ELECTROCHROMISM AND ELECTROCHROMIC DEVICES

Electrochromism has advanced greatly over the past decade with electrochromic substances – organic and/or inorganic materials and polymers – providing widespread applications in light-attenuation, displays and analysis.

Using reader-friendly electrochemistry, this book leads from electrochromic scope and history to new and searching presentations of optical quantification and theoretical mechanistic models. Non-electrode electrochromism and photo-electrochromism are summarised, with updated comprehensive reviews of electrochromic oxides (tungsten trioxide particularly), metal coordination complexes and metal cyanometallates, viologens and other organics; and more recent exotics such as fullerenes, hydrides and conjugated electroactive polymers are also covered. The book concludes by examining device construction and durability.

Examples of real-world applications are provided, including minimal-power electrochromic building fenestration, an eco-friendly application that could replace air conditioning; moderately sized electrochromic vehicle mirrors; large electrochromic windows for aircraft; and reflective displays such as quasi-electrochromic sensors for analysis, and electrochromic strips for monitoring of frozen-food refrigeration.

With an extensive bibliography, and step-by-step development from simple examples to sophisticated theories, this book is ideal for researchers in materials science, polymer science, electrical engineering, physics, chemistry, bioscience and (applied) optoelectronics.

P. M. S. MONK is a Senior Lecturer in physical chemistry at the Manchester Metropolitan University in Manchester, UK.

R. J. MORTIMER is a Professor of physical chemistry at Loughborough University, Loughborough, UK.

D. R. ROSSEINSKY, erstwhile physical chemist (Reader) at the University of Exeter, UK, is an Hon. University Fellow in physics, and a Research Associate of the Department of Chemistry at Rhodes University in Grahamstown, South Africa.

ELECTROCHROMISM AND ELECTROCHROMIC DEVICES

P. M. S. MONK,
Manchester Metropolitan University

R. J. MORTIMER
Loughborough University

AND

D. R. ROSSEINSKY
University of Exeter

CAMBRIDGE UNIVERSITY PRESS
Cambridge, New York, Melbourne, Madrid, Cape Town, Singapore, São Paulo

Cambridge University Press
The Edinburgh Building, Cambridge CB2 8RU, UK

Published in the United States of America by Cambridge University Press, New York

www.cambridge.org
Information on this title: www.cambridge.org/9780521822695

© P. M. S. Monk, R. J. Mortimer, D. R. Rosseinsky 2007

This publication is in copyright. Subject to statutory exception
and to the provisions of relevant collective licensing agreements,
no reproduction of any part may take place without
the written permission of Cambridge University Press.

First published 2007

Printed in the United Kingdom at the University Press, Cambridge

A catalogue record for this publication is available from the British Library

ISBN 978-0-521-82269-5 hardback

Cambridge University Press has no responsibility for
the persistence or accuracy of URLs for external or
third-party internet websites referred to in this publication,
and does not guarantee that any content on such
websites is, or will remain, accurate or appropriate.

Contents

Preface		page ix
Acknowledgements		xii
List of symbols and units		xiv
List of abbreviations and acronyms		xvii
1	Introduction to electrochromism	1
	1.1 Electrode reactions and colour: electrochromism	1
	1.2 Non-redox electrochromism	3
	1.3 Previous reviews of electrochromism and electrochromic work	6
	1.4 Criteria and terminology for ECD operation	7
	1.5 Multiple-colour systems: electropolychromism	17
	References	18
2	A brief history of electrochromism	25
	2.1 Bibliography; and 'electrochromism'	25
	2.2 Early redox-coloration chemistry	25
	2.3 Prussian blue evocation in historic redox-coloration processes	25
	2.4 Twentieth century: developments up to 1980	27
	References	30
3	Electrochemical background	33
	3.1 Introduction	33
	3.2 Equilibrium and thermodynamic considerations	34
	3.3 Rates of charge and mass transport through a cell	41
	3.4 Dynamic electrochemistry	46
	References	51
4	Optical effects and quantification of colour	52
	4.1 Amount of colour formed: extrinsic colour	52
	4.2 The electrochromic memory effect	53
	4.3 Intrinsic colour: coloration efficiency η	54
	4.4 Optical charge transfer (CT)	60
	4.5 Colour analysis of electrochromes	62
	References	71

5	Kinetics of electrochromic operation	75
	5.1 Kinetic considerations for type-I and type-II electrochromes: transport of electrochrome through liquid solutions	75
	5.2 Kinetics and mechanisms of coloration in type-II bipyridiliums	79
	5.3 Kinetic considerations for bleaching type-II electrochromes and bleaching and coloration of type-III electrochromes: transport of counter ions through solid electrochromes	79
	5.4 Concluding summary	115
	References	115
6	Metal oxides	125
	6.1 Introduction to metal-oxide electrochromes	125
	6.2 Metal oxides: primary electrochromes	139
	6.3 Metal oxides: secondary electrochromes	165
	6.4 Metal oxides: dual-metal electrochromes	190
	References	206
7	Electrochromism within metal coordination complexes	253
	7.1 Redox coloration and the underlying electronic transitions	253
	7.2 Electrochromism of polypyridyl complexes	254
	7.3 Electrochromism in metallophthalocyanines and porphyrins	258
	7.4 Near-infrared region electrochromic systems	265
	References	274
8	Electrochromism by intervalence charge-transfer coloration: metal hexacyanometallates	282
	8.1 Prussian blue systems: history and bulk properties	282
	8.2 Preparation of Prussian blue thin films	283
	8.3 Electrochemistry, *in situ* spectroscopy and characterisation of Prussian blue thin films	285
	8.4 Prussian blue electrochromic devices	289
	8.5 Prussian blue analogues	291
	References	296
9	Miscellaneous inorganic electrochromes	303
	9.1 Fullerene-based electrochromes	303
	9.2 Other carbon-based electrochromes	304
	9.3 Reversible electrodeposition of metals	305
	9.4 Reflecting metal hydrides	307
	9.5 Other miscellaneous inorganic electrochromes	309
	References	309
10	Conjugated conducting polymers	312
	10.1 Introduction to conjugated conducting polymers	312
	10.2 Poly(thiophene)s as electrochromes	318

	10.3	Poly(pyrrole)s and dioxypyrroles as electrochromes	327
	10.4	Poly(aniline)s as electrochromes	328
	10.5	Directed assembly of electrochromic electroactive conducting polymers	331
	10.6	Electrochromes based on electroactive conducting polymer composites	332
	10.7	ECDs using both electroactive conducting polymers and inorganic electrochromes	333
	10.8	Conclusions and outlook	334
		References	335
11	The viologens	341
	11.1	Introduction	341
	11.2	Bipyridilium redox chemistry	342
	11.3	Bipyridilium species for inclusion within ECDs	346
	11.4	Recent elaborations	360
		References	366
12	Miscellaneous organic electrochromes	374
	12.1	Monomeric electrochromes	374
	12.2	Tethered electrochromic species	387
	12.3	Electrochromes immobilised within viscous solvents	391
		References	391
13	Applications of electrochromic devices	395
	13.1	Introduction	395
	13.2	Reflective electrochromic devices: electrochromic car mirrors	395
	13.3	Transmissive ECD windows for buildings and aircraft	397
	13.4	Electrochromic displays for displaying images and data	401
	13.5	ECD light modulators and shutters in message-laser applications	404
	13.6	Electrochromic paper	405
	13.7	Electrochromes applied in quasi-electrochromic or non-electrochromic processes: sensors and analysis	406
	13.8	Miscellaneous electrochromic applications	407
	13.9	Combinatorial monitoring of multiples of varied electrode materials	409
		References	410
14	Fundamentals of device construction	417
	14.1	Fundamentals of ECD construction	417
	14.2	Electrolyte layers for ECDs	419
	14.3	Electrodes for ECD construction	422

	14.4	Device encapsulation	424
		References	425
15	Photoelectrochromism	433
	15.1	Introduction	433
	15.2	Direction of beam	433
	15.3	Device types	434
	15.4	Photochromic–electrochromic systems	438
		References	440
16	Device durability	443
	16.1	Introduction	443
	16.2	Durability of transparent electrodes	444
	16.3	Durability of the electrolyte layers	445
	16.4	Enhancing the durability of electrochrome layers	445
	16.5	Durability of electrochromic devices after assembly	446
		References	449
		Index	452

Preface

While the topic of electrochromism – the evocation or alteration of colour by passing a current or applying a potential – has a history dating back to the nineteenth century, only in the last quarter of the twentieth century has its study gained a real impetus. So, applications have hitherto been limited, apart from one astonishing success, that of the Gentex Corporation's self-darkening rear-view mirrors now operating on several million cars. Now they have achieved a telling next step, a contract with Boeing to supply adjustably darkening windows in a new passenger aircraft. The ultimate goal of contemporary studies is the provision of large-scale electrochromic windows for buildings at modest expenditure which, applied widely in the USA, would save billions of dollars in air-conditioning costs. In tropical and equatorial climes, savings would be proportionally greater: Singapore for example spends one quarter of its GDP (gross domestic product) on air conditioning, a *sine qua non* for tolerable living conditions there. Another application, to display systems, is a further goal, but universally used liquid crystal displays present formidable rivalry. However, large-scale screens do offer an attractive scope where liquid crystals might struggle, and electrochromics should almost certainly be much more economical than plasma screens. Numerous other applications have been contemplated. There is thus at present a huge flurry of activity to hit the jackpot, attested by the thousands of patents on likely winners. However, as a patent is *sui generis*, and we wish to present a scientific overview, we have not scanned in detail the patent record, which would have at least doubled the work without in our view commensurate advantages.

There are thousands of chemical systems that are intrinsically electrochromic, and while including explanatory examples, we incorporate here mostly those that have at least a promise of being useful. Our approach has been to concentrate on systems that colorise or change colour by electron transfer ('redox') processes, without totally neglecting other, electric-potential dependent, systems now particularly useful in applications to bioscience. The latter especially seem set to shine.

Several international gatherings have been convened to discuss electrochromism for devices. Probably the first was The Electrochemical Society meeting in 1989 (in Hollywood, Fl).[1] Soon afterwards was 'Fundamentals of Electrochromic Devices' organised by The American Institute of Chemical Engineers at their Annual Meeting in Chicago, 11–16 November 1990.[2] The following year, the authors of this present volume called a Solid-State Group (Royal Society of Chemisty) meeting in London. At the Electrochemical Society meeting in New Orleans (in 1994),[3] it was decided to host the first of the so-called International Meetings on Electrochromism, 'IME'. The first such meeting 'IME-1' met in Murano, Venice in 1994,[4] IME-2 in San Diego in 1996,[5] IME-3 was in London in 1998,[6] IME-4 in Uppsala in 2000,[7] IME-5 in Colorado in 2002 and IME-6 in Brno, Czech Republic in 2004.[8] Further electrochromics symposia occurred at Electrochemical Society meetings that took place at San Antonio, TX, in 1996[9] and Paris in 2003.[10]

The basis of the processes on which we concentrate is electrochemical, as is outlined in the first chapter. A historical outline is given in Chapter 2, and any reader not familiar with the electrochemistry presented here may find this explained sufficiently in Chapter 3. A fairly extensive presentation of twentieth-century electrochemistry in Chapter 3 seems necessary also to follow some later details of the exposition, and those familiar with this arcane science may choose to flip through a chapter largely comprising 'elderly electrochemistry', to quote from ref. 18 of Chapter 1.

Details of assessing coloration follow in Chapter 4, and in Chapter 5 attempts at theoretically modelling the electrochromic process in the most popular electrochromic material to date, tungsten trioxide, are outlined. In subsequent chapters, the work that has been conducted on a wide variety of materials follow, from metal oxides through complexed metals and metal-organic complexes to conjugated conductive polymers. Applications and tests finish the account. In order hopefully to make each chapter almost free-standing, we do quite frequently repeat the gist of some previous chapter(s).

A comment about the citations which end each chapter: early during our discussions of the book's contents, we decided to reproduce the full titles of each paper cited. Each title is cited as it appeared when first published. We have systematised capitalisation throughout (and corrected spelling errors in two papers).

In our account we have probably not succeeded in conveying all the aesthetic pleasure of studying aspects of colour and its creation, or the profound science-and-technology interest of understanding the reactions and of mastering the associated processes: this book does represent an attempt to spread these interests. However, further at stake is the prospect of controlling an important part of personal environments while economising on air-conditioning costs, thereby cutting down fuel consumption and lessening the human 'carbon footprint', to cite the mode words. There are the other perhaps lesser applications that are also promisingly useful. So, to a more controlled-colour future, read on.

DISCLAIMER: Superscripted reference citations in the text are, unusually, listed in full e.g. 1, 2, 3, 4 rather than the customary 1–4. The need arises from the parallel publication of this monograph as an e-book. In this version, 'each reference citation is hyper-linked to the reference itself, which requires that they be cited separately.'

References

1. Proceedings volume was *Electrochromic Materials*, Carpenter, M. K. and Corrigan, D. A. (eds.), **90–2**, Pennington, NJ, Electrochemical Society, 1990.
2. Proceedings of the Annual Meeting of the American Institute of Chemical Engineers, published in *Sol. Energy Mater. Sol. Cells*, **25**, 1992, 195–381.
3. Proceedings volume was *Electrochromic Materials II*, Ho, K.-C. and MacArthur, D. A. (eds.), **94–2**, Pennington, NJ, Electrochemical Society, 1994.
4. Proceedings volume was *Sol. Energy Mater. Sol. Cells*, 1995, **39**, issue 2–4.
5. Proceedings volume was *Sol. Energy Mater. Sol. Cells*, 1998, **54**, issue 1–4.
6. Proceedings volume was *Sol. Energy Mater. Sol. Cells*, 1999, **56**, issue 3–4.
7. Proceedings volume was *Electrochim. Acta*, 2001, **46**, issue 13–14.
8. Proceedings volume was *Sol. Energy Mater. Sol. Cells*, 2006, **90**, issue 4.
9. Proceedings volume was *Electrochromic Materials III*, Ho, K. C., Greenberg, C. B. and MacArthur, D. M. (eds.), **96–24**, Pennington, NJ, Electrochemical Society, 1996.
10. Proceedings volume was *Electrochromic Materials and Applications*, Rougier, A., Rauh, D. and Nazri, G. A. (eds.), **2003–17**, Pennington, NJ, Electrochemical Society, 2003.

Acknowledgements

We are indebted to numerous colleagues and correspondents who have collaborated in research or in providing information.

PMSM wishes to thank: Professor Claus-Gören Granqvist of Uppsala University, Professor Susana de Córdoba, Universidade de São Paulo, Brazil, Professor L. M. Loew of the University of Connecticut Health Center and Dr Yoshinori Nishikitani of the Nippon Mitsubishi Oil Corporation. Also, those on the computer helpdesk at MMU who helped with the scanning of figures.

RJM wishes to thank: Dr Joanne L. Dillingham, (Ph.D, Loughborough University), Dr Steve J. Vickers erstwhile of the Universities of Birmingham and Sheffield, Dr Natalie M. Rowley of the University of Birmingham; Dr Frank Marken, University of Bath; Professor Paul D. Beer, University of Oxford; Professor John R. Reynolds, University of Florida; Aubrey L. Dyer, University of Florida; Dr Barry C. Thompson, University of California, Berkeley – erstwhile of the Reynolds group at the University of Florida; Professor Mike D. Ward, University of Sheffield, and Professor Steve Fletcher of Loughborough University.

DRR wishes to thank: Bill Freeman Esq. then of Finisar Corp., now heading Thermal Transfer (Singapore), Dr Tom Guarr of Gentex, Dr Andrew Glidle now of Glasgow University, Dr Richard Hann of ICI, the late Dr Brian Jackson of Cookson Ltd, Professor Hassan Kellawi of Damascus University, Mr (now Captain) Hanyong Lim of Singapore, graduate of Carnegie-Mellon University, Professor Paul O'Shea of Nottingham University, Ms Julie Slocombe, erstwhile of Exeter University, and Dr Andrew Soutar and Dr Zhang Xiao of SIMTech in Singapore.

We also wish to thank the following for permission to reproduce the figures (in alphabetical order): The American Chemical Society, The American Institute of Physics, The Electrochemical Society, Elsevier Science, The Royal

Society of Chemistry (RSC), The Japanese Society of Physics, The Society of Applied Spectroscopy, and the Society for Photo and Information Engineering. In collecting the artwork for figures, we also acknowledge the kind help of the following: Dr Charles Dubois, formerly of the University of Florida.

From the staff of Cambridge University Press, we wish to thank Dr Tim Fishlock (now at the RSC in Cambridge, UK), who first commissioned the book, and his successor Dr Michelle Carey, and Assistant Editor, Anna Littlewood, together with Jo Bottrill of the production team for their help; and a particularly big thank you to the copy-editor Zoë Lewin for her consistent good humour and professionalism.

We owe much to our families, who have enabled us to undertake this project. We apologise if we have been preoccupied or merely absent when you needed us.

We also thank the numbers of kindly reviewers of our earlier book (and even the two who commented adversely) and much appreciate passing comment in a paper by Dr J. P. Collman and colleagues.

Though obvious new leaders exploring different avenues are currently emerging, if one individual is to be singled out in the general field, Claes-Goran Granqvist of the Ångstrom Laboratory, Uppsala, has to be acknowledged for the huge input into electrochromism that he has sustained over decades.

We alone are responsible for the contents of the book including the errors.

Symbols and units

A	ampere, area
Abs	optical absorbance
$c(y,t)$	time-dependent concentration of charge at a distance of y into a solid thin film
c_m	maximum concentration of charge in a thin film
c_0	initial concentration of charge in a thin film
D	diffusion coefficient
$\overline{D}$	chemical diffusion coefficient
d	thickness of a thin film
e	charge on an electron
e^-	electron
$\mathcal{E}$	energy
E	potential
E_a	activation energy
$E_{(appl)}$	applied potential
$E_{(eq)}$	equilibrium potential
E_{pa}	potential of anodic peak
E_{pc}	potential of cathodic peak
$E^{\ominus}$	standard electrode potential
eV	electron volt
F	Faraday constant
Hz	hertz
i	current density
i	subscripted, represents component 1 or 2 ...
i_b	bleaching current density
i_c	coloration current density
i_o	exchange current density
$\mathcal{J}$	imaginary part of impedance
J_o	charge flux (rate of passage of electrons or ionic species)
K	equilibrium constant
K_a	equilibrium constant of acid ionisation

K_{sp}	equilibrium constant of ionic solubility ('solubility product')
$l(t)$	time-dependent thickness of a narrow layer of the WO_3 film adjacent to the electrolyte (during electro-bleaching)
M	mol dm^{-3}
n	number in part of iterative calculation
n	number of electrons in a redox reaction
p	volume charge density of protons in the H_0WO_3
p	the operator $-\log_{10}$
Pa	pascal
q	charge per unit volume
Q	charge
R	gas constant
$\mathcal{R}$	real component of impedance
r	radius of sphere (e.g. of a solid, spherical grain)
S	Seebeck coefficient
s	second
T	thermodynamic temperature
t	time
v	scan rate
V	volt
V	volume
V_a	applied potential
W	Wagner enhancement factor ('thermodynamic enhancement factor')
x	insertion coefficient
$x_{(critical)}$	insertion coefficient at a percolation threshold
x_1	constant (of value ≈ 0.1)
x_o	proton density in a solid thin film
x, y, z, w or c	subscripted, non-integral composition indicators, in non-stoichiometric materials
Z	impedance
γ	gamma photon
ε	extinction coefficient ('molar absorptivity')
η	coloration efficiency
η_o	coloration efficiency of an electrochromic device
η_p	coloration efficiency of primary electrochrome
η_s	coloration efficiency of secondary electrochrome
η	overpotential

λ	wavelength
λ_{max}	wavelength maximum
Λ	ionic molar conductivity
μ	mobility, chemical potential
$\mu_{(ion)}$	mobility of ions
$\mu_{(electron)}$	mobility of electrons
ν	frequency of light
ρ	density of atoms in a thin film
ρ_0	constant equal to ($2\,e\,\rho\,d\,i_0$)
σ	electronic conductivity
τ_D	'characteristic time' for diffusion
φ_s	membrane surface potential
υ	kinematic viscosity
$\bar{\upsilon}$	velocity of solution flow
ω	frequency of ac signal

Abbreviations and acronyms

a	amorphous
ac	alternating current
AEIROF	anodically electrodeposited iridium oxide film
AES	atomic emission spectroscopy
AFM	atomic force microscopy
AIROF	anodically formed iridium oxide film
AMPS	2-acrylamido-2-methylpropanesulfonic acid
ANEPPS	3-{4-[2-(6-dibutylamino)-2-naphthyl]-*trans*-ethenyl pyridinium} propane sulfonate
aq	aqueous
AR	anti reflectance
ASSD	all-solid-state device
ATO	antimony–tin oxide
BEDOT	2,2′-bis(3,4-ethylenedioxythiophene)
BEDOT-*N*MeCz	3,6-bis[2-(3,4-ethylenedioxythiophene)]-*N*-alkylcarbazole
bipy	2,2′-bipyridine
bipm	4,4′-bipyridilium
c	crystalline
CAT	catecholate
CCE	composite coloration efficiency
CE	counter electrode
ChLCs	cholesteric liquid crystals
CIE	Commission Internationale de l'Eclairage
cmc	critical micelle concentration
CPQ	cyanophenyl paraquat [1,1′-bis(*p*-cyanophenyl)-4,4′-bipyridilium]
CRT	cathode-ray tube
CT	charge transfer

CTEM	conventional transmission electron microscopy
CuHCF	copper hexacyanoferrate
CVD	chemical vapour deposition
dc	direct current
DDTP	2,3-di(thien-3-yl)-5,7-di(thien-2-yl)thieno[3,4-*b*]pyrazine
DEG	diethyleneglycol
DMF	dimethylformamide
DMSO	dimethyl sulfoxide
EC	electrochromic
EC	electrode reaction followed by a chemical reaction
ECB	electrochromic battery
ECD	electrochromic device
ECM	electrochromic material
ECW	electrochromic window
EDAX	energy dispersive analysis of X-rays
EDOT	3,4-(ethylenedioxy)thiophene
EIS	electrochemical impedance spectroscopy
EQCM	electrochemical quartz-crystal microbalance
FPE	fluoresceinphosphatidyl-ethanolamine
FTIR	Fourier-transform infrared
FTO	fluorine[-doped] tin oxide
GC	glassy carbon
HCF	hexacyanoferrate
HOMO	highest occupied molecular orbital
HRTEM	high-resolution transmission electron microscopy
HTB	hexagonal tungsten bronze
HV	heptyl viologen (1,1'-di-*n*-heptyl-4,4'-bipyridilium)
IBM	Independent Business Machines
ICI	Imperial Chemical Industries
IR	infrared
ITO	indium–tin oxide
IUPAC	International Union of Pure and Applied Chemistry
IVCT	intervalence charge transfer
LB	Langmuir–Blodgett
LBL	layer-by-layer [deposition]
LCD	liquid crystal display
LED	light-emitting diode
LFER	linear free-energy relationships

LPCVD	liquid-phase chemical vapour deposition
LPEI	linear poly(ethylene imine)
LUMO	lowest unoccupied molecular orbital
MB	Methylene Blue
MLCT	metal-to-ligand charge transfer
MOCVD	metal-oxide chemical vapour deposition
MV	methyl viologen (1,1'-dimethyl-4,4'-bipyridilium)
nc	naphthalocyanine
NCD	nanochromic display
Ni HCF	nickel hexacyanoferrate
NMP	N-methylpyrrolidone
NRA	nuclear reaction analysis
NREL	National Renewable Energy Laboratory, USA
NVS©	Night Vision System©
OD	optical density
OEP	octaethyl porphyrin
OLED	organic light-emitting diode
OTE	optically transparent electrode
OTTLE	optically transparent thin-layer electrode
pa	peak anodic
PAA	poly(acrylic acid)
PAH	poly(allylamine hydrochloride)
PANI	poly(aniline)
PB	Prussian blue
PBEDOT-B(OC$_{12}$)$_2$	poly{1,4-bis[2-(3,4-ethylenedioxy)thienyl]-2,5-didodecyloxybenzene}
PBEDOT-N-MeCz	poly{3,6-bis[2-(3,4-ethylenedioxy)thienyl]-N-methylcarbazole}
PBEDOT-Pyr	poly{3,6-bis[2-(3,4-ethylenedioxy)thienyl] pyridine}
PBEDOT-PyrPyr(Ph)$_2$	poly{5,8-bis(3-dihydro-thieno[3,4-b]dioxin-5-yl)-2,3-diphenyl-pyrido[3,4-b]pyrazine}
PBuDOP	poly[3,4-(butylenes dioxy)pyrrole]
pc	peak cathodic
Pc	dianion of phthalocyanine
PC	propylene carbonate
PCNFBS	poly{cyclopenta[2,1-b;4,3-b']dithiophen-4-(cyanononafluorobutylsulfonyl)methylidene}
PdHCF	palladium hexacyanoferrate

PDLC	phase-dispersed liquid crystals
PEDOP	poly[3,4-(ethylenedioxy)pyrrole]
PEDOT	poly[3,4-(ethylenedioxy)thiophene]
PEDOT-S	poly{4-(2,3-dihydrothieno[3,4-*b*]-[1,4]dioxin-2-yl-methoxy}-1-butanesulfonic acid, sodium salt
PEO	poly(ethylene oxide)
PET	poly(ethylene terephthalate)
PG	Prussian green
PITT	potentiostatic intermittence titration technique
PMMA	poly(methyl methacrylate)
PMT	polaromicrotribometric
PP	plasma polymerised
PP	poly(1,3,5-phenylene)
PProDOP	poly[3,4-(propylenedioxy)pyrrole]
PProDOT	poly(3,4-propylenedioxythiophene)
PSS	poly(styrene sulfonate)
PTPA	poly(triphenylamine)
PVA	poly(vinyl acrylate)
PVC	poly(vinyl chloride)
PVD	physical vapour deposition
PW	Prussian white
PX	Prussian brown
Pyr	pyridine
Q	Quinone
RE	reference electrode
rf	radio frequency
RP	ruthenium purple: iron(III) hexacyanoruthenate(II)
RRDE	rotated ring-disc electrode
s	solid
s. soln	solid solution
SA	sacrificial anode
SCE	saturated calomel electrode
SQ	semi quinone
SEM	scanning electron microscopy
SHE	standard hydrogen electrode
SI	Système internationale
SIMS	secondary ion mass spectroscopy
SIROF	sputtered iridium oxide film
soln	solution

SPD	suspended particle device
SPM	solid paper matrix
STM	scanning tunnelling microscopy
TA	thiazine
TCNQ	tetracyanoquinodimethane
TGA	thermogravimetric analysis
THF	tetrahydrofuran
TMPD	tetramethylphenylenediamine
Tp*	hydrotris(3,5-dimethylpyrazolyl)borate
TTF	tetrathiafulvalene
UCPC	user-controllable photochromic [material]
UPS	ultraviolet photoelectron spectroscopy
VDU	visual display unit
VHCF	vanadium hexacyanoferrate
WE	working electrode
WPA	tungsten phosphoric acid
XAS	X-ray absorption spectroscopy
XPS	X-ray photoelectron spectroscopy
XRD	X-ray diffraction
XRG	xerogel

1
Introduction to electrochromism

1.1 Electrode reactions and colour: electrochromism

The terminology and basis of the phenomenon that we address are briefly outlined in this chapter. Although there are several usages of the term 'electrochromism', several being summarised later in this chapter, 'electrochromes' later in the present text are always 'electroactive', as follows. An electroactive species can undergo an electron uptake, i.e. 'reduction', Eq. (1.1), or electron release, i.e. 'oxidation', the reverse of Eq. (1.1) in a 'redox' reaction that takes place at an electrode. An electrode basically comprises a metal or other conductor, with external connections, in contact with forms O and R of an 'electroactive' material, and can be viewed as a 'half-cell':

$$\text{oxidised form, O} + \text{electron(s)} \rightarrow \text{reduced form, R.} \qquad (1.1)$$

Though in strict electrochemical parlance all the components, O and R and *the metallic or quasi-metallic conductor*, comprise 'the electrode', we and others often depart from this complete definition when we imply that 'the electrode' comprises the just-italicised component, which conforms with the following definition: 'An electrode basically comprises a metal or metallic conductor or, especially in electrochromism, an adequately conductive semiconductor often as a thin film on glass.' We thus usually refer to the 'electrode substrate' for the metal or metal-like component to make the distinction clear. Furthermore, in Chapter 3 it is emphasised that any electrode in a working system must be accompanied by a second electrode, with intervening electrolyte, in order to make up a cell allowing passage of current, in part comprising the flow of just those electrons depicted in Eq. (1.1).

An electroactive material may be an atom or ion, a molecule or radical, sometimes multiply bonded in a solid film, and must be in contact with the electrode substrate prior to successful electron transfer. It may be in solution – solvated and/or complexed – in which case it must approach sufficiently closely to the electrode substrate and undergo the adjustments that contribute to the (sometimes low) activation energy accompanying electron transfer. In other systems, the electroactive material may be a solid or dispersed within a solid matrix, in which case that proportion of the electrochrome physically in contact with the electrode substrate undergoes the redox reaction most rapidly, the remainder of the electroactive material less so. The underlying theory of electrochemical electron-transfer reactions is treated elsewhere.[1]

That part of a molecular system having or imparting a colour is termed a chromophore. White light comprises the wavelengths of all the colours, and colour becomes evident when photons from part of the spectrum are absorbed by chromophores; then the colour seen is in fact the colour *complementary* to that absorbed. Thus, for example, a blue colour is reflected (hence seen) if, on illumination with white light, the material absorbs red. Light absorption enables electrons to be promoted between quantised (i.e. wave-mechanically allowed) energy levels, such as the ground and first excited states. The wavelength of light absorbed, λ, is related to the magnitude of the energy gap E between these levels according to the Planck relation, Eq. (1.2):

$$E = h\nu = \frac{hc}{\lambda}, \qquad (1.2)$$

where ν is the frequency, h is the Planck constant and c the speed of light in vacuo. The magnitude of E thus relates to the colour since, when λ is the wavelength at the maximum (usually denoted as λ_{max}) of the absorption band observed in the spectrum of a chromophore, its position in the spectrum clearly governs the observed colour. (To repeat, the colour arises from the non-absorbed wavelengths.) Most electrochromes colourise by reflection, as in displays; transmission-effective systems, as in windows, follow a corresponding mechanism.

Electroactive species comprise different numbers of electrons before and after the electron-transfer reaction (Eq. (1.1) or its reverse), so different redox states will necessarily exhibit different spectroscopic transitions, and hence will require different energies E for electron promotion between the ground and excited states. Hence all materials will undergo change of spectra on redox change.

However, the colours of electroactive species only *may* be different before and after electron transfer because often the changes are not visible (except by suitable spectrometry) when the wavelengths involved fall outside the visible range. In other words, the spectral change accompanying a redox reaction is visually indiscernible if the optical absorptions by the two redox states fall in the ultraviolet (UV) or near infrared (NIR). When the change is in the visible region, then a pragmatic definition of electrochromism may be formulated as follows. 'Electrochromism is a change, evocation, or bleaching, of colour as effected either by an electron-transfer (redox) process or by a sufficient electric potential. In many applications it is required to be reversible.' However, regarding intensity-modulation filters for, say, IR message-laser pulses in optical fibres, such terms as 'electrochromic switching or modulation' are increasingly being used for such invisible effects.

Visible electrochromism is of course only ever useful for display purposes if one of the colours is markedly different from the other, as for example when the absorption band of one redox state is in the visible region while the other is in the UV. If the colours are sufficiently intense and different, then the material is said to be electrochromic and the species undergoing change is usefully termed an 'electrochrome'.[2]

Simple laboratory demonstrations of electrochromism are legion.[3,4] The website in ref. 5 contains a video sequence clearly demonstrating electrochromic coloration, here of a highly conjugated poly(thiophene) derivative. Many organic and inorganic materials are electrochromic; and even some biological species exhibit the phenomenon:[6] *Bacteriorhodopsin* is said to exhibit very strong electrochromism with a colour change from bright blue to pale yellow.[6]

The applications of electrochromism are outlined in Chapter 13 and the general criteria of device fabrication are outlined in Chapter 14.

1.2 Non-redox electrochromism

The word 'electrochromism' is applied to several, disparate, phenomena. Many are not electrochromic in the redox sense defined above.

Firstly, charged species such as 3-{4-[2-(6-dibutylamino-2-naphthyl)-*trans*-ethenyl] pyridinium} propane sulfonate ('di-4-ANEPPS') (**I**), called 'electrochromic probes', are employed in studies of biological membrane potentials.[7] (A similar-looking but intrinsically different mechanism involving deprotonation is outlined below.) For a strongly localised system, such as a protein system where electron-donor and -acceptor sites are separated by large distances, the potential surfaces involved in optical electron excitation (see Eq. (1.2)) become highly asymmetrical.[7] For this reason, the electronic spectrum of (**I**) is extraordinarily sensitive to its environment, demonstrating large solvent-dependent 'solvatochromic' shifts,[8] so much information can be gained by quantitative analysis of its UV-vis spectra. In effect, it is possible to image the electrical activity of a cell membrane.[7] Loew *et al.* first suggested this use of such electrochromism in[9] 1979; they pointed out how the best species for this type of work are compounds like (**I**), its 8-isomer, or nitroaminostilbene,[10] both of which have large non-linear second-harmonic effects.[9] In consequence, significant changes are induced by the environment in the dipole moment so on excitation from the ground to the excited states, different colours result.

[Structure of compound **I**]

I

This application is not electrochromism as effected by redox processes of the kind we concentrate on in the present work, but can alternatively be viewed as a molecular Stark effect[11] in which some of the UV-vis bands of polarisable molecules evince a spectroscopic shift in the presence of a strong electric field. Vredenberg[12] reviewed this aspect of electrochromism in 1997. Such a Stark effect was the original sense implied by 'electrochromism' when the word was coined in 1961.[13]

While many biological and biochemical references to 'electrochromism' mean a Stark effect of this type, some are electrochromic in the redox sense. For example, the (electrochromic) colours of quinone reduction products have been used to resolve the respective influences of electron and proton transfer processes in bacterial reactions.[14,15,16] In some instances, however, this electrochromic effect is unreliable.[17]

A valuable electrochromic application has been employed by O'Shea[18] to probe local potentials on *surfaces* of biological cell membranes. The effect of electric potential on acidity constants is employed: weak acids in solution are partly ionised into proton and ('base') residue to an extent governed ordinarily by the equilibrium constant particular to that acid, its acidity constant K_a. However, if the weak acid experiences an extraneous electric potential, the extent of ionisation is enhanced by further molecular scission (i.e. proton release) resulting from the increased stabilisation of the free-proton charge. With 'p' representing (negative) decadic logarithms, the outcome may be represented by the equation $pK_a(\varphi_s) = pK_a(0) - F\varphi_s/RT(\ln 10)$ where φ_s is the membrane surface potential.[18] This result (a close parallel of the observed 'second Wien effect' in high-field conductimetry on weak acids) arises from combining the Boltzmann equation with the Henderson–Hasselbalch equation. The application proceeds as follows. A fluorescent molecule is chosen, that is a proton-bearing acid of suitable K_a, with only its deprotonated moiety

1.2 Non-redox electrochromism

showing visible fluorescence, and then only when the potential experienced is high enough. The probe molecules are inserted by suitable chemistry into the surface of the cell membrane. Then it will fluoresce, in areas of sufficiently high electric potential, thus illuminating such areas of φ_s, and monitoring even rapid rates of change as can result from say cation acquisition by the surface. Suitable probe molecules are[18,19] fluoresceinphosphatidyl-ethanolamine (FPE) and[20,21,22] 1-(3-sulfonatopropyl)-4-[p[2-(di-n-octylamino)-6-naphthyl]-vinyl]-pyridinium betaine. To quote,[23] 'Probe molecules such as FPE have proved to be particularly versatile indicators of the electrostatic nature of the membrane surface in both artificial and cellular membrane systems.' This ingenious probe of electrical interactions underlying biological cell function thus relies unusually not on electron transfer but on proton transfer as effected by electric potential changes.

Secondly, the adjective 'electrochromic' is often applied to a widely differing variety of fenestrative and device applications. For example, a routine web search using the phrase 'electrochromic window' yielded many pages describing a suspended-particle-device (SPD) window. Some SPD windows are also termed 'Smart Glass'[24] – a term that, until now, has related to genuine electrochromic systems. On occasion (as occurs also in some patents) a lack of scientific detail indicates that the claims of some manufacturers' websites are perhaps excessively ambitious – a practice that may damage the reputation of electrochromic products should a device fail to respond to its advertised specifications.

Also to be noted, 'gasochromic windows' (also called gasochromic smart-glass windows) are generally not electrochromic, although sometimes described as such, because the change in colour is wholly attributable to a direct chemical gas + solid redox reaction, with no externally applied potential, and no measurable current flow. The huge complication of the requisite gaseous plumbing is rarely addressed, while electrochromic devices require only cables. (The most studied gasochromic material is, perhaps confusingly, tungsten oxide, which is also a favoured electrochrome.) The gasochromic devices in refs. 25,26,27,28,29,30,31,32 are not electrochromic in the sense adopted by this book.

Thirdly, several new products are described as 'electrochromic' but are in fact electrokinetic–colloidal systems, somewhat like SPDs with micro-encapsulation of the active particles. A good example is Gyricon 'electrochromic paper',[33] developed by Xerox. Lucent and Philips are developing similar products. Such paper is now being marketed as 'SmartPaper^{TM}'. Gyricon is intended for products like electronic books, electronic newspapers, portable signs, and foldable, rollable displays. It comprises two plastic sheets, each of thickness *ca.* 140 μm, between which are millions of 'bichromal' (i.e. two colour) highly dipolar spheres of diameter 0.1 μm, and are suspended within minute oil-filled pockets. The spheres rotate following exposure to an electric field, as from a 'pencil' tip attached to a battery also connected to a metallically conductive backing sheet;[34] the spheres rotate fully to display either black or white, or partially (in response to weaker electrical pulses), to display a range of grey shades.[33] Similar mechanisms operate in embedded sacs of sol in which charged black particles are 'suspended' (when in the colourless state) but on application of a potential by an 'electric pencil', black particles visibly deposit on the upper surface of the sacs. Some of these systems being deletable and re-usable promise substantial saving of paper.

Note that the NanoChromics™ paper described on page 347, marketed by NTera of Eire, is genuinely electrochromic in the redox sense.

1.3 Previous reviews of electrochromism and electrochromic work

The broadest overview of all aspects of redox electrochromism is *Electrochromism: Fundamentals and Applications*, by Monk, Mortimer and Rosseinsky.[2] It includes criteria for electrochromic application, the preparation of electrochromes and devices, and encompasses all types of electrochromic materials considered in the present book, both organic and inorganic. A major review of redox electrochromism appears in *Handbook of Inorganic Electrochromic Materials* by Granqvist,[35] a thorough and detailed treatise covering solely inorganic materials.

Other reviews of electrochromism appearing within the last fifteen years include (in alphabetical order) those of: Agnihotry[36] in 1996, Bange et al.[37] in 1995, Granqvist (sometimes with co-workers) in 1992,[38] 1993,[39,40] 1997,[41,42] 1998,[43,44,45] 2000,[46] 2003,[47,48] and 2004,[49] Green[50] in 1996, Greenberg in 1991[51] and 1994,[52] Lampert in 1998,[53] 2001[54] and 2004,[55] Monk in 2001[56] and 2003,[57] Mortimer[58] in 1997, Mortimer and Rosseinsky[59] in 2001, Mortimer and Rowley[60] in 2002, Mortimer, Dyer and Reynolds[61] in 2006, Scrosati, Passerini and Pileggi[62] and Scrosati,[63] 1992, Somani and Radhakrishnan[64] in 2003 and Yamamoto and Hayashida[65] in 1998.

Bamfield's book[8] *Chromic Phenomena*, published in 2001, includes a substantial review of electrochromism. Non-English reviews include that by Volke and Volkeova[66] (in Czech: 1996). McGourty (in 1991),[67] Hadfield (in 1993),[68] Hunkin (in 1993)[69] and Monk, Mortimer and Rosseinsky (in 1995)[70] have all written 'popular-science' articles on electrochromism.

Bowonder et al.'s 1994 review[71] helps frame electrochromic displays within the wider corpus of display technology. Lampert's[55] 2004 review 'Chromogenic materials' similarly helps place electrochromism within the wider scope of other forms of driven colour change, such as thermochromism. Lampert's review, shorter, crammed with acronyms but more up-to-date, includes other forms of display device, such as liquid crystal displays (LCDs), phase-dispersed liquid crystals (PDLCs), cholesteric liquid crystals (ChLCs) and suspended particle devices (SPDs).

There are also many dozen reviews concerning specific electrochromes, electrochromic-device applications and preparative methodologies, which we cite in relevant chapters. The now huge numbers of patents on materials, processes or devices are usually excluded, the reliability – often just the plausibility – of patents being judged by different, not always scientific, criteria.

1.4 Criteria and terminology for ECD operation

The jargon used in discussions of the operation of electrochromic devices (ECD) is complicated, hence the criteria and terminology cited below, necessarily abridged, might aid clarification. The terms comply with the 1997 IUPAC recommended list of terms on chemically modified electrodes (CMEs). A CME is[72]

an electrode made up of a conducting or semi conducting material that is coated with a selected monomolecular, multimolecular, ionic or polymeric film of a chemical modifier and that, by means of faradaic ... reactions or interfacial potential differences ... exhibits chemical, electrochemical and/or optical properties of the film.

Chemically modified electrodes are often referred to as being derivatised, especially but not necessarily when the modifier is organic or polymeric. *All* electrochromic electrodes comprise some element of modification, but are rarely referred to as CMEs; this is simply to be understood.

1.4.1 Electrochrome type

In the early days of ECD development, the kinetics of electrochromic coloration were discussed in terms of 'types' as in the seminal work of Chang, Sun and Gilbert[73] in 1975. Such types are classified in terms of the phases, present initially and thence formed electrochemically, which dictate the precise form of the current–time relationships evinced during coloration, and thus affect the coloration–time relationships. While the original classifications are somewhat dated, they remain useful and are followed here throughout. A **type-I** electrochrome is soluble, and remains in solution at all times during electrochromic usage. A good example is aqueous methyl viologen

(1,1′-dimethyl-4,4′-bipyridilium – **II**), which colours during a reductive electrode reaction, Eq. (1.3):

$$MV^{2+}(aq) + e^- \rightarrow MV^{+\bullet}(aq). \quad (1.3)$$
colourless   intense blue

$$H_3C-N^+\!\!=\!\!\langle\ \rangle\!-\!\langle\ \rangle\!=\!N^+-CH_3$$
$$2X^-$$
II

X^- can be a halide or complex anion such as BF_4^-. The cation is abbreviated to MV^{2+}. Other type-I electrochromes include any viologen often soluble in aqueous solution, or a phenathiazine (such as Methylene Blue), in non-aqueous solutions.

Type-II electrochromes are soluble in their colourless forms but form a coloured *solid* on the surface of the electrode following electron transfer. This phase change increases the write–erase efficiency and speeds the response time of the electrochromic bleaching. A suitable example of a type-II system is cyanophenyl paraquat (**III**), again in water,[74,75,76] Eq. (1.4):

$$CPQ^{2+}(aq) + e^- + X^-(aq) \rightarrow [CPQ^{+\bullet}\ X^-](solid). \quad (1.4)$$
colourless   olive green

$$NC-\langle\ \rangle\!-\!N^+\!=\!\langle\ \rangle\!-\!\langle\ \rangle\!=\!N^+\!-\!\langle\ \rangle\!-\!CN$$
III

The solid material here is a salt of the radical cation product[74] (the incorporation of the anionic charge X^- ensures electro-neutrality within the solid product).

Other type-II electrochromes commonly encountered include aqueous viologen systems such as heptyl or benzyl viologens,[77] or methoxyfluorene compounds in acetonitrile solution.[78] Inorganic examples include the solid products of electrodeposited metals such as bismuth (often deposited as a finely divided solid), or a mirror of metallic lead or silver (Section 9.3), in which the electrode reaction is generally reduction of an aquo ion or of a cation in a complex with attached organic or inorganic moieties ('ligands').

Type-III electrochromes remain solid at all times. Most inorganic electrochromes are type III, e.g. for metal oxides, Eq. (1.5),

$$\text{MO}_y(s) + x(\text{H}^+(\text{soln.}) + e^-) \rightarrow \text{H}_x\text{MO}_y(s), \quad (1.5)$$

$$\text{colourless} \qquad\qquad\qquad \text{intense colour}$$

where the metal M is most commonly a d-block element such as Mo, Ni or W, and the mobile counter ion (arbitrarily cited here as the proton) could also be lithium; $y=3$ is commonly found, and WO$_3$ has been the most studied. The parameter x, the 'insertion coefficient', indicates the proportion of metal sites that have been electro-reduced. The value of x usually lies in the approximate range $0 \leq x < 0.3$.

Other inorganic type-III electrochromes include phthalocyanine complexes and metal hexacyanometallates such as Prussian blue. Organic type-III systems are typified by electroactive conducting polymers. The three groups of polymer encountered most often in the literature of electrochromism are generically termed poly(pyrrole), poly(thiophene) or poly(aniline) and relate to the parenthesised monomer from which the electrochromic solid is formed by electro-polymerisation, as discussed below.

1.4.2 Contrast ratio CR

The contrast ratio CR is a commonly employed measure denoting the intensity of colour formed electrochemically, as seen by eye, Eq. (1.6):

$$CR = \left(\frac{R_o}{R_x}\right), \quad (1.6)$$

where R_x is the intensity of light reflected diffusely though the coloured state of a display, and R_o is the intensity reflected similarly but from a non-shiny white card.[79] The ratio CR is best quoted at a specific wavelength – usually at $\lambda_{\max}$ of the coloured state. As in practice, a CR of less than about 3 is almost impossible to see by eye. As high a value as possible is desirable.

The CR is commonly expressed as a ratio such as 7:1. A CR of 25:1 is cited for a type-II display involving electrodeposited bismuth metal,[80] and as high[81] as 60:1 for a system based on heptyl viologen radical cation, electrodeposited from aqueous solution with a charge[82] of 1 mC cm^{-2}, and 10:1 for the cell WO$_3$|electrolyte|NiO.[83]

More elaborate measures of coloration are outlined in Chapter 4.

1.4.3 Response time τ

The response time τ is the time required for an ECD to change from its bleached to its coloured state (or vice versa). It is generally unlikely that $\tau_{\text{(coloration)}} = \tau_{\text{(bleach)}}$. At present, there are few *reliable* response times in the literature since there is no consistency in the reporting and determination of cited data, and especially in the way different kinetic criteria are involved when determining τ. For example, τ may represent the time required for some fraction of the colour (defined or arbitrary) to form, or it may relate to the time required for an amount of charge (again defined or arbitrary) to be consumed in forming colour at the electrode of interest.

While most applications do not require a rapid colour change, some such as for electrochromic office windows actually require a very slow response, as workers can feel ill when the colour changes too rapidly.[84] For example, a film of WO_3 (formed by spray pyrolysis of a solution generated by dissolving W powder in H_2O_2) became coloured in 15 min, and bleached in 3 min,[85] but the choice of both potential and preparative method was made to engender such slowness. In contrast, a film of sol–gel-derived titanium dioxide is coloured by reductive insertion of Li^+ ions at a potential of about –2 V with a response time of about 40 s.[86]

However, applications such as display devices require a more rapid response. To this end, Sato[87] reports an anodically formed film of iridium oxide with a response time of 50 ms; Canon[88] made electrochromic oxide mixtures that undergo absorbance changes of 0.4 in 300 ms. Reynolds *et al.*[89] prepared a series of polymers based on poly(3,4-alkylenedioxythiophene) 'PEDOT' (**IV**); multiple switching studies, monitoring the electrochromic contrast, showed that films of polymer of thickness *ca.* 300 nm could be fully switched between reduced and oxidised forms in 0.8–2.2 s with a modest transmittance change of 44–63%. Similarly, a recently fabricated electrochromic device was described as 'ultra fast', with a claimed[90] τ of 250 ms; the viologen bis(2-phosphonoethyl)-4,4'-bipyridilium (**V**), with a coloration efficiency η of 270 $cm^2\ C^{-1}$ was employed as chromophore.

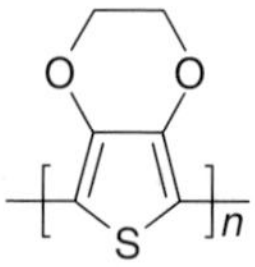

IV

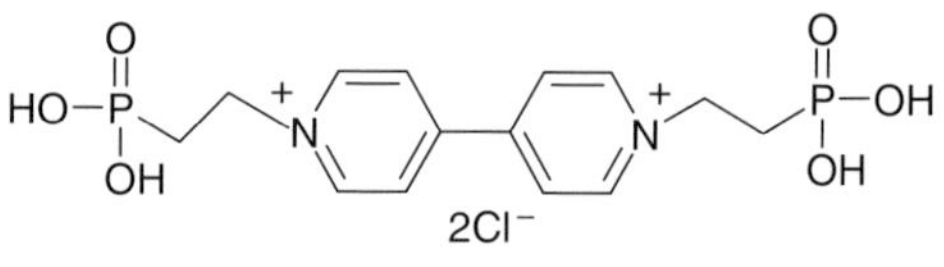

V

1.4 Criteria and terminology for ECD operation

Furthermore, the electrochrome–electrolyte interface has a capacitance C. Such capacitances are well known in electrochemistry to arise from ionic 'double layer' effects in which the field at (or charge on) the electrode attracts a 'layer' – really just an excess – of oppositely charged electrolyte ions from the bulk solution. The so-called 'rise time' of any electrochemical system denotes the time needed to set up (i.e. fully charge) this interfacial capacitance prior to successful transfer of electronic charge across the interface. Coloration will not commence between instigation of the colouring potential and completion of the rise time, a time that may be tens of milliseconds.

Applying a *pulsed* potential has been shown[91,92,93,94,95,96,97,98] to enhance significantly the rate at which electrochromic colour is generated, relative to potential-jump (or linear potential-increase) coloration. Although a quantitative explanation is not readily formulated, in essence the pulsing modifies the mass transport of electrochrome, eliminating kinetic 'bottle-necks', as outlined in Chapter 5. Pulsing is reported to speed up the response of viologen-based displays, enhancing the rate of electrochromic colour formation for 'viologens',[91] methyl,[92] heptyl[93] and aryl-substituted viologens;[94] pulsing also enhances the rates of electro-coloration ECDs based on TiO_2,[95] WO_3[96,97] and 'oxides'.[98] The Donnelly mirror in ref. 97 operates with a pulse sequence of frequency 10–20 Hz.

Substrate resistance

The indium–tin oxide (ITO) electrode substrate in an ECD has an appreciable electrical resistance R, although its effects will be ignored here. References 98 and 99 present a detailed discussion of the implications.

In many chemical systems, the uncoloured form of the electrochrome also has a high resistance R: poly(thiophene), poly(aniline), WO_3 and MoO_3 are a few examples. Sudden decreases in R during electro-coloration can cause unusual effects in the current time profiles.[100,101]

1.4.4 Write–erase efficiency

The write–erase efficiency is the fraction (percentage) of the originally formed coloration that can be subsequently electro-bleached. The efficiency must approach 100% for a successful display, which is a stringent test of design and construction.

The write–erase efficiency of an ECD of aqueous methyl viologen MV^{2+} as the electrochrome will always be low on a realistic time scale owing to the slowness of diffusion to and from the electrode through solution. The kinetics of electrochrome diffusion here are complicated since this electrochrome is extremely soluble in all applicable solvents for both its dicationic (uncoloured) and radical-cation (coloured) forms. Electrochrome diffusion is discussed in Chapters 4 and 5.

The simplest means of increasing the write–erase efficiency is to employ a type-II or type-III electrochrome, since between the write and erase parts of the coloration cycle the coloured form of the electrochrome is not lost from the electrode by diffusion. The write–erase efficiency of a type-I ECD may be improved by retarding the rate at which the solution-phase electrochrome can diffuse away from the electrode and into the solution bulk. Such retardation is achieved either by *tethering* the species to the surface of an electrode (then termed a 'derivatised' electrode), with, e.g., chemical bonding of viologens to the surface of particulate[102] TiO_2, or by immobilising the viologen species within a semi-solid electrolyte such as poly(AMPS). This is amplified in Section 14.2. Such modified type-I systems are effectively 'quasi type-III' electrochromes. While embedding in this way engenders an excellent long-term write–erase efficiency and a good electrochromic memory, it will also cause all response times to be extremely slow, perhaps unusably so.

1.4.5 Cycle life

An adjunct to the write–erase efficiency is the electrochromic device's cycle life which represents the number of write–erase cycles that can be performed by the ECD before any significant extent of degradation has occurred. (Such a write–erase cycle is sometimes termed a 'double potential step'.) The cycle life is therefore an experimental measure of the ECD durability. Figure 1.1 shows such a series of double potential steps, describing the response of hydrous nickel oxide immersed in KOH solution (0.1 mol dm^{-3}). The effect of film degradation over an extended time is clear. However, a 50% deterioration is often tolerable in a display.

Since ECDs are usually intended for use in windows or data display units, deterioration is best gauged by eye and with the same illumination, environment and cell driving conditions, that would be employed during normal cell operation. While it may seem obvious that the cycle life should be cited this way, many tests of cell durability in the literature of electrochromism involve cycles of much shorter duration than the ECD response time τ. Such partial tests are clearly of dubious value, but studies of cycle life are legion. Some workers have attempted to address this problem of variation in severity of the cycle test by borrowing terminology devised for the technology of battery discharge and describing a write–erase cycle as 'deep' or 'shallow' (i.e. the cycle length being greater than τ or less than τ, respectively).

1.4 Criteria and terminology for ECD operation

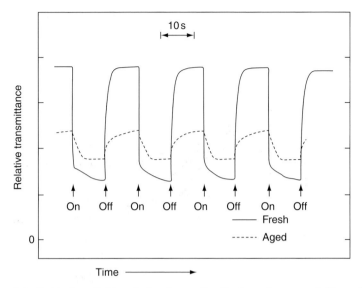

Figure 1.1 Optical switching behaviour of a fresh and an aged film of NiO electrodeposited onto ITO. The potential was stepped between 0 V (representing 'off') and 0.6 V (as 'on'). The aged film had undergone about 500 write–erase cycles. (Figure reproduced from Carpenter, M. K., Conell, R. S. and Corrigan, D. A. 'The electrochromic properties of hydrous nickel oxide'. *Sol. Energy Mater.* **16**, 1987, 333–46, by permission of Elsevier Science.)

The maximising of the cycle life is an obvious aim of device fabrication. A working minimum of about 10^5 is often stipulated.

There are several common reasons why devices fail: the conducting electrodes fail, the electrolyte fails, or one or both of the electrochromic layers fail. The electrolyte layers are discussed in Section 14.2, and overall device stability is discussed in Chapter 16. An individual device may fail for any or all of these reasons. Briefly, the most common causes of low cycle life are photodegradation of organic components within a device, either of the solvent or the electrochrome itself; and also the repeated recrystallisations within solid electrochromes associated with the ionic ingress and egress[99] that necessarily accompany redox processes of type-II and -III electrochromes.

1.4.6 Power consumption

An electrochromic display consumes no power between write or erase cycles, this retention of coloration being called the 'memory effect'. The intense colour of a sample of viologen radical cation remains undimmed for many months in the absence of chemical oxidising agents, such as molecular oxygen.

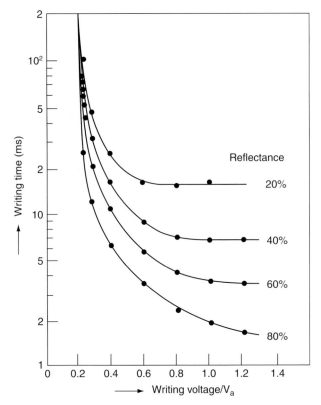

Figure 1.2 Calibration curve of electrochromic response time τ against the potentiostatically applied 'writing' potential V_a (cited against SCE) for heptyl viologen dibromide (**VI**) (0.1 mol dm^{-3}) in aqueous KBr (0.3 mol dm^{-3}). It is assumed that $\tau_{\text{(bleaching)}} = \tau_{\text{(coloration)}}$. (Figure reproduced from Schoot, C. J., Ponjeé, J. J., van Dam, H. T., van Doorn, R. A. and Bolwijn, P. J. 'New electrochromic memory device.' *Appl. Phys. Lett.*, **23**, 1973, 64, by permission of the American Institute of Physics.)

However, no-one has ever invented a perfect battery of infinite shelf life, and any ECD (all of which follow battery operation) will eventually fade unless the colour is renewed by further charging.

The charge consumed during one write–erase cycle is a function of the amount of colour formed (and removed) at an electrode during coloration (and decoloration). Schoot *et al.*[103] state that a contrast ratio of 20:1 may be achieved with a device employing heptyl viologen (1,1'-di-*n*-heptyl-4,4'-bipyridilium dibromide, **VI**) with a charge of 2 mC cm^{-2}, yielding an optical reflectance of 20%. Figure 1.2 shows a plot of response time for electrochromic coloration for HV^{2+} 2Br$^-$ in water as a function of electrochemical driving voltage.

1.4 Criteria and terminology for ECD operation

$$H_{15}C_7-\overset{+}{N}\diagdown\diagup-\overset{+}{N}-C_7H_{15}$$

2Br⁻

VI

Displays operating via cathode ray tubes (CRTs) and mechanical devices consume proportionately much more power than do ECDs. The amount of power consumed is so small that a solar-powered ECD has recently been reported,[104] the driving power coming from a single small cell of amorphous silicon. Such photoelectrochromic systems are discussed further in Chapter 15.

The power consumption of light-emitting diodes (LEDs) is relatively low, usually less than that of an ECD. Furthermore, ECDs consume considerably more power than liquid crystal displays, although a LCD-based display requires an applied field at all times if an image is to be permanent, i.e. it has no 'memory effect.' For this reason, Cohen asserts that ECD power consumption rivals that of LCDs;[105] he cites 7 or 8 mC cm^{-2} during the short periods of coloration or bleaching, and a zero consumption of charge during the longer periods when the optical density remains constant. This last criterion is overstated: a miniscule current is usually necessary to maintain the coloured state against the 'self-bleaching' processes mentioned earlier, comparable to battery deterioration (see p. 54).

1.4.7 Coloration efficiency η

The amount of electrochromic colour formed by the charge consumed is characteristic of the electrochrome. Its value depends on the wavelength chosen for study. The optimum value is the absorbance formed per unit charge density measured at λ_{max} of the optical absorption band. The coloration efficiency η is defined according to Eq. (1.7):

$$Abs = \eta Q, \qquad (1.7)$$

where *Abs* is the absorbance formed by passing a charge density of Q. A graph of *Abs* against Q accurately gives η as the gradient. For a detailed discussion of the way such optical data may be determined; see Section 4.3.

The majority of values cited in the literature relate to metal oxides; few are for organic electrochromes. A comprehensive list of coloration efficiencies is included in Section 4.3; additional values are sometimes cited in discussions of individual electrochromes.

1.4.8 Primary and secondary electrochromism

To repeat the definition, a cell comprises two half-cells. Each half-cell comprises a redox couple, needing the second electrode to allow the passage of charge through cell and electrodes. As ECDs are electrochemical cells, so each ECD requires a minimum of two electrodes. The simplest electrochromic light modulators have two electrodes directly in the path of the light beam. Solid-state electrochromic displays are, in practice, multi-layer devices (often called 'sandwiches'; see Chapter 14). If both electrodes bear an electrochromic layer, then the colour formation within the two should operate in a *complementary* sense, as illustrated below using the example of tungsten and nickel oxides. The WO_3 becomes strongly blue-coloured during reduction, while being effectively colourless when oxidised. However, sub-stoichiometric nickel oxide is dark brown-black when oxidised and effectively colourless when reduced.

When an ECD is constructed with these two oxides – each as a thin film (see Chapter 15) – one electrochrome film is initially reduced while the other is oxidised; accordingly, the operation of the device is that portrayed in Eq. (1.8):

$$\underbrace{WO_3 \quad + \quad M_xNiO_{(1-y)}}_{\text{bleached}} \rightarrow \underbrace{M_xWO_3 \quad + \quad NiO_{(1-y)}}_{\text{coloured}}. \qquad (1.8)$$

pale yellow colourless dark blue brown-black

The tungsten oxide in this example is the more strongly coloured material, so is termed the primary electrochrome, and the $NiO_{(1-y)}$ acts as the secondary (or counter) electrode layer. Ideally, the secondary electrochrome is chosen in order to complement the primary electrochrome, one colouring on insertion of counter ions while the other *loses* that ionic charge (or gains an oppositely charged ion) concurrently with its own coloration reaction, i.e. their respective values of η are of different sign. Note the way that charge passes through the cell from left to right and back again during electrochromic operation – 'electrochromism via the rocking chair mechanism', an uninformative phrase coined by Goldner et al. in 1984.[106] Nuclear-reaction analysis (NRA) is said to confirm this mechanistic mode,[107] but it is difficult to conceive of any other mechanism. In perhaps a majority of recent investigations, tungsten trioxide has been the primary electrochrome chosen owing to its high coloration efficiency, while the secondary layer has been an oxide of, e.g., iridium, nickel or vanadium.

The second electrode need not acquire colour at all. So-called 'optically passive' materials (where 'passive' here implies visibly non-electrochromic) are often the choice of counter electrode for an ECD. Examples of optically passive oxide layers include indium–tin oxide and niobium pentoxide. In an unusual design, if the counter electrode is a mirror-finish metal that is very thin and porous to ions, then ECDs can be made with one electroactive layer *behind* this electrode. In such a case, the layer behind the mirror electrode can be either strongly (but ineffectively) coloured or quite optically passive. Chapter 14 cites examples of such counter electrodes.

In devices operating in a complementary sense, both electrodes form their colour concurrently, although it is often impossible to deconvolute the optical response of a whole device into those of the two constituent electrochromic couples. When the electrochrome is a permanently solid in both forms (that is, type III), an approximate deconvolution is possible. This requires sophisticated apparatus such as *in situ* ellipsometry[108] and accompanying mathematical transformations. Recently, however, the group of Hagen and Jelle[109,110,111,112,113,114,115,116] have devised an ingenious and valuable means of overcoming this fundamental problem of distinguishing the optical contributions of each electrode. Devices were fabricated in which each constituent film had a narrow 'hole' (a bare area) of diameter *ca.* 5 mm, the hole in each film being positioned at a different portion of each film. By careful positioning of a narrow spectrometer beam through the ECD, the optical response of each individual layer is obtainable, while simultaneously the electrochemical response of the *overall* ECD is obtained concurrently via chronoamperometry in real time. This simple yet powerful 'hole' method has led to otherwise irresolvable analyses of these complicated, multi layer systems. For optimal results, the holes should not exceed about one hundredth of the overall active electrode area.

1.5 Multiple-colour systems: electropolychromism

While single-colour electrochromic transformations are usually considered elsewhere in this book, applications may be envisaged in which one electrochrome, or more together, evince a whole series of different colours, each coloured state generated at a characteristic applied potential. For a single-species electrochrome, a series of oxidation states, or charge states – each with its own colour – could be produced. Each state forms at a particular potential if each such state can be sustained, that is, if the species is 'multi-valent' in chemical parlance. Such systems should be called electropolychromic (but 'polyelectrochromic' prevails). A suitable example is methyl viologen, which is colourless as a dication, MV^{2+} (**II**), blue as a radical cation, and red–brown as a di-reduced neutral species, as described in Chapter 11. Electrochromic viologens with as many as six colours have been synthesised.[117]

Other systems that are electropolychromic are actually mixtures of several electrochromes. An example is Yasuda and Seto's[118] trichromic device comprising individual pixels addressed independently, each encapsulated to contain a different electrochrome. For example, the red electrochrome was 2,4,5,7-tetranitro-9-fluorenone (**VII**); a product from 2,4,7-trinitro-9-fluorenylidene malononitrile (**VIII**) is green, and reduction of TCNQ (tetracyanoquinodimethane, **IX**) yields the blue radical anion TCNQ⁻•. The chromophores in this system always remained in solution, i.e. were type I.

The colour evinced is a simple function of the potential applied, provided that each chromophore generates colour at a different potential (i.e. differs in $E^\ominus$ value: see Chapter 3) and there is no chemical interaction (that can be prevented by encapsulation).

References

1. Bard, A. J. and Faulkner, L. R. *Electrochemical Methods: Fundamentals and Applications* (2nd edn.), New York, Wiley, 2002.
2. Monk, P. M. S., Mortimer, R. J. and Rosseinsky, D. R. *Electrochromism: Fundamentals and Applications*, Weinheim, VCH, 1995.
3. [Online] at jchemed.chem.wisc.edu/Journal/Issues/1997/Aug/abs962.html (accessed 27 January 2006).
4. Forslund, B., A simple laboratory demonstration of electrochromism, *J. Chem. Ed.*, **74**, 1997, 962–3. The demonstration employed tungsten trioxide, electrodeposited from aqueous sodium tungstate onto SnO_2 coated electrodes.
5. [Online] at www.ifm.liu.se/biorgel/research/div/electrochromic.html (accessed 27 January 2006).

6. [Online] at www.aps.org/apsnews/0697/11962g.html (accessed 27 January 2006).
7. Murga, L. F. and Ondrechen, M. J. Theory of the Stark Effect in protein systems containing an electron donor–acceptor couple. *J. Inorg. Biochem.*, **70**, 1998, 245–52.
8. Bamfield, P. *Chromic Phenomena: Technological Applications of Colour Chemistry*, Cambridge, Royal Society of Chemistry, 2001.
9. Loew, L. M., Scully, L., Simpson, L. M. and Waggoner, A. S. Evidence for a charge shift electrochromic mechanism in a probe of membrane potential. *Nature (London)*, **281**, 1979, 497–9.
10. Huesmann, H., Gabrielli, G. and Caminati, G. Monolayers and Langmuir–Blodgett films of the electrochromic dye Di-8-ANEPPS. *Thin Solid Films*, **327–329**, 1998, 804–7.
11. Professor L. M. Loew, 1999, personal communication.
12. Vredenberg, W. J. Electrogenesis in the photosynthetic membrane: fields, facts and features. *Bioelectrochem. Bioenergy*, **44**, 1997, 1–11.
13. Platt, J. R. Electrochromism, a possible change of color producible in dyes by an electric field. *J. Chem. Phys.*, **34**, 1961, 862–3.
14. Tiede, D. M., Utschig, L., Hanson, D. K. and Gallo, D. M. Resolution of electron and proton transfer events in the electrochromism associated with quinone reduction in bacterial reaction centers. *Photosyn. Res.*, **55**, 1998, 267–73.
15. Tiede, D. M., Vazquez, J., Cordova, J. and Marone, P. A. Structural and function changes in photosynthetic bacterial reaction center proteins induced by incorporating different metal ions. *Biochemistry*, **35**, 1996, 10763–73.
16. Miksovska, J., Maróti, P., Tandori J., Schiffer, M., Hanson, D. K. and Sebban, P. Modulation of the free energy level of Q_A^- by distant electrostatic interactions in the photosynthetic reaction center. *Biochemistry*, **35**, 1996, 15411–17.
17. Crimi, M., Fregni, V., Altimari, A. and Melandri, B. A. Unreliability of carotenoid electrochromism for the measure of electrical potential differences induced by ATP hydrolysis in bacterial chromatophores. *FEBS Lett.*, **367**, 1995, 167–72.
18. O'Shea, P. Physical landscapes in biological membranes. *Philos. Trans. R. Soc. London, Ser. A Math. Phys. Eng. Sci.*, **363**, 2005, 575–88.
19. Asawakarn, T., Cladera, J. and O'Shea, P. Effects of the membrane dipole potential on the interaction of Saquinavir with phospholipid membranes and plasma membrane receptors of Caco-2 cells. *J. Biol. Chem.*, **276**, 2001, 38457–63.
20. Cladera, J. and O'Shea, P. Generic techniques for fluorescence measurements of protein–ligand interactions: real-time kinetics and spatial imaging. In Harding S. E. and Chowdhry, B. (eds.), *Protein–Ligand Interactions*, Oxford, Oxford University Press, 2001, pp. 169–200.
21. Ross, E., Bedlack, R. S. and Loew, L. M. Dual-wavelength ratiometric fluorescence measurement of the membrane dipole potential. *Biophys. J.*, **67**, 1994, 208–16.
22. Montana, V., Farkas, D. L. and Loew, L. M. Dual-wavelength ratiometric fluorescence measurements of membrane potential. *Biochemistry*, **28**, 1989, 4536–9.
23. Wall, J. S., Golding, C., van Veen, M. and O'Shea, P. S. The use of fluoresceinphosphatidylethanolamine as a real-time probe for peptide–membrane interactions. *Mol. Memb. Biol.*, **12**, 1995, 181–90.
24. [Online] at www.glass.ie/ (accessed 27 January 2006).
25. Georg, A., Graf, W., Neumann, R. and Wittwer, V. The role of water in gasochromic WO_3 films. *Thin Solid Films*, **384**, 2001, 269–75.

26. Georg, A. Graf, W. Neumann, R. and Wittwer, V. Stability of gasochromic WO_3 films, *Sol. Energy Mater. Sol. Cells*, **63**, 2000, 165–176.
27. Georg, A., Graf, W., Neumann, R. and Wittwer, V. Mechanism of the gasochromic coloration of porous WO_3 films. *Solid State Ionics*, **127**, 2000, 319–328.
28. Georg, A., Graf, W., Schweiger, D.,Wittwer, V., Nitz, P. and Wilson, H. R. Switchable glazing with a large dynamic range in total solar energy transmittance (TSET). *Sol. Energy*, **62**, 1998, 215–228.
29. Opara Krašovec, U., Orel, B., Georg, A. and Wittwer, V. The gasochromic properties of sol–gel WO_3 films with sputtered Pt catalyst. *Sol. Energy*, **68**, 2000, 541–551.
30. Schweiger, D., Georg, A., Graf, W. and Wittwer, V. Examination of the kinetics and performance of a catalytically switching (gasochromic) device. *Sol. Energy Mater. Sol. Cells*, **54**, 1998, 99–108.
31. Shanak, H., Schmitt, H., Nowoczin, J. and Ziebert, C. Effect of Pt-catalyst on gasochromic WO_3 films: optical, electrical and AFM investigations. *Solid State Ionics*, **171**, 2004, 99–106.
32. Wittwer, V., Datz, M., Ell, J., Georg, A., Graf, W. and Walze, G. Gasochromic windows. *Sol. Energy Mater. Sol. Cells*, **84**, 2004, 305–14.
33. [Online] at www.gyriconmedia.com/smartpaper/faq.asp (accessed 27 January 2006).
34. [Online] at www2.parc.com/dhl/projects/gyricon (accessed 27 January 2006).
35. Granqvist, G. C. *Handbook of Inorganic Electrochromic Materials*, Amsterdam, Elsevier, 1995.
36. Agnihotry, S. A. Electrochromic devices: present and forthcoming technology. *Bull. Electrochem.*, **12**, 1996, 707–12.
37. Bange, K., Gambke, T. and Sparschuh, G. Optically active thin-film coatings. In Hummel, R. E. and Guenther, K. H. (eds.), *Handbook of Optical Properties*, Boca Raton, FL, CRC Press, 1995, pp. 105–34.
38. Granqvist, C. G. Electrochromism and smart window design. *Solid State Ionics*, **53–6**, 1992, 479–89.
39. Granqvist, C. G. Electrochromic materials: microstructure, electronic bands, and optical properties. *Appl. Phys. A*, **56**, 1993, 3–12.
40. Granqvist, C. G. Electrochromics and smart windows. *Solid State Ionics*, **60**, 1993, 213–14.
41. Granqvist, C. G. Electrochromic materials and devices. *Proc. SPIE*, **2968**, 1997, 158–166.
42. Granqvist, C. G. Electrochromism and electrochromic devices. In Gellings, P. J. and Bouwmeester, H. J. M. (eds.), *The CRC Book of Solid State Electrochemistry*. Boca Raton, FL, CRC Press, 1997, pp. 587–615.
43. Granqvist, C. G. Progress in solar energy materials: examples of work at Uppsala University. *Renewable Energy*, **15**, 1998, 243–250.
44. Granqvist, C. G., Azens, A., Hjelm, A., Kullman, L., Niklasson, G. A., Rönnow, D., Strømme Mattson, M., Veszelei, M. and Vaivers, G. Recent advances in electrochromics for smart windows applications, *Sol. Energy*, **63**, 1998, 199–216.
45. Granqvist, C. G. and Wittwer, V. Materials for solar energy conversion: an overview. *Sol. Energy Mater. Sol. Cells*, **54**, 1998, 39–48.
46. Granqvist, C. G. Electrochromic tungsten oxide films: review of progress 1993–1998. *Sol. Energy Mater. Sol. Cells*, **60**, 2000, 201–62.
47. Granqvist, C. G., Avendaño, E. and Azens, A. Electrochromic coatings and devices: survey of some recent advances. *Thin Solid Films*, **442**, 2003, 201–11.

48. Granqvist, C. G. Solar energy materials *Adv. Mater.*, **15**, 2003, 1789–1803.
49. Granqvist, C. G., Avendaño, E. and Azens, A. Advances in electrochromic materials and devices. *Mater. Sci. Forum*, **455–456**, 2004, 1–6.
50. Green, M. The promise of electrochromic systems. *Chem. Ind.*, 1996, 641–4.
51. Greenberg, C. B. Chromogenic materials: electrochromic. In Krosch, J. I. (ed.), *Kirk-Othmer Encyclopedia of Chemical Technology* (fourth edn.), New York, Wiley, 1991, vol. 6, pp. 312–21.
52. Greenberg, C. B. Optically switchable thin films: a review. *Thin Solid Films*, **251**, 1994, 81–93.
53. Lampert, C. M. Smart switchable glazing for solar energy and daylight control. *Sol. Energy Mater. Sol. Cells*, **52**, 1998, 207–21.
54. Lampert, C. M. Progress in switching mirrors. *Proc. SPIE*, **4458**, 2001, 95–103.
55. Lampert, C. M. Chromogenic smart materials. *Materials Today*, **7**, 2004, 28–35.
56. Monk, P. M. S. Electrochromism and electrochromic materials for displays. In Nalwa, H. S. (ed.), *Handbook of Advanced Electronic and Photonic Materials*, San Diego, Academic Press, 2001, vol. 7, pp. 105–59.
57. Monk, P. M. S. Electrochromism and electronic display devices. In Nalwa, H. S. and Rohwer, L. S. (eds.), *Handbook of Luminescent Display Materials and Devices*, San Diego, Academic Press, 2002, vol. 3, pp. 261–370.
58. Mortimer, R. J. Electrochromic materials. *Chem. Soc. Rev.*, **26**, 1997, 147–56.
59. Rosseinsky, D. R. and Mortimer, R. J. Electrochromic systems and the prospects for devices. *Adv. Mater.*, **13**, 2001, 783–93.
60. Rowley, N. M. and Mortimer, R. J. New electrochromic materials. *Sci. Prog.*, **85**, 2002, 243–62.
61. Mortimer, R. J., Dyer, A. L. and Reynolds, J. R. Electrochromic organic and polymeric materials for display applications. *Displays*, **27**, 2006, 1–18.
62. Passerini, S., Pileggi, R. and Scrosati, B. Laminated electrochromic devices: an emerging technology. *Electrochim. Acta*, **37**, 1992, 1703–6.
63. Scrosati, B. Properties of selected electrochromic materials. In Chowdari, B. V. R. and Radharkrishna, S. (eds.), *Proceedings of the International Seminar on Solid State Ionic Devices*, Singapore, World Publishing Co., 1992, pp. 321–36.
64. Somani, P. R. and Radhakrishnan, S. Electrochromic materials and devices: present and future. *Mater. Chem. Phys.*, **77**, 2003, 117–33.
65. Yamamoto, T. and Hayashida, N. π-Conjugated polymers bearing electronic and optical functionalities: preparation, properties and their applications. *Reactive and Functional Polymers*, **37**, 1998, 1–17.
66. Volke, J. and Volkeova, V. Electrochromismus a zavádení elektrochromní techniky ['Electrochromism and electrochromic technology']. *Chem. Listy*, **90**, 1996, 137–46 [in Czech: the abstract and title are in English].
67. McGourty, C. 'Thinking' windows cut the dazzle. *Daily Telegraph*, 2 April 1991.
68. Hadfield, P. Tunable sunglasses that can fade in the shade. *New Scientist*, 22 March 1993, 22.
69. Hunkin, T. Just give me the fax. *New Scientist*, 13 February 1993, 33–7.
70. Monk, P. M. S., Mortimer, R. J. and Rosseinsky, D. R. Through a glass darkly. *Chem. Br.*, **31**, 1995, 380–382.
71. Bowonder, B., Sarnot, S. L., Rao, M. S. and Rao, D. P. Electronic display technologies – state of the art, *Electron. Inform. Plan.*, **21**, 1994, 683–746.
72. Durst, R. A., Baumner, A. J., Murray, R. W., Buck, R. P. and Andrieux, C. P. Chemically modified electrodes: recommended terminology and definitions. *Pure Appl. Chem.*, **69**, 1997, 1317–23.

73. Chang, I. F., Gilbert, B. L. and Sun, T. I. Electrochemichromic systems for display applications. *J. Electrochem. Soc.*, **122**, 1975, 955–62.
74. Compton, R. G., Waller, A. M., Monk, P. M. S. and Rosseinsky, D. R. Electron paramagnetic resonance spectroscopy of electrodeposited species from solutions of 1,1′-bis (*p*-cyanophenyl)-4,4′-bipyridilium (cyanophenylparaquat, CPQ). *J. Chem. Soc., Faraday Trans.*, **86**, 1990, 2583–6.
75. Rosseinsky, D. R. and Monk, P. M. S. Electrochromic cyanophenylparaquat (CPQ: 1,1′-bis-cyanophenyl-4,4′-bipyridilium) studied voltammetrically, spectroelectrochemically and by ESR. *Sol. Energy Mater. Sol. Cells*, **25**, 1992, 201–10.
76. Rosseinsky, D. R., Monk, P. M. S. and Hann, R. A. Anion-dependent aqueous electrodeposition of electrochromic 1,1′-*bis*-cyanophenyl-4,4′-bipyridilium (cyanophenylparaquat) radical cation by cyclic voltammetry and spectroelectrochemical studies. *Electrochim. Acta*, **35**, 1990, 1113–23.
77. Monk, P. M. S. *The Viologens: Physicochemical Properties, Synthesis and Applications of the Salts of 4,4′-Bipyridine*, Chichester, Wiley, 1998.
78. Grant, B., Clecak, N. J. and Oxsen, M. Study of the electrochromism of methoxyfluorene compounds. *J. Org. Chem.*, **45**, 1980, 702–5.
79. Faughnan, B. W. and Crandall, R. S. Electrochromic devices based on WO_3, in Pankove J. L. (ed.), *Display Devices*, Berlin, Springer-Verlag, 1980, pp. 181–211.
80. Ziegler, J. P. and Howard, B. M. Applications of reversible electrodeposition electrochromic devices. *Sol. Energy Mater. Sol. Cells*, **39**, 1995, 317–31.
81. Barclay, D. J., Dowden, A. C., Lowe, A. C. and Wood, J. C. Viologen-based electrochromic light scattering display. *Appl. Phys. Lett.*, **42**, 1983, 911–13.
82. Barclay, D. J., Bird, C. L., Kirkman, D. H., Martin, D. H. and Moth, F. T. An integrated electrochromic data display. *SID Digest*, 1980, 124–5.
83. Mathew, J. G. H., Sapers, S. P., Cumbo, M. J., O'Brien, N. A., Sargent, R. B., Raksha, V. P., Lahaderne, R. B. and Hichwa, B. P. Large area electrochromics for architectural applications. *J. Non-Cryst. Solids*, **218**, 1997, 342–6.
84. Siddle, J., Pilkington PLC, personal communication, 1991.
85. Munro, B., Kramer, S., Zapp, P., Krug, H. and Schmidt, H. All sol–gel electrochromic system for plate glass. *J. Non-Cryst. Solids*, **218**, 1997, 185–8.
86. Özer, N. Reproducibility of the coloration processes in TiO_2 films. *Thin Solid Films*, **214**, 1992, 17–24.
87. Sato, Y. Characterization of thermally oxidized iridium oxide films. *Vacuum*, **41**, 1990, 1198–200.
88. Canon, K. K. Electrochromic device, Jpn. Kokai Tokkyo Koho, Japanese Patent JP 6,004,925, as cited in *Chem. Abs.* **102**: P 212,797, 1985.
89. Welsh, D. M., Kumar, A., Morvant, M. C. and Reynolds, J. R. Fast electrochromic polymers based on new poly(3,4-alkylenedioxythiophene) derivatives. *Synth. Met.*, **102**, 1999, 967–8.
90. Cummins, D., Boschloo, G., Ryan, M., Corr, D., Rao, S. N. and Fitzmaurice, D. Ultrafast electrochromic windows based on redox-chromophore modified nanostructured semiconducting and conducting films. *J. Phys. Chem. B*, **104**, 2000, 11449–59.
91. Knapp, R. C., Turnbull, R. R. and Poe, G. B., Gentex Corporation. Reflectance control of an electrochromic element using a variable duty cycle drive. US Patent 06084700, 2000.
92. Monk, P. M. S., Fairweather, R. D., Ingram, M. D. and Duffy, J. A. Pulsed electrolysis enhancement of electrochromism in viologen systems: influence of comproportionation reactions. *J. Electroanal. Chem.*, **359**, 1993, 301–6.

93. Electrochromic displays. In Howells, E. R. (ed.), *Technology of Chemicals and Materials for the Electronics Industry*, Chichester, Ellis Horwood, 1984, pp. 266–76.
94. Protsenko, E. G., Klimisha, G. P., Krainov, I. P., Kramarenko, S. F. and Distanov, B. G. *Deposited Doc.*, SPSTL 971, Khp-D81, 1981, as cited in *Chem. Abs.* **98**: 170,310.
95. Ottaviani, M., Panero, S., Morizilli, S., Scrosati, B. and Lazzari, M. The electrochromic characteristics of titanium oxide thin film. *Solid State Ionics*, **20**, 1986, 197–202.
96. Monk, P. M. S., Duffy, J. A. and Ingram, M. D. Pulsed enhancement of the rate of coloration for tungsten trioxide based electrochromic devices. *Electrochim. Acta*, **43**, 1998, 2349–57.
97. Schierbeck, K. L., Donnelly Corporation. Digital electrochromic mirror system. US Patent, 06089721, 2000.
98. Statkov, L. I. Peculiarities of the mechanism of the electrochromic coloring of oxide films upon pulsed electrochemical polarization. *Russ. J. Appl. Chem.*, **70**, 1997, 653–4.
99. Ho, K.-C., Singleton, D. E. and Greenberg, C. B. The influence of terminal effects on the performance of electrochromic windows. *J. Electrochem. Soc.*, **137**, 1990, 3858–64.
100. Aoki, K. and Tezuki, Y. Chronoamperometric response to potentiostatic doping at polypyrrole-coated microdisk electrodes. *J. Electroanal. Chem.*, **267**, 1989, 55–66.
101. Ingram, M. D., Duffy, J. A. and Monk, P. M. S. Chronoamperometric response of the cell ITO | H_xWO_3 | PEO–H_3PO_4 (MeCN) | ITO. *J. Electroanal. Chem.*, **380**, 1995, 77–82.
102. Cinnsealach, R., Boschloo, G., Nagaraja Rao, S. and Fitzmaurice, D. Electrochromic windows based on viologen-modified nanostructured TiO_2 films. *Sol. Energy Mater. Sol. Cells*, **55**, 1998, 215–23.
103. Schoot, C. J., Ponjeé, J. J., van Dam, H. T., van Doorn, R. A. and Bolwijn, P. J. New electrochromic memory device. *Appl. Phys. Lett.*, **23**, 1973, 64–5.
104. Bullock, J. N., Bechinger, C., Benson, D. K. and Branz, H. M. Semi-transparent *a*-SiC:H solar cells for self-powered photovoltaic-electrochromic devices. *J. Non-Cryst. Solids*, **198–200**, 1996, 1163–7.
105. Cohen, C. Electrochromic display rivals liquid crystals for low-power needs. *Electronics*, **11**, 1981, 65–6.
106. Goldner, R. B., Arntz, F. O., Dickson, K., Goldner, M. A., Haas, T. E., Liu, T. Y., Slaven, S., Wei, G., Wong, K. K. and Zerigian, P. Some lessons learned from research on a thin film electrochromic window. *Solid State Ionics*, **70–71**, 1994, 613–18.
107. Goldner, R. B., Haas, T., Arntz, F. O., Slaven, S. and Wong, G. Nuclear reaction analysis profiling as direct evidence for lithium ion mass transport in thin film 'rocking chair' structures. *Appl. Phys. Lett.*, **62**, 1993, 1699–701.
108. Bader, G., Ashrit, P. V. and Truong, V.-V. Transmission and reflection ellipsometry of thin films and multilayer systems. *Appl. Opt.*, **37**, 1998, 1146–1151.
109. Jelle, B. P. and Hagen, G. Transmission spectra of an electrochromic window based on polyaniline, Prussian blue, and tungsten oxide. *J. Electrochem. Soc.*, **140**, 1993, 3560–5.
110. Jelle, B. P., Hagen, G., Hesjevik, S. M. and Ødegård, R. Transmission through an electrochromic window based on polyaniline, tungsten oxide and a solid polymer electrolyte. *Mater. Sci. Eng. B*, **13**, 1992, 239–41.

111. Jelle, B. P., Hagen, G. and Nodland, S. Transmission spectra of an electrochromic window consisting of polyaniline, Prussian blue and tungsten oxide, *Electrochim. Acta*, **38**, 1993, 1497–500.
112. Jelle, B. P., Hagen, G. and Ødegård, R. Transmission spectra of an electrochromic window based on polyaniline, tungsten oxide and a solid polymer electrolyte. *Electrochim. Acta*, **37**, 1992, 1377–80.
113. Jelle, B. P., Hagen, G., Sunde, S. and Ødegård, R. Dynamic light modulation in an electrochromic window consisting of polyaniline, tungsten oxide and a solid polymer electrolyte. *Synth. Met.*, **54**, 1993, 315–20.
114. Jelle, B. P. and Hagen, G. Performance of an electrochromic window based on polyaniline, prussian blue and tungsten oxide, *Sol. Energy Mater. Sol. Cells*, **58**, 1999, 277–86.
115. Jelle, B. P. and Hagen, G. Electrochemical multilayer deposition of polyaniline and Prussian blue and their application in solid state electrochromic windows. *J. Appl. Electrochem.*, **28**, 1998, 1061–65.
116. Jelle, B. P., Hagen, G. and Birketveit, O. Transmission properties for individual electrochromic layers in solid state devices based on polyaniline, Prussian Blue and tungsten oxide. *J. Appl. Electrochem.*, **28**, 1998, 483–9.
117. Rosseinsky, D. R. and Monk, P. M. S. Studies of tetra-(bipyridilium) salts as possible polyelectrochromic materials. *J. Appl. Electrochem.*, **24**, 1994, 1213–21.
118. Yasuda, A. and Seto, J. Electrochemical studies of molecular electrochromism. *Sol. Energy Mater. Sol. Cells*, **25**, 1992, 257–68.

2
A brief history of electrochromism

2.1 Bibliography; and 'electrochromism'

Brief histories of electrochromism have been delineated by Chang[1] (in 1976), Faughnan and Crandall[2] (in 1980), Byker[3] (in 1994) and Granqvist[4] (in 1995). Other published histories rely very heavily on these sources. The additional histories of Agnihotry and Chandra[5] (in 1994) and Granqvist et al.[6] (in 1998) chronicle further advances in making electrochromic devices for windows. The first books on electrochromism were those of Granqvist,[4] and Monk, Mortimer and Rosseinsky,[7] which were both published in 1995.

Platt[8] coined the term 'electrochromism' in 1961 to indicate a colour generated via a molecular Stark effect (see page 4) in which orbital energies are shifted by an electric field. His work follows earlier studies by Franz and Keldysh in 1958,[9,10] who applied huge electric fields to a film of solid oxide causing spectral bands to shift. These effects are not the main content of this book.

2.2 Early redox-coloration chemistry

In fact, redox generation of colour is not new – twentieth-century redox titration indicators come to the chemist's mind ('redox', Section 1.1, implying electron transfer). However, as early as 1815 Berzelius showed that pure WO_3 (which is pale yellow) changed colour on reduction when warmed under a flow of dry hydrogen gas,[11] and in 1824, Wöhler[12] effected a similar chemical reduction with sodium metal. Section 1.4 and Eq. (1.5), and Eq. (2.5) below, indicate the extensive role of WO_3 in electrochromism, amplified further in Section 6.2.1.

2.3 Prussian blue evocation in historic redox-coloration processes

An early form of photography devised in 1842 by Sir John Frederick William Herschel[13] is a ubiquitous example of a photochromic colour change involving electron transfer, devised for a technological application. Its inventor was a friend of Fox Talbot, who is credited with inventing silver-based photography, of like mechanism, in 1839. Herschel's method produced photographs and diagrams by generating Prussian blue $KFe^{III}[Fe^{II}(CN)_6](s)$ from moist paper pre-impregnated with ferric ammonium citrate and potassium ferricyanide, forming yellow Prussian brown $Fe^{3+}[Fe(CN)_6]^{3-}$ or $Fe^{III}[Fe^{III}(CN)_6]$ (for Prussian blue details see reaction (3.12) and p. 282 ff.; for oxidation-state representation by Roman numerals; see p. 35). Wherever light struck the photographic plate, photo reduction of Fe^{III} yielded Fe^{II} in the complex, hence Prussian blue formation; see eq. (2.1):

$$Fe^{3+}[Fe^{III}(CN)_6]^{3-}(s) + K^+(aq) + e^-(h\nu) \rightarrow KFe^{III}[Fe^{II}(CN)_6](s), \quad (2.1)$$

where $e^-(h\nu)$ represents an electron photolysed from water or other ambient donor, a process often oversimplified as resulting from reduction of Fe^{3+} by the photolysed e^-:

$$[H_2O + h\nu \rightarrow e^- + \{H_2O^+\}; Fe^{3+}(aq) + e^- \rightarrow Fe^{2+}(aq)]; \quad (2.2)$$

followed by

$$K^+(aq) + Fe^{2+}(aq) + [Fe^{III}(CN)_6]^{3-}(aq) \rightarrow KFe^{III}[Fe^{II}(CN)_6](s), \quad (2.3)$$

where $\{H_2O^+\}$ represents water-breakdown species. Herschel called his process 'cyanotype'. By the 1880s, so-called 'blueprint' paper was manufactured on a large scale as engineers and architects required copies of architectural drawings and mechanical plans. This widespread availability revived cyanotype, as a photographic process for large reproductions, to late in the twentieth century, under the common name of 'blueprint'. This word has become an English synonym for 'plan'.

Soon after Herschel, in 1843 Bain patented a primitive form of fax transmission that again relied on the generation of a Prussian blue compound.[14,15] It involved a stylus of pure soft iron resting on damp paper pre-impregnated with potassium ferrocyanide. In an electrical circuit, electro-oxidation of the (positive) iron tip formed ferric ion from the metal, which consumes the iron as it combines with ferrocyanide ion to produce a very dark form of insoluble Prussian blue. Thus the iron electrode generates a track of darkly-coloured deposit wherever the positive stylus touches the paper.

2.4 Twentieth century: developments up to 1980

Probably the first suggestion of an electrochromic *device* involving electrochemical formation of colour is presented in a London patent of 1929,[16] which concerns the electrogeneration of molecular iodine from iodide ion. Such molecular I_2 then effects the chemical oxidation of a dye precursor, thus forming a bright colour. This example again represents an electrochromic reaction. However, the proneness of iodide to photo-oxidation is discouraging to any further development.

In 1962, Zaromb published now-neglected studies of electrodepositing silver in desired formats from aqueous solutions of Ag^+ [17,18] or complexes thereof.[19] Electro-reduction of Ag(I) ion yields a thin layer of metallic silver that reflects incident light if continuous, or is optically absorbent if the silver is particulate. Zaromb called his system an 'electroplating light modulator', and explicitly said it represented a 'viable basis for a display'. His work was not followed up until the mid 1970s, e.g. by the groups of Camlibel[20] and of Ziegler, who deposited metallic bismuth.[21,22]

The first recorded colour change following electrochemical reduction of a solid, tungsten trioxide, was that of Kobosew and Nekrassow[23] in 1930. The colour generation reaction (*cf.* Section 9.2.1) followed Eq. (2.4):

$$WO_3(s) + x(H^+ + e^-) \rightarrow H_xWO_3(s). \tag{2.4}$$

Their WO_3 was coated on an electrode, itself immersed in aqueous acid. The electrode substrate is unknown, but presumably inert.

By 1942 Talmay[24,25] had a patent for electrochromic printing – he called it 'electrolytic writing paper' – in which paper was pre-impregnated with particulate MoO_3 and/or WO_3. A blue–grey image forms following an electron-transfer reaction: in effect, the electrode acted as a stylus, forming colour wherever the electrode traversed the paper. The electrochromic coloration reaction followed Eq. (2.4) above, and the proton counter ion came from the ionisation of the water in the paper.

In 1951, Brimm *et al.*[26] extended the work of Kobosew and Nekrassow to effect *reversible* colour changes, for Na_xWO_3 immersed in aqueous acid (sulfuric acid of concentration 1 mol dm^{-3}). A little later, in 1953, Kraus of Balzers in Lichtenstein[27] advocated the reversible colour–bleach behaviour of WO_3 (again immersed in aqueous H_2SO_4) *as a basis for a display*: this work was regrettably never published.

Probably the first company to seek commercial exploitation of an electrochromic product was the Dutch division of Philips, again in the early 1960s.

Their prototype device utilised an aqueous organic viologen (see Chapter 11), heptyl viologen (HV: 1,1'-n-heptyl-4,4'-bipyridilium) as the bromide salt. Their first patent dates from 1971,[28] and their first academic paper from 1973.[29]

At much the same time, Imperial Chemical Industries (ICI) in Britain initiated a far-reaching program to develop an electrochromic device. Like Philips, they first analysed the response of heptyl viologen in water but quickly decided its coloration efficiency η was too low, and changed to the larger viologen cyanophenyl paraquat [CPQ: 1,1'-bis(1-cyanophenyl)-4,4'-bipyridilium] as the sulfate salt. Their first patent dates from May 1969.[30] By early 1970, ICI was seeking tenders to commercialise a CPQ-based device.[31,32]

Other devices based on heptyl viologen were being investigated by Barclay's group at Independent Business Machines (IBM),[33] and by Texas Instruments in Dallas, although their work was not published until after their programme was discontinued.[34]

As none of these studies attracted much attention, probably most workers now attribute the first widely accepted suggestion of an electrochromic device to Deb (then at Cyanamid in the USA) in 1969,[35] following a technical report from the previous year.[36] Deb formed electrochromic colour by applying an electric field of 10^4 V cm^{-1} across a thin film of dry tungsten trioxide vacuum deposited on quartz: he termed the effect 'electrophotography'. (This wording may reflect his earlier work dating from 1966, when he analysed thin-film vacuum-deposited MoO_3 on quartz, which acquired colour following UV irradiation.[37]) Figure 2.1 shows a schematic representation of his cell. In fact, Deb's film of WO_3 was open to the air rather than immersed in ion-containing electrolyte solutions, suggesting that the mobile counter cations might have come from simultaneous ionisation of interstitial and/or adsorbed water. At the time, Deb suggested the colour arose from F-centres, much like the colour formed by heating or irradiating crystals of metal halides in a field. The background to Deb's work was recounted much later,[38] in 1995.

In 1971, Blanc and Staebler[39] produced an electrochromic effect superior to most previously published. They applied electrodes to the opposing faces of doped, crystalline $SrTiO_3$ and observed an electrochromic colour move into the crystal from the two electrodes. The charge carriers are (apparently) oxide ions, which migrate through the crystal in response to redox changes at the electrodes. Their work has not been followed, probably because no viable device was likely to ensue as their crystal had to be heated to *ca.* 200 °C.

In 1972, Beegle developed a display of WO_3 having identical counter and working electrodes, with an intervening opaque layer.[40]

2.4 Twentieth century: developments up to 1980

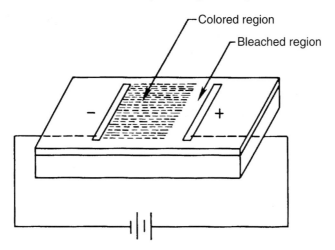

Figure 2.1 Electrocoloration of thin-film WO$_3$ film using a surface electrode geometry. (Figure reproduced from Deb, S. K. 'Reminiscences on the discovery of electrochromic phenomena in transition-metal oxides'. *Sol. Energy Mater. Sol. Cells*, **39**, 1995, 191–201.)

Nowadays most workers cite Deb's later paper,[41] which dates from 1973, as the true birth of electrochromic technology. It is often said that this seminal paper describes the first 'true' electrochromic device, with a film of WO$_3$ immersed in an ion-containing electrolyte. In fact ref. 41 does not mention aqueous electrolytes at all but rather, another film of WO$_3$ vacuum evaporated onto a substrate of quartz. Deb does correctly identify the ionisation of water as the source of the protons necessary for Eq. (2.4), but suggests oxide ions extracted from the WO$_3$ lattice, rather than proton insertion, for the coloration mechanism.[42]

Within a year of Deb's 1973 paper, Green and Richman[43] in London proposed a system based on WO$_3$ in which the mobile ion was Ag$^+$. In 1975, Faughnan *et al.* of the RCA Laboratories in Princeton, New Jersey, in a pivotal review,[44] reported WO$_3$ undergoing reversible electrochromic colour changes while immersed in aqueous sulfuric acid. Faughnan *et al.* analysed the speed of colour change in terms of Butler–Volmer electrode dynamics, establishing a pioneering model of electro-bleaching[45] and electro-coloration[46] that is still relevant now.

Mohapatra of the Bell Laboratories in New Jersey published the first description of the reversible electro-insertion of lithium ion, Eq. (2.5), in 1978:[47]

$$WO_3(s) + x(Li^+(aq) + e^-) \rightarrow Li_xWO_3(s). \qquad (2.5)$$

Meanwhile, the electrochromism of organic materials also developed momentum. In 1974, Parker et al.[48] prepared methoxybiphenyl species, the electrogenerated radical cations of which are intensely coloured (see p. 379). While he nowhere employs the word 'electrochromism' or its cognates, his paper, displaying acute awareness of the technological scope of such colour changes, cited values of λ_{max} for the several radical cations.

Later, Kaufman et al. of IBM in New York published the first report of an electrochromic polymer comprising an alkyl-chain backbone with pendant electroactive species[49,50] (see Section 10.2). The details in his preliminary report[51] are as indistinct as are many patents, but his later work reveals that his electrochromes were based on tetrathiafulvalene and quinone moieties.[49] In 1979 came the first account of an electrochromic conducting polymer, when Diaz et al.[52] (also of IBM in San Jose, California), announced the electrosynthesis of thin-film poly(pyrrole); see Section 10.3.

The electrochemical literature of the twentieth century will undoubtedly provide further early reports of electrochromism.

References

1. Chang, I. F. Electrochromic and electrochemichromic materials and phenomena. In Kmetz, A. R. and von Willisen, F. K. (eds.), *Non-emissive Electrooptical Displays*, New York, Plenum Press, 1976, pp. 155–96.
2. Faughnan, B. W. and Crandall, R. S. Electrochromic displays based on WO_3. In Pankove, J. I. (ed.), *Display Devices*, Berlin, Springer Verlag, 1980, ch. 5, pp. 181–211.
3. Byker, H. J. Commercial developments in electrochromics. *Proc. Electrochem. Soc.*, **94–2**, 1994, 1–13.
4. Granqvist, G. C. *Handbook of Inorganic Electrochromic Materials*, Amsterdam, Elsevier, 1995.
5. Agnihotry, S. A. and Chandra, S. Electrochromic devices: present and forthcoming technology, *Indian J. Eng. Mater. Sci.*, **1**, 1994, 320–34.
6. Granqvist, C. G., Azens, A., Hjelm, A., Kullman, L., Niklasson, G. A., Rönnow, D., Strømme Mattson, M., Veszelei, M. and Vaivers, G. Recent advances in electrochromics for smart window applications, *Sol. Energy*, **63**, 1998, 199–216.
7. Monk, P. M. S., Mortimer, R. J. and Rosseinsky, D. R. *Electrochromism: Fundamentals and Applications*, Weinheim, VCH, 1995.
8. Platt, J. R. Electrochromism, a possible change of color producible in dyes by an electric field. *J. Chem. Phys.*, **34**, 1961, 862–3.
9. Franz, W. *Z. Naturforsch*, **13A**, 1958, 44, as cited in ref. 7.
10. Keldysh, L. V. *Zh. Eksp. Teor. Fiz.*, **34**, 1958, 1138, as cited in ref. 7.
11. Berzelius, J. J. Afhandlingar i fysik. *Kemi Och Mineralogie*, **4**, 1815, 293, as cited in ref. 4.
12. F. Wöhler. *Ann. Phys.*, **2**, 1824, 350, as cited in ref. 4.
13. See the article 'True Blue (cyanotype) part 2: blue history' by Peter Marshall, available [online] at photography.about.com/library/weekly/aa061801b.htm (accessed 26 January 2006).

14. Bain, A., UK Patent, 27 May 1843, as cited in ref. 4.
15. Hunkin, T. Just give me the fax. *New Scientist*, 13 February 1993, 33–7.
16. Smith, F. H., British Patent, 328,017, 1929, as cited in ref. 4.
17. Zaromb, S. Theory and design principles of the reversible electroplating light modulator. *J. Electrochem. Soc.*, **109**, 1962, 903–12.
18. Zaromb, S. Geometric requirements for uniform current densities at surface-conductive insulators of resistive electrodes. *J. Electrochem. Soc.*, **109**, 1962, 912–18.
19. Mantell, J. and Zaromb, S. Inert electrode behaviour of tin oxide-coated glass on repeated plating–deplating cycling in concentrated NaI–AgI solutions. *J. Electrochem. Soc.*, **109**, 1962, 992–3.
20. Camlibel, I., Singh, S., Stocker, H. J., Van Ultert, L. G. and Zydzik, G. I. An experimental display structure based on reversible electrodeposition, *Appl. Phys. Lett.*, **33**, 1978, 793–4.
21. Howard, B. M. and Ziegler, J. P. Optical properties of reversible electrodeposition electrochromic materials, *Sol. Energy Mater. Sol. Cells*, **39**, 1995, 309–16.
22. Ziegler, J. P. Status of reversible electrodeposition electrochromic devices, *Sol. Energy Mater. Sol. Cells*, **56**, 1995, 477–93.
23. Kobosew, N. and Nekrassow, N. I. *Z. Electrochem.*, **36**, 1930, 529, as cited in ref. 4.
24. Talmay, P. US Patent 2,281,013, 1942, as cited in ref. 4.
25. Talmay, P. US Patent 2,319,765, 1943, as cited in ref. 4.
26. Brimm, E. O., Brantley, J. C., Lorenz, J. H. and Jellinek, M. H. *J. Am. Chem. Soc.*, **73**, 1951, 5427, as cited in ref. 4.
27. Kraus, T. Laboratory report: Balzers AG, Lichtenstein, entry date 30 July 1953, as cited in ref. 4.
28. Philips Electronic and Associated Industries Ltd. Image display apparatus. British Patent 1,302,000, 4 Jan 1973. [The patent was first filed on 24 June 1971.]
29. Schoot, C. J., Ponjée, J. J., van Dam, H. T., van Doorn, R. A. and Bolwijn, P. T. New electrochromic memory display. *Appl. Phys. Lett.*, **23**, 1973, 64–5. [The paper was first submitted in April 1973.]
30. Short, G. D. and Thomas, L. Radiation sensitive materials containing nitrogenous cationic materials, British Patent 1,310,813, published 21 March 1973. [The patent was first filed on 28 May 1969.]
31. J. G. Allen, ICI Ltd. Personal communication, 1987.
32. Kenworthy, J. G., ICI Ltd. Variable light transmission device. British Patent 1,314,049, 18 April 1973. [The patent was first filed on 8 Dec 1970.]
33. Barclay, D. J., Bird, C. L., Kirkman, D. K., Martin, D. H. and Moth, F. T. An integrated electrochromic data display, *SID 80 Digest*, 1980, abstract 12.2, 124.
34. For example, see Jasinski, R. J. N-Heptylviologen radical cation films on transparent oxide electrodes. *J. Electrochem. Soc.*, **125**, 1978, 1619–23.
35. Deb, S. K. A novel electrophotographic system, *Appl. Opti.*, **Suppl. 3**, 1969, 192–5.
36. Van Ruyven, L. J. The role of water in vacuum deposited electrochromic structures. *Cyanamid Technical Report*, **14**, 1968, 187. As cited in Giglia, R. D. and Haake, G. Performance achievements in WO_3 based electrochromic displays. *Proc. SID*, **12**, 1981, 76–81.
37. Deb, S. K. and Chopoorian, J. A. Optical properties and color-center formation in thin films of molybdenum trioxide. *J. Appl. Phys.*, **37**, 1966, 4818–25.
38. Deb, S. K. Reminiscences on the discovery of electrochromic phenomena in transition-metal oxides, *Sol. Energy Mater. Sol. Cells*, **39**, 1995, 191–201.

39. Blanc, J. and Staebler, D. L. Electrocoloration in $SrTiO_3$: vacancy drift and oxidation–reduction of transition metals. *Phys. Rev. B.*, **4**, 1971, 3548–57.
40. Beegle, L. C. Electrochromic device having identical display and counter electrodes. US Patent 3,704,057, 28 November 1972.
41. Deb, S. K. Optical and photoelectric properties and colour centres in thin films of tungsten oxide. *Philos. Mag.*, **27**, 1973, 801–22. [The paper was submitted in November 1972.]
42. Deb, S. K. Some aspects of electrochromic phenomena in transition metal oxides. *Proc. Electrochem. Soc.*, **90–92**, 1990, 3–13.
43. Green, M. and Richman, D. A solid state electrochromic cell – the $RbAg_4 I_5/WO_3$ system. *Thin Solid Films*, **24**, 1974, S45–6.
44. Faughnan, B. W., Crandall, R. S. and Heyman, P. M. Electrochromism in WO_3 amorphous films. *RCA Rev.*, **36**, 1975, 177–97.
45. Faughnan, B. W., Crandall, R. S. and Lampert, M. A. Model for the bleaching of WO_3 electrochromic films by an electric field. *Appl. Phys. Lett.*, **27**, 1975, 275–7.
46. Crandall, R. S. and Faughnan, B. W. Dynamics of coloration of amorphous electrochromic films of WO_3 at low voltages. *Appl. Phys. Lett.*, 1976, **28**, 95–7.
47. Mohapatra, S. K. Electrochromism in WO_3. *J. Electrochem. Soc.*, **285**, 1978, 284–8. [The paper was submitted for publication in April 1977.]
48. Ronlán, A., Coleman, J., Hammerich, O. and Parker, V. D. Anodic oxidation of methoxybiphenyls: the effect of the biphenyl linkage on aromatic cation radical and dication stability. *J. Am. Chem. Soc.*, **96**, 1974, 845–9.
49. Kaufman, F. B., Schroeder, A. H., Engler, E. M. and Patel, V. V. Polymer-modified electrodes: a new class of electrochromic materials. *Appl. Phys. Lett.*, **36**, 1980, 422–5.
50. Kaufman, F. B. and Engler, E. M. Solid-state spectroelectrochemistry of cross-linked donor bound polymer films. *J. Am. Chem. Soc.*, **101**, 1979, 547–9.
51. Kaufman, F. B. New organic materials for use as transducers in electrochromic display devices. Conference Record., 1978 Biennial Display Research Conference, Publ. IEEE, 23–4.
52. Kanazawa, K. K., Diaz, A. F., Geiss, R. H., Gill, W. D., Kwak, J. F., Logan, J. A., Rabolt, J. F. and Street, G. B. 'Organic metals': polypyrrole, a stable synthetic 'metallic' polymer. *J. Chem. Soc., Chem. Commun.*, 1979, 854–5.

3
Electrochemical background

3.1 Introduction

This chapter introduces the basic elements of the electrochemistry encompassing the redox processes that are the main subject of this monograph. Section 3.2 describes the fundamentals, starting with the origin of the cell *emf* (the electric potential across it), introducing the use of electrode potentials, and their determination in equilibrium conditions within simple electrochemical cells. In the first example (with electroactive species that resemble type-I electrochromes), the reactants are all ions in solution. In the second example, the cell assembly comprises two electrodes, each a metal in contact with a solution of its own ions, somewhat resembling type-II electrochromes. Though electrochromic electrodes are intrinsically more complicated than the two examples cited here, they follow just the principles established. Details of fabrication for electrochromic devices (ECDs) appear in Chapter 14. Section 3.3 exemplifies the kinetic features underlying electrochromic coloration. In it, the rates of mass transport and those of electron transfer, the three rate-limiting (thus current-limiting) processes encountered during the electrochemistry, are described. Diffusion of both electrochrome and counter ions is discussed more fully in Chapter 5, to illustrate the way charge-carrier movement limits the rate of the coloration/bleaching redox processes within ECDs.

Section 3.4 covers electrochemical methods involving dynamic electrochemistry, particularly cyclic voltammetry, which is important in studying electrochromism; three-electrode systems are required here.

More comprehensive treatments of electrochemical theory will be found elsewhere.[1,2,3]

3.2 Equilibrium and thermodynamic considerations

3.2.1 A cell with dissolved ions as reactants: the Gibbs energy and electromotive force

The fundamental origin of an electrochemical *emf* ('electromotive force') in a cell sometimes seems obscure. Basically it arises from the energy of a chemical reaction involving electron transfer (exactly, the Gibbs free energy change for unit amount of reaction). The simplest example involves solely ions in water, such as the reaction that occurs on mixing the ions:

$$Fe^{2+} + Mn^{3+} \rightleftharpoons Fe^{3+} + Mn^{2+}. \tag{3.1}$$

This electron transfer reaction is known to proceed from left to right spontaneously, effectively to completion, and quite rapidly. If Fe^{2+} and Fe^{3+} were contained in one solution and Mn^{2+} and Mn^{3+} in another, and the two solutions were connected via a tube containing a salt solution, there would be no way for the reaction to proceed, although a 'cell' would have been partly created. If however two inert wires, of say Pt, were inserted into each of the metal-ion solutions, then on connecting the wires, Fe^{2+} would transfer electrons e^- to the Pt so becoming Fe^{3+}, while at the other Pt, Mn^{3+} would gain e^- becoming Mn^{2+}. Thus the reaction proceeds as it would on directly mixing the reactants, but now via the electrode processes, each at its own rate, with rate constants k_{et}. The flow of electrons in the wire is accompanied by net ionic motion through the solutions: a current flows through the cell and in the wire.

If, instead of connecting the Pt wires, a meter, or opposing voltage, were connected, so frustrating the electrode processes, these would indicate the voltage (the cell *emf*, $E_{(cell)}$) evoked by the tendency of the reactions of the ions to proceed as stated, owing to the Gibbs free energy change ΔG that would accompany direct reaction. The connection of $E_{(cell)}$ with the thermodynamics of the cell reaction then follows from the identification

$$\Delta G = -nFE_{(cell)} \tag{3.2}$$

as a charge nF traverses a potential $E_{(cell)}$ in a (virtual) occurrence of the cell reaction, Eq. (3.1). Here n is the number of electrons transferred in the written reaction (1 in this example), and F is the Faraday constant, the charge on 6.022×10^{23} electrons, i.e. the charge involved in unit-quantity (a mole) of a complete reaction where $n = 1$.

In general, an electrochemical cell comprises a minimum of two electrodes, each made up of two different 'charge states' of a particular chemical. For inorganic species, the charge state is more properly the oxidation state or (colloquially) redox state, which is shown in superscripted Roman numerals by the element symbol, thus Fe^{II}, Fe^{III} and Mn^{II}, Mn^{III} in the initial examples, sometimes as Fe(II), Fe(III), and so on. This is a widely used 'chemical-accountancy' abbreviation ploy based on summarily assigning a charge of 2^- to the oxide ion, e.g. as in $W^{VI}O_3$. Here the precise charge distribution will differ considerably from the conventional, assigned, oxidation state. (The use of Roman numerals for oxidation states in chemistry differs from that used for gaseous species by spectroscopists, who write an *atom* as MI, a singly charged ion M^+ as MII, M^{2+} as MIII, etc., the numerals here being on par and unparenthesised.)

3.2.2 Individual electrode processes

Consider what happens at the electrodes individually. At an electrode, the two states stay in equilibrium (i.e. constant in composition) at only one potential, the 'equilibrium potential', applied to this electrode. A comparable statement is also true for the other electrode. 'Applying a potential' always requires the presence in the cell of the second electrode also connected to the source, say a battery, of the external potential. If the potential applied to the electrode, when in contact with both redox states, is different from this equilibrium electrode potential, then one of two 'redox' reactions (or 'half-reactions') can occur: electron gain – reduction, Eq. (3.3):

$$O + n e^- \rightarrow R, \tag{3.3}$$

or electron loss – oxidation, the reverse of Eq. (3.3) – which will alter compositions at the electrode. O and R, like Mn^{3+}, Mn^{2+} or Fe^{3+}, Fe^{2+} are called a 'redox couple'. (Equation (3.3), itself abbreviated to 'O,R', is sometimes loosely referred to as 'the O,R electrode'.)

3.2.3 Electrode potentials defined and illustrated

The potential of an unreactive metal in contact, and in equilibrium, with the two redox states, is termed the electrode potential $E_{O,R}$ or, colloquially, the 'redox potential'. When just this value of potential is applied from a battery or voltage source, no overall composition change occurs via Eq. (3.3), but electron transfer does persist because in these conditions the forward and reverse processes in Eq. (3.3) are conduced to proceed at the same rate.

If no external potential is applied, the O,R species at their particular concentrations control the energy (and hence the potential) of the electrons in the metal contact, thereby allowing electrical communication to a meter. (Measurement of this energy in a single electrode can be contemplated in principle but is difficult in practice and will henceforth be viewed as impossible.) While a value of $E_{O,R}$ for the (O,R) half cell cannot be determined independently, only *differences* in electric potential between two sites being ordinarily accessible by communication to a meter, the usual cell construction comprising two electrodes intrinsically avoids this problem. Assigning an arbitrary value to $E_{O,R}$ for *one* O,R couple (the H^+/H_2 couple) then establishes, for all other couples, values of their electrode potentials as appear in tabulations. This is amplified below.

Only redox couples (i.e. 'electroactive materials') that can transfer electrons with reasonable rapidity can set up stable redox potentials for measurement. The application of a potential greater or less than the equilibrium value (see 'Overpotentials' on p. 42 below) can effect desired composition changes in either direction by *driving* the electron process in either direction, to the required extent. Only with fast redox couples can the composition be rapidly governed by an applied voltage.

For rapidly reacting redox couples, the (equilibrium) electrode potential $E_{O,R}$ is governed by the ratio of the respective O,R concentrations (which are related to their 'activities' – a thermodynamic concept, see next paragraph) by a form of the Nernst equation:

$$E_{O,R} = E_{O,R}^{\ominus} + \frac{RT}{nF}\ln\left(\frac{a(O)}{a(R)}\right), \tag{3.4}$$

where $E^{\ominus}$ is the *standard electrode potential* (see below), the terms a are the activities, R is the gas constant, F the Faraday constant, T the thermodynamic temperature and n is the number of electrons in the electron-transfer reaction in Eq. (3.3).

The two oxidation states O and R can be solid, liquid, gaseous or dissolved. Dissolved states can comprise either liquids or solids as solvent. Activity may be described as the 'thermodynamically perceived concentration'. The relationship between concentration c and activity a is: $a = (c/c_{std}) \times \gamma$, where γ is the dimensionless activity coefficient representing interactions with ambient ions, and c_{std} is best set at unity in the chosen concentration units. Observed values of γ for ions are somewhat less than 1 in moderately dilute aqueous solutions. Here just for illustration we take activities of ions (or other solutes) in liquid solution as being the ionic concentrations (which is empirically true if always in a maintained excess of inert salt). Activities of gases are closely enough their pressures, while activities of pure solids – those that remain unaffected in composition by possible redox reactions, thus being always constant in composition – are assigned the value unity. However, when solid electrode material undergoes a redox reaction where the product forms a solid solution within the reactant, the respective activities are represented by mole fractions x; but if the result of redox reaction is a mixture of two pure bulk solids, then each is represented as being of unit activity.

The term $E_{O,R}^{\ominus}$ is the *standard electrode potential* defined as the electrode potential $E_{O,R}$ measured at a standard pressure of 0.1013 MPa and designated temperature, with both O and R (and any other ion species in the redox reaction) present at unit activity. Fundamentally, the value of $E_{O,R}^{\ominus}$ is determined by the effective condensed-phase electron affinity of O (or, equivalently, the effective condensed-phase ionisation potential of R) on a relative scale. This scale of $E_{O,R}^{\ominus}$ values was established by assigning a particular value to one selected redox system, by convention zero for H^+/H_2, as detailed below.

3.2.4 A cell with metal electrodes in contact with ions of those metals

Figure 3.1 shows an electrochemical cell that comprises our second simple example. The left-hand electrode is a zinc rod immersed in an aqueous solution containing Zn^{2+}; the two redox states Zn^{2+}, Zn comprise the redox couple and, when connected to an external wire, make up the redox electrode. As in Fig. 3.1, one of the redox species in Eq. (3.3) also functions as the contact electrode by which E may be monitored, since zinc metal is a good conductor,

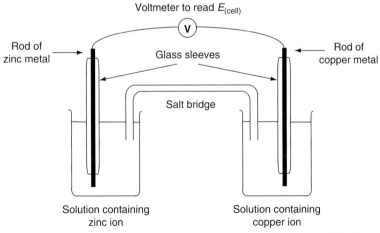

Figure 3.1 Schematic of the primitive cell $Zn(s)|Zn^{2+}(aq)||Cu^{2+}(aq)|Cu(s)$ for equilibrium measurements. Each metal rod is immersed in a solution of its own ions: the two half cells are Zn^{2+}, Zn and Cu^{2+}, Cu.

as is copper. Both electrodes, however, need to be connected to the same 'inert' conducting material in connections between the cell and meter; Pt is often used, as in the introductory example. For other redox couples the inert metal is not written but taken as understood. The 'inert electrodes' – better, inert *contacts* – do not contribute to the electrode reaction. They comprise an inert metal such as platinum or gold in contact with two oxidation states O,R of a chemical species dissolved either in water or other solvent, or in solid solution, or otherwise from gaseous, insoluble-salt, or pure-liquid components.

The spontaneous reaction in the cell depicted in Figure 3.1 is the following:

$$Cu^{2+}(aq) + Zn(s) \rightarrow Cu(s) + Zn^{2+}(aq), \qquad (3.5)$$

where (s) denotes 'solid' and (aq) is aqueous (alternatively here, one uses (soln) for general solvent, or specifies which solvent by suitable abbreviation). The suffixes (l) and (g) are for 'liquid' and 'gas,' and there is a need for (s. soln) meaning 'solid solution', that of one species within another, forming a solid. The Nernst equation for the whole cell is:

$$E_{(cell)} = E^{\ominus}_{(cell)} - \frac{RT}{nF} \ln \frac{[Cu(s)][Zn^{2+}]}{[Cu^{2+}][Zn(s)]} = E^{\ominus}_{(cell)} - \frac{RT}{nF} \ln \frac{[Zn^{2+}]}{[Cu^{2+}]}, \qquad (3.6)$$

where the square brackets [ ] represent concentrations (better, activities), but the fictional values for the metals are conventionally represented by unit activity as in the right-hand form of the equation here.

Comparably with our first example, the cell depicted would therefore spontaneously produce current if the electrodes were connected externally with a conducting wire, the 'applied potential' then obviously being zero and not $E_{(cell)}$. This reaction proceeds via the two reactions, $Cu^{2+} + 2e^- \rightarrow Cu$ and $Zn \rightarrow Zn^{2+} + 2e^-$ at the two respective electrodes. The resultant flow of electrons e^- is discernible as an external current I in the wire. Concomitant ion motion occurs within the solution phase in attempting to maintain electrical neutrality throughout the cell. The direction of the reaction is reflected in the relative values of the two electrode potentials E, evaluated as outlined below. The magnitude of I depends on the net rate of reaction (3.3) or its reverse, when applicable, at the more slowly operating electrode.

The electrode reactions are shown above as simple processes though in detail comprising a complicated series of steps. To exemplify, aqueous Cu^{2+} has hexacoordinated water molecules, two on longer 'polar' bonds than the other four 'equatorial' waters. All these have to be shed, in obscure steps; meanwhile Cu^{2+} becomes Cu^+ then Cu^0 atoms, then metal-lattice components. So in such an apparently simple process, appreciable mechanistic

3.2 Equilibrium and thermodynamic considerations

complexity underlies the simplified reaction cited. Thus even greater complexity can be expected in the chemically more intricate electrochromic systems dealt with later.

3.2.5 The cell emf and the electrode potentials: the hydrogen scale

The amount of Zn^{2+} in solution will remain constant, that is, at equilibrium, only when the potential applied to the Zn equals the electrode potential $E_{Zn^{2+},Zn}$ and, simultaneously, the copper redox couple (right-hand side of the cell) is only at equilibrium when the potential applied to the copper is $E_{Cu^{2+},Cu}$. Neither electrode potential as explained above is known as an absolute or independent value: only the difference between the two, that is, $E_{(cell)}$, is the measurable quantity. Then

$$E_{(cell)} = E_{(\text{right-hand side})} - E_{(\text{left-hand side})} + E_j = E_{Cu^{2+},Cu} - E_{Zn^{2+},Zn} + E_j, \quad (3.7)$$

where E_j is a junction potential at the contact between the solutions about the two electrodes, usually minimised. Further detail concerning cell notation is set out in ref. 1. (E_j is usually of unknown magnitude but approaches zero when the two solutions are nearly similar in composition. Alternatively, precautions can be taken to minimise the value of E_j via, e.g., a 'salt bridge', a tube containing suitable electrolyte, between the two solutions. Often an inert electrolyte uniformly distributed throughout the cell suffices.)

$E_{(cell)}$ is then the observed electrical potential difference to be applied across the cell to effect zero current flow, i.e. to prevent thereby any redox reaction at either electrode, so 'preserving equilibrium', and is simply the difference between the electrode potentials:

$$E_{(cell)} = E_{(\text{right-hand side})} - E_{(\text{left-hand side})}. \quad (3.8)$$

This statement is obviously applicable to all electrochemical cells operating 'reversibly' (i.e. rapidly).

$E_{(cell)}$ may be measured on a voltmeter by allowing a negligibly small (essentially zero) current to flow through the voltmeter, but applying a measured potential from an external source that exactly opposes $E_{(cell)}$ is the precision choice. At zero current, $E_{(cell)}$ is the electromotive force ('*emf*') of the cell. When we wish to emphasise that the electrodes are being kept at equilibrium by an externally applied potential, we shall write $E_{(eq)}$ instead of $E_{(cell)}$.

For many redox couples, an electrode-potential scale has been devised. After measurement of $E_{(cell)}$, if one of the electrode potentials which comprise

$E_{(cell)}$ is summarily assigned a value, then the other is predetermined, following Eq. (3.7). In order to establish this formal scale, the half cell

$$\text{Pt} \mid \text{H}_2(\text{g})(1\,\text{atm}), \text{H}^+(\text{aq, unit activity})$$

is assigned an electrode potential $E^\ominus$ of zero for all temperatures. This is the standard hydrogen electrode (SHE), in which the electrode reaction is

$$\text{H}^+(\text{aq}) + \text{e}^- = \tfrac{1}{2}\,\text{H}_2(\text{g}). \tag{3.9}$$

It is the standard *reference* electrode: from comparisons made with cells in which one of the electrodes is a SHE, all standard electrode potentials are cited with respect to it. (Since no single ionic species like H^+ can make up a solution, to emulate the extreme dilutions that approximate to single-ion conditions, Nernst-equation extrapolation procedures can correct for finite-concentration effects. These considerations apply also to Eq. (3.4). This 'activity-coefficient' factor is henceforth supererogatory for our purposes.) Unless stated otherwise, the solvent is water. Any change of solvent changes the values of $E^\ominus$ and, in general, alters the sequence of $E^\ominus$ values somewhat. Note that in tabulations,[2,3,4,5] the half reactions (putatively taking place in 'half cells') to which these $E^\ominus$ refer, are formally written as reduction reactions with the electron e^- on the left-hand side.

The SHE is the *primary* reference electrode, but is thought cumbersome and care is needed handling H_2. Thus other, 'secondary', reference electrodes are preferred. The most common are the saturated calomel electrode (SCE) and the silver–silver chloride electrode. Quasi-reference electrodes are also admissible, the most common being a bare silver wire, presumably bearing traces of silver oxide to complete the redox couple. Potentials cited in this text have been converted to the saturated calomel electrode (SCE) potential scale, when aqueous electrolyte solution was used. (This attempt at uniformity will have involved cumbersomely reversing the procedures followed by some authors, of *citing* potentials with respect to zero for a SHE, for values *measured* with respect to an SCE, then 'corrected' to the hydrogen scale. We have used the value of 0.242 V for the SCE on the hydrogen scale.[6])

3.2.6 Electrochromic electrodes

To link the introductory electrochemical examples above with electrochromic systems, we cite the widely studied tungsten trioxide electrode:

$$\text{W}^{\text{VI}}\text{O}_3(\text{s}) + \text{e}^- \rightarrow \text{W}^{\text{V}}\text{O}_3(\text{s}). \tag{3.10}$$

This is an idealisation of the reaction that in practice proceeds only fractionally to the extent of the insertion coefficient x ($x < 1$ and in many cases $\ll 1$):

$$W^{VI}O_3(s) + xe^- + xM^+(soln) \rightarrow M_x(W^V)_x(W^{VI})_{1-x}O_3(s.\,soln), \quad (3.11)$$

where the product is a solid solution with mole fractions x incorporating an unreactive electrolyte cation M^+, often Li^+, but sometimes H^+. The counter cations may not always be unreactive. Further detail follows in Section 6.4.

Another oft-studied electrochrome is Prussian blue (PB) that undergoes the half-reaction, here represented in the reductive bleaching process in Eq. (3.12), the blue pigment PB on the left being decolourised:

$$M^+Fe^{3+}[Fe^{II}(CN)_6]^{4-}(s) + e^- + M^+(soln) \rightarrow (M^+)_2Fe^{2+}[Fe^{II}(CN)_6]^{4-}(s).$$
$$\text{blue} \qquad\qquad\qquad\qquad\qquad\qquad\qquad \text{white (clear)}$$
$$(3.12)$$

In the formulae, each CN is actually CN^- and M^+ is usually K^+. The oxidation-state notation allows a shorthand version of the essential reaction,

$$Fe^{III}[(Fe^{II}(CN)_6] + e^- \rightarrow Fe^{II}[(Fe^{II}(CN)_6], \quad (3.13)$$

where only the actual chromophore segment can thus be shown.

3.3 Rates of charge and mass transport through a cell: overpotentials

To reiterate, an electrochromic device is fundamentally an electrochemical cell. Applying a potential $V_a \neq E_{(cell)}$ across the cell causes charge to flow, and hence effects electrochromic operation. As just outlined, these charges enforce the consumption and generation of redox materials within the cell. Above a particular applied potential V_a, the reaction in the cell will proceed oxidatively at one electrode and reductively at the other and below it the electrochemical reactions at the electrodes are the reverse of these. At only one applied potential is the *current* through the cell zero: we call this potential the equilibrium potential $E_{(eq)} = E_{(cell)}$. A steady state exists at $E_{(eq)}$ and no charge is consumed at either electrode To elaborate, considering the electrodes separately, above a certain potential applied to a particular electrode, the reaction there within the cell is an oxidation reaction, and below it the electrode reaction is reduction. Complementary processes must occur at the partner electrode. Considering both electrodes, at only one potential applied across the cell is the current through the cell zero: at this equilibrium potential, $E_{(eq)} = E_{(cell)}$.

As before, we concentrate attention on one electrode. The charge that flows is measured per unit time as current I, which is clearly proportional to the rate at which electronic charge Q at an electrode is consumed by the electroactive species, or generated from it, by reduction or oxidation, respectively,

$$I = \frac{dQ}{dt}. \tag{3.14}$$

If the redox (electroactive) species are in solution, the magnitude of an electrochemical current is a function of three rates at that electrode: (i) the rate of electron transport through the materials comprising the electrode; (ii) the rate of electron movement across the electrode–solution interface, and (iii) the rate at which the electroactive material (ion, atom or molecule) moves through solution prior to a successful electron-transfer reaction (also, in the case of solid electroactive materials, involving the movement of non-electroactive ions if they are taken up or lost by electroactive solids). Processes (i) and (ii) are termed charge transfer (or charge transport); process (iii) involves mass transfer or transport.

When net (observable) current flows, the slowest of the three rates is '(overall) rate limiting', governing the overall rate of charge movement in a device or electrode process. Rate (i) is determined by the magnitude of the electronic conductivity σ of the material from which the electrode is constructed, when one or both components of the redox couple are solid. Electrodes comprising platinum, gold or glassy carbon contacts possess high electronic conductivities σ so rate (i) is rarely rate limiting with such substrates. For transparent electrode systems fluoride-doped tin oxide or ITO act the role, in the place of metals, of the 'inert contact' to the redox species. Their conductivities are both low relative to true metals, so rate (i) can apply in such systems.

The magnitude of rate (ii) is 'activated', that is, the system must surmount an energy barrier prior to electron transfer. The magnitude of rate (ii) is governed by the rate constant of the electron-transfer process k_{et}, and is dictated by the *overpotential* η of the electrode, defined by Eq. (3.15):

$$\eta = V_a - E_{(eq)}. \tag{3.15}$$

The rate constants k_{et} are potential dependent, the 'constancy' appellation referring to concentration dependences at a predetermined potential. Thus k_{et} is a curious rate constant dependent on the overpotential η, a complication dealt with below in the Butler–Volmer treatment. (In the literature, overpotential and coloration efficiency are unfortunately represented by the same symbol η. In later chapters, overpotential will be spelt out, and the symbol η alone will mean only coloration efficiency.)

Overpotential has sign as well as magnitude. More usefully, it is applied to just one electrode. By definition, an overpotential of zero indicates equilibrium, and hence zero current, i.e. no conversion of electrochrome to form its coloured state, and hence no electrochromic operation. Provided the overpotential applied is sufficiently large, k_{et} will be high and therefore rate (ii) will not be rate limiting. Applying an overpotential (i.e. forcing the potential of the electrode away from $E_{(eq)}$) causes a current I to flow, which is related to overpotential η by Eq. (3.16), a form of Tafel's law:[7,8]

$$\eta = a + b \ln I;$$

that is,

$$I \propto \text{exponential}(\eta/b), \quad (3.16)$$

where a and b are constants particular to the system (see 'Butler–Volmer kinetics' towards the end of the chapter, p. 46).

Occasionally, the overpotential η needs to be relatively small to prevent electrolytic side reactions, in which case rate (ii) may be rate limiting.

Rate (iii) is rate limiting in a number of electrochromic devices; but while electrons may be intuitively adjudged the fast movers in the processes with rates (i) and (ii), this is by no means always so. In type-I systems, the electrochrome must come into contact with the electrode before a successful electron-transfer reaction can occur. Since a type-I electrochrome is evenly distributed throughout the solution before the device is switched on, most of the electrochrome is distributed in the solution bulk, and must move toward the electrode interphase until sufficiently close for the electron transfer to take place. (The term interphase here is preferred to 'interface' to emphasise the number and diverse nature of the many layers between bulk electrochrome and bulk solvent, including, on the liquid side, potential-distributed ions, oriented molecules and adsorbed species, as well as the outermost solid surface, that always differs from bulk solid.)

3.3.1 Mass transport mechanisms

The process by which the electroactive material moves from the solution bulk toward the electrode, mass transport, proceeds via three separate mechanisms: migration, convection and diffusion. Mass transport is formally defined as the flux J_i of electroactive species i, that is, the number of i reaching the solution–electrode interphase per unit time, as defined in the Nernst–Planck equation, Eq. (3.17):[9]

$$J_i = -\frac{z_i F}{RT} c_i \left(\frac{\partial \phi(x)}{\partial x}\right) + c_i \,\overline{v}_i(x) - D_i \left(\frac{\partial c_i(x)}{\partial x}\right), \qquad (3.17)$$
migration　　convection　　diffusion

where $\phi(x)$ is the strength of the electric field along the x-axis, $\overline{v}_i$ is the velocity of solution (as a vector, where applicable), and D_i and c_i are respectively the diffusion coefficient and concentration of species i in solution. (Strictly, the equation describes one-dimensional mass transfer along the x-axis.)

The three transport modes operate in an additive sense. Convection is the physical movement of the solution. Deliberate stirring of the solution is termed 'forced' convection; density differences of the solution adjacent to the electrode cause 'natural' convection. Both forms of convection can be assumed absent in electrochromic cells, or at least of a negligible extent. Convection will not be discussed in any further detail since it is irrelevant for solid electrolytes and otherwise uncontrolled in other ECDs.

3.3.2 Migration

Migration represents the movement of ions in response to an electric field in accord with Ohm's Law: positive electrodes obviously attract negatively charged anions, negatively charged electrodes attracting cations. Migration may be neglected for liquid electrolytes containing 'swamping' excess of unreactive ionic salt (often termed a 'supporting electrolyte'), as excess concentrations of inert cations or anions that accumulate about their respective electrodes effectively inhibit continued migration.

However, solid polymer electrolytes or solid-state electrochromic layers experience a significant extent of migration since the transport numbers of (i.e. fractions of total current borne by) the electroactive species or of mobile counter ions become appreciable.

In the absence of both convection and migration, diffusion becomes the sole means of mass transport, delivering electroactive species to the electrode. Migration is still important in liquid-phase systems such as that in the *Gentex* mirror, described in Sections 11.1 (Fig. 11.3) and 13.2.

3.3.3 Diffusion

The most important mode of mass transport in electrochromism is usually diffusion, which ideally follows Fick's laws. The first law defining the flux J_i (the amount of diffusant traversing unit area of a cross-section in the solution normal to the direction of motion per unit time) is:

3.3 Rates of charge and mass transport through a cell

$$J_i = -D_i\left(\frac{\partial c_i}{\partial x}\right), \tag{3.18}$$

where D_i is the diffusion coefficient of the species i, and $(\partial c_i/\partial x)$ is the change in concentration c of species i per unit distance x (i.e. the concentration gradient). The concentration gradient $(\partial c_i/\partial x)$ arises in any electrochemical process with current flow because some of the electroactive species is consumed around the electrode, this depletion causing the concentration gradient. Diffusion results from a natural minimising of the magnitude of internal concentration gradients.

Fick's second law describes the time dependence (rate) of such diffusion, Eq. (3.19):

$$\left(\frac{\partial c_i}{\partial t}\right) = D_i\left(\frac{\partial^2 c_i}{\partial x^2}\right), \tag{3.19}$$

where t is time and i denotes the ith species in solution. The required integration of this second-order differential equation often leads to difficulty in accurately modelling a diffusive system. A rough-and-ready but useful version gives the approximate relation, Eq (3.20):

$$l \approx (Dt)^{1/2}, \tag{3.20}$$

where l is the distance travelled by species with diffusion coefficient D in time t. The implications of diffusive control are discussed below.

Movement of type-I and type-II electrochromes toward an electrode during coloration (see Sections 3.4 and 3.5 below) represents true diffusion of electrochrome. By contrast, electro-*bleaching* of a type-II electrochrome and coloration and bleaching of type-III electrochromes are all processes involving solids. Such diffusional movement is complicated by concomitant migration. For this reason, the 'diffusion' of a charged species through a solid is characterised by the so-called 'chemical diffusion coefficient' $\bar{D}$. The kinetics of bleaching in a type-II system, and either coloration or bleaching kinetics for a type-III electrochrome, will be characterised by the chemical, rather than the normal, diffusion coefficient D.

The implications for electrochromic coloration of straightforward diffusion are discussed in Section 5.1, and the kinetic distinctions between D and $\bar{D}$ are discussed in depth in Section 5.2.

Faradaic and non-faradaic currents

The contribution to any current that results in a redox (electron-transfer) reaction is termed 'faradaic' – that is, it obeys Faraday's laws – whereas that part arising solely from ionic motion without such accompanying redox, such

46   *Electrochemical background*

as in the formation of the ionic double layer, is 'non-faradaic'. Faraday's laws specifically relate to material deposition or dissolution effected by redox reactions, and, by extension, to redox transformation of dissolved species.

3.4 Dynamic electrochemistry

3.4.1 Butler–Volmer kinetics of electrode reactions

It is noted in Section 3.2 (page 39 above) that the (net) zero current at an electrode, when an external applied potential is equal to the electrode potential E, is the resultant of two opposing currents I_{cath} (cathodic, when electrons e^- are relinquished from the electrode) and I_{an} (anodic, the e^- are acquired by the electrode). At E these are equal in magnitude. We write that at E

$$I = I_{cath} + I_{an} = 0, \qquad (3.21)$$

where implied signs attach to the individual currents. (In this outline the O and R species are both in solution, as with type-I electrochromes. Minor elaborations are needed for type-II systems and major ones for type-III, but the underlying physics is identical throughout.) Details are in ref. 3 and works cited therein.

When, from Eq. (3.15), the applied potential differs by η (the overpotential) from E, I is non-zero and one or other of the individual currents dominates, depending on whether the electrode is positive or negative of E.

The (net) rate of the electrode reaction is defined as:

$$\text{rate} = \frac{I}{nFA} = \frac{i}{nF}, \qquad (3.22)$$

where n is the number of electrons involved in the reaction, F is the Faraday constant and A is the area of the electrode.

Rate constants covering concentration dependences on c_O and c_R for the reactions at a particular potential are defined in Eq. (3.23):

$$I_{cath} = -nFA\, k_{cath} c_O \text{ and } I_{an} = nFA\, k_{an} c_R. \qquad (3.23)$$

As Tafel's law states, Eq. (3.16), log I is linear with η, but this holds only when one of the individual ('cath' or 'an') currents dominates to the exclusion of the other; it therefore fails ever more seriously for decreasing η because when near to or approaching the electrode potential E (η small), I becomes small (both 'cath' and 'an' currents are appreciable), then $I \to 0$. (The law also fails for very large values of η, when the at-electrode concentrations of reactant decreases from the bulk values owing to the high consumption rates prevailing as follows, and replenishment by diffusion controls the current.)

3.4 Dynamic electrochemistry

The rate constants k_{cath} and k_{an} (for general reference we call either k_{et}) are both dependent on η. A zero value of η implies an applied potential equal to E, and a net current of zero. To obtain the current values I_{cath} and I_{an} applicable at E, these need to be obtained from extrapolation back to E of observed ($\ln I$) values vs. η, from linear Tafel's-law regions of η (Eqs. (3.15) and (3.16)). Here, at $\eta = 0$, the extrapolated values of each of the opposing currents I_{cath} and I_{an} pertain (and cancel) at E. When $E = E^{\ominus}$ (that is, with $c_O = c_R$), this procedure results in the requisite values of the electrode rate parameters. These are

$|I_{cath}| = |I_{an}| = I_0$, the *standard exchange current*;
$i_0 = I_0/A$ is the *standard exchange current density*;

$$k_{cath}(E^{\ominus}) = k_{an}(E^{\ominus}) = k^{\ominus}$$

where the parenthesised '$(E^{\ominus})$' denotes 'pertaining at $E^{\ominus}$', and $k^{\ominus}$ is the *standard electron transfer rate constant* for the electrode reaction.

Now $k^{\ominus}$, along with other rate constants, includes an exponential activation-energy term for the activation barrier to be surmounted in the electron transfer, which is intrinsic to the particular reaction involved. Then that activation energy is diminished by the energy supplied via η, some of which favours one direction of reaction, some the reverse; how much depends on the detail of the energy barrier, which if symmetrical results in a fraction $\alpha = \frac{1}{2}$ of the supplied energy for each direction. When not equal to $\frac{1}{2}$, α is usually found experimentally to be between 0.4 and 0.6, from Tafel-law slopes. The value $\frac{1}{2}$ is reasonably assumed in straightforward cases when not otherwise readily available.

The activation energy term $\exp(-E_a/RT)$, that arises from the barrier to electron transfer, is implicit within $k^{\ominus}$, hence the counter (driving) energy deriving from η will likewise comprise an exponential factor in the k_{et} expressions, with overpotential contributions in straightforward cases weighted as α and $1 - \alpha$ for opposing directions:

$$k_{cath} = k^{\ominus} \exp\left(-\frac{\alpha nF\eta}{RT}\right) \text{ and } k_{an} = k^{\ominus} \exp\left(\frac{(1-\alpha)nF\eta}{RT}\right). \quad (3.24)$$

This equation leads to the final Butler–Volmer form, holding until η is made so large that reactant consumption becomes great (from the high prevailing k_{et} values), this depletion therefore bringing in diffusion control.

Hence, Eq. (3.25) is obtained:

$$i = i_0 \left\{ \exp\left(-\frac{\alpha nF\eta}{RT}\right) - \exp\left(\frac{(1-\alpha)nF\eta}{RT}\right) \right\}, \quad (3.25)$$

where the overpotential η is negative when the electrode is made cathodic but positive with electrode anodic.

Wider expositions follow different sign conventions and include special cases, but the essence of the kinetics is as outlined here. Advanced theories, besides indicating probable α values, show that the linearity of the Tafel region is not necessarily general, but it is certainly found to hold for the vast majority of reactions examined.

3.4.2 Cyclic voltammetry

Current flow through a cell alters the potentials at both electrodes, in accord with Eq. (3.16) which holds with different intrinsic parameters for each electrode. In order to isolate the processes at one electrode, the effects at the other are ignored (and this 'counter electrode' can then be chosen merely for convenience: Pt, electrolysing solvent water, for example; any unwanted byproducts are segregated within a sinter-separated compartment). The potential at the 'working electrode' (WE) is then measured not via the potential applied across the cell, but by measuring the potential between the WE and a closely juxtaposed reference electrode (RE) like the SCE (see Section 3.3). No net current flows through the SCE so its potential may be regarded as constant, while the WE bears a variable current and shows a true, measurable, potential. The cyclic-voltammetry experiment involves applying a potential smoothly varying with time t, over a range including the electrode potential $E_{O,R}$ of the WE and observing the resultant current, which will peak (with value I_p) near $E_{O,R}$. At the end of the chosen range the potential is reversed, to change at the same rate as for the forward 'potential sweep'. The control device (a *potentiostat*) in fact drives a current across the cell of such (changing) magnitude as to effect the desired steady potential change at a desired rate; at any *instant* of time the potential is in fact constant and known, hence the name of the control device. A record of the potential with time will show a saw-tooth trace of this 'potential ramping'. The so-called scan- or 'sweep'-rate (the rate of potential variation) ν can be varied to give desiderata like diffusion coefficients (see Chapter 5). Alternative procedures employ potentials varying as sine waves, rather than the saw-tooth mode described. Each voltammetric scan of an electrochromic electrode thus represents an on/off switching cycle and can be used to estimate survival times of such electrodes if allowed to run for a sufficiently long time.

Figure 3.2 (a) depicts a schematic circuit for cyclic voltammetric analyses, indicating the nature of the connections between the three electrodes. Figure 3.2 (b) shows a schematic cyclic voltammogram (CV).

3.4 Dynamic electrochemistry

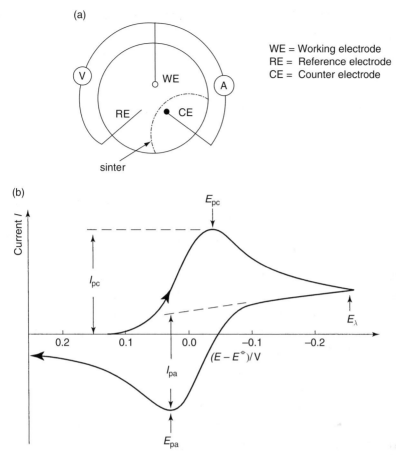

Figure 3.2 (a) Schematic cell (depicted within a circular vessel) for obtaining a cyclic voltammogram, showing connections between the three electrodes. The sinter prevents the products of electrode reactions at the counter electrode diffusing into the studied solution. (b) Schematic cyclic voltammogram for a simple, reversible, one-electron redox couple, in which all species remain in solution.

The controlling device can (or should be able to) measure total charge passed at each stage of the sweep, and with prolonged examination any loss or decomposition of electrochrome becomes apparent from observable diminution of cycle charge. Optical/spectroscopic examination of the electrode can be undertaken concomitantly. Other modifications of measurement are used, such as continuous pulses of potential, which trace versus time a series of square-well potentials above and below an average.

A widely used application involves the Randles–Sevčik equation linking the peak current I_p with concentration c, v and the diffusion coefficient D, from a solution of Fick's laws. D is dealt with in further detail in Chapter 5:

$$I_{(\lim,t)} = -0.4463\, nF\, A \left(\frac{nF}{RT}\right)^{\frac{1}{2}} D^{\frac{1}{2}}\, c\, v^{\frac{1}{2}}. \qquad (3.26)$$

The other symbols have already been defined.

3.4.3 Impedance spectroscopy

Electrochemical impedance spectroscopy (EIS) is summarised here, in an outline employing the familiar concepts of resistance and capacitance. Thus, one can measure the resistance of a circuit element, such as a redox electrode, and its apparent dependence on the frequency of the potential applied, together with the capacitance and *its* frequency dependence, and directly convert these data into the real $\mathcal{R}$ and imaginary $\mathcal{J}$ parts of the impedance Z. Plots of $\mathcal{J}$ against $\mathcal{R}$ or of either against applied frequency, or of other functions against either quantity, can yield useful rate parameters for electrode processes.[10] There are in fact four ways in which what can be thought of as basically resistance and capacitance measurements can be represented, each providing different weightings with respect to frequency. For example, the inverse of impedance is the quantity called admittance. All such treatments are called immitance measurements.

3.4.4 Ellipsometry

Ellipsometry is an optical technique that employs polarised light to study thin films. In this context, 'thin' means films ranging from essentially zero thickness to several thousand Ångstroms, although this upper limit can sometimes be extended. The technique has been known for almost a century, and today has many standard applications, including the measurement of film thicknesses and probing dielectric properties. It is mainly used in semiconductor research and fabrication to determine properties of layer stacks of thin films and the interfaces between the layers.

In the ellipsometry technique, linearly polarised light of known orientation strikes on the surface of a sample at an oblique angle of incidence. The reflected light is then polarised elliptically (hence 'ellipsometry'). The shape and orientation of this ellipse depends on the angle of incidence, the direction of the polarisation of the incident light, and the reflective properties of the surface. An ellipsometer quantifies the changes in the polarisation state of light as it reflects from a sample, as a function of these variables.

If the thin-film sample undergoes changes, for example its thickness alters, then its reflection properties will also change. More importantly to electrochromism, applying a potential across an electroactive film changes the optical properties of the film, and hence the polarisation of the reflected light. Therefore, by monitoring the polarisation of the reflected light while changing the applied potential ('*in-situ* electrochemical ellipsometry') and subsequently manipulating the resultant optical data, it is possible to deduce much concerning the electrochromic layers, such as any changes in film thickness with potential (called 'electrostriction') and the formation of concentration gradients within the film.[11,12]

References

1. Bard, A. J. and Faulkner, L. R. *Electrochemical Methods: Fundamentals and Applications*, 2nd edn, New York, Wiley, 2001, pp. 2–3, 48–9, 51–2.
2. A. J. Bard (ed.). *Encyclopedia of Electrochemistry of the Elements*, New York, Marcel Dekker, 1973–1986.
3. Antelman, M. S. *Encyclopedia of Chemical Electrode Potentials*, New York, Plenum, 1982.
4. David R. Lide (ed.). *The CRC Handbook of Chemistry & Physics*, 86th edn, Boca Raton, FL, CRC Press, 2005.
5. Bard, A. J. and Faulkner, L. R. *Electrochemical Methods: Fundamentals and Applications*, 2nd edn, New York, Wiley, 2001, pp. 808–12.
6. Hitchcock, D. I. and Taylor, A. C. The standardization of hydrogen ion determinations, I: hydrogen electrode measurements with a liquid junction. *J. Am. Chem. Soc.*, **59**, 1937, 1813–18.
7. Tafel, J. *Z. Physik. Chem.*, **50A**, 1905, 641, as cited in Bard and Faulkner.
8. Bard, A. J. and Faulkner, L. R. *Electrochemical Methods: Fundamentals and Applications*, 2nd edn, New York, Wiley, 2001, pp. 102 ff.
9. Bard, A. J. and Faulkner, L. R. *Electrochemical Methods: Fundamentals and Applications*, 2nd edn, New York, Wiley, 2001, p. 29.
10. Macdonald, D. D. *Transient Techniques in Electrochemistry*, New York, Plenum, 1977.
11. Tompkins, H. G. and McGahan, W. A. *Spectroscopic Ellipsometry and Reflectometry: A User's Guide*, New York, Wiley, 1999.
12. [Online] at www.beaglehole.com/elli_intro/elli_intro.html (accessed 15 November 2005).

4

Optical effects and quantification of colour

4.1 Amount of colour formed: extrinsic colour

The coloured form of the electrochrome is produced by electrochemical reaction(s) at the electrode, Eq. (4.1) and its reverse (see Section 3.2). At the electrode, each redox centre of the electroactive species can accept or donate electrons from or to an external metal connection, one centre being formed per n electrons, where n is usually one or two according to the balanced redox reaction, Eq. (4.1):

$$\text{oxidised form, O} + \text{electron\{s\}} \rightarrow \text{reduced form, R.} \qquad (4.1)$$

In the simplest cases, the number of colour centres formed by the electrode reaction, and hence the change in absorbance $\Delta(Abs)$, is in direct proportion to the electrochemical charge passed Q, following Faraday's first law, 'The amount of (new) material formed at an electrode is proportional to the electrochemical charge passed':

$$\Delta(Abs) \propto Q. \qquad (4.2)$$

The term 'electrochemical' charge here implies that no unwanted side reactions involving electron transfer occur at the electrode during electrochromic colour change, i.e. that the relevant reaction is 100% efficient. The component of the total charge passed that is directly involved in forming the desired product is termed the faradaic charge for that process (but redox side reactions involving unwanted electrochemical products also involve faradaic charge). If the total charge passed is greater than the faradaic charge, then the difference is termed 'non-faradaic'. This represents processes like 'parasitic' current leakage possibly resulting from undesirable electronic current such as through the intra-electrode cell materials (electrolyte), or double-layer charging in the electrolyte adjacent to the electrode, an effect emulating the charging of an electrolytic capacitor.

The magnitude of the optical absorbance change obviously follows the (ideal) faradaic charge Q governing the amount of coloured material formed. The Beer–Lambert law – Eq. (4.3) – relates the optical absorbance Abs proportionally to the concentration of a chromophore:

$$Abs = \varepsilon l c, \qquad (4.3)$$

where ε is the extinction coefficient or molar absorptivity, c is the concentration of the coloured species and l the spectroscopic path length in the sample; l could be the thickness of a thin solid film of electrochrome, or the thickness of a liquid layer containing a dissolved chromophore. In the case of electrochemically generated colour, $\Delta(Abs)$ is the change in the optical absorbance and, from Eq. (4.3), is related by Eq. (4.4) to Δc, the change in the concentration of chromophore generated by the electrochemical charge passed:

$$\Delta(Abs) = \varepsilon l \Delta c. \qquad (4.4)$$

Even when the electrode efficiency is 100%, the relationship $\Delta Abs \propto Q$ in Eq. (4.2) will only hold if the absorbance is determined at fixed wavelength. However, many solid-state electrochromic systems do not follow the relation $\Delta(Abs) \propto Q$ because both the shape of the major absorption band and the wavelength maximum can change somewhat with the extent of charge insertion (i.e. of electrochemical change as gauged by the insertion coefficient x) and hence, of course, with concentration of coloured species. This deviation can result from changes in the molecular environment about the colorant with amount of colorant produced.

4.2 The electrochromic memory effect

A liquid-crystal display (LCD) is field-responsive, while electrochromic devices are potential-responsive. As colour generated in an ECD results from the application of a voltage across it that causes charge to flow, in an ECD therefore the colour intensity can be readily modulated between 'negligible' at the one extreme (all electroactive sites being in a non- or weakly absorbing redox state), and 'intense' at the other (all electroactive sites being in the coloured redox state). In brief, light intensity is modulated by varying the amount of charge passed.

Exemplifying, Figure 4.1 shows an electrochromic figure '**3**', the image being formed at those separate and insulated electrodes to which a suitable potential is applied, where charge therefore flows and coloration ensues. The electrochromic colour is removed (bleached) by applying a potential now with the

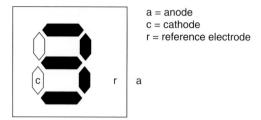

Figure 4.1 Schematic representation of an electrochromic alphanumeric character comprising seven separate electrodes.

polarity reversed, thereby reversing the electron-transfer process in Eq. (4.1). The ECD on/off operation thus relies on the reversible redox reaction at an electrochromic electrode,

$$\text{oxidation} + ne^- \rightleftarrows \text{reductant}, \qquad (4.1)$$

as discussed more fully in connection with Eq. (3.1) and elaborated in Section 3.2. With a second electrode plus interposed electrolyte, ECDs behave just like rechargeable (i.e. 'secondary') batteries, but in thin-film form; the similarities are explored by Heckner and Kraft.[1]

Since the perception of ECD colour arises from formation of a coloured *chemical*, rather than from a light-emitting or interference effect, the colour in solid-state electrochromes – i.e. type III – will persist after the current has ceased to flow. This persistence of colour leads to the useful property of ECDs, the so-called 'memory effect'. Such memory is occasionally referred to as being 'non-volatile'. However, since nearly all redox states are somewhat reactive, unwanted redox reactions can occur within devices after colour formation, thus, in the sense that no storage battery is ever perfect, most ECDs do not retain their colour indefinitely. Furthermore, type-I all-solution electrochromes diffuse from the solid contact, then being decolorised by reactions in mid-solution, and a maintaining current is necessary for colour persistence. In practice then, the memory is never permanent. Organic electrochromes in particular can also photodegrade. Such colour loss is often termed 'self bleaching'; see p. 15. Device durability is addressed in Chapter 16.

4.3 Intrinsic colour: coloration efficiency η

Although the number of colour centres formed is a function of the electrochemical charge passed, the observed intensity of colour will also depend on the specific electrochrome, some electrochromes being intensely coloured, others only feebly so. The optical absorption of an electrochrome is related

4.3 Intrinsic colour: coloration efficiency η

to the inserted charge per unit area Q (the 'charge density') by an expression akin to the Beer–Lambert law (Eq. (4.3) above), since Q is proportional to the number of colour centres formed, Eq. (4.5):

$$Abs = \log\left(\frac{I_o}{I}\right) = \eta Q. \qquad (4.5)$$

Here the proportionality factor η, the 'coloration efficiency', is a quantitative measure of the electrochemically formed colour. For an ECD in transmission mode, η is measured as the *change* in optical absorbance $\Delta(Abs)$ evoked by the electrochemical charge density Q passed, Eq. (4.6):

$$\eta = \frac{\Delta(Abs)}{Q}. \qquad (4.6)$$

The proportionality factor η is clearly independent of the optical pathlength l within the sample.

The coloration efficiency can be thought of as an electrochemical equivalent of the more familiar extinction coefficient ε (*cf.* Eq. (4.3) above), which characterises a chromophore *in solution* (in a particular solvent); η thus represents the area of electrochrome on which colour is intensified, in absorbance units per coulomb of charge passed. In many electrochromic studies it is (erroneously) expressed in cm^2, rather than area per unit charge, for example cm^2 C^{-1}.

Needless to say, values of η should thus be maximised for most efficient device operation. A compendium of η for metal oxide electrochromes is given in Table 4.1, and for organic species in Table 4.2. Many additional values are available in refs. 2 and 3; and many other values are cited elsewhere in this work. The obviously larger values of η for organic species owes largely to enhanced quantum-mechanical properties governing the probability of electronic transitions responsible for coloration (see p. 60 ff.).

Since the optical absorbance Abs depends on the wavelength of observation, η must be determined at a fixed, cited, wavelength; η is defined as positive if colour is generated cathodically, but negative if colour is generated anodically (in accordance with the IUPAC definitions: anodic currents are deemed negative, cathodic currents positive). A negative for anodic coloration is not always stated, however, so care is needed here.

Values of η are clearly smaller for metal oxides than for all other classes of electrochrome, but this has not deterred most investigators from studying the electrochromic properties of oxides (see Chapter 6).

Table 4.1. *Coloration efficiencies η for thin films of metal-oxide electrochromes. Positive values denote cathodically formed colour, negative values denote anodic coloration.*

Oxide	Morphology	Preparative method[a]	[b]$\eta/cm^2 C^{-1}$	Ref.
Anodically colouring oxides				
FeO	Polycrystalline	CVD	−6.0	4
FeO	Polycrystalline	Sol–gel	−28	5
FeO	Polycrystalline	Electrodeposition	−30	6
IrO_x	Polycrystalline	rf sputtering	−15 (633)	7
IrO_x	Amorphous	Anodic deposition	−30	8
NiO	Polycrystalline	dc sputtering	−41 – 25	9
NiO	Amorphous	Dipping technique	−35	10
NiO	Amorphous	Electrodeposition	−20	11
NiO	Polycrystalline	rf sputtering	−36 (640)	12
NiO	Polycrystalline	Spray pyrolysis	−37	13
NiO	Amorphous	Vacuum evaporation	−32 (670)	14
Rh_2O_5	Amorphous	Anodic deposition	−20 (546)	8
V_2O_5	Polycrystalline	rf sputtering	−35 (1300)	15
Cathodically colouring oxides				
Bi_2O_3	Amorphous	Sputtering	3.7 (650)	16
CoO	Polycrystalline	CVD	21.5	17
CoO	Amorphous	Electrodeposited	24	18,19
CoO	Polycrystalline	Sol–gel	25	20
CoO	Polycrystalline	Spray pyrolysis	12 (633)	21
CoO	Amorphous	Thermal evaporation	20–27	22
MoO_3	Amorphous	Ther. evap. of Mo(s)	19.5 (700)	23
MoO_3	Polycrystalline	Oxidation of MoS_3	35 (634)	24
MoO_3	Amorphous	Thermal evaporation	77 (700)	12
$Mo_{0.008}W_{0.992}O_3$	Amorphous	Thermal evaporation	110 (700)	25
Nb_2O_5	Polycrystalline	rf sputtering	12 (800)	26
Nb_2O_5	Polycrystalline	Sol–gel	38 (700)	27
Ta_2O_5	Polycrystalline	rf sputtering	5 (540)	26
TiO_2	Amorphous	Thermal evaporation	7.6	28
TiO_2	Polycrystalline	rf sputtering	8 (546)	29
TiO_2	Amorphous	Thermal evaporation	8 (646)	30
TiO_2	Polycrystalline	Sol–gel	50	31
WO_3	Amorphous	Thermal evaporation	115 (633)	32
WO_3	Amorphous	Electrodeposition	118 (633)	33
WO_3	Amorphous	Electrodeposition	62–66 (633)	34
WO_3	Amorphous	Thermal evaporation	79 (800)	35
WO_3	Polycrystalline	rf sputtering	21	36
WO_3	Polycrystalline	Spin-coated gel	64 (650)	37
WO_3	Amorphous	Dip-coating[c]	52	38
WO_3	Polycrystalline	Spray pyrolysis	42	39
WO_3	Polycrystalline	Sol–gel	36 (630)	40
WO_3	Polycrystalline	CVD	38–41	41
WO_3	Polycrystalline	dc sputtering	109 (1400)	26

[a] 'CVD' = chemical vapour deposition; 'dc sputtering' = dc magnetron sputtering.
[b] Wavelength (λ/nm) used for measurement in parentheses.

Table 4.2. *Coloration efficiencies η for organic electrochromes. Positive values denote cathodically formed colour, negative values denote anodic coloration. (Table reproduced from Rauh, R. D., Wang, F., Reynolds, J. R. and Meeker, D. L. 'High coloration efficiency electrochromics and their application to multi-color devices'. Electrochim. Acta, **46**, 2001, 2023–2029, by permission of Elsevier Science.)*

Electrochrome	$\lambda_{(max)}$/ nm	$^a\eta$/ cm^2 C^{-1}
Monomeric organic redox dyes		
Indigo Blue	608	−158
Toluylene Red	540	−150
Safranin O	530	−274
Azure A	633	−231
Azure B	648	−356
Methylene Blue	661	−417
Basic Blue 3	654	−398
Nile Blue	633	−634
Resazurin	598	−229
Resorufin	573	−324
Methyl viologen	604	176
Conducting polymers		
Poly(3,4-ethylenedioxythiophenedidodecyloxybenzene)	552	−1240^b
	730	650^c
Poly(3,4-propylenedioxypyrrole)	480	−520
Poly(3,4-propylenedioxythiophene), PProDOT	551	−275

^{a} Values were *calculated* from data published in ref. 43; ^{b} reduced form; ^{c} oxidised form.

4.3.1 Intrinsic colour: composite *coloration efficiency (CCE)*

Although measuring values of η is important for assessing the power requirements of an electrochrome, Reynolds *et al.*[44] emphasise that the methods chosen for measurement often vary between research groups which causes difficulty in comparing values for different electrochromes. A general method for effectively and consistently measuring *composite* coloration efficiencies (CCEs) (see below) has been proposed,[44] and applied to measurements on electrochromic films of conductive polymers[44,45,46] and the mixed-valence inorganic complex, Prussian blue – PB, iron(III) hexacyanoferrate(II):[47] PB is reducible to the clear Prussian white – PW, iron(II) hexacyanoferrate(II). Such measurements have also been applied to conductive polymers[48] but performed with reflected light as opposed to the usual transmitted light.

A tandem chronocoulometry–chronoabsorptometry method is employed to measure composite coloration efficiencies, with CCEs being calculated at *specific percentage transmittance changes*, at the λ_{max} of the appropriate absorbance band. To illustrate this approach, Figure 4.2(a) shows the absorbance during the dynamic measurement of a film of Prussian blue (PB) at 686 nm, to effect the electrochromic transition. A square wave pulse was switched between +0.50 V (PB, of high absorbance) and −0.20 V (PW, of low absorbance); these potentials are cited against a Ag|AgCl wire in KCl solution (0.2 mol dm^{-3}). For the PB → PW transition, the electrochromic

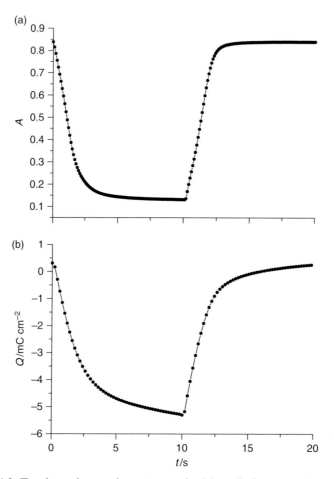

Figure 4.2 Tandem chronoabsorptometric (a) and chronocoulometric (b) data for a PB|ITO|glass electrode in aqueous KCl (0.2 mol dm^{-3}) supporting electrolyte, on square-wave switching between +0.50 V (PB, high absorbance) and −0.20 V (PW, low absorbance) vs. Ag|AgCl. (Figure reproduced from Mortimer, R. J. and Reynolds, J. R. '*In situ* colorimetric and composite coloration efficiency measurements for electrochromic Prussian blue'. *J. Mater. Chem.*, **15**, 2005, 2226–33, with permission from The Royal Society of Chemistry.)

Table 4.3. *Optical and electrochemical data collected for coloration efficiency measurements. Prussian blue is reviewed in Chapter 7, and PEDOT in Chapter 10. (Table reproduced with permission of The Royal Society of Chemistry, from: Mortimer, R. J. and Reynolds, J. R. 'In situ colorimetric and composite coloration efficiency measurements for electrochromic Prussian blue.' J. Mater. Chem., 15, 2005, 2226–33.)*

Transition	% of full switch	$\Delta(\%T)$	ΔA	$Q/\text{mC cm}^{-2}$	$\eta/\text{cm}^2\text{C}^{-1}$	t/s	Ref.
PB → PW	90	53.8	0.673	4.49	−150	3.4	
PB → PW	**95**	**56.6**	**0.691**	**4.85**	**−143**	**4.4**	47
PB → PW	98	58.3	0.701	5.18	−135	6.0	
PW → PB	90	52.9	0.564	3.85	−147	1.9	
PW → PB	**95**	**55.9**	**0.632**	**4.21**	**−150**	**2.2**	47
PW → PB	98	57.5	0.675	4.54	−149	2.6	
PEDOT	90	48	0.48	2.49	192	0.33	
PEDOT	**95**	**51**	**0.49**	**2.68**	**183**	**0.36**	44
PEDOT	98	53	0.50	3.04	165	0.45	

The bold figures represent the authors' preferred reference percentage.

contrast at 686 nm was 60% of the total *transmittance* ($\Delta\%T$), calculated from the maximum and minimum absorbance values. The charge measurements recorded simultaneously with the absorbance data are given in Figure 4.2(b).

In the *composite* coloration efficiency method, to provide points of reference with which to compare the CCE values of various electrochromes, values of η are *calculated* at a specific transmittance change, as a percentage of the total $\Delta(\%T)$. Table 4.3 shows data for 90, 95 and 98% changes, for both reduction of PB to form PW, and the reverse process, oxidation of PW to re-form PB.

Although the chronocoulometric data in Table 4.3 were corrected for background charging, as were the measurements with conducting polymer films,[44] the η values for the reduction process are seen to decrease slightly with increases in optical change. This decrease demonstrates the importance of measuring the charge passed at a very specific transmittance value and not simply to divide the total absorbance change by the maximum charge passed. This practice is important in considering the reduction of PB to PW, because PW is a good catalyst for the reduction of oxygen: molecular O_2 may diffuse into the cuvette during long measurement times, resulting in an erroneously high charge measurement.

It should be noted that in the original publication[44] that introduced composite coloration efficiency measurements, the calculated values of η were described as being at 90, 95 and 98% of the total *optical density* change [$\Delta OD\ (=\Delta Abs)$], at

λ_{max}. In view of the fundamental definition of ΔOD, this choice of variable represents a mis-statement and all composite coloration efficiencies recorded in Table 4.3, and previously,[44] were determined using the ΔOD at 90, 95 and 98% of $\Delta(\%T)$. Although as observed above, inorganic materials typically exhibit lower η values than conducting polymers, it is interesting to note from Table 4.3 how the carefully measured values calculated here are comparable to those for films of poly(3,4-ethylenedioxythiophene) – PEDOT – (at a film thickness of 150 nm), although switching times are longer for the PB–PW transition. The η values are similar for both the reduction of PB to PW, and for the re-oxidation of PW back to PB, although the switching times for the latter process are slightly shorter. To preserve the electroneutrality of the solid electrochrome, uptake or loss of potassium ions must accompany the colour-transforming electron transfer; see Chapter 8. The difference in switching times probably arises from different rates of ingress or egress of potassium ions in these films.

4.4 Optical charge transfer (CT)

Films of solid electrochrome are comparatively thin, usually sub-micron in thickness, and thus comprise very little material; and solution-phase electrochromes are enclosed in ECDs within small volumes of solvent, typically of maximum optical path length 1 mm. An electrochromic colour that is intense enough to observe under normal illumination will therefore require a spectroscopic transition that is very intense, i.e. having a very high extinction coefficient ε.

Of the organic electrochromes, the most intense absorptions are encountered with systems having an extended conjugation system, such as cyanines and conductive polymers, or a large extent of internal conjugation such as radicals of the viologens (see Chapter 11). As an example, the radical cation of CPQ, cyanophenyl paraquat (**I**) (formally 1,1'-bis(p-cyanophenyl)-4,4'-bipyridilium) in acetonitrile has an intense green colour:[49] at $\lambda_{(max)} = 674$ nm its ε is 83 300 dm^3 mol^{-1}cm^{-1}, cf. ε for the aqueous MnO$_4^-$ ion (which is generally thought to be intensely coloured) of only[50] 2400 dm^3 mol^{-1} cm^{-1}.

NC—⌬—N⁺—⌬—⌬—N⁺—⌬—CN

I

The metal-oxide system to have received the most attention for electrochromic purposes is tungsten trioxide, WO$_3$ (Section 6.2). The bulk trioxide is pale

4.4 Optical charge transfer (CT)

yellow in colour and transparent as a thin film, but forms a blue colour on reduction. In metal-oxide systems, the source of the required intense electrochromic colour is usually an *intervalence* optical charge-transfer (CT) transition,[51,52] where the term 'intervalence' implies here that the two atoms or ions are of the same element.

In colourless WO_3, all tungsten sites have a common oxidation state of $+VI$. Reductive electron transfer to a W^{VI} site forms W^V, and the blue form of the electrochrome becomes evident from the optical CT. This blue form is commonly called a 'bronze' (see Chapter 6), although strictly, tungsten *bronzes* are characterised by metallic conductivity, and have compositions M_xWO_3 where x is typically greater than about 0.3. A WO_3-based electrochrome (rather than a bronze), as used in an ECD, must be restricted to a lower value of x in order to preserve switchability, and is thus a semiconductor.

The optical intervalence CT of this sort is usually regarded as the major cause of the electrochromic colour in many inorganic systems. Other mechanisms such as the Stark effect are briefly dealt with in Chapter 1. In a CT-based system, following photon absorption an electron is optically excited from an orbital on the donor species in the ground-state (pre-transfer) electronic configuration of the system, to a vacant electronic orbital on an adjacent ion or atom, producing an excited state. The blue colour is caused by red light being absorbed to effect the intervalence transition between adjacent ('A' and 'B') W^{VI} and W^V centres, Eq. (4.7):

$$W^{VI}_{(A)} + W^{V}_{(B)} + h\nu \rightarrow W^{V}_{(A)} + W^{VI}_{(B)}. \qquad (4.7)$$

The product species, which are hence in an excited state, subsequently lose the excess energy acquired from the absorbed photon by thermal dissipation to surrounding structures. (Close examination of the PB–PW structures shows that the photo-effected product distribution unusually involves an intrinsic chemical change absent in the Eq. (4.7) transition for $W^{VI/V}$, ferric ferrocyanide being chemically different from photo-product ferrous ferricyanide, in contrast with the transition depicted in Eq. (4.7).) These intervalence transitions are characterised by broad, intense and relatively featureless absorption bands in the UV, visible or near IR, with molar absorptivities (extinction coefficients) of useful magnitudes. As an example, ε for the $W^{V,VI}$ oxide system in Eq. (4.7) lies in the range[53] 1400–5600 $dm^3\,mol^{-1}\,cm^{-1}$, the value decreasing with increasing insertion coefficient x. (The optical properties of WO_3 are discussed in Chapter 6, Section 6.4 on p. 140 ff.)

4.5 Colour analysis of electrochromes

Colour is a very subjective phenomenon, causing its *description* or, for example, the *comparison* of two colours, to be quite difficult. However, a new method of colour analysis, *in situ colorimetric analysis* has recently been developed.[54] It is based on the CIE (Commission Internationale de l'Eclairage (the 'International Commission on Illumination')) system of colorimetry, which is elaborated below. The CIE method has been applied[44,45,54,55,56,57,58,59,60,61] to the quantitative colour measurement of conducting electroactive polymer and other electrochromic films on optically transparent electrodes (OTEs) under electrochemical potential control in a spectroelectrochemical cell. Experimentally, the method is straightforward in operation: a spectroelectrochemical cell is assembled within a light box, and a commercial portable colorimeter (such as the *Minolta* CS-100 Chroma Meter), mounted on a tripod, measures changes in the electrochromic film during transformations performed under potentiostatic control. This method allows the *quantitative* colour description of electrochromes, as perceived by the human eye, in terms of hue, saturation and luminance (that is, relative transmissivity). Such colour analyses provide a more precise way to define colour[62,63] than more familiar forms of spectrophotometry. Rather than simply measuring spectral absorption bands, in colour analysis the human eye's sensitivity to light across the whole visible spectral region is measured and a numerical description of a particular colour is given.

This approach, which has been applied to electrochromic conducting electroactive polymer films and, more recently, to Prussian blue films,[47] is likely to be applicable to a wide range of both organic and inorganic electrochromes. There are three main advantages to *in situ* colorimetric analysis. First, by acquiring a quantitative measure of the colour, it is possible to report accurately the colour of new materials. Second, by utilising colorimetric analysis, it is possible to represent graphically the path of an electrochrome's colour change. Third, the method can ultimately function as a valuable tool in the construction of electrochromic devices. Beyond these practical considerations, colorimetric analyses can also provide valuable information about the optical and electrochemical processes in electrochromes. The approach is exemplified in Figures 4.5 and 4.6 for PB, and elaborated below.

4.5.1 A brief synopsis of colorimetric theory

Colour is described by three attributes. The first identifies a colour by its location in the spectral sequence, i.e. the wavelength associated with the colour. This is known as the *hue, dominant wavelength* or *chromatic colour,* and is the wavelength where maximum contrast occurs. It is this aspect which is commonly, but mistakenly, referred to as colour.

The second attribute relates to the relative levels of white and/or black, and is known as *saturation, chroma, tone, intensity* or *purity*.

The third attribute is the *brightness* of the colour, and is also referred to as *value, lightness* or *luminance*. Luminance provides information about the perceived transparency of a sample over the entire visible range.

Using the three attributes of hue, saturation and luminance, any colour can be both described and actually quantified. In order to assign a quantitative scale to colour measurement, the hue, saturation and luminance must be defined numerically in a given colour system. The most well known and most frequently used colour system is that developed by the Commission Internationale de l'Eclairage, commonly known as the CIE system of colorimetry. It was first devised in 1931, and is based on a so-called '2° Standard Observer', that is, a system characterised by the result of tests in which people had to visually match colours in a 2° field of vision.[64]

Thus the CIE system is based on how the 'average' person subjectively sees colours, and thus simulates mathematically how people perceive colours. The original CIE experiments resulted in the formulation of colour-matching functions, which were based on the individual's response to various colour stimuli. There are three modes by which the eye is stimulated when viewing a colour, hence the CIE system is expressed in terms of a 'tristimulus'. These colour matching functions are used to calculate such tristimulus values (symbolised as X, Y and Z), which define the CIE system of colorimetry. Once obtained, values of X, Y and Z allow the definition of all the CIE recommended colour spaces, where the phrase 'colour space' implies a method for expressing the colour of an object or a light source using some kind of notation, such as numbers. The concept for the XYZ tristimulus values is based on the three-component theory of colour vision, which states that the eye possesses three types of cone photoreceptors for three primary colours (red, green and blue) and that all colours are seen as mixtures of these three primary colours.

Colour spaces are usually defined as imaginary geometric constructs, containing all possible colour perceptions, and represented in a systematic manner according to the three attributes. Colour spaces are the means by which the information of the X, Y and Z tristimulus values is represented graphically, either in two- or three-dimensional space. Actually the tristimulus values themselves constitute a colour space, although the three-dimensional vectoral nature of the comprehensive system makes it quite unwieldy for presenting data. Colour is a three-dimensional phenomenon, so it is not easily represented quantitatively.

Colour quantification is more easily visualised if separated into the two attributes, lightness and chromaticity. The 'lightness' describes how light or dark a colour is, and 'chromaticity' (representing hue and chroma) can be shown two-dimensionally.

The CIE has defined numerous colour spaces based on various criteria. The three most commonly used are the CIE 1931 Yxy colour space, the CIE 1976 $L^*u^*v^*$ colour space, and the CIE 1976 $L^*a^*b^*$ colour space. The latter is also referred to as CIELAB. The evolution of the CIE criteria is now outlined.

The colour sensitivity of the eye changes according to the angle of view. In 1931, the CIE proposed its first recommended colour space based on the X, Y and Z tristimulus values and a 2° field of view, hence the name '2° Standard Observer'. In this system, the tristimulus value Y is retained as a direct measure of the brightness or luminance of the colour. The two-dimensional graph obtained with such data is Cartesian – an xy graph – and known as the 'xy chromaticity diagram'. From this diagram, respective values of x and y are calculated from the X, Y and Z tristimulus values via Eq. (4.8) and Eq. (4.9):

$$x = \frac{X}{X+Y+Z}, \quad (4.8)$$

$$y = \frac{Y}{X+Y+Z}. \quad (4.9)$$

On the graph represented in Figure 4.3, the line surrounding the horse-shoe-shaped area is called the 'spectral locus', which shows the wavelengths of light in the visible region. Colour Plate 1 shows a colour representation of this figure.

The line connecting the longest and shortest wavelengths contains the non-spectral purples, and is therefore known as the 'purple line'. Surrounded by the spectral locus and the purple line is the region known as the 'colour locus', which contains every colour that can exist. The point (labelled as **W** in Figure 4.3) within this locus is known as the white point and its location is dependent on the light source. The CIE has several recommended light sources (so-called 'illuminants'), such as the D_{50} (5000 K) constant-temperature daylight simulating light source. The location of a point in the xy diagram then gives the hue and chroma of the colour. The hue is determined by drawing a straight line through the point representing 'white' and the point of interest to the spectral locus thus obtaining the dominant wavelength of the colour.

To exemplify, Figure 4.4 shows the determination of the dominant wavelength (~550 nm) for 'sample B'; and to reiterate terminology, the spectral locus refers only to the horse-shoe-shaped curve and not the purple line which

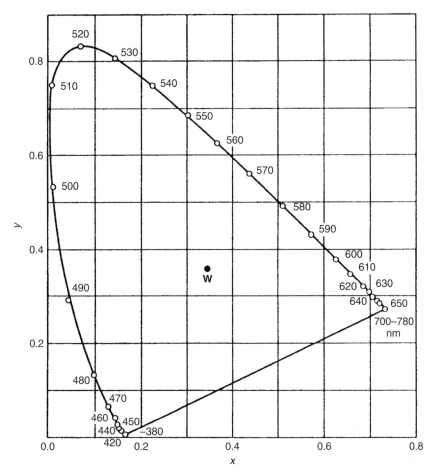

Figure 4.3 CIE 1931 *xy* chromaticity diagram with labelled white point (**W**).

is defined by non-spectral purples. For placing a wavelength dependence on samples such as 'sample A' that are found along the purple line, a complementary wavelength can be expressed by drawing a straight line from the sample coordinate through the white point to the spectral locus. Indeed a complementary wavelength can be expressed for any sample with which this procedure can be applied. The purity (or saturation) as expressed by the relation in the figure is a measure of the intensity of specific hue, with the most intense (or saturated) colours lying closest to the spectral locus.

The most saturated colours lie along the spectral locus. It is important, however, to realise that the CIE does not associate any given colour with any point on the diagram: if colours are ever included on a diagram, they are only an artist's representation of what colour a region is *most likely* to represent. The reason that colours cannot be specifically associated with a given pair of

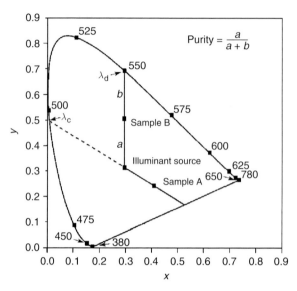

Figure 4.4 CIE 1931 *xy* chromaticity diagram showing the determination of the complementary wavelength of a sample with *xy* coordinates of arbitrary sample A, and the dominant wavelength and purity of a sample with *xy* coordinates of arbitrary sample B. (Figure reproduced from DuBois, Jr, C. J. 'Donor-acceptor methods for band gap control in conjugated polymers'. Ph.D. Thesis, Department of Chemistry, University of Florida, 2003, p. 21, by permission of the author.)

xy coordinates is because the third dimension of colour, lightness, is not included in the diagram. The relative lightness or darkness of a colour is very important in how it is perceived. The brightness is usually presented as a percentage, as expressed in Eq. (4.10):

$$\% Y = \frac{Y}{Y_0} \times 100, \qquad (4.10)$$

in which Y_0 is the background luminance and Y is the luminance measured for the sample. In the corresponding dome-shaped three-dimensional diagram, it is recognised that the highest purity or saturation can only be achieved when the luminance or lightness of the colour is at a low value.[63]

In 1976, the CIE proposed two new colour spaces, $L^*u^*v^*$ and $L^*a^*b^*$, in order to correct flaws in the earlier proposed systems. Both were defined as uniform colour spaces, which are geometrical constructs containing all possible colour sensations. This new system is formulated in such a way that equal distances correspond to colours that are perceptually equidistant. The main reason for designing such systems was to provide an accurate means of representing and calculating colour difference.

4.5 Colour analysis of electrochromes

The CIE $L^*u^*v^*$ colour space is a uniform colour space based on the X, Y and Z tristimulus values defined in 1931. The L^* value measures the lightness; chroma and hue are defined in terms of u^* and v^*. The CIE $L^*u^*v^*$ system has a corresponding two-dimensional chromaticity diagram known as the $u'v'$ UCS ('uniform colour space'), which is very similar to the 1931 xy chromaticity diagram. The $L^*u^*v^*$ colour space is now used as a standard in television, video and the display industries.

In a further development the $L^*a^*b^*$ colour space is also a uniform colour space defined by the CIE in 1976. The L^* value represents the same quantity as in CIE $L^*u^*v^*$ and hue and saturation bear similar relationships to a^* and b^*. The CIE $L^*a^*b^*$ space is a standard commonly used in the paint, plastic and textile industries.

The values of L^*, a^* and b^* are defined as in Equations (4.11)–(4.13):

$$L^* = 116 \times \left(\frac{Y}{Y_n}\right)^{1/3} - 16; \tag{4.11}$$

$$a^* = 500 \times \left[\left(\frac{X}{X_n}\right)^{1/3} - \left(\frac{Y}{Y_n}\right)^{1/3}\right]; \tag{4.12}$$

$$b^* = 200 \times \left[\left(\frac{Y}{Y_n}\right)^{1/3} - \left(\frac{Z}{Z_n}\right)^{1/3}\right]; \tag{4.13}$$

where X_n, Y_n and Z_n are the tristimulus values of a perfect reflecting diffuser (as calculated from the background measurement). In the $L^*a^*b^*$ chromaticity diagram, $+a^*$ relates to the red direction, $-a^*$ is the green direction, $+b^*$ is the yellow direction, and $-b^*$ is the blue direction. The centre of the chromaticity diagram (0, 0) is achromatic; as the values of a^* and b^* increase, the saturation of the colour increases.

None of the systems is perfect, but the 1931 xy chromaticity diagram is probably the best known and most widely recognised way to represent a colour. The diagram conveys information in a straightforward manner and hence is very easy to use and understand. In addition, the CIE 1931 system is useful in that it can be used to analyse colour in many different ways; notably, the system can be used to predict the outcome of mixing colour. The result of mixing two colours is known to lie along the straight line on the xy chromaticity diagram connecting the points representing the colours of the pure components in the mixture. The position on this line representing the actual chromicity depends on the ratio of the amounts of the two mixed colours.

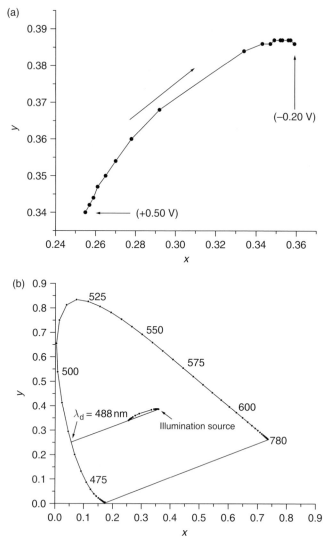

Figure 4.5 CIE 1931 xy chromaticity diagrams for a Prussian blue (PB)|ITO|glass electrode in aqueous KCl (0.2 mol dm^{-3}) supporting electrolyte. (a) The potential (vs. Ag/AgCl) was decreased, in the steps indicated in Table 4.4, from the coloured PB (+0.50 V) to the transparent Prussian white (PW) (−0.20 V) redox states. (b) The xy coordinates are plotted onto a diagram that shows the locus coordinates, with labelled hue wavelengths, and the evaluation of the dominant wavelength (488 nm) of the PB redox state. (Figure reproduced from Mortimer, R. J. and Reynolds, J. R. 'In situ colorimetric and composite coloration efficiency measurements for electrochromic Prussian blue'. J. Mater. Chem., 15, 2005, 2226–33, with permission from The Royal Society of Chemistry.)

4.5 Colour analysis of electrochromes

The advantage of the CIE $L^*u^*v^*$ and CIE $L^*a^*b^*$ colour spaces is that they are 'uniform', i.e. equal distances on the graph represent equal perceived colour differences; the $L^*u^*v^*$ and $L^*a^*b^*$ systems therefore resolve a major drawback of the earlier 1931 system, correcting a defect of the latter which was that equal distances on the graph did not represent equal perceived colour differences.

As uniform colour spaces, CIE $L^*u^*v^*$ and CIE $L^*a^*b^*$ allow the accurate representation and calculation of colour differences. In addition, calculations can be performed to conclude whether differences in colour are due to differences in lightness, hue or saturation. The only difference between the $L^*u^*v^*$

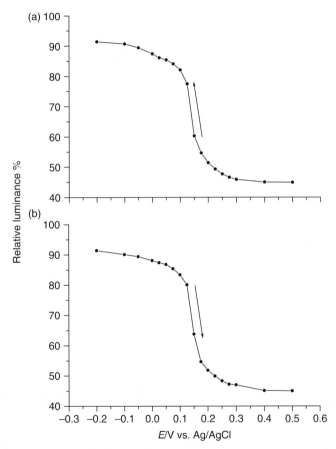

Figure 4.6 Relative luminance (%), vs. applied potential (E/V vs. Ag/AgCl), for a PB|ITO|glass electrode in aqueous KCl (0.2 mol dm^{-3}) as supporting electrolyte. The potential was decreased (a) and then increased (b), in the same steps as used for Figure 4.5, between the coloured PB (+ 0.50 V) and the transparent PW (− 0.20 V) redox states. (Figure reproduced from Mortimer, R. J. and Reynolds, J. R. 'In situ colorimetric and composite coloration efficiency measurements for electrochromic Prussian blue'. J. Mater. Chem., **15**, 2005, 2226–33, with permission from The Royal Society of Chemistry.)

Table 4.4. *Coordinates for reduction of Prussian blue to Prussian white as a film on an ITO|glass substrate in aqueous KCl (0.2 mol dm^{-3}) supporting electrolyte. Data come from ref. 47. (Table reproduced with permission of The Royal Society of Chemistry, from: Mortimer, R. J. and Reynolds, J. R. In situ colorimetric and composite coloration efficiency measurements for electrochromic Prussian blue. J. Mater. Chem., 15, 2005, 2226–33.)*

E/V vs. Ag/AgCl	%Y	x	y	L*	a*	b*
0.500	**44.9**	**0.255**	**0.340**	**73**	**−26**	**−33**
0.400	45.0	0.255	0.340	73	−26	−33
0.300	45.9	0.257	0.342	73	−26	−32
0.275	46.6	0.259	0.344	74	−26	−31
0.250	47.7	0.261	0.347	75	−27	−30
0.225	49.3	0.265	0.350	76	−26	−29
0.200	51.4	0.270	0.354	77	−26	−27
0.175	54.7	0.278	0.360	79	−25	−24
0.150	60.3	0.292	0.368	82	−22	−19
0.125	77.5	0.334	0.384	91	−10	−6
0.100	82.1	0.343	0.386	93	−7	−3
0.075	84.1	0.347	0.386	93	−5	−2
0.050	85.4	0.349	0.387	94	−5	−1
0.025	86.1	0.352	0.387	94	−3	−1
0.000	87.4	0.353	0.387	95	−3	0
−0.050	89.4	0.356	0.387	96	−2	0
−0.100	90.7	0.357	0.387	96	−1	1
−0.200	**91.4**	**0.359**	**0.386**	**97**	**0**	**1**

and $L^*a^*b^*$ colour spaces is that the $L^*a^*b^*$ lacks a two-dimensional diagram, which is probably its only major drawback.

The $u'v'$ uniform colour space diagram only functions as a *uniform* colour space when the plotted points lie in a plane of constant luminance. Therefore, the graphical representation of colour for materials with widely varying luminance, causes the $u'v'$ chromaticity diagram to lose all advantage over the 1931 xy chromaticity diagram.

Considering all the assets and drawbacks of these three different colour spaces, generally *in situ* colorimetric results are expressed graphically in the CIE 1931 Yxy colour space system. (In addition, due to the common use of the $L^*a^*b^*$ system, values of $L^*a^*b^*$ are also often reported.) By way of illustration, Figures 4.5 and 4.6 show sample colour coordinates and luminance data on switching between the (oxidised) blue and (reduced) colourless ('bleached') states of the electrochrome Prussian blue.

In this example, sharp changes in hue, saturation and luminance take place, with an exact coincidence of data in the reverse (colourless to blue) direction. Table 4.4

shows the Yxy coordinates, together with the calculated $L^*a^*b^*$ coordinates. Comparing the PB $L^*a^*b^*$ coordinates with those of the blue states for a range of different electrochromic conducting polymer films[54] shows the distinct nature of the blue colour provided by PB. For example, the $L^*a^*b^*$ coordinates for the (deep blue) neutral form of PEDOT are 20, 15, and −43 respectively,[54] while for PB they are 73, −26 and −33.

References

1. Heckner, K.-H. and Kraft, A. Similarities between electrochromic windows and thin film batteries. *Solid State Ionics*, **152–153**, 2002, 899–905.
2. Granqvist, G. C. *Handbook of Inorganic Electrochromic Materials*, Amsterdam, Elsevier, 1995.
3. Lev, O., Wu, Z., Bharathi, S., Glezer, V., Modestov, A., Gun, J., Rabinovich, L. and Sampath, S. Sol–gel materials in electrochemistry. *Chem. Mater.*, **9**, 1997, 2354–75.
4. Maruyama, T. and Kanagawa, T. Electrochromic properties of iron oxide thin films prepared by chemical vapor deposition. *J. Electrochem. Soc.*, **143**, 1996, 1675–8.
5. Özer, N. and Tepehan, F. Optical and electrochemical characterisation of sol–gel deposited iron oxide films. *Sol. Energy Mater. Sol. Cells*, **56**, 1999, 141–52.
6. Zotti, G., Schiavon, G., Zecchin, S. and Casellato, U. Electrodeposition of amorphous Fe_2O_3 films by reduction of iron perchlorate in acetonitrile. *J. Electrochem. Soc.*, **145**, 1998, 385–9.
7. Sato, Y., Ono, K., Kobayashi, T., Watanabe, H. and Yamanoka, H. Electrochromism in iridium oxide films prepared by thermal oxidation of iridium–carbon composite films. *J. Electrochem. Soc.*, **134**, 1987, 570–5.
8. Dautremont-Smith, W. C. Transition metal oxide electrochromic materials and displays, a review. Part 2: oxides with anodic coloration. *Displays*, **3**, 1982, 67–80.
9. Scarminio, J., Gorenstein, A., Decker, F., Passerini, S., Pileggi, R. and Scrosati, B. Cation insertion in electrochromic NiO_x films. *Proc. SPIE*, **1536**, 1991, 70–80.
10. Fantini, M. C. A., Bezerra, G. H., Carvalho, C. R. C. and Gorenstein, A. Electrochromic properties and temperature dependence of chemically deposited $Ni(OH)_x$ thin films. *Proc., SPIE*, **1536**, 1991, 81–92.
11. Carpenter, M. K., Conell, R. S. and Corrigan, D. A. The electrochromic properties of hydrous nickel oxide. *Sol. Energy Mater. Sol. Cells*, **16**, 1987, 333–46.
12. Kitao, M. and Yamada, S. Electrochromic properties of transition metal oxides and their complementary cells. In Chowdari, B. V. R. and Radharkrishna, S. (eds.), *Proceedings of the International Seminar on Solid State Ionic Devices*, Singapore, World Scientific Publishing Co., 1988, 359–78.
13. Kadam, L. D. and Patil, P. S. Studies on electrochromic properties of nickel oxide thin films prepared by spray pyrolysis technique. *Sol. Energy Mater. Sol. Cells*, **69**, 2001, 361–9.
14. Velevska, J. and Ristova, M. Electrochromic properties of NiO_x prepared by low vacuum evaporation. *Sol. Energy Mater. Sol. Cells*, **73**, 2002, 131–9.
15. Cogan, S. F., Nguyen, N. M., Perrotti, S. J. and Rauh, R. D. Optical properties of electrochromic vanadium pentoxide, *J. Appl. Phys.*, **66**, 1989, 1333–7.
16. Shimanoe, K., Suetsugu, M., Miura, N. and Yamazoe, N. Bismuth oxide thin film as new electrochromic material. *Solid State Ionics*, **113–115**, 1998, 415–19.

17. Maruyama, T. and Arai, S. Electrochromic properties of cobalt oxide thin films prepared by chemical vapour deposition. *J. Electrochem. Soc.*, **143**, 1996, 1383–6.
18. Polo da Fonseca, C. N., De Paoli, M.-A. and Gorenstein, A. The electrochromic effect in cobalt oxide thin films. *Adv. Mater.*, **3**, 1991, 553–5.
19. Polo da Fonseca, C. N., De Paoli, M.-A. and Gorenstein, A. Electrochromism in cobalt oxide thin films grown by anodic electroprecipitation, *Sol. Energy Mater. Sol. Cells*, **33**, 1994, 73–81.
20. Svegl, F., Orel, B., Hutchins, M. G. and Kalcher, K. Structural and spectroelectrochemical investigations of sol–gel derived electrochromic spinel Co_3O_4 films. *J. Electrochem. Soc.*, **143**, 1996, 1532–9.
21. Kadam, L. D., Pawar, S. H. and Patil, P. S. Studies on ionic intercalation properties of cobalt oxide thin films prepared by spray pyrolysis technique. *Mater. Chem. Phys.*, **68**, 2001, 280–2.
22. Svegl, F., Orel, B., Bukovec, P., Kalcher, K. and Hutchins, M. G. Spectroelectrochemical and structural properties of electrochromic Co(Al)-oxide and Co(Al, Si)-oxide films prepared by the sol–gel route. *J. Electroanal. Chem.*, **418**, 1996, 53–66.
23. Bica De Moraes, M. A., Transferetti, B. C., Rouxinol, F. P., Landers, R., Durant, S. F., Scarminio, J. and Urbano, B. Molybdenum oxide thin films obtained by hot-filament metal oxide deposition technique. *Chem. Mater.*, **163**, 2004, 513–20.
24. Laperriere, G., Lavoie, M. A. and Belenger, D. Electrochromic behavior of molybdenum trioxide thin films, prepared by thermal oxidation of electrodeposited molybdenum trisulfide, in mixtures of nonaqueous and aqueous electrolytes. *J. Electrochem. Soc.*, **143**, 1996, 3109–17.
25. Faughnan, B. W. and Crandall, R. S. Optical properties of mixed-oxide WO_3/MoO_3 electrochromic films. *Appl. Phys. Lett.*, **31**, 1977, 834–6.
26. Cogan, S. F., Anderson, E. J., Plante, T. D. and Rauh, R. D. Materials and devices in electrochromic window development, *Proc. SPIE*, **562**, 1985, 23–31.
27. Ohtani, B., Masuoka, M., Atsui, T., Nishimoto, S. and Kagiya, N. Electrochromism of tungsten oxide film prepared from tungstic acid. *Chem. Express*, **3**, 1988, 319–22.
28. Yonghong, Y., Jiayu, Z., Peifu, G., Xu, L. and Jinfa, T. Electrochromism of titanium oxide thin films. *Thin Solid Films*, **298**, 1997, 197–9.
29. Dyer, C. K. and Leach, J. S. Reversible optical changes within anodic oxide films of titanium and niobium. *J. Electrochem. Soc.*, **125**, 1978, 23–9.
30. Bange, K. and Gambke, T. Electrochromic materials for optical switching devices. *Adv. Mater.*, **2**, 1992, 10–16.
31. Lee, G. R. and Crayston, J. A. Sol-gel processing of transition-metal alkoxides for electronics. *Adv. Mater.*, **5**, 1993, 434–42.
32. Faughnan, B. W., Crandall, R. S. and Heyman, P. M. Electrochromism in WO_3 amorphous films. *RCA Rev.*, **36**, 1975, 177–97.
33. Deepa, M., Srivastava, A. K., Singh, S. and Agnihotry, S. A. Structure–property correlation of nanostructured WO_3 thin films produced by electrodeposition. *J. Mater. Res.*, **19**, 2004, 2576–85.
34. Pauporté, T. A simplified method for WO_3 electrodeposition. *J. Electrochem. Soc.*, **149**, 2002, C539–45.
35. Cogan, S. F., Plante, T. D., Anderson, E. J. and Rauh, R. D. Materials and devices in electrochromic window development. *Proc. SPIE*, **562**, 1985, 23–31.
36. Hutchins, M., Kamel, N. and Abdel-Hady, K. Effect of oxygen content on the electrochromic properties of sputtered tungsten oxide films with Li^+ insertion. *Vacuum*, **51**, 1998, 433–9.

37. Özkan, E., Lee, S.-H., Liu, P., Tracy, C. E., Tepehan, F. Z., Pitts, J. R. and Deb, S. K. Electrochromic and optical properties of mesoporous tungsten oxide films. *Solid State Ionics*, **149**, 2002, 139–46.
38. Park, N.-G., Kim, M. W., Poquet, A., Campet, G., Portier, J., Choy, J. H. and Kim, Y. I. New and simple method for manufacturing electrochromic tungsten oxide films. *Active and Passive Electronic Components*, **20**, 1998, 125–33.
39. Arakaki, J., Reyes, R., Horn, M. and Estrada, W. Electrochromism in NiO_x and WO_x obtained by spray pyrolysis. *Sol. Energy Mater. Sol. Cells*, **37**, 1995, 33–41.
40. Bessière, A., Badot, J.-C., Certiat, M.-C., Livage, J., Lucas, V. and Baffier, N. Sol–gel deposition of electrochromic WO_3 thin film on flexible ITO/PET substrate. *Electrochim. Acta*, **46**, 2001, 2251–6.
41. Davazoglou, D., Donnadieu, A. and Bohnke, O. Electrochromic effect in WO_3 thin films prepared by CVD. *Sol. Energy Mater.*, **16**, 1987, 55–65.
42. Rauh, R. D., Wang, F., Reynolds, J. R. and Meeker, D. L. High coloration efficiency electrochromics and their application to multi-color devices. *Electrochim. Acta*, **46**, 2001, 2023–9.
43. Green, F. J. *The Sigma–Aldrich Handbook of Stains, Dyes and Indicators*, Milwaukee, WI, Aldrich Chemical Company, Inc., 1990.
44. Gaupp, C. L., Welsh, D. M., Rauh, R. D. and Reynolds, J. R. Composite coloration efficiency measurements of electrochromic polymers based on 3,4-alkylenedioxythiophenes. *Chem. Mater.*, **14**, 2002, 3964–70.
45. Cirpan, A., Argun, A. A., Grenier, C. R. G., Reeves, B. D. and Reynolds, J. R. Electrochromic devices based on soluble and processable dioxythiophene polymers. *J. Mater. Chem.*, **13**, 2003, 2422–8.
46. Reeves, B. D., Grenier, C. R. G., Argun, A. A., Cirpan, A., McCarley, T. D. and Reynolds, J. R. Spray coatable electrochromic dioxythiophene polymers with high coloration efficiencies. *Macromolecules*, **37**, 2004, 7559–69.
47. Mortimer, R. J. and Reynolds, J. R. *In situ* colorimetric and composite coloration efficiency measurements for electrochromic Prussian blue. *J. Mater. Chem.*, **15**, 2005, 2226–33.
48. Aubert, P.-H., Argun, A. A., Cirpan, A., Tanner, D. B. and Reynolds, J. R. Microporous patterned electrodes for color-matched electrochromic polymer displays. *Chem. Mater.*, **16**, 2004, 2386–93.
49. Compton, R. G., Waller, A. M., Monk, P. M. S. and Rosseinsky, D. R. Electron paramagnetic resonance spectroscopy of electrodeposited species from solutions of 1,1'-*bis*(*p*-cyanophenyl)-4,4'-bipyridilium (cyanophenyl paraquat, CPQ). *J. Chem. Soc., Faraday Trans.*, **86**, 1990, 2583–6.
50. Duffy, J. A. *Bonding, Energy Levels and Inorganic Solids*, London, Longmans, 1990.
51. Robin, M. B. and Day, P. Mixed valence chemistry – a survey and classification. *Adv. Inorg. Chem. Radiochem.*, **10**, 1967, 247–422.
52. Brown, D. B. (ed.), *Mixed Valence Compounds (NATO Conference)*, 1980, London, D. Reidel.
53. Baucke, F. G. K., Duffy, J. A. and Smith, R. I. Optical absorption of tungsten bronze thin films for electrochromic applications. *Thin Solid Films*, **186**, 1990, 47–51.
54. Thompson, B. C., Schottland, P., Zong, K. and Reynolds, J. R. In situ colorimetric analysis of electrochromic polymers and devices. *Chem. Mater.*, **12**, 2000, 1563–71.
55. Thompson, B. C., Schottland, P., Sönmez, G. and Reynolds, J. R. In situ colorimetric analysis of electrochromic polymer films and devices. *Synth. Met.*, **119**, 2001, 333–4.

56. Schwendeman, I., Hickman, R., Sönmez, G., Schottland, P., Zong, K., Welsh, D. M. and Reynolds, J. R. Enhanced contrast dual polymer electrochromic devices. *Chem. Mater.*, **14**, 2002, 3118–22.
57. Sönmez, G., Schwendeman, I., Schottland, P., Zong, K. and Reynolds, J. R. N-Substituted poly(3,4-propylenedioxypyrrole)s: high gap and low redox potential switching electroactive and electrochromic polymers. *Macromolecules*, **36**, 2003, 639–47.
58. Sönmez, G., Meng, H. and Wudl, F. Organic polymeric electrochromic devices: polychromism with very high coloration efficiency. *Chem. Mater.*, **16**, 2004, 574–80.
59. Thomas, C. A., Zong, K., Abboud, K. A., Steel, P. J. and Reynolds, J. R. Donor-mediated band gap reduction in a homologous series of conjugated polymers. *J. Am. Chem. Soc.*, **126**, 2004, 16440–50.
60. Sönmez, G., Shen, C. K. F., Rubin, Y. and Wudl, F. A red, green, and blue (RGB) polymeric electrochromic device (PECD): the dawning of the PECD era. *Angew. Chem., Int. Ed. Engl.*, **43**, 2004, 1498–502.
61. Sönmez, G. and Wudl, F. Completion of the three primary colours: the final step towards plastic displays. *J. Mater. Chem.*, **15**, 2005, 20–2.
62. Berns, R. S. *Billmeyer and Saltzman's Principles of Color Technology*, 3rd edn, New York, J. Wiley & Sons, 2000.
63. Wyszecki, G. and Stiles, W. S. *Color Science: Concepts and Methods, Quantitative Data and Formulae*, 2nd edn, New York, J. Wiley & Sons, 1982.
64. [online] at www.efg2.com/Lab/Graphics/Colors/Chromaticity.htm (accessed 4 January 2006).

5
Kinetics of electrochromic operation

5.1 Kinetic considerations for type-I and type-II electrochromes: transport of electrochrome through liquid solutions

Type-I and type-II electrochromes are dissolved in solution prior to the electron-transfer reaction that results in colour. Such electron-transfer reactions are said to be 'nernstian' or 'reversible' when uncomplicated and fast and in accord with the Nernst equation (Eq. (3.1), Chapter 3). When two conditions regarding the motions of electroactive species (or indeed other participant species) are met, there is a particular means, that needs definition, whereby the key electroactive species arrives at the electrode. These conditions are: the absence both of convection (i.e. the solution unstirred, 'still'), and also of electroactive-species migration.[a] Then 'mass transport' (directional motion) of any electroactive species is constrained to occur wholly by diffusion. On the one hand, the rate of forming coloured product can be dictated by the rate of electron transfer with rate constant k_{et}, which if low may render the electrode response non-nernstian (the electrode potential $E_{O,R}$ diverges from the Nernst equation (3.1) in terms of bulk electroactive concentrations), and furthermore, the rate of the process governed by k_{et} largely determines the current. On the other hand, if k_{et} is high, then electroactive/electrode electron transfer is not the rate- and current-controlling bottleneck, and the overall rate of colour formation is dictated by the rate of mass transport of electroactive species toward the electrode.

---

[a] To recapitulate Section 3.3, 'migration' here means charge motion resulting in *ohmic* conduction of current. This migration is subtly prevented when the solution contains an excess of inert ('swamping') electrolyte ions that themselves cannot conduct, because, being inert (i.e. redox-unreactive), such ions, on contact with the appropriate electrode, cannot undergo the electron transfer required to complete the conduction process. Excess ionic charge of these species accumulates up to a potential-determined limit. Huge applied potentials can in some cases subvert 'inertness'.

The experimental context of these considerations arises as follows. An electrochromic cell is primed for use ('polarised') by applying an overpotential (Section 3.3, Chapter 3). Polarising the cell ensures that, if in solution, *some* of the electrochrome impinging on the electrode will undergo an electron-transfer reaction. However, *all* of the electrochrome reaching the electrode is electro-modified if the overpotential is sufficiently large, in which case the current becomes directly proportional to the concentration of electrochrome, a result that arises from Fick's laws of diffusion[1] (Chapter 3). The current is then said to have its *limiting* value $I_{(lim)}$, i.e. increases in the applied overpotential will not increase the magnitude of the current. The value of $I_{(lim)}$ decreases slowly with time (with electrode and solution motionless), as outlined below. A large positive value of overpotential generates a limiting anodic (oxidative) current, while a large negative value of overpotential results in a limiting cathodic (reductive) current.

Because the amount of colour formed in a given time is by definition proportional to the rate of charge passage at the electrode, as high a current as possible is desirable for rapid device operation, i.e. if possible, a limiting current is enforced. (If the current *I* is made too high, however, deleterious side reactions may occur at the electrode, as discussed below. The current that yields electrochemical reaction is termed 'faradaic', but current otherwise utilised say in solely ionic movement is 'non-faradaic'– Section 3.4, Chapter 3.)

The current is thus best increased by enhancing the rates of mass transport to the electrode. In a laboratory cell, stirring the solution will maximise the current since convection (Section 3.3) is the most efficient form of mass transport. However, in a practicable ECD this expedient will always be impossible, and natural convection, as e.g. caused by localised heating of the solution at the electrode, can also be dismissed.

If migration is also minimised because an excess of inert 'swamping' electrolyte has been added to the solution (Section 3.3 and footnote to previous page), then the time-dependence of the limiting current, $I_{(lim, t)}$ owing to electrode reaction of the ion *i* is given by the Cottrell equation, Eq. (5.1):

$$I_{(lim, t)} = nFA c_i \sqrt{\frac{D_i}{\pi t}}, \quad (5.1)$$

where F is the Faraday constant, c_i is the concentration of the electroactive species *i*, *n* is the number of electrons involved in the electron-transfer reaction, Eq. (1.1), and A is the electrode area. The derivation of the Cottrell equation presupposes semi-infinite linear diffusion toward a planar electrode, and more complicated forms of the Cottrell equation have been derived for the thin-layer

5.1 Transport of electrochromes through solutions

Table 5.1. *Diffusion coefficients D of solvated cations moving through solution prior to reductive electron transfer.*

Diffusing entity	$D/\text{cm}^2\ \text{s}^{-1}$	Diffusion medium	Ref.
Fe^{3+}	5×10^{-6}	Water	2
Methyl viologen	8.6×10^{-6}	Water	3
Cyanophenyl paraquat	2.1×10^{-6}	Propylene carbonate	4

cells[2] that are used for type-I ECDs. Table 5.1 lists a few values of diffusion coefficient D obtained from Cottrell analyses.

Equation (5.1) predicts that the magnitude of the current – and hence the rate at which charge is consumed in forming the coloured form of the electrochrome – is not constant, but decreases monotonically with a $t^{-1/2}$ dependence in a diffusion-controlled electrochemical system. This kinetic result is indeed found until quite long times (>10 s after the current flow commences). Figure 5.1 shows such a plot of current I against time $t^{-1/2}$ during the electro-oxidation of aqueous *o*-tolidine (3,3′-dimethyl-4,4′-diamino-1,1′-biphenyl) (**I**), which, being a kinetically straightforward ('nernstian') system,[5] conforms with the analysis.

I

The rate of coloration is obviously a linear function of the rate of electron uptake, $I = \mathrm{d}Q/\mathrm{d}t$. Accordingly, for optical absorbance *Abs* (which is $\propto Q$), the rate of colour formation $\mathrm{d}(Abs)/\mathrm{d}t$ (which is $\propto I$, Eq. (1.7)) ought also to have the time dependence of $t^{-1/2}$ according to the Cottrell relation, Eq. (5.1). Integration hence predicts $Abs \propto t^{+1/2}$ and for (**I**) in water; the test plot, Figure 5.2, is satisfactorily linear.[5] Support for a diffusion-controlled mechanism is thus demonstrated.

The slope of Figure 5.2 should be independent of the concentrations of the electroactive species, as is shown in Figure 5.3. Here, slopes of *Abs* versus $t^{1/2}$ plots at various concentrations and currents are plotted against I for the electro-oxidation of *o*-tolidine (**I**) in water,[5] and they superimpose regardless of concentration, as expected. However, the plot Figure 5.3 should *not* be linear, as $\mathrm{d}(Abs)/\mathrm{d}t^{1/2}$ is clearly not linear with I, which can be inferred from Eq. (1.7), and the spurious straight line shown results largely from employing restricted ranges of the variables.

Absorbance–time relationships like these have seldom been used as tests (presumably discouraged by confusion arising from the apparent irrationality

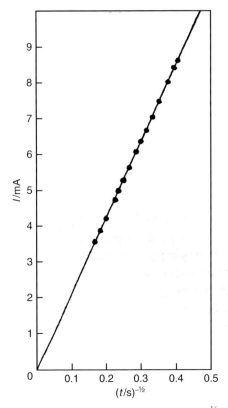

Figure 5.1 Cottrell plot of limiting current I against $t^{-1/2}$ during the electro-oxidation of o-tolidine (3,3'-dimethyl-4,4'-diamino-1,1'-biphenyl) in aqueous solution at a ITO electrode polarised to 1.5 V vs. SCE. (Figure reproduced from Hansen, W. N., Kuwana, T. and Osteryoung, R. A. 'Observation of electrode–solution interface by means of internal reflection spectrometry'. *Anal. Chem.*, **38**, 1966, 1810–21, by permission of The American Chemical Society.)

of the Figure 5.3-type plots) but in 1995 Tsutsumi et al.[6] emulated the tests of these relations for electrogenerating the aromatic radical anion of p-diacetylbenzene (**II**) with similar success.

Such diffusion control is expected during coloration for all type-I electrochromes, while type-II electrochromes should evince the same behaviour at very short times. Deviations must occur at longer times because the transferring

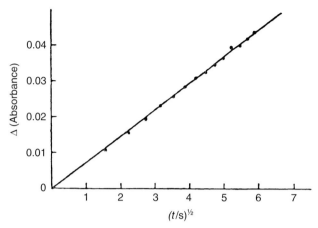

Figure 5.2 Plot of the change of optical absorbance Abs against $t^{1/2}$ during the electro-oxidation of o-tolidine (3,3'-dimethyl-4,4'-diamino-1,1'-biphenyl) in aqueous solution at a ITO. (Figure reproduced from Hansen, W. N., Kuwana, T. and Osteryoung, R. A. 'Observation of electrode–solution interface by means of internal reflection spectrometry'. *Anal. Chem.*, **38**, 1966, 1810–21, by permission of The American Chemical Society.)

electron needs to traverse a layer of solid coloured product, with concomitant complication of the analysis.

5.2 Kinetics and mechanisms of coloration in type-II bipyridiliums

As the details of the coloration mechanisms are, exceptionally, so specific to the chemistry of this group of type-II electrochromes, where the uncoloured reactant is dissolved but the coloured form becomes deposited as a solid film, the complications of the chemistry are dealt with in Chapter 11, on the bipyridiliums. Sections 11.2 and 11.3 specifically are devoted to these aspects.

5.3 Kinetic considerations for bleaching type-II electrochromes and bleaching and coloration of type-III electrochromes: transport of counter ions through solid electrochromes

Type-II electrochromes such as heptyl viologen (see Chapter 11) are solid prior to bleaching. Type-III electrochromes remain solid during oxidation and reduction reactions. The majority of studies relating to the kinetic aspects of electrochromic operation of solid materials relate to tungsten oxide as a thin film. With suitable and probably slight modification, the theories below relating to solid WO_3 will equally apply to many other solid electrochromes, such

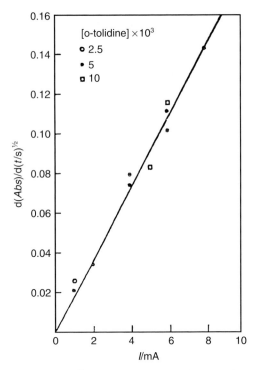

Figure 5.3 Plot of $\mathrm{d}(Abs)/\mathrm{d}t^{1/2}$ against current I for the electro-oxidation of o-tolidine (3,3'-dimethyl-4,4'-diamino-1,1'-biphenyl) in aqueous solutions at a ITO electrode polarised to 1.5 V vs. SCE. The concentrations of electrochrome are o 2.5×10^{-3} mol dm^{-3}, • 5 mmol dm^{-3}, and □ 10 mmol dm^{-3}. The straight line is spurious – see text. (Figure reproduced from Hansen, W. N., Kuwana, T. and Osteryoung, R. A. 'Observation of electrode-solution interface by means of internal reflection spectrometry'. *Anal. Chem.*, **38**, 1966, 1810–21, by permission of The American Chemical Society.)

as the other metal oxides in Chapter 6. Some of the results may also apply straightforwardly to the inherently conducting polymers in Chapter 10.

Even a brief survey of the literature on tungsten trioxide shows a number of competing models in circulation for the coloration and decoloration processes. As already noted, the most by far of reported kinetic studies of electrochromism relate to solid tungsten trioxide. Its coloration reaction is summarised in Eq. (5.2) (which is actually 'a gross over-simplification',[7] since the initial solid almost invariably also involves water and hydroxyl ions):

$$WO_3 + x(M^+ + e^-) \rightarrow M_xWO_3. \quad (5.2)$$

Thus in the discussion below WO_3 is the paradigm, with the M^+ as an inert, i.e. electro-inactive ion, usually designated 'counter ion', that is entrained to

5.3 Transport of counter ions through solid systems

preserve or maximise electroneutrality within the solid oxide film. (Systems generally adjust, subject to electromagnetic, electrostatic and quantal laws, to minimise concentrations of charge and high potentials.) Other electrochromes, organic as well as inorganic, are mentioned here if data are available.

5.3.1 Kinetic background: preliminary assumptions

(i) Initial state: mass balance Prior to the application of the coloration potential V_a, solid films of WO_3 are assumed to contain no electro-inserted counter ions. However, an ellipsometric study by Ord et al.[8] apparently disproves this assumption. His thin-film WO_3, formed anodically on W metal immersed in acetic acid, was shown to contain protonic charge, but this charge had no optical effect: presumably acid had been unreactively absorbed by the solid.

Another source of charge inside a film is the ionisation of water: $H_2O \rightarrow H^+ + OH^-$ (or with sufficient H_2O about, $2H_2O \rightarrow H_3O^+ + OH^-$). Such water may be replenished during coloration and bleaching since there is evidence for movement of molecular water through transition-metal oxides during redox cycling, e.g. H_2O will be inserted into electrodeposited cobalt oxyhydroxide[9] or into vacuum-evaporated[10] WO_3 when the impressed potential is cathodic; and water will also move through polymers of organic viologen in response to redox cycling.

(ii) Electronic motion As we assume a particulate electron, the niceties of quantum-mechanical tunneling associated with wave properties will be glossed over. At low extent of reduction x, electron conduction probably occurs via activated site-to-site hopping rather than through occupied conduction bands, since most of these metal oxides when fully oxidised are, at best, poorly conducting semiconductors.[11] In accord, the electrical conductivity of fully oxidised WO_3 is extremely low, both as a solid and as a thin film. In contrast, the electronic conductivity of M_xWO_3 (where $M = H^+$, Li^+ or Na^+) is metallic for so-called 'bronzes'[b] of x greater than ca. 0.3. Figure 5.4 shows a plot of electronic conductivity in WO_3 as a function of insertion coefficient x. The WO_3 was prepared either by vacuum evaporation, that produces an *amorphous oxide*, denoted a-WO_3, or by sputtering, that produces a *crystalline oxide*,

---

[b] In this context, a 'bronze' is a solid with metallic or near-metallic conductivity. Below a metal-to-insulator transition, WO_3 is a semiconductor, but above it near-free electrons impart reflectivity. 'Free' here implies 'akin to conduction electrons in true metals'.

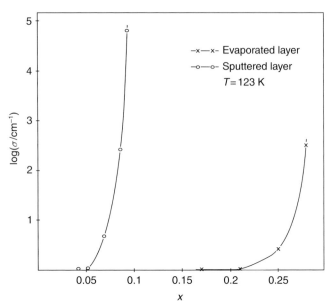

Figure 5.4 Plot of electronic conductivity σ of H_xWO_3 as a function of insertion coefficient x. Data determined at 123 K. (Figure reproduced from Wittwer, V., Schirmer, O. F. and Schlotter, P. 'Disorder dependence and optical detection of the Anderson transition in amorphous H_xWO_3 bronzes'. *Solid State Commun.*, **25**, 1978, 977–80, copyright (1978) with permission from Elsevier Science.)

denoted c-WO_3. *It should be noted that c-WO_3 is less electronically conductive than a-WO_3.*

Circumscribing the use of WO_3 in ECDs, the formation of the high-x bronzes M_xWO_3 ($x > 0.3$) is not reversible, so e.g. $Li_{0.4}WO_3$ cannot be electro-oxidised back to[12] WO_3. At high x values the transferred electrons, acquired in the electrochemical coloration process, are stabilised in an accessible conduction band largely comprising the tungsten d orbitals. Electrons from interphase redox reactions by external electroactive species, via a dissipating conduction through this band, may thwart the re-oxidative extraction of electrons from W^V by the electrode substrate. (Interphase rather than 'interface' is defined in Chapter 3, p. 43)

(iii) Motion of ions The solid electrochromic oxide, as a film on its electrode substrate, can be immersed in a solution containing a salt of the counter ions, such as H_2SO_4 for mobile protons, or $LiClO_4$ for Li^+ ion. During electro-coloration, electrons enter the film via the electrode substrate and, concurrently,

5.3 Transport of counter ions through solid systems

counter ions enter the film through the electrolyte-facing interphase of the WO_3 cathode. Bleaching entails a reversal of these steps.

So coloration or bleaching proceed with associated movements of both electrons and cations.[13] When the kinetics of electrochemical redox change are dictated by the motion of a species within the film, it is the slower, hence rate limiting, of the two charge carriers that is the determinant. The slower charge carrier is usually the ion because of its relatively large size. Indeed, the transport number t (= fraction of current borne) of ions can approach zero, then correspondingly the electron transport number $t_{(\text{electron})} \rightarrow 1$. Such dual motion is the cause of the curiously named 'thermodynamic enhancement' described by Weppner and Huggins,[14] as mentioned below.

A good gauge of rapidity of ion motion is its diffusion coefficient D. However, the movement of counter ions through solid WO_3 proceeds by both diffusion and migration. The two modes of mass transport operate additively, but the separate extents are usually not known. Exemplifying, Bell and Matthews[15] cite activation energies E_a for diffusion, varying in the range 56–70 kJ mol^{-1} (values that denote an appreciable temperature dependence): the spread of values arises from the pronounced curvature of an Arrhenius plot. True diffusion is an activated process and normally obeys the Arrhenius equation that gives a linear graph of ln D against $1/T$. In contrast, the temperature dependence of migration is relatively modest. As dual mechanisms with different activation energies often show curved $1/T$ plots of the rate-parameter logarithm, the non-linearity of Bell and Matthews' graph accordingly points to a significant extent of migration in the measured 'diffusion coefficient'. The latter is therefore unlikely to be a true diffusion coefficient but a combined-mechanism quantity $\bar{D}$, as defined below.

Diffusion coefficients are obtained from several measurements: impedance spectra, chronoamperometry, analysis of cyclic-voltammetric peak heights as a function of scan rate via the Randles–Sevčik equation, Eq. (3.12), and radiotracer methods.[16] Compendia from the literature of $\bar{D}$ values for mobile ions moving through WO_3 in refs. 12,17,18,19,20 provide the representative selection in Table 5.2, together with preparation method and insertion coefficient, x. For comparative purposes, values for mobile ions moving through other type-III electrochromes are listed in Table 5.3.

The variations in diffusion coefficient could reflect the disparity in rate between electrons and ions as they move through the solid. To minimise the charge imbalance during ion insertion or egress, the slower ions move faster and the fast electrons are slowed.[14] In this way, the overall rate is altered,[36] causing D to change by a factor of W, an enhancement factor. The factor W quantifies the extent of the so-called 'thermodynamic enhancement', and the resultant

Table 5.2. *Chemical diffusion coefficients $\bar{D}$ representing movement of lithium ions through tungsten trioxide: effect of preparative methodology and insertion coefficient. Measurements as in text, on three-electrode cells avoiding ECD complications.*

Morphology	x in Li_xWO_3	$\bar{D}$ /cm^2 s^{-1}	Ref.
(a) Effect of preparative methodology			
WO_3*,a,b	–	5×10^{-9}	21
WO_3*,b,d	–	1.6×10^{-12}	22
WO_3*,c,d	–	1.3×10^{-11}	23
WO_3*,e	–	2×10^{-11}	24
WO_3f	–	5×10^{-13}	25
(b) Effect of insertion coefficient x			
a-WO_3*,b,c	0.097	2.5×10^{-12}	21
a-WO_3*,b,c	0.138	4.9×10^{-12}	21
a-WO_3*,b,c	0.170	1.5×10^{-11}	21
a-WO_3*,b,c	0.201	2.6×10^{-11}	21
a-WO_3*,b,c	0.260	2.8×10^{-11}	21
$Li_{0.1}WO_3$f	0.1	1.7×10^{-9}	26
$Li_{0.37}WO_3$f	0.37	5.6×10^{-10}	26

*Thin film. [a] Sputtered film. [b] Impedance measurement. [c] Thermally evaporated sample. [d] Chronoamperometric measurement. [e] Electrodeposited film. [f] Film prepared from sol–gel intermediate.

diffusion coefficient is the '*chemical* diffusion coefficient'; W is also termed the 'Wagner factor'. The two diffusion coefficients D and $\bar{D}$ are related as:[14]

$$\bar{D} = WD. \tag{5.3}$$

In consequence, probably most of the 'diffusion coefficients' in the literature of solid-state electrochromism are *chemical* diffusion coefficients. The factor W was derived as:[14]

$$W = t_{(electron)} \left[\frac{\partial \ln a_{(ion)}}{\partial \ln c_{(ion)}} + z_{(ion)} \frac{\partial \ln a_{(electron)}}{\partial \ln c_{(ion)}} \right]. \tag{5.4}$$

Here the letters c and a are respectively concentration and activity, (see Chapter 3, p. 36); $z_{(ion)}$ is the charge on the mobile ion. The enhancement factor W can be[14] as great as 10^5, but is said to be 'about 10' for the motion of H^+ through WO_3.[12] In addition to morphological differences born of preparative

Table 5.3. *Chemical diffusion coefficients $\bar{D}$ of mobile ions through permanent, solid films of type-III electrochromes; diffusion of counter ion through the electrochromic layer. Methods as for Table 5.2.*

Compound	Ion : Solvent	$D/\text{cm}^2\,\text{s}^{-1}$	Ref.
Cerium(IV) oxide	Li^+:PC	5.2×10^{-13}	27
$(\text{F}_{16}\text{-pc})\text{Zn}^a$	TBAT:DMF	$1.6\text{--}8.0 \times 10^{-12}$	28
Lutetium bis(phthalocyanine)	Cl^- : H_2O	10^{-7b}	29
Nickel hydroxide	H^+:H_2O	2×10^{-7} to 2×10^{-9}	30,31
$\text{H}_{0.042}\text{Nb}_2\text{O}_5$	H^+	3.6×10^{-8}	32
$\text{H}_{0.08}\text{Nb}_2\text{O}_5$	H^+	5.2×10^{-7}	32
Poly(carbazole)	ClO_4^-:H_2O	10^{-11}	33
Poly(isothianaphthene)	BF_4^-:PC	10^{-14}	34
Tungsten(VI) trioxide^c	H^+:HCl(aq)	2×10^{-8d}	35
Tungsten(VI) trioxide^e	Li^+:PC	2.1×10^{-11f}	27
Vanadium(V) trioxide	Li^+:PC	3.9×10^{-11}	27

PC = Propylene carbonate. F_{16}-pc = perfluorinated phthalocyanine. ^{a}Value from analysis of a Randles–Sevčik graph. ^{b}Apparently calculated from chloride ion mobility. ^{c}Thermally-evaporated sample. ^{d}Chronoamperometric measurement. ^{e}Sputtered film. ^{f}Value determined from impedance measurement.

routes, variations in W are a likely reason for the wide differences in the $\bar{D}$ values listed in Tables 5.2 and 5.3.

Being fast, the transport number of the electron $t_{\text{(electron)}} \rightarrow 1$, hence the observed rate of transport through WO_3 is determined by the slower ions. Thus the expression for W can be simplified, Eq. (5.4) becoming:

$$W = \left[\frac{\partial \ln a_{\text{(ion)}}}{\partial \ln c_{\text{(ion)}}}\right]. \tag{5.5}$$

Substituting for W from Eq. (5.3) into Eq. (5.5) yields the so-called Darken relation:[14]

$$\bar{D} = D\left[\frac{\partial \ln a_{\text{(ion)}}}{\partial \ln c_{\text{(ion)}}}\right]. \tag{5.6}$$

It is assumed in Eqs. (5.3)–(5.6) above that only the counter ion is mobile since all other ions (e.g. oxide ions O^{2-} that are, more likely,[37] in the oxygen bridges –O–) are covalently bound or otherwise immobile. This tenable assumption has been verified in part by impedance spectroscopy.[38]

(iv) Energetic assumptions A relatively crude model of insertion has the counter ion entering or leaving the oxide layer after surmounting an activation barrier E_a associated with the WO_3–electrolyte interphase. For example, a recent Raman-scattering investigation of H_xWO_3 electro-bleached in aqueous H_2SO_4 is said to indicate, by analysis of the WO_3 vibrational modes, that the rate of electro-bleaching is dominated by proton expulsion from the H_xWO_3 as the H^+ traverses the electrochrome–solution interphase.[10]

There is also an activation barrier to electron insertion/egress from or to the electrode substrate, the barrier often being represented as the *resistance to charge transfer*, $R_{(CT)}$. Many of the measured values of '$R_{(CT)}$' may be composites of terms containing the interphase activation energy E_a (in an exponential) for ion insertion together with $R_{(CT)}$ for the electron transfer at the electrode substrate, with the former E_a effect being the larger. The motion of counter ions within the film may also contribute, and certainly play a role in the interpretative models considered below.

The motion of a (bare thus minute) proton will be the most rapid of all the cations, in moving within the oxide layer following insertion during coloration. Protons come to rest when the external potential is removed and when, in addition, they attain sites of lowest potential energy. On equilibration inside the oxide layer, the inserted ion is assumed in most models to be uniformly distributed throughout the film, perhaps with slight deviations in concentration at interphases due to the interactions born of surface states.[39] The discussion below indicates how this last assumption probably understates the role of interphases.

5.3.2 Kinetic complications

The complications caused by the innate resistance of the ITO, called 'terminal effects', can be largely bypassed (but see refs. 40, 41) by including an ultra-thin layer of metallic nickel between the electrochrome and ITO,[42] or an ultrathin layer of precious metal on the outer, electrolyte-facing, side of the electrochrome. Both apparently improve the response time τ.[43,44,45,46] The effect is elaborated in refs. 40 and 41.

(i) Crystal structure There are several distinct crystallographic phases notably monoclinic discernible in reduced crystalline tungsten oxide (c-WO_3) at low insertion coefficients ($0 < x \leq 0.03$).[47] Slight spatial rearrangement of atoms (i.e. local phase transitions from the predominantly monoclinic) in c-WO_3 are said to occur during reduction,[48] which may affect the electrochromic response

time of WO_3 for colouring or for bleaching. Such structural changes are sometimes believed to be the rate-limiting process during ion insertion into WO_3.[49,50]

The value of $\bar{D}$ increases slightly with increasing insertion coefficient x, as exemplified by the data of Ho et al.[21] in Table 5.2; Avellaneda and Bulhões find the same effect.[26]

Green[24] has stated that WO_3 expands by ca. 6% on reductive ion insertion; and Ord et al.[51] show by ellipsometry that V_2O_5 on reduction in acetic acid electrolyte also expands by 6%, despite the thicknesses of electrochromic oxides being somewhat diminished when a field is applied owing to *electrostriction*.[52,53]

Similarly, samples of c-WO_3, when injected with Li^+ ion at a continuous rate, were found to have a higher capacity for lithium ion than do otherwise identical samples that are charged fitfully.[54] It was argued[54] that this result demonstrates that the Li_xWO_3 product has sufficient time to change structure on a microscopic scale during the slower, stepwise, charging, thus impeding subsequent scope for reduction.

(ii) The effect of the size of the mobile ion Questions arise as to what counter ion is taken up during reduction, and which one provides the charge motion within the film, but the picture is not clear-cut. A general picture does emerge from envisaging the constraints on ionic motion and the experimental observations, but it is not always intrinsically consistent in detail. As ions that move through solid oxide experience obstruction within the channels, ionic size is expected to govern the values of D for different ions. A model for this process from which activation energies can be estimated is outlined later, on p. 112.

For rapid ECD coloration, ion size should be minimised, so protons are favoured for WO_3. Deuterons[55,56,57] are found to be somewhat slower than protons; and lithium ions are slower still (see Table 5.2). Though some workers have reversibly inserted Na^+,[58,59,60] and even reversible incorporation of Ag^+ has been reported,[61] most other cations are too slow to act in ECDs. (The sequence of cations follows the indications of the activation-energy model referred to.) The only *anion* small and mobile enough to be inserted into anodically colouring electrochromes is OH^-.

Scarminio[62] reported that the stress induced in a film is approximately proportional to the insertion coefficient, x. The film capacitance also increases linearly with x.[63] Scrosati et al.[48] used a laser-beam deflection method to assess the stresses from electro-inserting Li^+ and Na^+, finding that phase transitions were induced.

Counter-ion swapping can occur since WO_3 does entrain indeterminate amounts of water, even if prepared as an anhydrous film. Variable water content may be the cause of the great discrepancies between reported values of $\bar{D}$. Some chemical diffusion coefficients for the (nominally) slow lithium ion appear to be fairly high for motion through WO_3. This suggests diffusion of the more mobile proton (presumably taken up interstitially, or formed by ionisation of interstitial water), followed at longer times by exchange of Li^+ for H^+ as charge-carrier, which is illustrated in the electrochemical quartz-crystal microbalance (EQCM) study by Bohnke et al.[64,65,66,67] Such unexpected swapping is considered thermodynamically (specifically entropy) driven. In common with Bohnke et al., Babinec's EQCM study[68] also suggested swapping of Li^+ for the more mobile H^+, but also suggested *egress of hydroxide ions* from the film during coloration (from water within the film ionising to OH^- and H^+). A dual-cation mechanism is suggested by Plinchon et al.'s[69] mirage-effect experiment that implied dual insertion of H^+ and Li^+ during reduction of WO_3.

Kim et al.,[20,70] studying the dual injection of H^+ and Li^+ by impedance spectroscopy, report the process to be 'extremely complicated'. For a chemically different WO_3, the diffusion coefficient of lithium ions inserted into rf-sputtered WO_3 was found to decrease as the extent of oxygen deficiency increased.[71]

(iii) The effect of electrochrome morphology Diffusion through amorphous oxides is significantly faster than through those same oxides when crystalline.[24] Kubo and Nishikitani,[72] in a Raman spectral study of WO_3, cite polaron–polaron interactions within *clusters* of c-WO_3 embedded in amorphous material, as a function of cluster size, concluding that the coloration efficiency η increases as the cluster becomes larger. Also, since electrochromic films commonly comprise both amorphous and crystalline WO_3, the mobile ions tend to move through the amorphous material as a kinetic 'fast-track'. Indeed, diffusion through c-WO_3 is so slow by comparison with diffusion through amorphous tungsten oxide (a-WO_3) that the c-WO_3 need not even be considered during kinetic modelling of films comprising both amorphous and crystalline oxide;[24] see page 98. In similar vein, the value of D for Li^+ motion through a-WO_3 that is thermally annealed decreases by about 5% over annealing temperatures ranging from 300–400 °C; the decrease in D is ascribed to increased crystallinity.[73] Similarly, diffusion through the amorphous grain boundaries within polycrystalline NiO is faster than through the NiO crystallites.[74]

An additional means of increasing the electrochromic coloration rate is to increase the size of the channels through the WO_3 by introducing heteroatoms into the lattice. The incorporation of other atoms like Mo, to form e.g. $W_yMo_{(1-y)}O_3$, causes strains in the lattice which are relieved by increases in all the lattice constants.

(iv) The effect of water The presence of water can greatly complicate kinetic analyses intrinsically, and additionally, *adsorption* of water at the electrochrome–electrolyte interface can make some optical analyses quite difficult[75] since specular effects are altered. Even the coloration efficiency can change following such adsorption.[76]

Hurditch[77] has stated that electrochromic colour of H_xWO_3 will form only if films contain moisture and, similarly, Arnoldussen[78] states that MoO_3 is not electrochromic if its moisture content drops below a minimum level. Curiously, he also states that his MoO_3 was electro-coloured as a dry film in a vacuum. One concludes that water, presumably adsorbed initially, is essential in effecting the reductive coloration, either by ionising to H^+ and OH^- so providing the conductive protons, or by being reduced to H_2 (also with OH^-) which itself can effect *chemical* reduction.

Hygroscopicity Thin films of metal oxide are often somewhat hygroscopic,[79] although it has been concluded that the cubic phase of WO_3 prefers two H^+ to one water molecule.[80] Adsorbed water can be removed by heating[81] above *ca.* 190°C (but extensive film crystallisation will also occur at such temperatures; see p. 140). References 82 and 83 describe the depth-profile of H_2O in WO_3, as shown in Figure 5.5.

Proton conductivity through solid-state materials, and its measurement, have been reviewed by Kreuer.[84]

Aquatic degradation Excess moisture inside films (especially *evaporated* films) will cause much structural damage,[85] perhaps following the formation of soluble tungstate ions.[81] Faughnan and Crandall,[86] Arnoldussen[87] and Randin[88] have all discussed dissolution effects. Furthermore, the rate of WO_3 dissolution is promoted by aqueous chloride ion.[89]

Energetics The effect on stabilities resulting from the incorporation of water needs consideration. The forces exerted on an atom, ion or molecule during its movement through an oxide interior are determined by the microscopic environment through which it moves, and on the physical size of the channels through which it must pass. Ions undergo some or total desolvation during ion insertion from solution, i.e. when traversing the solution–electrochrome interphase into the lattice. The loss of solvation stabilisation can be partly compensated by interaction with lattice oxides or indeed occluded H_2O, but the former – in addition to lattice-penetration obstacles – could retard motion (EQCM studies[67] however show Li^+ to be unsolvated as it moves through

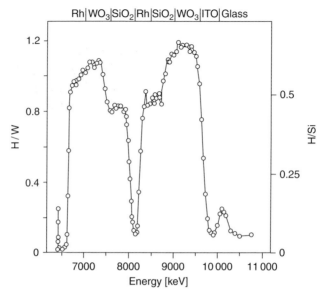

Figure 5.5 Hydrogen profile within the electrochromic cell at an applied voltage of 0 V: Rh|WO$_3$|SiO$_2$|Rh|SiO$_2$|WO$_3$|ITO|glass. The rhodium layers act as both a mirror and an ion-permeable layer. (Figure reproduced from Wagner, W., Rauch, F., Ottermann, C. and Bange, K. 'Hydrogen dynamics in electrochromic multilayer systems investigated by the ^{15}N technique'. *Nuc. Instr. Meth. Phys. Res. Sect. B*, **50**, 1990, 27–30, copyright (1990) with permission from Elsevier Science.)

WO$_3$). Proton motion through hydrated films is accordingly found to be much faster than through dry films,[90] the retarding proton/oxide interactions possibly being weaker than in dry oxides. Alternatively a Grötthus-type conduction process could be facilitating rapid proton conduction through hydrated oxide interiors.

Bohnke *et al.*[67] used data from EQCM studies to explain non-adherence to Nernst-type relations, postulating that adsorbed, unsolvated, anions are expelled from the surface of the WO$_3$ as cathodic coloration commences. The effects of interactions between inserted Li$^+$ and the lattice were also mentioned.

(iv) The effect of insertion coefficient on $\bar{D}$ Values of $\bar{D}$ can be obtained from the gradient of a graph of impedance vs. $\omega^{-1/2}$ as by Huggins and co-workers.[21] Three independent groups found that $\bar{D}$ decreased as the insertion coefficient of Li$^+$ in WO$_3$ increased;[60,71,91] Masetti *et al.*[60] found that $\bar{D}$ for Li$^+$ and Na$^+$ decreased by thirty-fold in WO$_3$ over the insertion coefficient range

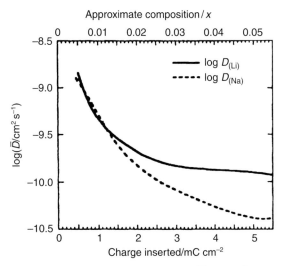

Figure 5.6 Plot of chemical diffusion coefficient $\bar{D}$ for Li$^+$ and Na$^+$ through WO$_3$ as a function of insertion coefficient. (Figure reproduced from Masetti, E., Dini, D. and Decker, F. 'The electrochromic response of tungsten bronzes M$_x$WO$_3$ with different ions and insertion rates'. *Sol. Energy Mater. Sol. Cells*, **39**, 301–7, copyright (1995) with permission from Elsevier Science.)

$0 < x < 0.05$; see Figure 5.6. By contrast, Huggins' results from an independent ac technique show the opposite trend, with $\bar{D}$ of Li$^+$ in WO$_3$ *increasing* as x increases. The sensitivity of motion parameters to preparative method has already been remarked on: fluctuations in $\bar{D}$ with x appear highly complicated, possibly too complicated to model at present.

5.3.3 Kinetic modelling of the electrochromic coloration process

For the electrochromic coloration reaction of WO$_3$ given in Eq. (5.2), each of the models below will be discussed, identifying M$^+$ as a proton unless specified otherwise. The distinctive features of the models discussed in the following sections are summarised in Table 5.4 overleaf.

Model of Faughnan and Crandall: potentiostatic coloration

Assumptions Faughnan and Crandall[86,92,93,94,95] provided a semi-empirical model for WO$_3$ coloration and bleaching, semi-empirical because they used data from measured values of the electrode potential E to provide empirical parameters used in their formulation. The main assumptions at the heart of the model[86,92,93,94,95] are the following.

Table 5.4. *Summary of the coloration models described on pages 91–104.*

Principal authors	Distinctive features	Refs.
Faughnan and Crandall	• No concentration gradients form within the film. • There is an H^+ injection barrier at the electrolyte–WO_3 interphase. • An empirically characterised back-potential acts at that interphase. • The back potential dominates the rate of coloration.	86,93
Green	• Concentration gradients of counter cations within the M_xWO_3 films were computed by analogy with heat flow through metal slabs. • The diffusing entity is uncharged so there are no effects owing to the electric field. • Hence the kinetic effects of cations and electrons are indistinguishable. • The H_xWO_3 adjacent to the inert electrode substrate remains $H_{\rightarrow 0}WO_3$.	100
Ingram, Duffy and Monk	• A percolation threshold sets in at $x = 0.03$. • When $x < 0.03$, rate-limiting species are electrons; at $x > 0.03$ counter-ion motions are rate limiting.	96
Bohnke	• Electrons and proton counter ions in the film form a neutral species $[H^+ e^-]$. • Reduction of WO_3 is a *chemical reaction*, effected by atomic hydrogen arising from this neutral.	101,102,103
Various	• W^{IV} species participate in addition to W^V and W^{VI}. • Reduction of WO_3 may be a two-electron process.	116,117,118,119, 120,121,122, 123, 124,125,126,127, 128,129,130,131,132

(i) The rate-limiting motion is always that of the proton as it enters the WO_3 from the electrolyte, in traversing the electrochrome–electrolyte interphase. The proton motion (intercalation) is rate limiting also because of assumption (ii).

(ii) A 'back potential' (Faughnan *et al.*, always call this potential a 'back *emf*') forms across the WO_3–electrolyte interphase during coloration, the potential increasing as the extent of insertion x increases.

From assumption (i), it is argued that, having entered the WO_3, the proton motion is relatively unhindered, apart from the restraint arising from the back

5.3 Transport of counter ions through solid systems

potential. Because the central kinetic determinant is the energy barrier to motion of protons into and out of the WO_3 layer via the WO_3–electrolyte interphase, a further assumption (iii) may be inferred.

(iii) The absence of concentration gradients of H^+ within the H_xWO_3 is implied, hence diffusion never directly controls current. Only the back potential – assumption (ii) – restrains proton motion and hence also the current flow[86] and the rate of increase in proton concentration.
(iv) The WO_3 film *initially* is free of W^V and hence of any initial complication from separate counter-cation charge (but this initial-state assumption – essentially a clarification – lacks mechanistic implication, and thus has no further role).

The unusual back potential – assumption (ii) – opposes the expected current flow.[86] It is invoked because the chemical potential of the inserted cation is increased (i.e. it is increasingly energetically disfavoured) as the proton concentration within the oxide[c] increases. The back potential then corresponds to the change in chemical potential of the proton that accompanies coloration. In essence, the developing back potential within the solid smooths out the usual requisite applied potentials (i.e. those sufficiently exceeding the electrode potential E so as to drive the coloration process) that would ordinarily result in a current 'jump' or peak associated with (i.e. effecting) oxidation-state change (Chapter 3). The involvement of the back potential is clearly seen in cyclic voltammograms (CVs) of WO_3, where there is no current peak directly associated with the reductive formation of colour. However, by contrast, CVs do show a peak associated with the (oxidative) electrochemical bleaching: see Figure 5.7. Clearly the back potential will oppose ion insertion during coloration but will aid ion egress (proton removal) during bleaching.[90,96]

The kinetics of the model Electro-coloration commences as soon as the potential is stepped from an initial value E_{in} at which reduction just starts to a second potential V_a. Since an equilibrium electrode potential E associated with the W^{VI}/W^V couple is set up following any reduction, the applied potential V_a is, in fact, an overpotential, so V_a is cited with respect to E (that is, V_a = applied potential minus E, where E changes with increase of W^V). Note that we now retain the symbolism of the original authors,[86,92,93,94,95] especially regarding V_a, which here, rather than the η of Chapter 3, denotes overpotential and not simply the applied voltage. Apart from this difference in meaning, the main

---

[c] Whether the increase in the protonic chemical potential with increase of its concentration is sufficient to produce an effective back potential could find independent support from a sufficiently detailed lattice-energy calculation, as has proved invaluable for comparable situations in other electrochromes: see ref. 41 of Chapter 8.

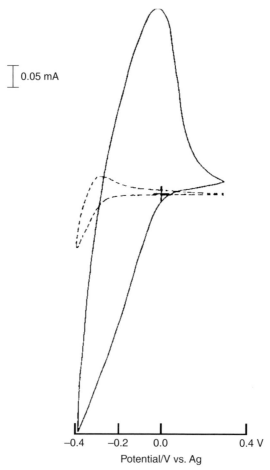

Figure 5.7 Typical cyclic voltammogram of an amorphous thin film of tungsten trioxide evaporated on ITO and immersed in PC–LiClO$_4$ (1 mol dm^{-3}) at 500 mV s^{-1} (solid line) and 50 mV s^{-1} (dotted line). (Figure reproduced from Kim, J. J., Tryk, D. A., Amemiya, T., Hashimoto, K. and Fujishima, F. 'Color impedance and electrochemical impedance studies of WO$_3$ thin films: H$^+$ and Li$^+$ transport'. *J. Electroanal. Chem.*, **435**, 31–8, copyright (1997) with permission from Elsevier Science.)

further change from the η of Chapter 3 is that now the overpotential V_a has simply a value *without sign*.

The chemical potential of H$^+$ was obtained from a statistical entropy-of-mixing term, together with empirical constants, as

$$\mu_{H^+} = A + 2Bx + nRT\ln\left(\frac{x}{1-x}\right), \tag{5.7}$$

where $n = 1$ and the A and B terms were derived from a plot of the observed *emf* E values versus x.

5.3 Transport of counter ions through solid systems

Taking into account the back potential induced within the WO$_3$, the magnitude of the current is governed by two energy barriers, each showing an exponential dependence on the applied potential. The first is influenced by the insertion coefficient x within the H$_x$WO$_3$, while the other is influenced by the barrier to ionic charge-transfer current flow across the WO$_3$–electrolyte interphase, owing to proton desolvation and the accompanying difficulty of intercalating into the lattice.

The basis to the development of the theory is to treat the proton uptake at the interphase as a conventional ion-uptake electrode process following the Tafel law (Eq. (3.16), Chapter 3). The kinetics that ensue then follow the Butler–Volmer development, where the effect of the driving potential V_a overcomes the intrinsic energy barrier by an extent αV_a where α is variously viewed as the symmetry or transmission coefficient and so represents the effectiveness of V_a. The α values found for various systems usually fall between 0.4 and 0.6, and ½ is often summarily assigned to it *faute de mieux*, as here. From this simplified Butler–Volmer viewpoint the observed current is hence expected to be proportional to $\exp(V_a e/2RT)$; the positive sign in the exponential arises because V_a *opposes* activation energies.

As colour forms with increasing x, so the current during coloration, i_c, decreases, from the back-potential influence. This current is a function of time t, decreasing because the back potential increases with time:[86,93]

$$i_c = i_o \left(\frac{1-x}{x} \right) \exp\left(-\frac{x}{x_1} \right) \exp\left(\frac{V_a e}{2RT} \right). \tag{5.8}$$

Faughnan *et al.*'s $x_1 = 0.1$ appears to be the extent of intercalation at which both assumptions (i) and (ii) are taken to be fulfilled. The term e is the electronic charge and i_o is the exchange current, itself a function of the coloration current and the extent of coloration, that needs to be established from the primed system at onset of operating:

$$i_o = i_e \frac{0.53 \, e \, x_o}{RT} \left(\frac{x_o}{1 - x_o} \right). \tag{5.9}$$

Here x_o is the mole fraction of protons within the film prior to the application of the voltage V_a and i_e is the current immediately on applying V_a. The numeral is an empirical value 0.53 V from a plot of *emf* E against x so relating to the back-potential effect invoked earlier.

Faughnan and Crandall introduce a 'characteristic time' (τ_D) for diffusion into the film, from an approximate solution of Fick's second law, akin to

Eq. (3.12), depending on the film thickness d and the proton diffusion coefficient D:

$$\tau_D = \frac{d^2}{4D}, \qquad (5.10)$$

i.e. time needed for the proton to penetrate to a representative point mid-film. (Note that the diffusion coefficient here was chosen to be D rather than the chemical diffusion coefficient $\bar{D}$.) This value is employed in arriving at a time-dependence for the effect of the back potential on current.

Combining these considerations[91] led to the equation:

$$i_c = i_o \left(\frac{\tau_o}{t}\right)^{1/2} \exp\left(\frac{V_a e}{4RT}\right), \qquad (5.11)$$

where τ_o is a constant equal to $(\rho d e / 2 i_o)$ in which ρ is the density of W sites within the film; incorporation of $[1/t \exp(V_a e/2RT)]^{1/2}$ (unsquaring d^2) underlies the form of the exponential factor in the equation. The coloration current predicted in Eq. (5.11) thus depends strongly on the applied voltage (overpotential) V_a. Furthermore, if diffusion through the film were alone responsible for the observed i–t behaviour, any potential dependence (above the redox-effecting value of V_a) would be absent.

Equation (5.11) has been verified experimentally for films of WO_3 on Pt immersed in liquid electrolytes and with either a proton[91] or a lithium ion[54] as the mobile counter ion. The equation has also been shown to apply to WO_3 on ITO in contact with solid electrolytes, the mobile cation being lithium[97,98] or the proton.[91] Equation (5.11) is obeyed only for limited ranges of x if the counter ion is the proton.

The kinetic treatment by Luo et al.[99] is somewhat similar to that above. Their principal divergence from Faughnan and Crandall is to suggest the magnitude of the bleaching voltage is unimportant below a certain critical value.

Model of Green: galvanostatic coloration

Assumptions An altogether different treatment is that of Green and co-workers.[100] In his model, coloration is effected galvanostatically, with the charge passed at low electric fields. The a priori conditions are that $dQ/dt = i =$ constant, therefore, from Faraday's laws, $dx/dt =$ constant, where x is the *average* insertion coefficient throughout the entire film of H_xWO_3.

Green assumes the following.

(i) All activity coefficients are the same.
(ii) The diffusing entity is uncharged so there are no effects owing to the electric field, i.e. migration is wholly absent, itself implying assumption (iii).

(iii) All diffusion coefficients are D rather than $\bar{D}$.
(iv) The WO_3 contains no mobile protons prior to the application of current.
(v) The film may or may not contain interstitial water.

Assumption (i) contradicts the derivation of the Weppner and Huggins' relations in Eqs. (5.3)–(5.6). Assumption (ii) can be classed as consistent with the model of Bohnke et al.,[101,102,103] as described below. The application of assumption (iv) is unlikely to affect significantly the utility of Green's model.

The kinetic features of the model The film of WO_3 has a thickness d. A constant ionic flux of J_o from the electrolyte layer reaches the solid WO_3 and thence penetrates to a distance y, where $0 < y < d$. The distance $y = d$ denotes the WO_3–electrolyte interphase. There is no ionic flux at the back-electrode (at $y = 0$).

By analogy with the conduction of heat through a solid slab positioned between two parallel planes,[104] Green obtained Eq. (5.12):

$$c(y,t) - c_o = \frac{J_o t}{d} + \frac{J_o d}{\bar{D}} \left\{ \frac{3y^2 - d^2}{6d^2} - \frac{2}{\pi^2} \sum_{n=1}^{\infty} \frac{(-1)^n}{n^2} \exp\left(-\pi^2 n^2 \frac{\bar{D}t}{d^2}\right) \cos\left(\frac{n\pi y}{d}\right) \right\},$$
(5.12)

or, in abbreviated form:

$$c(y,t) - c_o = \frac{J_o t}{d} + \frac{J_o d}{\bar{D}} F(y,t),$$
(5.13)

where $c(y,t)$ is the concentration of H^+ (possibly partially solvated) at a distance of y into the WO_3 film at the time t. Green omits specifying migration effects but does cite diffusion coefficients as $\bar{D}$. All the diffusion coefficients pertain to solid phase(s). In Green's notation, quantities c are number densities, and currents i represent numbers of ionic or electronic charges passing per unit time rather than, say, Ampères per unit area.

If $\bar{D}$ is large, then $c(y,t)$ is independent of y and so $c(y,t)$ increases linearly with $J_o t/d$, causing the concentration of H^+ *throughout* the film to be even, in agreement with assumption (iv) in the model of Faughnan and Crandall above. The second term on the right-hand side of Eq. (5.13) acts as a correction term to account for diffusion-limited processes in the solid.

Green has plotted curves of $F(y,t)$ against y/d for various values of $\bar{D}t/d^2$; see Figure 5.8. These show, at short times, that only the WO_3 adjacent to the electrolyte will contain any protonic charge, but the proton concentration gradient flattens out at longer times.

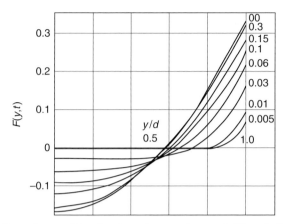

Figure 5.8 Green's model of coloration: values of $F(y, t)$ for a film of thickness d with no mass flow at $y = 0$ and constant flux J_o at $y = d$. The numbers on the curves are the values of Dt/d^2. (Figure reproduced from Green, M., Smith, W. C. and Weiner, J. A. 'A thin film electrochromic display based on the tungsten bronzes'. *Thin Solid Films*, **38**, 89–100, copyright (1976) permission from Elsevier Science.)

In a later development, Green[24] added into his model the effects on the concentration gradients of incorporating grain boundaries into his model. For simplicity the grains of c-WO_3 are assumed to be spherical. When such boundaries are considered, and assumed to be regions within the film acting as pathways for 'fast-track' diffusion, the second term on the right-hand side of Eq. (5.13) is simplified to $(J_o r^2)/(15 d \bar{D})$, where the sphere radius is r.

Green[24] concluded that for a response time of τ, the relationship

$$\bar{D}\tau \left(\frac{c_m}{n}\right)^2 \geq 1 \tag{5.14}$$

should be followed, where c_m is the maximum concentration of H^+ that arises (the number of H^+ equalling that of W^V), and n is the number of optically absorbing colour centres per unit area required to produce the required absorbance, equal to the number of H^+ per cm^2. The parenthesised term thus roughly represents the inverse of the average distance separating colour centres.

None of the concentration gradients predicted by Green's model have been measured.

The kinetic treatment of Seman and Wolden,[105] closely similar to that of Green, departs from Green's model in incorporating the back potential of Faughnan and Crandall.

The model of Ingram, Duffy and Monk:[96] an electronic percolation threshold

A percolation threshold is attained when previous directed electronic motions, proceeding by individual 'hops' from a small number of sites, during a steady increase in the number of occupied sites to a critical value, suddenly become profuse, because of the onset of multiple pathways through the increased number of occupied sites. In ordinary site-wise conductive systems this occurs when occupied sites become $\sim 15\%$ of the maximum.[106]

Assumptions

(i) The central assumption underlying the model of Ingram et al.[96] is that the motion of the *electron* is rate limiting below a percolation threshold, at $x_{(critical)}$, but electron movement is rapid when $x > x_{(critical)}$. Such a transition is documented[107,108] for WO_3.

(ii) Most of the assumptions and hence the theoretical elaboration of Faughnan and Crandall's model (see p. 91 ff.) are obeyed when $x > x_{(critical)}$.

The model It is already clear from Figure 5.4 and the discussions above that the electronic conductivity σ of pure WO_3 is negligibly low. The conductivity σ increases as x increases until, at ca. $x \approx 0.3$, the conductivity becomes metallic. The onset of metallicity is an example of a semiconductor-to-metal transition, an Anderson transition.[107] Then if the mobility $\mu_{(ion)}$ of ions is approximately constant, but the mobility of the electron $\mu_{(electron)}$ increases dramatically over the compositional range $0 \leq x < 0.3$ then, at a critical composition $x_{(critical)}$, the ionic and electronic mobilities will be equal: $\mu_{(ion)} = \mu_{(electron)}$. It follows then that $\mu_{(ion)} < \mu_{(electron)}$ when $x > x_{(critical)}$. Hence, at low x, the motion of the *electron* is rate limiting; and only above $x_{(critical)}$ will electron movement be the more rapid. It is shown in ref. 96 that Faughnan and Crandall's model (page 91 ff.) is obeyed extremely well when $x > x_{(critical)}$ but not at low values of x, below $x_{(critical)}$.

Ingram et al.[96] analysed the potentiostatic coloration of evaporated a-WO_3 on ITO, which involved obtaining transients of current i against time t during electroreduction. Such plots showed a peculiar current 'peak', see Figure 5.9, which was rationalised in terms of attaining a percolation threshold, with the electron velocity rising dramatically at $x \approx 0.03$.[d]

---

[d] The low value 0.03–0.04 claimed for the electron-conduction percolation limit may be understood as arising from a restricted electron delocalisation about a few neighbouring W^{VI}, that in effect extends the size of the 'sites' involved in allowing the onset of the critical percolation, which would hence lower the numbers of sites needed for criticality. The onset of metallicity at $x \sim 0.3$ results from the wave-mechanical overlap of conduction sites or bands with the valence bands, and the approximate correspondence here with the customary percolation value ~ 0.15 is then probably fortuitous.

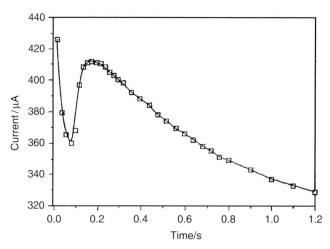

Figure 5.9 Chronoamperometric trace of current vs. time during the electro-coloration (reduction) of the cell ITO|WO$_3$|PEO–H$_3$PO$_4$|(H)ITO. The potential was stepped from a rest potential of about 0.0 V to –0.6 V at $t = 0$. Note the current peak at *ca.* 0.2 s. (Figure reproduced from Ingram, M. D., Duffy, J. A. and Monk, P. M. S. 'Chronoamperometric response of the cell ITO|H$_x$WO$_3$|PEO–H$_3$PO$_4$ (MeCN)|ITO'. *J. Electroanal. Chem.*, **380**, 1995, 77–82, with permission from Elsevier Science.)

Similar chronoamperometric plots of i against t which include a current peak have also been found by Armand and co-workers[97] and by Craig and Grant.[109] The value of x at the peak is also *ca.* 0.03 in ref. 97, as can be gauged by manual integration of the peak in the traces published. The percolation phenomenon was not seen by Ingram *et al.* when electro-colouring WO$_3$ with a small field, by applying a very small cathodic driving potential, perhaps because the transition was too slow to be noted.

Armand and co-workers[97] explain the peak in terms of the nucleation of hydrogen gas (via the electroreduction of H$^+$), possibly with the surface of the incipient H$_x$WO$_3$ acting as a catalyst:

$$2H^+ + 2\,e^- \rightarrow H_2. \tag{5.15}$$

While such nucleation phenomena can certainly cause strange current peaks in chronoamperometric traces, Armand's explanation may not be correct here since Craig and Grant,[109] who found a similar current peak, had inserted *lithium* ion into WO$_3$ from a super-dry PC-based electrolyte, i.e. an electrolyte free of mobile H$^+$: in this case Li^0 would have to be the corresponding reactant.

Also, in a different system, a current peak has been observed by Aoki and Tezuka[110] during the anodic electro-doping of poly(pyrrole), that was successfully modelled in terms of a percolation threshold. There was no mention of such a threshold in the study by Torresi and co-workers,[111] but their model of 'relaxation processes' in thin-film poly(aniline) does, again, suggest a *sudden* change in electronic conductivity with composition change.

In summary, Ingram, Duffy and Monk suggest that the kinetic behaviour of WO_3 in the insertion-coefficient range $0 \leq x < 0.03$–0.04 is dictated by slow electron motion; only after a percolation threshold at the upper coefficient limit here does ion motion become rate limiting. In contrast with the assumptions implicit in deriving Eq. (5.5), $t_{(electron)}$ does not tend toward 1, so values of $\bar{D}$ alter dramatically as the percolation threshold is reached.

In studies claiming *free* electronic motion, by Goldner et al.,[112] and Rauh and co-workers,[113] both groups employ Drude-type models (see p. 142) to describe the free-electron behaviour, but following Ingram et al., electrons are 'free' only above the percolation threshold.

Model of Bohnke: reduction of W^V via neutral *inserted species*

Assumptions The requirement of a new interpretation of the WO_3 coloration process was indicated by the need to explain the temporal relationships governing the *optical* data obtained during electrochromic coloration. Accordingly, the bases of most of the theories in the electrochemical models above are still regarded as valid (see discussion, below). The major divergence from the models above is the following.

(i) The rate-limiting process during electrochromism is the diffusion of an electron–ion pair (such as $[H^+ e^-]$), which may be atomic (as $H^\bullet$). Because the $[H^+ e^-]$ pair has no overall charge, the diffusion coefficient evinced by the system is D rather than $\bar{D}$. The meeting of H^+ and e^- is outlined below.

(ii) The rate of electrochromic colour formation is thus a *chemical* rather than an electrochemical reaction:

$$W^{VI} + [H^+ e^-]^0 \rightarrow W^V + H^+. \quad (5.16)$$

The proton product of Eq. (5.16) resides as a counter ion adjacent to the site of the chemical reduction reaction, i.e. to the W^V.

(iii) The chemical reduction in Eq. (5.16) occurs 'spontaneously' on the time scale for diffusion of the $[H^+ e^-]$ pair. From Eq (3.16), $d \sim (D\,t)^{1/2}$, inserting a reasonable assumed D of ca. $10^{-12}\,\text{cm}^2\,\text{s}^{-1}$ indicates that the mobile species would traverse a typical film in many seconds, lending reality to these suppositions, providing k_{et} is high enough (see point (iv), following).

(iv) The observed current is thus a function of the rate of forming $[H^+ e^-]$ pairs, but does not represent the formation rate of colour centres. The rate of forming colour is thus either a function of the rate of diffusion of the $[H^+ e^-]$ pair to available W^{VI} sites prior to 'instantaneous' electron transfer, Eq. (5.16), or, if the appropriate rate constant k_{et} is quite low, it is a function of the rate of the electron-transfer reaction itself, $W^{VI} + e^- \to W^V$. (The electrochromic colour in this model is still due to intervalence optical transitions between W^{VI} and W^V.)

The Model In contrast to the models above of Faughnan and Crandall, and of Green in which the motion of H^+ is rate limiting, or the model of Ingram *et al.* in which first the motion of the electron and then the motion of the proton is rate limiting, in Bohnke's model[101,102,103] the mobile diffusing species is suggested to be an electron–ion pair. Indeed, it is even possible that electron transfer has occurred within the pair, resulting in the formation of atomic hydrogen or lithium prior to coloration. On entering the WO_3, the inserted H^+ ion moves through the WO_3, probably moving only a very short distance within the WO_3 before encountering the faster electron from the electrode substrate. The charged species within the encounter pair then diffuse *together* as a neutral entity, or they react to form atomic hydrogen.

Furthermore, the model implies that the kinetics-controlling mobility, in moving through WO_3, of the $[H^+ e^-]$ pair that provides a quasi counter ion to W^V, will be simplified since migration effects, born of coulombic attractions, can be wholly neglected and, accordingly, the measured diffusion coefficient is better considered as D than as $\bar{D}$. In common with Faughnan and Crandall, and Ingram *et al.*, Bohnke acknowledges that the observed current–time behaviour is governed by the formation of a back potential, but parts from Faughnan and Crandall in asserting that concentration gradients are formed within the incipient H_xWO_3 during coloration. Bohnke's model is said[101,102,103] to be satisfactory in simulating the observed absorbance–time data except at short times, but is not applied in any detail to data for bleaching.

In support of the model, the rate of diffusion through Nb_2O_5 is similarly said to be dominated by 'redox pairs'.[114,115]

Recent developments: intervalence between W^{VI} and W^{IV}

Assumption A new view of the key tungsten species has emerged in the last decade. While broadly agreeing with the model of Faughnan and Crandall (above), Deb and co-workers[116] suggested in 1997 that the coloured form of the electrochrome is not $H_xW^{V,VI}O_3$ but $H_xW^{VI}_{(1-y)} W^{IV}_y O_{(3-y)}$, and hence that the optical intervalence transition is $W^{VI} \leftarrow W^{IV}$ rather than the hitherto widely accepted $W^{VI} \leftarrow W^{V}$.

The fully oxidised form of the trioxide (MoO_3 or WO_3) is confirmed to contain only the +VI oxidation state by studies with XPS[117,118,119] and ESCA.[120,121] Reduction during the coloration reaction $MO_3 + x(H^+ + e^-) \rightarrow H_xMO_3$ is expected to yield the +V oxidation state: but XPS shows that some of the +IV state is also formed during the reduction of Mo,[118,122,123,124] and of W.[117,118,119,125] Rutherford backscattering studies furthermore suggest that the amount of W^{IV} in nominal 'WO_3' is a function of the extent of oxygen deficiency.[126] Infrared[127] and Raman studies[128,129] also indicate the presence of W^{IV}. Indeed, Lee et al.[128] say that even as-deposited films contain appreciable amounts of W^{IV}. Additionally, it is notable that Sun and Holloway[130] (in 1983) and Bohnke and co-workers[131] (in 1991) both suggest that *reduction of WO_3 is a two-electron process*. Similarly, the electrochromic and photochromic properties of O-deficient WO_3 have also been found to depend on similar W^{IV} participation in both mechanisms.[132] Possibly the observed W^V is formed by comproportionation, as in Eq. (5.17):

$$W^{VI} + W^{IV} \rightarrow 2W^V. \tag{5.17}$$

Siokou et al.[118] suggest that the W^{IV} state 'plays a dominant role in deep coloration'.

Finally, de Wijs and de Groot deliberately omitted the involvement of W^{IV} in their recent wave-mechanical calculations.[133] Rather, from density-functional computations, they argue for W^V–W^V dimers rather than W^{IV} and W^{VI}.

The on-going growth of views on the roles played by the several W species, and their ultimate resolution, promises intriguing physicochemical developments for the near future.

Additional experimental results

(i) Coloration of non-stoichiometric 'bronzes' A non-stoichiometric reduced oxide has a non-integral ratio of oxygen and metal ions, e.g. $WO_{(3-y)}$, where y is likely to be small. Such materials are also called 'sub-stoichiometric'.

Zhang and Goto[71] found that D increased as the extent of sub-stoichiometry increased, i.e. as y in $Li_xWO_{(3-y)}$ increased; $WO_{(3-y)}$ is then in reality, $W^{VI}_{(1-y)}W^{IV}_y O_{(3-y)}$. Other materials of the type $WO_{(3-y)}$ are indeed also electrochromic, but trapping of electrons at shear planes and defect sites can be problematic for rapid, reversible electrochromic coloration.[134] For this reason, non-stoichiometry is best avoided, although note that[135] $MoO_{(3-y)}$

apparently electro-colours at a faster rate than does MoO_3 alone, and also has a superior contrast ratio CR. Nevertheless, such materials will not be considered further here because the additional complexities encountered with these systems, comparable to (but different from) those of the tungsten systems, do not yet lead to a clearer or general view of the mechanisms in electrochromic oxides.

(ii) Electrochemical titration In a brief study of galvanostatically injected lithium ion in[47] c-WO_3, the electrode potential E of the lithiated oxide was monitored as a function of x while a continuous (and constant) current was passed. It was found that dE/dx decreased suddenly at $x = 0.04$–0.05, close to the values of $x_{(critical)}$ noted above on page 99. In plots of *emf* against x, obtained during injection of Li^+ into, and removal from, c-WO_3, there is a considerable hysteresis between the E for reductive charge injection and that for oxidative Li^+ egress. This is a mobility-controlled kinetic phenomenon: on the time scales involved, there is a higher concentration of lithium on the surface of the particles than in the particle bulk.

(iii) Use of an interrupted current (from a 'pulsed' potential) The rate of electrochromic coloration of tungsten oxide-based ECDs may be enhanced considerably by applying a progression of potentiostatically controlled current pulses rather than enforcing a continuous current.[136] The rate of coloration depends strongly on the pulse length employed, the optimum pulse duration for a high $d(Abs)/dt$ also depending strongly on the pulse amplitude. However, according to the final paragraph of 'Kinetic complications: (i) crystal structure' above, p. 86, steady reduction does effect a greater capacity for Li^+ before bulk metallicity intervenes.

The effects of interrupting the current, by applying current pulses, is attributed to the formation of a thin layer of high-x bronze on the electrolyte-facing side of the WO_3. By interspersing the coloration currents with short periods of zero current, the steep concentration gradient associated with a high-x layer is allowed to dissipate into the film. The amount of charge that can be inserted per current pulse is thus greatly increased, as evidenced by increased peak currents.

An additional advantage of pulsing is to enhance the durability of electrochromic devices by decreasing the occurrence of undesirable electrolytic side reactions such as the formation of molecular hydrogen gas: it is likely that the catalytic properties of H_xWO_3 for H_2 generation are impaired. Several groups

Table 5.5. *Summary of the bleaching models described on pages 105–109.*

Principal authors	Distinctive features	Refs.
Faughnan and Crandall	• The bleaching current is primarily governed by a field-driven space-charge limited current of protons in the H_xWO_3 next to the electrolyte. • The activation energy to proton expulsion is slight. • No concentration gradients form within the film.	35,86
Green	• Concentration gradients of counter cations in M_xWO_3 films were computed from analogy with heat flow through metal slabs. • The kinetic effects of cations and electrons are indistinguishable.	100

have found that a pulsed potential enhances the rate of coloration and bleaching, and suppresses the extent of side reactions.[136,137,138,139,140,141,142,143]

Kinetic modelling of the electrochromic bleaching process

The process of film bleaching, Eq. (5.18), represents the reverse of Eq. (5.2) above:

$$H_xWO_3 \longrightarrow x(H^+ + e^-) + WO_3. \tag{5.18}$$

Bleaching is somewhat simpler than is coloration since the back potential contributes to, rather than acts against, the movement of the mobile counter ions.

Table 5.5 above summarises the various bleaching models cited in this section, citing the distinctive features of each.

Model of Faughnan and Crandall: potentiostatic bleaching

The potentiostatic removal of charge (i.e. bleaching of the electrochromic colour) of the WO_3 bronze has been modelled by Faughnan and Crandall.[35]

Assumptions
(i) The bleaching time of H_xWO_3 is primarily governed by a field-driven space-charge limited current of protons in the H_xWO_3 next to the electrolyte.
(ii) The resistance to charge transfer at the electrochrome–electrolyte interphase does not limit the magnitude of the bleaching current.

(iii) Ionic charge leaves the H_xWO_3 film during electro-bleaching, resulting in a layer of proton-depleted WO_3 at the electrolyte-facing side of the electrochrome. *All the voltage applied across the electrochrome layer film drops across this narrow layer of WO_3.* The layer has a time-dependent thickness termed $l(t)$.

(iv) There is a clear interface between H_xWO_3 and WO_3 layers within the electrochrome, the position of this interface moving into the oxide film from the electrolyte as the bleaching progresses, with $l(t)$ becoming thicker with time.

Since the back potential contributes toward the movement of the mobile charged species, rather than against it, the time-dependent bleaching current i_b shows a different response to the applied voltage V_a from that during coloration, according to Eq. (5.11): i_b now depends on the proton mobility μ_{H^+}:

$$i_b(t) = \frac{\varepsilon \mu_{H^+} V_a^2}{l(t)^3}. \tag{5.19}$$

where ε is the proper permittivity, and $l(t)$ is the time-dependent thickness of a narrow layer of the WO_3 film adjacent to the electrolyte. (Faughnan denotes this length x_I rather than $l(t)$ as here. Note that ε is not the molar absorptivity of Chapter 1.)

The thickness $l(t)$ is proportional to time, and is related to the initial proton concentration (number density) within the film c_o, such that[35,86] $l(t)^3 = J_o t/c_o e$. All the voltage applied to the ECD is assumed to occur across this thin layer, hence the observed i–V_a square law.

Solution of the differential equations for time-dependent diffusion across $l(t)$ during bleaching leads to an additional relationship:

$$i_b(t) = \frac{(p^3 \varepsilon \mu_{H^+})^{1/4} V_a^{1/2}}{(4t)^{3/4}}, \tag{5.20}$$

where p is the volume charge density of protons in the $H_{\rightarrow 0}WO_3$. The result in Eq. (5.20) assumes that bleaching occurs potentiostatically implying a fixed V_a across the *whole* of the WO_3 layer.

The current i_b decreases as $l(t)$ grows thicker, incurring a time dependence of $i_b \propto t^{-3/4}$. This i–t relationship has been verified often for WO_3 in contact with liquid electrolytes[35,54,98] and for WO_3 in contact with semi-solid polymeric electrolytes.[96,98] Figure 5.10 shows the logarithmic current–time response of H_xWO_3 bleached in $LiClO_4$–PC, clearly showing the expected gradient of $-3/4$ at intermediate times. A superior fit between experiment and theory is seen if the electrolyte is aqueous, as in Fig. 5.10 (a); Fig. 5.10 (b) is the analogous plot but for propylene carbonate as solvent.

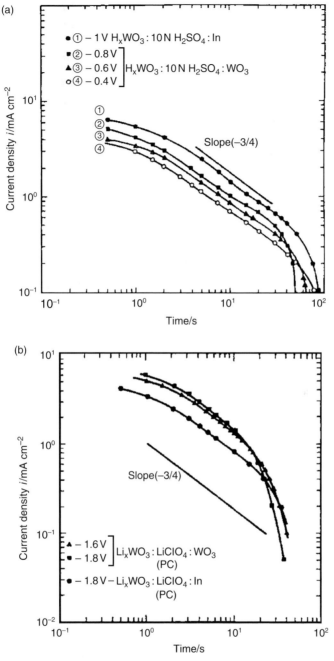

Figure 5.10 Current–time characteristics of H_xWO_3 during electrochemical bleaching as a function of potential: (a) H_xWO_3 in H_2SO_4 (15 mol dm^{-3}) and (b) H_xWO_3 in PC–LiClO$_4$ (1 mol dm^{-3}). The gradient of $-\tfrac{3}{4}$ predicted from Faughnan and Crandall's theory is indicated. (Figure reproduced from Mohapatra, S. K. 'Electrochromism in Li_xWO_3'. *J. Electrochem. Soc.*, **125**, 1978, 284–8, by permission of The Electrochemical Society, Inc.)

108   Kinetics of electrochromic operation

The time for complete bleaching to occur t_b (i.e. the time required for $l(t)$ to become the film thickness) is a function of film thickness d, proton mobility μ, permittivity ε of the film and the insertion coefficient:

$$t_b = \frac{e\rho d^4 x}{4\mu V_a^2 \varepsilon}, \tag{5.21}$$

where x is the insertion coefficient at the *commencement* of bleaching and ρ is the corresponding density of W atoms. Equation (5.21) fulfils expectation in indicating the longer time needed for a film to bleach if the sample is thick or is strongly coloured prior to bleaching.

Model of Green: potentiostatic bleaching

The potentiostatic bleaching of thin film WO_3 has also been modelled by Green.[100] In common with his model for coloration, the film thickness is d and the distance of a proton from the back inert-metal electrode is y. The time-dependent proton concentration is $c(y,t)$, and the initial concentration of H^+ c_o, both actually number densities.

Assumptions
(i) All H^+ ions reaching the electrochrome–electrolyte interphase are instantly removed, implying assumption (ii) below.
(ii) The activation energy for charge electron transfer across the interphase (ions and electrons) is slight. The best means of ensuring assumption (i) is to potentiostatically control the rates of charge movement, i.e. ensuring that assumption (ii) holds by applying a sufficiently large positive potential.
(iii) Accordingly, $c(y=d,t) = 0$ for all time $t > 0$.

The time- and thickness-dependent concentration of H^+ [$c(y,t)$] is then obtained as:

$$c(y,t) = \frac{4c_o}{\pi} \sum_{n=0}^{\infty} \frac{1}{2n+1} \exp\left(\frac{-\bar{D}(2n+1)^2 \pi^2 t}{d^2}\right) \sin\left\{\frac{(2n+1)\pi y}{d}\right\}. \tag{5.22}$$

Green[100] has again computed theoretical curves, in this instance of $c(t)/c_o$ against y/d, where $c(t)$ is the concentration of H^+ in the film at time t at a distance of $0 < y < d$. Such curves drawn for various Dt/d^2 are reproduced in Figure 5.11. As was the case during coloration, the condition for a rapidly responding ECD is $\bar{D}\tau(c_m/n)^2 \geq 1$; cf. Eq. (5.14).

The computed concentration gradients await experimental verification.

5.3 Transport of counter ions through solid systems

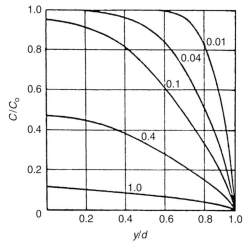

Figure 5.11 Green's model of bleaching: concentration c in the film $0 < y < d$; c_o is the initial concentration; at $t > 0$, $c(d) = 0$. The numbers on the curves are values of Dt/d^2. Note that $c(y = d, t)$ is 0 for all $t > 0$. (Figure reproduced from Green, M., Smith, W. C. and Weiner, J. A. 'A thin film electrochromic display based on the tungsten bronzes.' *Thin Solid Films*, **38**, 1976, 89–100, with permission from Elsevier Science.)

Additional experimental evidence for concentration gradients

Ellipsometry While the exact form of Green's computed concentration gradients need confirmation, other data suggest steep concentration gradients are likely. For example, *in situ* electrochemical ellipsometry – a non-destructive technique – has demonstrated a clear interface between the oxidised (colourless) and reduced (coloured) regions of thin films of vanadium oxide[51] or molybdenum oxide,[52,144] *cf.* assumption (iii) on page 108 above, but ellipsometry has so far failed to detect such an interface within films of WO_3 during reduction,[8] so weakening Faughnan and Crandall's assumption (iv). Within thin-film V_2O_5, this boundary separates the reduced (hydrogen-free) and oxidised (proton-containing) forms of the oxide. The interface was detected both during reduction and oxidation reactions of V_2O_5.

By contrast, the ellipsometric study by Duffy and co-workers[145] did find evidence that implied a surface layer of bronze *does* form during reduction, although note that in this latter study *dry* WO_3 was employed, then reduced *chemically* by gaseous $H_2 + N_2$. The surface of the bronze had a sufficiently high insertion coefficient to be metallic (implying that $x \geq 0.3$). Ingram and co-workers[96,136] also found evidence for surface layers of H_xWO_3 at very short, sub-millisecond, times; these latter studies involved *electrochemical* reduction. Results from inserting the relatively large Na^+ ion into WO_3 also suggest the formation of a high-x layer of Na$_x$WO$_3$ on the electrolyte-facing side of the WO$_3$ during reduction;[59,60] the slow motion of the entering Na$^+$ cation could accentuate incipient concentration gradients.

Nuclear reaction analysis While in some of the simpler models a constant concentration of inserted cation is assumed throughout the electrochromic film, several investigations afford compelling evidence of steep concentration gradients forming during electro-coloration and bleaching. For example, Bange and co-workers[82,83] measured proton densities with the ^{15}N technique (the 'nuclear reaction analysis', NRA):

$$^{15}N + {}^1H \rightarrow {}^{14}C + {}^4He + \gamma, \qquad (5.23)$$

in which, prior to analysis, a sample of WO$_3$ is electro-coloured normally with proton counter ion, and then bombarded with 'hot' ^{15}N atoms. The depth to which the ^{15}N atoms are inserted is varied by controlling their kinetic energy during bombardment. The emitted gamma rays are monitored as a function of energy, thus as a function of depth: the γ-ray count is taken to be directly proportional to the proton concentration. It has been thereby shown that a concentration gradient forms in a film of electrochromic oxide during coloration.

SIMS Secondary-ion mass spectroscopy (SIMS) was the technique of choice to study cation concentrations as a function of film thickness, exemplified by the work by Porqueras *et al.*,[146] Zhong *et al.*,[147] Deroo and co-workers[22] and Wittwer *et al.*[107] In each case, the surface of the film was slowly etched away, and the ablated material analysed. The last study[107] showed $\approx$50% change in cation across the WO$_3$ film. Again, a concentration gradient was clearly shown to form during electrochromic coloration.

However, both the NRA and SIMS techniques destroy the sample during measurement, allowing possible movement of mobile ions *during* measurement, so the results are not without qualification.

Discussion – coloration and bleaching

The back potential The theory of Faughnan and Crandall on p. 91 ff. is the most widely used in describing the coloration kinetics of thin-film electrochromic tungsten trioxide. It is now almost universally agreed that a back potential forms during coloration. Also, the relationships (from Eq. (5.11)) of $i_c \propto t^{-1/2}$ and $i_c \propto \exp(V_a)$ have often been verified experimentally during electro-coloration;

and the relationship $i_b \propto t^{-3/4}$, Eq. (5.20), has also been verified during the electro-bleaching of H_xWO_3, albeit over limited time scales in each case.

Concentration gradients The second area of consensus concerns concentration gradients: these are inferred from the ellipsometry, NRA, and SIMS analyses outlined above, that concentration gradients of H^+ form within the H_xWO_3 both during coloration (with a higher x at the *electrolyte*-facing side of the electrochrome) and during bleaching (with higher x at the *inert-electrode*-facing side of the electrochrome).

The existence of a concentration gradient in the electrochrome cannot be established directly, and can only be inferred. If they exist, they additionally contradict one of the few explicit assumptions of the model of Faughnan and Crandall, since in their theory the protonic charge is assumed effectively to be evenly distributed within the film at all times $t > \tau_D$ (where τ_D is the 'characteristic time' describing the temporal requirements for diffusion within the film, as defined in Eq. (5.10) above) thus implying $t \geq$ milliseconds. However, even in Faughnan's treatment, diffusion within the film arises at $t < \tau_D$, hence implying that concentration gradients enforcing Fick's laws do form within the film.

With the wide acceptance of the i–t–V_a characteristics predicted by Faughnan and Crandall's model (except where $x < 0.03$ in WO_3 reduction), Faughnan's central kinetic assumption of the interphase energy barrier that dictates the proton-insertion rate does appear tenable. The finding that concentration gradients are formed within the incipient H_xWO_3 does not contradict the model, but merely indicates that any contribution to an observed activation energy is small: the activation energy for diffusion are often not excessive (Table 5.6), thus any concentration gradients do not dominate

Table 5.6. *Activation energies for diffusion of mobile ions through a solid metal-oxide host.*

Host	Mobile ion	E_a/kJ mol^{-1}	Ref.
WO_3	Li^+	20–40[1]	91
a-WO_3	Li^+	50[2]	148
WO_3	Li^+	64[2]	149
WO_3	Li^+	20[2]	150
NiOH	H^+	7.0	30

[1] The value of E_a depends sensitively on x. [2] Converted from the original eV.

the observed kinetic laws. They might, however, influence the numerical magnitudes of the rates determined experimentally.

The activation energy E_a for ionic movement has been modelled by Anderson and Stuart;[151] M^{z+} (of charge z_+ and radius r_+) is transferred over a distance d from an oxygen ligand (of radius r_-), to a vacancy near a similar oxygen (each bearing a charge z_-). The activation energy E_a is then given in Eq. (5.24) as:[152,153]

$$E_a = \frac{B z_+ z_- e^2}{\varepsilon (r_+ + r_-)} - \frac{2 z_+ z_- e^2}{\frac{1}{2} d \varepsilon} + \frac{\Gamma \pi l (r_+ - r_d)^2}{2}, \quad (5.24)$$

where z_+ and z_- are the respective charge numbers on the cation and the non-bridging oxygen, and r_+ and r_- are the corresponding radii. The symbol ε here is the *relative* permittivity of the material, and B/ε is a form of effective Madelung constant, this term representing the loss of lattice stabilisation at onset of the ionic 'jump' from its initially stable lattice site. The second term is the coulombic stabilisation acquired by interaction with two oxygens at the mid-point of the jump, i.e. to $\frac{1}{2}d$, mid-way between these oxygens, the numerator '2' denoting interaction with both. The final term covers mechanical stress: Γ is the shear modulus, l is the jump length and r_d is half the distance between bridging-oxygen surfaces that form the 'doorway' needing enlargement to r_+ to enable M^{z+} to pass. The values of E_a calculated from Eq. (5.24) are 'about satisfactory' for Li^+ and Na^+.[154]

Energetics of diffusion Since concentration gradients in ECDs can only be inferred, they are too limited for precise measurement. In such attempts, on etching away the surface of a solid by SIMS, the energy required to remove the surface is sufficient to perturb the H^+ or Li^+ ions, not only volatilising many of the ions but also driving others into the remaining WO_3 during measurement.

Diffusion of neutral species The more recent novel model of Bohnke et al.,[101,102,103] encompassing ion–electron pairs, could have serious implications for many solid-state ionic devices in addition to those involving electrochromism: the good fit between her data and the model does invite attention. In contrast, in the thermodynamic-enhancement model of Huggins and Weppner,[14] differing rates are presupposed of ionic and electronic motion in the film, which appears impossible except prior to the meeting of an ion and electron, thereby forming a neutral pair. If Bohnke's model holds, all values of diffusion coefficient observed will be those of D rather than $\bar{D}$.

While none of the other authors' models comprise the concept of ion–electron pairs, Green[100] notably states that the kinetic behaviour of electrons and ions can be separated only under the influence of a high electric field, implying that the kinetics both of counter ions and electrons moving separately and in pairs could be identical for electrochromic coloration effected by applying a small electric potential. It may be that ion–electron pairs *are* present but not noted in other studies. Complementarily, perhaps the need to invoke their existence can be dismissed in studies employing higher electric fields.

However, several results contradict the Bohnke model: Hall-effect measurements on preprepared Li_xWO_3 and Na_xWO_3 show diffusion coefficients that are roughly proportional to the number of alkali-metal cations inserted and are independent of temperature,[155,156] thus demonstrating the complete dissociation of electrons and cations. (Conductivity σ as a function of x has also been measured by Bohnke *et al.* by microwave results coupled with electrochemical measurements, on an electrochemical cell containing lithium electrolyte.[157] The conductivity of a-WO_3 increased during insertion and decreased during extraction of Li^+ ions.)

Similarly, the Seebeck coefficient S is proportional to $x^{-2/3}$ (x being the insertion coefficient), which is consistent[158,159] 'with a *free* electron' moving through preprepared reduced oxide; and magnetic susceptibility data appear to show the same results.[160]

Insertion coefficients $0 \leq x \leq 0.03$ As mentioned on p. 99 ff. above, Faughnan and Crandall's i–t–V_a characteristics are poorly followed when $x < 0.03$. The near ubiquity of this value of insertion coefficient, in demarcating discontinuities in the physicochemical behaviour of H_xWO_3, has been ascribed by Ingram *et al.*[96] to reaching thence surpassing a percolation threshold. The invocation of a percolation threshold does answer a number of questions, such as the cause of the peculiar current peak in potential-step traces, and possibly the deviating at very low x of the Bohnke model.

The relationship between the extent of localisation and x may also be discerned from the electronic conductivity σ, which only becomes significant at $x = 0.05$ or so (Figure 5.4). Furthermore, the relationship between the extent of localisation and x may also be discerned optically since the molar absorptivity (extinction coefficient) ε is not constant but decreases as x increases,[161] summarised in Figure 5.12. Following the reduction of the WO_3, four separate absorbance–insertion coefficient domains may be discerned: $0 < x < 0.04$ (of extinction coefficient $\varepsilon_1 = 5600$ dm^3 mol^{-1} cm^{-1}); $0.04 < x < 0.28$ (of $\varepsilon_2 = 2800$ dm^3 mol^{-1} cm^{-1}); $0.28 < x < 0.44$ (of $\varepsilon_3 = 1400$ dm^3 mol^{-1} cm^{-1}), and $x > 0.4$ (of $\varepsilon_4 = 0$ dm^3 mol^{-1} cm^{-1}). The value of ε_4 probably means the current did not effect reduction of further tungsten sites. All data were obtained at

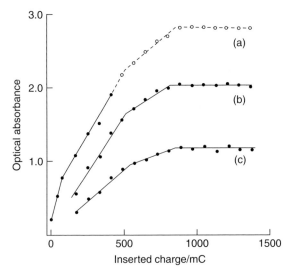

Figure 5.12 Increase in the absorbance of the intervalence charge-transfer band of H_xWO_3 as a function of charge passed: (a) at the wavenumber maximum of 9000 cm^{-1} (the points O were calculated from the curves in (b) and (c); (b) absorbances at 20 000 cm^{-1} and (c) absorbances at 16 000 cm^{-1}. (Figure reproduced from Baucke, F. G. K., Duffy, J. A. and Smith, R. I. 'Optical absorption of tungsten bronze thin film for electrochromic applications'. *Thin Solid Films*, **186**, 1990, 47–51, with permission from Elsevier Science.)

constant wavelength, λ. Similarly, Mohapatra[54] shows a plot of absorbance vs. inserted charge, where some traces are linear only in the range $0 < x < 0.033$, and Scrosati and co-workers[48] found ε of Li_xWO_3 and Na_xWO_3 differed significantly over the insertion coefficient range $0 < x \leq 1$. Values of x characterised by ε_1 were postulated[96] to represent single W^V species below the percolation threshold and, similarly, values of x for ε_4 represent metallic H_xWO_3 in which charge 'inserted' is *conducted* without any valence trapping (i.e. without reduction). The values of x represented by ε_2 and ε_3 are, in all probability, representations of different extents of electron delocalisation.

Notably, in the studies by Monk *et al.*[136] and by Siddle and co-workers,[162,163] the optical absorbance of the incipiently reduced oxide was observed to increase for a short time *after* the driving potential was removed. Since the extinction coefficient ε for H_xWO_3 is a function of insertion coefficient x, these observations can be taken as direct evidence for flattening-out of a concentration gradient in the absence of an applied field.

Toward a consensus model The evidence for each model seems quite convincing if taken in isolation and, as discussed above, some elements of the theories seem to fit all models: the positing of a back potential is a case in point. Only Faughnan and Crandall[92] dismiss the idea of concentration gradients within the incipient H_xWO_3.

A combined model describing the electro-coloration of thin-film tungsten trioxide would suggest that the kinetics are dominated by the formation of a back potential. Initially, when insertion coefficients are small, in the range $0 < x < 0.03$–0.05, the motion of electrons is rate limiting, but as the upper value of this x limit is passed, so the mobility of electrons increases and ionic motion becomes rate limiting. The percolation threshold x of 0.03 is sufficiently small that many workers may have missed anomalous properties at small x. Also, the mobility $\mu_{(electron)}$ is usually said to be higher in hydrated WO_3 – and perhaps also for reduced oxides immersed in electrolyte solutions – thus further masking the effects of low x. The percolation phenomenon was not seen by Ingram *et al.* when electro-colouring WO_3 with a very small field, where no current peak was observed. Green's observation, that the behaviour of electrons and ions cannot be separated except at high fields, may be sufficient explanation.

Bohnke's[101,102,103] assumption that $[H^+ \; e^-]$ pairs form during coloration has received no support from subsequent workers; but many of these studies may have been incapable of discerning such pairs. The data of Ingram *et al.*[96] and others suggest that in particular circumstances electrons and ions do indeed move autonomously, and no $[H^+ \; e^-]$ pairs are required to form at the higher fields.

5.4 Concluding summary

The electrochemical insertion and egress of counter ions into thin films of solid electrochrome is clearly a complicated process. While several new studies provide general views of intercalation, diffusion and migration (e.g. refs. 164,165,166), a complete mechanism describing the controlling redox processes and ionic motions in coloration and bleaching has not yet been established.

References

1. Bard, A. J. and Faulkner, L. R. *Electrochemical Methods: Fundamentals and Applications*, 2nd edn, New York, Wiley, 2002, pp. 148 ff.
2. Bard, A. J. and Faulkner, L. R. *Electrochemical Methods: Fundamentals and Applications* 2nd edn, New York, Wiley, 2002.
3. Steckhan, E. and Kuwana, T. Spectroelectrochemical study of mediators, I: bipyridilium salts and their electron transfer rates to Cytochrome *c*. *Ber. Bunsen-Ges. Phys. Chem.*, **78**, 1974, 253–9.

4. Rosseinsky, D. R. and Monk, P. M. S. Electrochromic cyanophenylparaquat (CPQ: 1,1′-bis-cyanophenyl-4,4′-bipyridilium) studied voltammetrically, spectroelectrochemically and by ESR. *Sol. Energy Mater. Sol. Cells*, **25**, 1992, 201–10.
5. Hansen, W. N., Kuwana, T. and Osteryoung, R. A. Observation of electrode-solution interface by means of internal reflection spectrometry. *Anal. Chem.*, **38**, 1966, 1810–21.
6. Tsutsumi, H., Nakagawa, Y. and Tamura, K. Single-film electrochromic devices with polymer gel films containing aromatic electrochromics. *Sol. Energy Mater. Sol. Cells*, **39**, 1995, 341–8.
7. Granqvist, C. G. Electrochromic tungsten oxide films: review of progress 1993–1998. *Sol. Energy Mater. Sol. Cells*, **60**, 2000, 201–62.
8. Ord, J. L., Pepin, G. M. and Beckstead, D. J. An optical study of hydrogen insertion in the anodic oxide of tungsten. *J. Electrochem. Soc.*, **136**, 1989, 362–8.
9. Noshino, T. and Baba, N. Characterization and properties of electrochromic cobalt oxide thin film prepared by electrodeposition. *Sol. Energy Mater. Sol. Cells*, **39**, 1995, 391–7.
10. Wadayama, T., Wako, H. and Hatta, A. Electrobleaching of WO_3 as probed by Raman scattering. *Mater. Trans. JIM*, **37**, 1996, 1486–91.
11. Goodenough, J. B. Transition metal oxides with metallic conductivity. *Bull. Soc. Chim. Fr.*, **4**, 1965, 1200–7. The introduction includes a comprehensive list.
12. Whittingham, M. S. The formation of tungsten bronzes and their electrochromic properties. In Chowdari, B. V. R. and Radhakrishna, S. (eds.), *Proceedings of the International Seminar on Solid State Ionic Devices*, Singapore, World Publishing Company, 1988, pp. 325–40.
13. Hersch, H. N., Kramer, W. E. and McGee, J. K. Mechanism of electrochromism in WO_3. *Appl. Phys. Lett.*, **27**, 1975, 646–8.
14. Weppner, W. and Huggins, R. A. Determination of the kinetics parameters of mixed conducting electrodes and application to the system Li_3 Sb. *J. Electrochem. Soc.*, **124**, 1977, 1569–78.
15. Bell, J. M. and Matthews, J. P. Temperature dependence of kinetic behaviour of sol–gel deposited electrochromics. *Sol. Energy Mater. Sol. Cells*, **68**, 2001, 249–63.
16. Tuck, B. *Atomic Diffusion in III–V Semiconductors*, Bristol, Adam Hilger, 1988; e.g. see ch. 2 'Elements of diffusion', pp. 9–45.
17. Monk, P. M. S., Mortimer, R. J. and Rosseinsky, D. R. *Electrochromism: Fundamentals and Applications*, Weinheim, VCH, 1995.
18. Granqvist, G. C. *Handbook of Inorganic Electrochromic Materials*, Amsterdam, Elsevier, 1995.
19. Goldner, R. B. Some aspects of charge transport in electrochromic films. In Chowdari, B. V. R. and Radhakrishna, S. (eds.), *Proceedings of the International Seminar on Solid State Ionic Devices*, Singapore, World Publishing Company, 1988, pp. 351–8.
20. Kim, J. J., Tryk, D. A., Amemiya, T., Hashimoto, K. and Fujishima, A. Color impedance and electrochemical impedance studies of WO_3 thin films: H^+ and Li^+ transport *J. Electroanal. Chem.*, **435**, 1997, 31–8.
21. Ho, C.-K., Raistrick, I. D. and Huggins, R. A. Application of AC-techniques to the study of lithium diffusion in tungsten trioxide thin-films. *J. Electrochem. Soc.*, **127**, 1980, 343–50.
22. Baudry, P., Aegerter, M. A., Deroo, D. and Valla, B. Electrochromic window with lithium conductive polymer electrolyte. *Proc. Electrochem. Soc.*, **90–2**, 1990, 274–87.

23. Kamimori, T., Nagai, J. and Mizuhashi, M. Electrochromic devices for transmissive and reflective light control. *Sol. Energy Mater.*, **16**, 1987, 27–38.
24. Green, M. Atom motion in tungsten bronze thin films. *Thin Solid Films*, **50**, 1978, 148–50.
25. Xu, G. and Chen, L. Lithium diffusion in WO_3 films. *Solid State Ionics*, **28–30**, 1988, 1726–8.
26. Avellaneda, C. O. and Bulhões, L. O. S. Intercalation in WO_3 and WO_3:Li films. *Solid State Ionics*, **165**, 2003, 59–64.
27. Baudry, P., 1989. Ph.D thése nouveau regime. Grenoble, France.
28. Hesse, K. and Schlettwein, D. Spectroelectrochemical investigations on the reduction of thin films of hexadecafluorophthalocyaninatozinc (F_{16} PcZn). *J. Electroanal. Chem.*, **476**, 1999, 148–58.
29. Nicholson, M. M. and Pizzarello, F. Galvanostatic transients in lutetium diphthalocyanine. *J. Electrochem. Soc.*, **127**, 1980, 821–7.
30. MacArthur, D. M. The proton diffusion coefficient for the nickel hydroxide electrode. *J. Electrochem. Soc.*, **117**, 1970, 729–32. MacArthur consistently in this paper talks of 'ΔH for diffusion', but in fact the data from his Arrhenius-type graphs yield E_A.
31. Lukovtsev, P. D. and Slaidin, G. J. Proton diffusion through nickel oxide. *Electrochim. Acta*, **6**, 1962, 17–21.
32. Gomes, M. A. B. and Bulhões, L. O. S. Diffusion coefficient of H^+ at Nb_2O_5 layers prepared by thermal oxidation of niobium. *Electrochim. Acta*, **35**, 1990, 765–8.
33. Tran-Van, F., Henri, T. and Chevrot, C. Synthesis and electrochemical properties of mixed ionic and electronic modified polycarbazole. *Electrochim. Acta*, **47**, 2002, 2927–36.
34. Yashima, H., Kobayashi, M., Lee, K.-B., Chung, D., Heeger, A. J. and Wudl, F. Electrochromic switching of the optical properties of polyisothianaphthene. *J. Electrochem. Soc.*, **134**, 1987, 46–52.
35. Faughnan, B. W., Crandall, R. S. and Lampert, M. A. Model for the bleaching of WO_3 electrochromic films by an electric field. *Appl. Phys. Lett.*, **27**, 1975, 275–7.
36. Wagner, C. *Z. Phys. Chem., Abs. B*, **21**, 1933, 25, as cited in ref. 14 here.
37. Bisquert, J. and Vikhrenko, V. S. Analysis of the kinetics of ion intercalation: two state model describing the coupling of solid state ion diffusion and ion binding processes. *Electrochim. Acta*, **47**, 2002, 3977–88.
38. Bisquert, J. Analysis of the kinetics of ion intercalation: ion trapping approach to solid-state relaxation processes. *Electrochim. Acta*, **47**, 2002, 2435–49.
39. Yoshimura, T., Watanabe, M., Koike, Y., Kiyota, K. and Tanaka, M. Effect of surface states on WO_3 on the operating characteristics of thin film electrochromic devices. *Thin Solid Films*, **101**, 1983, 141–51.
40. Ho, K.-C., Singleton, D. E. and Greenberg, C. B. The influence of terminal effects on the performance of electrochromic windows. *J. Electrochem. Soc.*, **137**, 1990, 3858–64.
41. Kaneko, H. and Miyake, K. Effects of transparent electrode resistance on the performance characteristics of electrochemichromic cells. *Appl. Phys. Lett.*, **49**, 1986, 112–14.
42. Jeong, D. J., Kim, W.-S. and Sung, Y. E. Improved electrochromic response time of nickel hydroxide thin films by ultra-thin nickel metal underlayer. *Jpn. J. Appl. Phys.*, **40**, 2001, L708–10.
43. He, T., Ma, Y., Cao, Y., Yang, W. and Yao, J. Enhanced electrochromism of WO_3 thin film by gold nanoparticles. *J. Electroanal. Chem.*, **514**, 2001, 129–32.

44. Yao, J. N., Yang, Y. A. and Loo, B. H. Enhancement of photochromism and electrochromism in MoO_3/Au and MoO_3/Pt thin films. *J. Phys. Chem. B*, **102**, 1998, 1856–60.
45. Haranahalli, A. R. and Holloway, P. H. The influence of metal overlayers on electrochromic behavior of tungsten trioxide films. *J. Electron. Mater.*, **10**, 1981, 141–72.
46. Haranahalli, A. R. and Dove, D. B. Influence of a thin gold surface layer on the electrochromic behavior of WO_3 films. *Appl. Phys. Lett.*, **36**, 1980, 791–3.
47. Cheng, K. H. and Whittingham, M. S. Lithium incorporation in tungsten oxides. *Solid State Ionics*, **1**, 1980, 151–61.
48. Dini, D., Passerini, S., Scrosati, B. and Decker, F. Stress changes in electrochromic thin film electrodes: laser beam deflection method (LBDM) as a tool for the analysis of intercalation processes. *Sol. Energy Mater. Sol. Cells*, **56**, 1999, 213–21.
49. Berezin, L. Y. and Malinenko, V. P. Electrochromic coloration and bleaching of polycrystalline tungsten trioxide. *Pis'ma. Zh. Tekh. Fiz.* **13**, 1987, 401–4 [in Russian], as cited in *Chem. Abs.* **107**: 449,382t.
50. Berezin, L. Y., Aleshina, L. A., Inyushin, N. B., Malinenko, V. P. and Fofanov, A. D. Phase transitions during electrochromic processes in tungsten trioxide. *Fiz. Tverd Tela (Leningrad)*, **31**, 1989, 41–9 [in Russian], as cited in *Chem. Abs.* **112**: 225,739.
51. Ord, J. L., Bishop, S. D. and DeSmet, D. J. Hydrogen insertion into anodic oxide films on vanadium. *Proc. Electrochem. Soc.*, **90–2**, 1990, 116–24.
52. Ord, J. L. and DeSmet, D. J. Optical anisotropy and electrostriction in the anodic oxide of molybdenum. *J. Electrochem. Soc.*, **130**, 1983, 280–4.
53. Ord, J. L. and Wang, W. P. Optical anisotropy and electrostriction in the anodic oxide of tantalum. *J. Electrochem. Soc.*, **130**, 1983, 1809–14.
54. Mohapatra, S. K. Electrochromism in Li_xWO_3. *J. Electrochem. Soc.*, **125**, 1978, 284–8.
55. Shiyanovskaya, I. Isotopic effect in evolution of structure and optical gap during electrochromic coloration of $WO_3 \cdot 1/3(H_2O)$ films. *Mikrochim. Acta*, **S14**, 1997, 819–22.
56. Kurita, S., Nishimura, T. and Taira, K. Proton injection phenomena in WO_3-electrolyte electrochromic cells. *Appl. Phys. Lett.*, **36**, 1980, 585–7.
57. Shiyanovskaya, I. and Hepel, M. Isotopic effects in cation-injected electrochromic films. *J. Electrochem. Soc.*, **145**, 1998, 1023–8.
58. Kang, K. and Green, M. Solid state electrochromic cells: optical properties of the sodium tungsten bronze system. *Thin Solid Films*, **113**, 1984, L29–32.
59. Dini, D., Decker, F. and Masetti, E. A comparison of the electrochromic properties of WO_3 films intercalated with H^+, Li^+ and Na^+. *J. Appl. Electrochem.*, **26**, 1996, 647–53.
60. Masetti, E., Dini, D. and Decker, F. The electrochromic response of tungsten bronzes M_xWO_3 with different ions and insertion rates. *Sol. Energy Mater. Sol. Cells*, **39**, 1995, 301–7.
61. Green, M. and Richman, D. A solid state electrochromic cell: the $RbAg_4I_5|WO_3$ system. *Thin Solid Films*, **24**, 1974, S45–6.
62. Scarminio, J. Stress in photochromic and electrochromic effects on tungsten oxide film. *Sol. Energy Mater. Sol. Cells*, **79**, 2003, 357–68.
63. García-Cañadas, J., Mora-Seró, I., Fabregat-Santiago, F., Bisquert, J. and Garcia-Belmonte, G. Analysis of cyclic voltammograms of electrochromic

a-WO_3 films from voltage-dependent equilibrium capacitance measurements, *J. Electroanal. Chem.*, **565**, 2004, 329–334.

64. Bohnke, O., Bohnke, C., Robert, G. and Carquille, B. Electrochromism in WO_3 thin films, I: $LiClO_4$–propylene carbonate-water electrolytes. *Solid State Ionics*, **6**, 1982, 121–8.

65. Bohnke, C. and Bohnke, O. Impedance analysis of amorphous WO_3 thin films in hydrated $LiClO_4$–propylene carbonate electrolytes. *Solid State Ionics*, **39**, 1990, 195–204.

66. Bohnke, O., Vuillemin, B., Gabrielli, C., Keddan, M., Perrot, H., Takenouti, H. and Torresi, R. An electrochemical quartz crystal microbalance study of lithium insertion into thin films of tungsten trioxide, I: modeling of the ionic insertion mechanism. *Electrochim. Acta*, **40**, 1995, 2755–64.

67. Bohnke, O., Vuillemin, B., Gabrielli, C., Keddam, M. and Perrot, H. An electrochemical quartz crystal microbalance study of lithium insertion into thin films of tungsten trioxide, II: experimental results and comparison with model calculations. *Electrochim. Acta*, **40**, 1995, 2765–73.

68. Babinec, S. J. A quartz crystal microbalance analysis of ion insertion into WO_3. *Sol. Energy Mater. Sol. Cells*, **25**, 1992, 269–91.

69. Plinchon, V., Giron, J.-C., Deloulbe, J. P. and Lerbet, F. Detection by mirage effect of the counter-ion flux between an electrochrome and a liquid electrolyte: application to WO_3, Prussian blue and lutetium diphthalocyanine film. *Proc. SPIE*, **1536**, 1991, 37–47.

70. Kim, J. J., Tryk, D. A., Amemiya, T., Hashimoto, K. and Fujishima, A. Color impedance and electrochemical impedance studies of WO_3 thin films: behavior of thinner films in non-aqueous electrolyte. *J. Electroanal. Chem.*, **433**, 1997, 9–17.

71. Zhang, L. and Goto, K. S. Measurement of Li^+ diffusivity in thin films of tungsten troxide with oxygen deficiency. *Proc. Electrochem. Soc.*, **90–2**, 1990, 23–39.

72. Kubo, T. and Nishikitani, Y. Deposition temperature dependence of optical gap and coloration efficiency spectrum in electrochromic tungsten oxide films. *J. Electrochem. Soc.*, **145**, 1998, 1729–35.

73. Wang, J., Bell, J. M. and Skryabin, I. L. Kinetics of charge injection in sol–gel deposited WO_3. *Sol. Energy Mater. Sol. Cells*, **56**, 1999, 465–75.

74. Chen, X., Hu, X. and Feng, J. Nanostructured nickel oxide films and their electrochromic properties. *Nanostruct. Mater.*, **6**, 1995, 309–12.

75. Shamritskaya, I. G., Lazorenko-Manevich, R. M. and Sokolova, L. A. Effects of anions on the electroreflectance spectra of anodically oxidized iridium in aqueous solutions. *Russ. J. Electrochem.*, **33**, 1997, 645–52.

76. Yoshiiki, N. and Kondo, S. Electrochemical properties of $WO_3.x\ H_2O$, I: the influence of water adsorption and hydroxylation. *J. Electrochem Soc.*, **130**, 1983, 2283–7.

77. Hurditch, R. Electrochromism in hydrated tungsten-oxide films. *Electron. Lett.*, **11**, 1975, 142–4.

78. Arnoldussen, T. C. Electrochromism and photochromism in MoO_3 films. *J. Electrochem. Soc.*, **123**, 1976, 527–31.

79. Holland, L. *Vacuum Deposition of Thin Films*, London, Chapman and Hall, 1956.

80. Hjelm, A., Granqvist, C. G. and Wills, J. M. Electronic properties and optical properties of WO_3. $LiWO_3$, $NaWO_3$ and HWO_3. *Phys. Rev. B*, **54**, 1996, 2436–45.

81. Yishiike, N. and Kondo, S. Electrochemical properties of $WO_3.x$ (H_2O), II: the influence of crystallization as hydration. *J. Electrochem. Soc.*, **131**, 1984, 809–13.
82. Wagner, W., Rauch, F., Ottermann, C. and Bange, K. Hydrogen dynamics in electrochromic multilayer systems investigated by the ^{15}N technique. *Nucl. Instr. Meth. Phys. Res. B.*, **50**, 1990, 27–30.
83. Wagner, W., Bange, K., Rauch, F. and Ottermann, C. In-depth profiling of hydrogen in oxide multi-layer systems. *Surf. Sci. Anal.*, **16**, 1990, 331–4.
84. Kreuer, K. D. Proton conductivity: materials and applications. *Chem. Mater.*, **8**, 1996, 610–41.
85. Duffy, J. A., Ingram, M. D. and Monk, P. M. S. The effect of moisture on tungsten oxide electrochromism in polymer electrolyte devices. *Solid State Ionics*, **58**, 1992, 109–14.
86. Faughnan, B. W. and Crandall, R. S. Electrochromic devices based on WO_3. In Pankove, J. L. (ed.), *Display Devices*, Berlin, Springer-Verlag, 1980, pp. 181–211.
87. Arnoldussen, T. C. A model for electrochromic tungsten oxide microstructure and degradation. *J. Electrochem. Soc.*, **128**, 1981, 117–23.
88. Randin, J.-P. Chemical and electrochemical stability of WO_3 electrochromic films in liquid electrolytes. *J. Electron. Mater.*, **7**, 1978, 47–63.
89. Hefny, M. M., Gadallah, A. G. and Mogoda, A. S. Some electrochemical properties of the anodic oxide film on tungsten. *Bull. Electrochem.*, **3**, 1987, 11–14.
90. Reichman, B. and Bard, A. J. The electrochromic process at WO_3 electrodes prepared by vacuum evaporation and anodic oxidation of W. *J. Electrochem. Soc.*, **126**, 1979, 583–91.
91. Kumagai, N., Abe, M., Kumagai, N., Tanno, K. and Pereria-Ramos, J. P. Kinetics of electrochemical insertion of lithium into WO. *Solid State Ionics*, **70–71**, 1994, 451–7.
92. Crandall, R. S. and Faughnan, B. W. Electronic transport in amorphous H_xWO_3. *Phys. Rev. Lett.*, **39**, 1977, 232–5.
93. Crandall, R. S. and Faughnan, B. W. Dynamics of coloration of amorphous electrochromic films of WO_3 at low voltages. *Appl. Phys. Lett.*, **28**, 1976, 95–7.
94. Crandall, R. S., Wojtowicz, P. J. and Faughnan, B. W. Theory and measurement of the change in chemical potential of hydrogen in amorphous H_xWO_3 as a function of the stoichiometric parameter x. *Solid State Commun.*, **18**, 1976, 1409–11.
95. Crandall, R. S. and Faughnan, B. W. Measurement of the diffusion coefficient of electrons in WO_3 films. *Appl. Phys. Lett*, **26**, 1975, 120–1.
96. Ingram, M. D., Duffy, J. A. and Monk, P. M. S. Chronoamperometric response of the cell ITO | H_xWO_3 | PEO–H_3PO_4 (MeCN) | ITO. *J. Electroanal. Chem.*, **380**, 1995, 77–82.
97. Pedone, P., Armand, M. and Deroo, D. Voltammetric and potentiostatic studies of the interface WO_3/polyethylene oxide–H_3PO_4. *Solid State Ionics*, **28–30**, 1988, 1729–32.
98. Nishikawa, M., Ohno, H., Kobayashi, T., Tsuchida, E. and Hirohashi, R. All solid-state electrochromic device containing poly[oligo(oxyethylene) methylmethacrylate]/$LiClO_4$ hybrid polymer ion conductor. *J. Soc. Photoagr. Sci. Technol. Jpn.*, **81**, 1988, 184–90 [in Japanese].
99. Luo, Z., Ding, Z. and Jiang, Z. Electrochromic kinetics of amorphous WO_3 films. *J. Non-Cryst. Solids*, **112**, 1989, 309–13.

100. Green, M., Smith, W. C. and Weiner, J. A. A thin film electrochromic display based on the tungsten bronzes. *Thin Solid Films*, **38**, 1976, 89–100.
101. Bohnke, O. and Vuillermin, B. Proton insertion into thin films of amorphous WO_3: kinetics study. In Balkanski, M., Takahashi, T. and Tuller, H. L. (eds.), *Solid State Ionics*, Amsterdam, Elsevier, 1992, pp. 593–8.
102. Bohnke, O. and Vuillermin, B. Proton insertion into thin films of amorphous WO_3: kinetics study. *Mater. Sci. Eng. B*, **13**, 1992, 243–6.
103. Bohnke, O., Rezrazi, M., Vuillermin, B., Bohnke, C., Gillet, P. A. and Rousellot, C. *In situ* optical and electrochemical characterization of electrochromic phenomena into tungsten trioxide thin films. *Sol. Energy Mater. Sol. Cells*, **25**, 1992, 361–74.
104. Carslaw, H. S. and Jaeger, J. C. *Conduction of Heat in Solids*, 2nd edn, Oxford, Oxford University Press, 1959.
105. Seman, M. and Wolden, C. A. Characterization of ion diffusion and transient electrochromic performance in PECVD grown tungsten oxide thin films. *Sol. Energy Mater. Sol. Cells*, **82**, 2004, 517–30.
106. Stauffer, D. *Introduction to Percolation Theory*, London, Taylor and Francis, 1985.
107. Wittwer, V., Schirmer, O. F. and Schlotter, P. Disorder dependence and optical detection of the Anderson transition in amorphous H_xWO_3 bronzes. *Solid State Commun.*, **25**, 1978, 977–80.
108. Likalter, A. A. Impurity states and insulator–metal transition in tungsten bronzes. *Physica B*, **315**, 2002, 252–60.
109. Craig, J. B. and Grant, J. M. Kinetic of electrochromic processes in tungsten oxide films. *J. Mater. Chem.*, **2**, 1992, 521–8.
110. Aoki, K. and Tezuka, Y. Chronoamperometric response to potentiostatic doping at polypyrrole-coated microdisk electrodes. *J. Electroanal. Chem.*, **267**, 1989, 55–66.
111. Malta, M., Gonzalez, E. R. and Torresi, R. M. Electrochemical and chromogenic relaxation processes in polyaniline films. *Polymer*, **43**, 2002, 5895–901.
112. Goldner, R. B., Norton, P., Wong, G., Foley, E. L., Seward, G. and Chapman, R. Further evidence for free electrons as dominating the behaviour of electrochromic polycrystalline WO_3 films. *Appl. Phys. Lett.*, **47**, 1985, 536–8.
113. Cogan, S. F., Plante, T. D., Parker, M. A. and Rauh, R. D. Free-electron electrochromic modulation in crystalline Li_xWO_3. *J. Appl. Phys.*, **60**, 1986, 2735–8.
114. Maranhão, S. L. D. A. and Torresi, R. M. Electrochemical and chromogenics kinetics of lithium intercalation in anodic niobium oxide films. *Electrochim. Acta*, **43**, 1998, 257–64.
115. Maranhão, S. L. D. A. and Torresi, R. M. Filmes de óxidos anódicos de nióbio: efeito eletrocrômico e cinética da reação de eletro-intercalação. *Quim. Nova*, **21**, 1998, 284–8.
116. Zhang, J. G., Benson, D. K., Tracy, C. E., Deb, S. K., Czanderna, A. W. and Bechriger, C. Chromic mechanism in amorphous WO_3 films. *J. Electrochem. Soc.*, **144**, 1997, 2022–6.
117. Leftheriotis, G., Papaefthimiou, S., Yianoulis, P. and Siokou, A. Effect of the tungsten oxidation states in the thermal coloration and bleaching of amorphous WO_3 films. *Thin Solid Films*, **384**, 2001, 298–306.
118. Siokou, A., Leftheriotis, G., Papaefthimiou, S. and Yianoulis, P. Effect of the tungsten and molybdenum oxidation states on the thermal coloration of amorphous WO_3 and MoO_3 films. *Surf. Sci.*, **482–5**, 2001, 294–9.

119. Wang, X. G., Jang, Y. S., Yang, N. H., Yuan, L. and Pang, S. J. XPS and XRD study of the electrochromic mechanism of WO_x films. *Surf. Coat. Technol.*, **99**, 1998, 82–6.
120. Temmink, A., Anderson, O., Bange, K., Hantsche, H. and Yu, X. Optical absorption of amorphous WO_3 and binding state of tungsten, *Thin Solid Films*, **192**, 1990, 211–18.
121. Temmink, A., Anderson, O., Bange, K., Hantsche, H. and Yu, X. 4f level shifts of tungsten and colouration state of *a*-WO_3. *Vacuum*, **41**, 1990, 1144–6.
122. Wang, B. X., Hu, G., Liu, B. F. and Dong, S. J. Electrochemical preparation of microelectrodes modified with non-stoichiometric mixed-valent molybdenum oxides. *Acta Chim. Sinica*, **54**, 1996, 598–604 [in Chinese]. (Abstract available on Web of Science website.)
123. Fleisch, T. H. and Mains, G. J. An XPS study of the UV reduction and photochromism of MoO_3 and WO_3. *J. Chem. Phys.*, **76**, 1982, 780–6.
124. Cruz, T. G. S., Gorenstein, A., Landers, R., Kleiman, G. G. and deCastro, S. C. Electrochromism in MoO_x films characterized by X-ray electron spectroscopy. *J. Electron. Spectrosc. Rel. Phenom.*, **101–3**, 1999, 397–400.
125. Papaefthimiou, S., Leftheriotis, G. and Yianoulis, P. Study of electrochromic cells incorporating WO_3, MoO_3, WO_3–MoO_3 and V_2O_5 coatings. *Thin Solid Films*, **343–344**, 1999, 183–6.
126. Bohnke, O., Frand, G., Fromm, M., Weber, J. and Greim, O. Depth profiling of W, O and H in tungsten trioxide thin films using RBS and ERDA techniques. *Appl. Surf. Sci.*, **93**, 1996, 45–52.
127. Antonaia, A., Santoro, M. C., Fameli, G. and Polichetti, T. Transport mechanism and IR structural characterisation of evaporated amorphous WO_3 films. *Thin Solid Films*, **426**, 2003, 281–7.
128. Lee, S.-H., Cheong, H. M., Tracy, C. E., Mascarenhas, A., Benson, D. K. and Deb, S. K. Raman spectroscopic studies of electrochromic *a*-WO_3. *Electrochim. Acta*, **44**, 1999, 3111–15.
129. Lee, S.-H., Cheong, H. M., Zhang, J.-G., Mascarenhas, A., Benson, D. K. and Deb, S. K. Electrochromic mechanism in WO_{3-y} thin films. *Appl. Phys. Lett.*, **74**, 1999, 242–4.
130. Sun, S.-S. and Holloway, P. H. Modification of vapor-deposited WO_3 electrochromic films by oxygen backfilling. *J. Vac. Sci. Technol. A.*, **1**, 1983, 529–33.
131. Rezrazi, M., Vuillemin, B. and Bohnke, O. Thermodynamic study of proton insertion into thin films of *a*-WO_3. *J. Electrochem. Soc.*, **138**, 1991, 2770–4.
132. Bechinger, C., Burdis, M. S. and Zhang, J.-G. Comparison between electrochromic and photochromic coloration efficiency of tungsten oxide thin films. *Solid State Commun.*, **101**, 1997, 753–6.
133. de Wijs, G. A. and de Groot, R. A. Amorphous WO_3: a first-principles approach. *Electrochim. Acta*, **46**, 2001, 1989–93.
134. Green, M. and Pita, K. Non-stoichiometry in thin film dilute tungsten bronzes: $M_x WO_{3-y}$. *Sol. Energy Mater. Sol. Cells*, **43**, 1996, 393–411.
135. Gorenstein, A., Scarminio, J. and Lourenço, A. Lithium insertion in sputtered amorphous molybdenum thin films. *Solid State Ionics*, **86–88**, 1996, 977–81.
136. Monk, P. M. S., Duffy, J. A. and Ingram, M. D. Pulsed enhancement of the rate of coloration for tungsten trioxide based electrochromic devices. *Electrochim. Acta*, **43**, 1998, 2349–57.

137. Knapp, R. C., Turnbull, R. R. and Poe, G. B. (Gentex Corporation). Reflectance control of an electrochromic element using a variable duty cycle drive. US Patent 06084700, 2000.
138. Monk, P. M. S., Fairweather, R. D., Ingram, M. D. and Duffy, J. A. Pulsed electrolysis enhancement of electrochromism in viologen systems: influence of comproportionation reactions. *J. Electroanal. Chem.*, **359**, 1993, 301–6.
139. Barclay, D. J. and Martin, D. H. Electrochromic displays. In Howells, E. R. (ed.), *Technology of Chemicals and Materials for the Electronics Industry*, Chichester, Ellis Horwood, 1984, pp. 266–76.
140. Protsenko, E. G., Klimisha, G. P., Krainov, I. P., Kramarenko, S. F. and Distanov, B. G. *Deposited Doc.*, 1981, SPSTL 971, Khp-D81. *Chem. Abs.* **98**: 170, 310 (1983).
141. Schierbeck, K. L. (Donnelly Corporation). Digital electrochromic mirror system. US Patent 06089721, 2000.
142. Statkov, L. I. Peculiarities of the mechanism of the electrochromic coloring of oxide films upon pulsed electrochemical polarization, *Russ. J. Appl. Chem.*, **70**, 1997, 653–4.
143. Ottaviani, M., Panero, S., Morizilli, S., Scrosati, B. and Lazzari, M. The electrochromic characteristics of titanium oxide thin film. *Solid State Ionics*, **20**, 1986, 197–202.
144. DeSmet, D. J. and Ord, J. L. An optical study of hydrogen insertion in the anodic oxide of molybdenum. *J. Electrochem. Soc.*, **134**, 1987, 1734–40.
145. Duffy, J. A., Baucke, F. G. K. and Woodruff, P. R. Optical properties of tungsten bronze surfaces. *Thin Solid Films*, **148**, 1987, L59–61.
146. Porqueras, I., Viera, G., Marti, J. and Bertran, E. Deep profiles of lithium in electrolytic structures of ITO/WO_3 for electrochromic applications. *Thin Solid Films*, **343–4**, 1999, 179–82.
147. Zhong, Q., Wessel, S. A., Heinrich, B. and Colbow, K. The electrochromic properties and mechanism of H_3WO_3 and Li_xWO_3. *Sol. Energy Mater.*, **20**, 1990, 289–96.
148. Kamimori, T., Nagai, J. and Mizuhashi, M. Transport of Li^+ ions in amorphous tungsten oxide films. *Proc. SPIE*, **428**, 1983, 51–6.
149. Matthews, J. P., Bell, J. M. and Skryabin, I. L. Effect of temperature on electrochromic device switching voltages. *Electrochim. Acta*, **44**, 1999, 3245–50.
150. Bell, J. M., Matthews, J. P. and Skryabin, I. L. Modelling switching of electrochromic devices – a route to successful large area device design. *Solid State Ionics*, **152–3**, 2002, 853–60.
151. Anderson, O. L. and Stuart, D. A. *J. Am. Ceram. Soc.*, **37**, 1954, 573, as cited in Elliott, S. R., *Physics of Amorphous Materials*, Harlow, Longman, 1990.
152. Strømme Mattson, M., Niklasson, G. A. and Granqvist, C. G. Diffusion of Li, Na, and K in fluorinated Ti dioxide films: applicability of the Anderson–Stuart model. *J. Appl. Phys.*, **81**, 1997, 2167–72.
153. Krasnov, Y. S., Sych, O. A., Patsyuk, F. N. and Vas'ko, A. T. Electrochromism and diffusion of charge carriers in amorphous tungsten trioxide, taking into account the electron capture on localized sites. *Elektrokhimiya*, **24**, 1988, 1468–1474 [in Russian], as cited in *Chem Abs.* **1110**: 1447,1513z.
154. Rosseinsky, D. R. and Mortimer, R. J. Electrochromic systems and the prospects for devices. *Adv. Mater.*, **13**, 2001, 783–93.
155. Gardner, W. R. and Danielson, G. C. Electrical resistivity and Hall coefficient of sodium tungsten bronze. *Phys. Rev.*, **93**, 1954, 46–51.

156. Jones Jr., W. H., Garbaty, E. A. and Barnes, R. G. Nuclear magnetic resonance in metal tungsten bronzes. *J. Chem. Phys.*, **36**, 1962, 494–9.
157. Bohnke, O., Gire, A. and Theobald, J. G. *In situ* detection of electrical conductivity variation of an a-WO_3 thin film during electrochemical reduction and oxidation in $LiClO_4$ (M)–PC electrolyte. *Thin Solid Films*, **247**, 1994, 51–5.
158. Muhlestein, L. D. and Danielson, G. C. Effects of ordering on the transport properties of sodium tungsten bronze. *Phys. Rev.*, **158**, 1967, 825–32.
159. Muhlestein, L. D. and Danielson, G. C. Seebeck effect in sodium tungsten bronze. *Phys. Rev.*, **160**, 1967, 562–7.
160. Wolfram, T. and Sutcu, L. x Dependence of the electronic properties of cubic Na_xWO_3. *Phys. Rev. B*, **31**, 1985, 7680–7.
161. Baucke, F. G. K., Duffy, J. A. and Smith, R. I. Optical absorption of tungsten bronze thin films for electrochromic applications. *Thin Solid Films*, **186**, 1990, 47–51.
162. Burdis, M. S. and Siddle, J. R. Observation of non-ideal lithium insertion into sputtered thin films of tungsten oxide. *Thin Solid Films*, **237**, 1994, 320–5.
163. Batchelor, R. A., Burdis, M. S. and Siddle, J. R. Electrochromism in sputtered WO_3 thin films. *J. Electrochem. Soc.*, **143**, 1996, 1050–5.
164. Montella, C. Discussion on permeation transients in terms of insertion reaction mechanism and kinetics. *J. Electroanal. Chem.*, **465**, 1999, 37–50.
165. Diard, J.-P., Le Gorrec, B. and Montella, C. Logistic differential equation: a general equation for electrointercalation processes? *J. Electroanal. Chem.*, **475**, 1999, 190–2.
166. Torresi, R. M., Córdoba de Torresi, S. I. and Gonzalez, E. R. On the use of the quadratic logistic differential equation for the interpretation of electrointercalation processes. *J. Electroanal. Chem.*, **461**, 1999, 161–6.

6
Metal oxides

6.1 Introduction to metal-oxide electrochromes

Metal oxides as thin films feature widely in the literature, in large part owing to their photochemical stability (see Section 6.1.2); by contrast, most, if not all, organic electrochromes may be susceptible to photochemical degradation.[1]

The oxides of the following transition metals are electrochromic: cerium, chromium, cobalt, copper, iridium, iron, manganese, molybdenum, nickel, niobium, palladium, praseodymium, rhodium, ruthenium, tantalum, titanium, tungsten and vanadium. Most of the electrochromic colours derive from intervalence charge-transfer optical transitions, as described in Section 4.4. The intervalence coloured forms of most transition-metal oxide electrochromes are in the range blue or grey through to black; it is much less common for transition-metal oxides to form other colours by intervalence transitions (see Table 6.1).

The oxides of tungsten, molybdenum, iridium and nickel show the most intense electrochromic colour changes. Other metal oxides of lesser colourability are therefore more useful as optically passive, or nearly passive, counter electrodes; see Section 1.4 on 'secondary electrochromism'.

At least one redox state of each of the oxides IrO_2, MoO_3, Nb_2O_5, TiO_2, NiO, RhO_2 and WO_3 can be prepared as an essentially colourless thin film, so allowing the electrochromic transition *colourless (clear)* $\rightleftarrows$ *coloured*. This property finds application in on–off or light-intensity modulation roles. Other oxides in Section 6.2 demonstrate electrochromism differently by showing two colours, i.e. switching as *colour 1* $\rightleftarrows$ *colour 2*, one of these colours often being much more intense than the other. Display-device applications can be envisaged for the latter group of electrochromes.

Granqvist[2] describes how the solid-state crystals of all of the well-known electrochromic metal oxides Ce, Co, Cr, Cu, Ir, Ni, Mo, Nb, Ni, Mo, Ta, Ti, V,

Table 6.1. *Summary of the colours of metal-oxide electrochromes.*

Metal	Oxidised form[a] of oxide	Reduced form[a] of oxide	Balanced redox reaction for electrochromic operation
Bismuth	Bi_2O_3 Transparent	$Li_xBi_2O_3$ Dark brown	(6.16)
Cerium	CeO_2 Colourless	M_xCeO_2 Colourless	(6.17)
Cobalt	CoO Pale yellow	Co_3O_4 Dark brown	(6.19)
	$LiCoO_2$ Pale yellow–brown	M_xLiCoO_2 (M $\neq$ Li) Dark brown	(6.20)
Copper	CuO Black	Cu_2O Red–brown	(6.22)
Iridium	$Ir(OH)_3$ Colourless	$IrO_2 \cdot H_2O$ Blue–grey	(6.11) or (6.12)
Iron	$FeO \cdot OH$ Yellow–green	$Fe(OH)_2$ Transparent	(6.24)
	Fe_2O_3 Brown	Fe_3O_4 Black	(6.25)
	Fe_3O_4 Black	FeO Colourless	(6.26)
	Fe_2O_3 Brown	$M_xFe_2O_3$ Black	(6.27)
	FeO Colourless	Fe_2O_3 Brown	(6.28)
Manganese	MnO_2 Dark brown	Mn_2O_3 Pale yellow	(6.29)
	MnO_2 Brown	$MnO_{(2-x)}(OH)_x$ Yellow	(6.30)
	MnO_2 Brown	M_xMnO_2 Yellow	(6.31)
Molybdenum	MoO_3 Colourless	M_xMoO_3 Intense blue	(6.9)
Nickel	$Ni^{II}O_{(1-y)}H_z$ Brown–black	$Ni^{II}_{(1-x)}Ni^{III}_xO_{(1-y)}H_{(z-x)}$ Colourless	(6.13)
Niobium	Nb_2O_5 Colourless	$M_xNb_2O_5$ Blue	(6.33)
Praseodymium	$PrO_{(2-y)}$ Dark orange	$M_xPrO_{(2-y)}$ Colourless	(6.34)
Rhodium	Rh_2O_3 Yellow	RhO_2 Dark green	(6.35)
Ruthenium	RuO_2 Blue–brown	Ru_2O_3 Black	(6.36)
Tantalum	Ta_2O_5 Colourless	TaO_2 Very pale blue	(6.37)

Table 6.1.(cont.)

Metal	Oxidised form[a] of oxide	Reduced form[a] of oxide	Balanced redox reaction for electrochromic operation
Tin	SnO_2 Colourless	Li_xSnO_2 Blue–grey	(6.38)
Titanium	TiO_2 Colourless	M_xTiO_2 Blue–grey	(6.39)
Tungsten	WO_3 Very pale yellow	M_xWO_3 Intense blue	(6.8)
Vanadium	V_2O_5 Brown–yellow	$M_xV_2O_5$ Very pale blue	(6.40)

[a] The counter cation M is lithium unless stated otherwise.

W, are composed of MO_6 octahedra arranged in a variety of corner-sharing and edge-sharing arrangements, and emphasises that these structural units persist in electrochromic films. Furthermore, he explains how the coordination of the ions leads to electronic bands that are able to explain the presence or absence of cathodic and anodic electrochromism in the numerous defect perovskites, rutiles and layer structures adopted by these oxides.

Solid-state electrochromism as in metal oxides requires the following.

(i) Bonding in structures whose electron orbital energies (or where applicable, band energies) allow of electron uptake or loss from an inert contact, i.e. 'redox switchability';
(ii) During the redox coloration process, a uniformity-conferring charge dispersibility via electron hopping or conduction bands, and complementary ion motion;
(iii) Subsequent photon-effected electronic transitions involving the redox-altered species, that are responsible for colour evocation or colour change.

The electron-hopping in (ii) is sometimes deemed to be small-polaron motion.

That transition energies in (iii) comprise a spread around a most probable value is shown in spectroscopy by absorption bands having an appreciable *width*. The optical charge transfers in (iii) can either involve discrete sites of the *same* element in different charge states, (different 'oxidation states'), in *homo*nuclear intervalence charge transfer ('IVCT'), or between sites occupied by *different* elements, in *hetero*nuclear IVCT. The former often (perhaps usually) holds in single-metal oxides, though optical charge transfer between a metal and an oxide ion is also a possibility. In binary-metal oxides, homonuclear or heteronuclear transfer between the metals, or metal/oxide-ion electron transfer, are possible. (All of the several possibilities here could in principle occur together but no corresponding totality of discrete bands has been so assigned). Intra-atomic or inter-band transitions (resulting from the redox-effected changes) can also – perhaps less usually – confer some colour, the former rarely being intense.

Most of the electrochromic oxides above are compounds of d-block metals. Some oxides of p-block elements – bismuth oxide, tin oxide, or mixed-cation such as indium–tin oxide (ITO) – likewise show a new colour (i.e. absorption band) on electro-reduction.

6.1.1 Bibliography

The literature describing the electrochromism of metal oxides is extensive. Granqvist's[3] 1995 book *Handbook of Inorganic Electrochromic Materials* provides the standard text. There is also a chapter on 'metal oxides' in *Electrochromism: Fundamentals and Applications* (1995).[4] Early reviews on cathodic coloration[5] and on anodic coloration[6] (both 1982) are still informative, as are those on WO_3 amorphous films[7] (1975) and WO_3 displays[7] (1980).

'Tungsten bronzes, vanadium bronzes and related compounds'[8] is the most thorough survey, despite its date (1973), of the electronic and structural properties of compounds of interest such as $M'_x MO_3$ where M is W, V or Mo, and M' represents a wide range of metal cations. The description 'bronzes' should strictly apply to metallically reflective, quite highly reduced, oxides, but the term is widely used in the literature for the moderately reduced non-metallic regimes also.

6.1.2 Stability and durability of oxide electrochromes

Metal-oxide electrochromes are studied for their relative photolytic stability, and ease of deposition in thin, even films over large-area electrodes (Section 6.1.3, below). However, four main disadvantages are detailed below. Firstly, the metal oxides can be somewhat unstable chemically, particularly to the presence of moisture. Secondly, while more photostable than organic electrochromes, many do evince some photoactivity. Thirdly, the metal oxides are inherently brittle. And finally, many oxides achieve only low coloration efficiencies.

Reaction with moisture and chemical degradation Most studies of ECDs suggest that chemical degradation is the principal cause of poor durability.

Thus, some workers believe that the thin-film ITO used to manufacture optically transparent electrodes (OTEs) is so moisture sensitive, particularly in its partially reduced form M_xITO, that all traces of moisture should be excluded from ITO-containing ECDs.[9,10] Similarly, the avoidance of water is sometimes advised[11] if ECDs contain either Ni(OH)$_2$ or NiO·OH. Tungsten oxide is said to be particularly prone to dissolution in water and aqueous acid,[12,13,14] particularly if the film is prepared by evaporation in vacuo;[15] see p. 150.

Photochemical stability The photochemical stability of metal oxides surpasses that of organic systems like polymers and viologens, or metallo-organic systems such as the phthalocyanines. Nevertheless, the metal oxides are not wholly photo-inert. For example, titanium dioxide is notably photoactive, particularly in its anatase allotrope, although in different applications like catalysing the photodecomposition of organic materials, such a high photoactivity is extremely desirable. Irradiating TiO$_2$ generates large numbers of positively charged holes, which are particularly reactive toward organic materials. Hence no electrochromic device should comprise thin-film TiO$_2$ in intimate contact with an organic electrolyte. Other metal oxides show photoactivity such as photochromism in a few cases. Photo-*electro*chromism is discussed in Chapter 15.

The following electrochromic oxides show photoactivity such as photochromism or photovoltaism in thin-film form: iridium (in its reduced state),[16] nickel,[17,18] molybdenum,[19,20,21,22,23] titanium,[24,25,26] and tungsten.[27,28,29,30,31,32,33,34]

Mechanical stability Like most solid-state crystalline structures, thin films of metal oxide are fragile. Bending or mechanical shock can readily cause insulating cracks and dislocations. Cracking is particularly problematic if the electrolyte layer(s) also comprise metal oxide, like Ta$_2$O$_5$. Some recent electrochromic devices have been developed in which the substrate is ITO deposited on PET or other polyester (see Section 14.3) in the fabrication of flexible ECDs, although their life expectancy is unlikely to be high because of fragility to bending.

Mechanical breakdown also occurs because the films swell and contract with the chemical changes taking place during electrochromic coloration and bleaching. Stresses arise from changes in the lattice constants, that adjust to the insertion and egress of charged counter ions, and also to the change of charge on the central metal cations. Green[35] and Ord *et al.*[36] show that WO$_3$[35] and V$_2$O$_5$[36] expand by about 6% during ion insertion. The oxide film cracks then disintegrates after repeated write–erase cycles if no accommodation or compensation is allowed for these stresses; see below.

Amongst many probes, stresses from electrochromic cycling can be sensitively monitored by the laser-deflection method: a laser beam impinges on the outer surface of the electrochrome, and analysis of the way its trajectory is deflected during redox cycling provides data that allow quantification of these mechanical stresses. In this way Scrosati and co-workers[37] found a linear dependence between the amount of charge inserted into WO_3 and the induced stress, when the inserted ions were H^+, Li^+ and Na^+. The linearity held only for small[37] amounts of inserted charge. Their correlation also suggests this induced stress is relieved in *direct* proportion to the extent of ion egress. Above certain values of x, though, new (unnamed) crystal phases were formed, particularly when the inserted ions were Li^+ or Na^+, that caused the loss of reversibility. Laser-beam deflection has been used to monitor electrochromic transitions in the oxides of iridium,[38] nickel[39] and tungsten.[30,40,41]

Alternative methods of analysing electrochromically induced stresses include electrochemical quartz-crystal microbalance (EQCM) studies, as described in Section 3.4. The stresses in oxide films of nickel,[11,42,43] titanium,[44] and tungsten[45,46,47,48] have been analysed thus.

Information on electrochemically induced stresses can also be inferred from X-ray diffraction, e.g. in oxides of nickel[49] and vanadium,[50] while those in molybdenum oxide have been studied by Raman vibrational spectroscopy.[51]

Employing an elastomeric polymer electrolyte largely accommodates the ion volume changes occurring during redox cycling: Goldner *et al.*[52] says 'nearly complete stress-change compensation' can be achieved by this method, for switching electrochromic windows. Other methods include adding small amounts of other metal oxides to the film: these minor built-in distortions introduce some mechanical 'slack' into the crystal lattices. For example, adding about 95% nickel oxide to WO_3 greatly enhances its cycle life.[53]

Chapter 16 contains an assessment of the durability of assembled electrochromic devices, and how such durability is tested.

6.1.3 The preparation of thin-film oxide electrochromes

In ECDs the metal-oxide electrochrome must be deposited on an electrode substrate as a thin, even film of sub-micron thickness, typically in the range 0.2–0.5 μm. Such thin films are either amorphous or polycrystalline, sometimes both admixed, the morphology depending strongly on the mode of film preparation. (i) Amorphous layers result from electrodeposition or thermal evaporation in vacuo. (ii) Other methods, sputtering for example, tend to form layers that are polycrystalline (microcrystalline or 'nanocrystalline').

Methods such as CVD or sol–gel generally proceed in two stages: the first-formed amorphous layer needs to be subsequently annealed ('curing,' 'sintering' or 'high-temperature heating'). Annealing assists the phase transition *amorphous* → *polycrystalline*, which greatly extends the growth of crystalline material within the amorphous.[54]

Such crystallisation is sometimes called [55] a 'history effect', thereby alluding to the extent of crystallinity, which depends largely on whether the sample was previously warmed or not. The crystallites formed can remain embedded in amorphous material, which could have serious implications for the speed of electrochromic operation; see p. 98. The number and size distribution of the crystallites depends on the temperature and duration of the annealing process.[56]

There are no reviews dedicated solely to the deposition of metal oxides, although many authors have reviewed one or more specific deposition methods: Granqvist's book[3] gives extensive detail on the preparation of metal-oxide films. Granqvist's review[57] 'Electrochromic tungsten oxide films: review of progress 1993–1998' provides further detail, as does Kullman's book, *Components of Smart Windows: Investigations of Electrochromic Films, Transparent Counter Electrodes and Sputtering Techniques*[58] (published in 1999). Finally, Venables'[59] book *Introduction to Surface and Thin Film Processes* (published in 2000) contains some useful comments about these preparations.

Deposition methods are outlined below, in alphabetical order.

Chemical vapour deposition (CVD)

In the CVD technique, a volatile precursor is introduced into the vacuum deposition chamber, and decomposes on contact with a heated substrate. Such volatiles commonly include metal hexacarbonyls or alkoxides and hexafluorides. For example, $W(CO)_6$ decomposes according to Eq. (6.1):

$$W(CO)_6\,(g) \rightarrow W\,(s) + 6CO\,(g). \tag{6.1}$$

The carbon monoxide waste byproduct is extracted by the vacuum system. The solid tungsten product is finely divided, approaching the atomic level. Annealing at high temperature in an oxidising atmosphere yields the required oxide. The films are made polycrystalline by the annealing process. Chemical vapour deposition with carbonyl precursors has provided thin oxide films of both molybdenum[60,61,62,63] and tungsten.[62,63,64,65,66,67,68,69,70,71]

An alternative precursor, a metal alkoxide such as $Ta(OC_2H_5)_5$,[72] is allowed into the deposition chamber at a low partial pressure. Decomposition occurs at the surface of a heated substrate (in this example[72] the temperature was 620 °C) to effect the reaction in Eq. (6.2):

$$2\text{Ta}(\text{OC}_2\text{H}_5)_5\,(\text{g}) + 5\text{O}_2\,(\text{g}) \to \text{Ta}_2\text{O}_5\,(\text{s}) + \text{products}\,(\text{g}). \qquad (6.2)$$

The resulting oxide film is heated for a further hour at 750 °C in an oxygen-rich atmosphere.[72] Vanadium oxide can similarly be prepared from the volatile alkoxide, $\text{VO}(\text{O}^i\text{Pr})_3$.[71]

If the CVD precursor does not decompose completely, the resultant films may contain carbon and hydrogen impurities, or other elements if different precursors are employed. The impurities either form gas-filled insulating voids in the oxide film, or their trace contamination adversely affects the electronic and optical properties of the electrochrome.

Other metallo-organic precursors have been used, e.g. Watanabe et al.[73] employed the two volatile materials tris(acetylacetonato)indium and di(pivaloylmethanato)tin to make ITO. Furthermore, precursors can be wholly inorganic, such as TaCl_5.[74]

Electrodeposition

Virtually all electrochromic films made by electrodeposition are amorphous prior to annealing.[75] Transition-metal oxides other than W or Mo are easily electrodeposited from aqueous solutions of metal nitrates, the lowest metal oxidation state usually being employed if there is a choice. Electrochemical reduction of aqueous nitrate ion generates hydroxide ion[76,77,78] according to Eq. (6.3):

$$\text{NO}_3^-\,(\text{aq}) + 7\,\text{H}_2\text{O} + 8\,\text{e}^- \to \text{NH}_4^+\,(\text{aq}) + 10\,\text{OH}^-\,(\text{aq}). \qquad (6.3)$$

The electrogenerated hydroxide ions diffusing away from the electrode associate with metal ions in solution. Subsequent precipitation then forms an insoluble layer of metal oxide as in Eq. (6.4):

$$\text{M}^{n+}\,(\text{aq}) + n\text{OH}^-\,(\text{aq}) \to [\text{M}(\text{OH})_n]\,(\text{s}), \qquad (6.4)$$

followed by dehydration during heating according to Eq. (6.5):

$$[\text{M}(\text{OH})_n]\,(\text{s}) \to \tfrac{1}{2}\,[\text{M}_2\text{O}_n]\,(\text{s}) + n/2\,\text{H}_2\text{O}. \qquad (6.5)$$

Dehydration as in Eq. (6.5) is usually incomplete, so the electrochrome comprises both oxide and hydroxide, often termed 'oxyhydroxide' and given the formulae $\text{MO}\cdot\text{OH}$ or $\text{MO}\cdot(\text{OH})_x$. Hence most electrogenerated films of 'oxide' are oxyhydroxide of indeterminate composition unless sufficient annealing followed the electrodeposition. Electrodeposition from nitrate-containing solutions has produced oxide (and oxyhydroxide) films of cobalt[79,80,81] and nickel.[77,78,79,82,83,84,85,86,87]

The mechanism of WO_3 electrodeposition is discussed at length by Meulenkamp.[75] Tungsten- or molybdenum-containing films can be electrodeposited from aqueous solutions of tungstate or molybdate ions, but good-quality oxide films are prepared from a solute obtained by oxidative dissolution of powdered metal in H_2O_2. This generates a peroxometallate species of uncertain composition, but the dissolution may proceed according to Eq. (6.6), as depicted for tungsten:

$$2W\,(s) + 6H_2O_2 \rightarrow 2H^+[(O_2)_2(O)W\text{–}O\text{–}W(O)(O_2)_2]^{2-} \text{ (aq)} \\ + H_2O + 4H_2\,(g). \tag{6.6}$$

Such peroxo species are also employed in the sol–gel deposition method described below.

The counter cations in Eq. (6.6) are either protons (as shown here), or they could be uncomplexed metal cations.[88] Excess peroxide is removed when the reactive dissolution is complete, usually by catalytic decomposition at an immersed surface coated with Pt-black. While still relatively unstable, dilution with an H_2O–EtOH mixture (volume ratio 1:1) confers greater long-term stability until used.[89] Marginal ethanol incorporation in the electroformation of WO_3[89] and NiO[90] films has been investigated.

Oxide films of cobalt,[80,81] molybdenum,[91,92] tantalum,[93] tungsten[56,89,94,95,96,97] and vanadium[98] have been made by electrodeposition from similar solutions.

It is difficult to tailor the composition of films comprising *mixtures* of metal oxide since the ratio of metals in the resultant film is not always determined by the cation ratio in the precursor solution. This divergence in composition arises from thermodynamic speciation. When the deposition solution contains more than one cation, the electrogenerated hydroxide must partition between all the metal cations in solution, each involving the consumption of hydroxide ions as governed by both the kinetics and/or equilibria associated with the formation of each particular hydroxo complex. As the mixing of the precursor cations in solution occurs on the molecular level, the final mixed-metal oxide can be homogeneous and even. The mole fractions x of each metal oxyhydroxide in the deposit can be tailored by using both pre-determined compositions and potentiostatically applied voltages V_a.[81,91,96,99] Alternatively, applying a limiting current by imposing a large electro-deposition overpotential (Section 3.3) yields a film with a composition approximating that of the deposition solution.[79,80,81,91,97,100,101,102,103,104,105] Computer-based speciation analyses have been demonstrated that describe the product distribution during the electrodeposition of such mixed-metal depositions.[105,106]

In a modification, electrochromes derived from $Ni(OH)_2$ and $Co(OH)_2$ are electrodeposited while the precursor solution is sonicated.[107] The main difference from conventional electrodeposition is the way sonication causes the formation, growth and subsequent collapse of microscopic bubbles. The bubble collapse takes place in less than 1 ns when the size is maximal, at which time the local temperature can be as high as 5000–25 000 K. After collapse, the local rate of cooling is about 10^{11} K s^{-1}, leading to crystallisation and reorganisation of the solute.[108] The reasons for the differences in the nanoproducts formed using this method are somewhat controversial; Gedanken and co-workers[108] suggest it obviates the need for particles to grow at finite rates.[107]

The method has been used to make thin films of Ni(OH)$_2$[107,109] and Co(OH)$_2$.[107,110] Córdoba de Torresi and co-workers report[107] that the method yields electrochromes with significantly higher coloration efficiencies η.

Sol–gel techniques

Regarding present terminology, 'colloid' is a general term denoting any more-or-less subdivided phase determined by its surface properties, 'sol' denotes sub-micron or nano particles visible only by the scattering of a parallel visible light beam (the so-called Tyndall effect); while 'gel' denotes linked species forming a three-dimensional network, sometimes including a second species within the minute enclosures.[111] The sol–gel method involves decomposing a precursor (one chosen from often several candidates) in a liquid, to form a sol, which, on being allowed to stand, is further spontaneously transformed into a gel.

The sol–gel method is an attractive route to preparing large-area films, as outlined in the review (2001) by Bell *et al.*[112] Many reviews of sol–gel chemistry include electrochromism: for example, Lakeman and Payne:[113] 'Sol–gel processing of electrical and magnetic ceramics' (1994); 'The hydrothermal synthesis of new oxide materials' (1995) by Whittingham *et al.*;[114] Alber and Cox,[115] (1997) 'Electrochemistry in solids prepared by sol–gel processes'; Lev *et al.*,[116] 'Sol–gel materials in electrochemistry' (1997), 'Electrochemical synthesis of metal oxides and hydroxides' (2000) by Therese and Kamath;[117] 'Electrochromic thin films prepared by sol–gel process' (2001) by Nishio and Tsuchiya;[118] 'Anti-reflection coatings made by sol–gel processes: a review' (2001) by Chen;[119] 'Sol–gel electrochromic coatings and devices: a review' (2001) by Livage and Ganguli,[120] and 'Electrochromic sol–gel coatings' by Klein (2002).[121]

As indicated by the number of literature citations, the preferred sol–gel precursors are metal alkoxides such as M(OEt)$_3$.[122] Many alkoxides react with water, so adding water to, say, Nb(OEt)$_5$ yields colloidal (sol) Nb$_2$O$_5$[123] according to Eq. (6.7):

$$2\text{Nb(OEt)}_5\,(l) + 5\text{H}_2\text{O}\,(l) \rightarrow \text{Nb}_2\text{O}_5\,(\text{sol}) + 10\text{EtOH}\,(aq), \qquad (6.7)$$

which, on standing, expands to form the gel.

6.1 Introduction to metal-oxide electrochromes

The other favoured sol–gel precursor is the peroxometallate species formed by oxidative dissolution of the respective metal in hydrogen peroxide (Eq. (6.6) above). Thus appropriate peroxo precursors have yielded electrochromic oxide films of cobalt,[80,81,99] molybdenum,[91,124,125] nickel,[126] titanium,[127,128] tungsten[99,123,127,129,130,131] and vanadium.[124,132,133] A similar peroxo species is formed by dissolving a titanium alkoxide $Ti(OBu)_4$ in H_2O_2.[128,134]

Whatever the preparative method, the gel is then applied to an electrode substrate, as below.

Spray pyrolysis The simplest method of applying a gelled sol involves spraying it onto the hot substrate, often in a relatively dilute 'suspension'.[135,136] This method, sometimes called 'spray pyrolysis', has been used to make electrochromic oxides of cerium,[137] cobalt,[138,139] nickel[140,141,142] and tungsten.[143,144,145,146] It is especially suitable for making mixtures, since the stoichiometry of the product accurately reproduces that of the precursor solution. The coated electrode is annealed at high temperature in an oxidising atmosphere, as for CVD-derived films, to give a polycrystalline electrochrome. Burning away the organic components is more problematic than for CVD since the proportion of carbon and other elements in the gel is usually higher, with concomitant increases in impurity levels.

Dip coating 'Dip coating' is comparable to spraying: the conductive substrate (inert metal; ITO on glass, etc.) is fully immersed in the gel then removed slowly to leave a thin adherent film. The process may be repeated many times when thicker layers are desired. The film is then annealed in an oxidising atmosphere. The method has produced oxide films of cerium,[147] nickel,[148,149,150,151,152] iridium,[153] iron,[154] niobium,[147,155,156,157,158,159,160,161,162,163,164,165,166,167,168,169,170,171,172,173] titanium,[174] tungsten[29,129,130,131,175,176,177,178,179,180,181] and vanadium.[182] Being particularly well suited to making mixed oxides, it has been used extensively for mixtures of precisely defined compositions such as indium tin oxide (ITO).[183]

Spin coating A further modification of dip coating is the 'spin coating' method: the solution or gel is applied to a spinning substrate, and excess is flung away by centrifugal motion. Film thickness is controlled by altering solution viscosity, temperature and spinning rate. Many oxide films have been made this way: cerium,[184] cobalt,[185] ITO,[186,187] iron,[188] molybdenum,[189,190] niobium,[191,192] tantalum,[193] titanium,[128] tungsten[129,190,194,195,196,197,198] and vanadium.[132,133,199,200] Once formed, such films are annealed in an oxidising atmosphere.

Spin coating is one of the preferred ways of forming thin-film metal-oxide *mixtures*, again producing precisely defined final compositions.[124,201,202,203,204,205]

Other methods: sputtering in vacuo

Sputtering techniques detailed below generally yield polycrystalline material[206] since the high temperatures within the deposition chamber effectively anneals the incipient film, thereby facilitating the crystallisation process *amorphous → polycrystalline*. Thin films of sputtered electrochrome are formed by three comparable techniques: dc magnetron sputtering, electron-beam sputtering and rf sputtering.

In *dc magnetron sputtering*, a target of the respective metal is bombarded by energetic ions from an ion gun aimed at it at an oblique angle. The ion of choice is Ar^+, which is both ionised and accelerated by a high potential comprising the 'magnetron'. The high-energy ions smash into the target in inelastic collisions that cause small particles of target to be dislodged by ablation. The atmosphere within the deposition chamber contains a small partial pressure of oxygen, so the ablated particles are oxidised: ablated tungsten becomes WO_3. The substrate is positioned on the far side of the target. The oxidised, ablated material impinges on it and condenses, releasing much energy. The substrate thus has to be water-cooled to prevent its melting, especially if it is ITO on glass.

Granqvist's 1995 book[3] and 2000 review[57] describe in detail how the experimental conditions, such as the partial pressures, substrate composition, sputtering energiser and impact angle, affect the properties of deposited films. As an example, Azens *et al*.[207] made films of W–Ce oxide and Ti–Ce oxide by co-sputtering from two separate targets of the respective metals. Such targets are typically 5 cm in diameter and have a purity of 99.9%. The deposition chamber contained a precisely controlled mixture of Ar and O_2, each of purity 99.998%. This sputter-gas pressure was maintained at 5–40 mTorr, the operating power varying between 100 and 250 W. The ratio of gaseous O_2 to Ar was adjusted from 1, to produce pure WO_3 and TiO_2 oxides, to 0.05 when pure Ce oxide was required. The deposition substrates were positioned 13 cm from the target. Deposition rates (from sputter time and ensuing film thicknesses as recorded by surface profilometry) were typically $0.4\,nm\,s^{-1}$.

Such reactive dc magnetron sputtering has been used to make oxide films of ITO,[208] molybdenum,[209] nickel,[210,211,212,213,214,215,216] niobium,[192,217,218] praseodymium,[219] tantalum,[220] tungsten[221,222] and vanadium.[50,223,224,225,226]

Electron-beam sputtering Here an impinging electron beam generates a vapour stream from the target for condensation on the substrate. This technique, also called 'reactive electron-beam evaporation,' has been used to prepare thin films of ITO,[227,228,229,230] MnO_2,[231] MoO_3[232] and V_2O_5.[233]

Radio-frequency (rf) sputtering Like dc sputtering, a target of the respective metal is bombarded with reactive atoms (argon or oxygen) at low pressure. The required thin film of metal oxide forms by heating the ablated material in an oxidising atmosphere. In the rf variant, the target-vaporising energy is derived from a beam of reactive atoms, generated at an rf frequency.

The rf-sputtering technique is often employed for making metal oxides, and yields good-quality films which are flat and even. No post-deposition treatment is needed, since the high temperatures within the deposition chamber yield samples that are already polycrystalline. The technique has been used to make oxide films of: iridium,[234,235] lithium cobalt oxide,[236,237,238] ITO,[239,240,241,242,243,244,245] manganese,[246,247] nickel,[86,248,249,250,251,252,253,254,255,256] tantalum,[257,258,259,260] titanium,[261] tungsten[262] and vanadium.[206,263,264,265,266,267]

Thermal deposition in vacuo The oxides of tungsten, molybdenum and vanadium are highly cohesive solids with extensive intra-lattice bonding, which require high temperatures for vaporisation when heated in vacuo. The vapour consists of molecular species (oligomers) such as the tungsten oxide trimer $(WO_3)_3$.[268] (Arnoldussen suggests that these trimers persist in the solid state.[14]) A pressure of about 10^{-5} Torr is maintained during the deposition process. Thin films of metal oxide form when the sublimed vapour condenses on a cooled substrate. In practice, a small quantity of powdered oxide is placed in an electrically heated boat, typically of sheet molybdenum. Molybdenum or tungsten oxides can be prepared thus, although small amounts of elemental molybdenum can sublime and contaminate the electrochromic film.[269]

The electrochromic properties of films deposited in vacuo are usually highly dependent on the method and conditions employed. Higher temperatures may cause slight decomposition in transit between the evaporation boat and substrate. Hence, evaporated tungsten oxide is often oxygen deficient to an extent y, in $WO_{(3-y)}$. Deb[270] suggests $y = 0.03$ but Bohnke and Bohnke[271] quote 0.3. The extent of oxygen deficiency will depend on the temperature of the evaporation boat and/or of the substrate target.[272] Nickel oxide formed by thermal deposition is generally of poor quality, since the high temperatures needed for sublimation cause loss of oxygen, resulting in sub-stoichiometric films $NiO_{(1-y)}$, where the extent of oxygen deficiency $y \ll 1$, so good-quality NiO is best made by sputtering methods. Thermal evaporation is often used to make the oxides of molybdenum,[23,273,274,275,276] tantalum,[220] tungsten[15,277,278,279] and vanadium.[233,277]

138 *Metal oxides*

Vacuum Deposition of Thin Films[280] by Holland (1956), though old, remains a valued text on thermal evaporation in the preparation of thin films.

Langmuir–Blodgett deposition

Langmuir–Blodgett methodology for preparing films of metal-oxide electrochromes was reviewed in 1994 by Goldenberg.[281] In essence, by the arcane methods of Langmuir–Blodgettry employing an appropriately constructed bath, an electrochrome precursor in a solvent is laid down on the surface of another, non-dissolving, liquid in monolayer form. This can then be drawn onto the (say metal or ITO-glass) substrate by slow immersion then emersion of the latter, suitably repeated for multi-layers. Conversion to the required oxide follows one of the routes described above.

6.1.4 Electrochemistry in electrochromic films of metal oxides

To add detail to the electrochemistry outlined in Chapter 3, electrochromic coloration of metal-oxide systems proceeds via the dual insertion of electrons (that effect redox change) and ions (that ensure the *ultimate* overall charge neutrality of the film). The dual charge injection is shown in Figure 6.1: the thin film of electrochrome concurrently accepts or loses electrons through the electrochrome–metal-electrode interface while ions enter or exit through the outer, electrochrome–electrolyte, interface. *Thus a considerable electric field is set up initially across the film before these separate charges reach their*

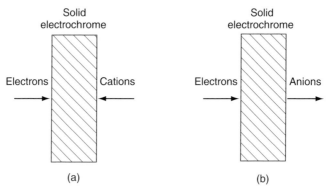

Figure 6.1 Schematic representation of 'double charge injection', depicted for a reduction reaction: (a) cations as mobile ion, and (b) anions as mobile ion. The charge carriers move in their opposite directions during oxidation. Note the way that equal amounts of ionic and electronic charge move into or out from the film in order to maintain charge neutrality within the solid layer of electrochrome, though separation can occur, causing potential gradients.

ultimate, equilibrium, distributions. An important aspect of mechanistic studies concerns whether the ionic motion or the electronic is the slower, because the outcome often decides what determines the rate of coloration (*cf.* Chapter 5). A simple but not unexceptionable surmise would impute faster electronic motion to predominant crystallinity, but ionic rapidity to predominant amorphism (for the same material).

The conductive electrode substrate can be either a metal or semiconductor; a highly-doped ITO or FTO film on glass usually acts as a transparent inert quasi-metal. The solid electrode assembly is in contact with a solution (solid or liquid) containing mobile counter ions (the ion source being termed 'electrolyte' hereafter). The mobile ion we imply to be lithium unless otherwise stated, though the proton also is often used thus. Anions are only occasionally employed as mobile ion, usually being the hydroxide ion OH^-.

While the following sections inevitably represent but an excerpt from the huge literature available, Granqvist's monograph[3] (1995) is comprehensive to that date. Tungsten trioxide is treated first in Section 6.2 because it has been investigated more fully than the other highly colourant metal-oxide electrochromes. Other oxide electrochromes are reviewed subsequently in Section 6.3. Finally, mixtures of oxide electrochromes are discussed in Section 6.4, including metal-oxide mixtures with noble metals and films of metal oxyfluoride.

For ECD usage, amorphous films are generally preferred for superior coloration efficiency η and response times. Polycrystalline films, by contrast, generally are more chemically durable. For this reason, studies have employed both amorphous and polycrystalline materials.

6.2 Metal oxides: primary electrochromes

6.2.1 Tungsten trioxide

Selected biblography

There are many reviews in the literature. The most comprehensive is: 'Case study on tungsten oxide' in Granqvist's 1995 book.[3] Also by Granqvist is: 'Electrochromic tungsten-oxide based thin films: physics, chemistry and technology',[282] (1993); 'Progress in electrochromism: tungsten oxide revisited'[283] (1999); and 'Electrochromic tungsten oxide films: review of progress 1993–1998' (2000).[57] Also useful are the reviews by Azens *et al.*:[284] 'Electrochromism of W-oxide-based thin films: recent advances'(1995); and by Monk:[285] 'Charge movement through electrochromic thin-film tungsten oxide' (1999).

Finally in this section the reader is referred to reviews by Bange:[286] 'Colouration of tungsten oxide films: a model for optically active coatings' (1999) and Faughnan and Crandall:[7] 'Electrochromic devices based on WO_3' (1980).

Morphology

The structure of WO_3 is based on a defect perovskite.[2,287,288,289,290] An XRD crystallographic study of thick and thin films from screen-printed WO_3 established that WO_3 nanopowder has two monoclinic phases of space groups $P2_1/n$ and Pc.[291] Metal dopants (see Section 6.4) such as In, Bi and Ag have different influences on the phase ratio $P2_1/n$ to Pc. Cell parameters and crystallite sizes (about 50 nm) were marginally affected by these inclusions and, in detail, depended on the dopant.

Tungsten trioxide as a thin film can be amorphous or microcrystalline, a-WO_3 or c-WO_3, or indeed a mixture of phases and crystal forms. The preparative method dictates the morphology, the amorphous form resulting from thermal evaporation in vacuo and electrodeposition, the microcrystalline from sputtering or from thermal annealing of a-WO_3. X-Ray diffraction showed Deb's[270] evaporated WO_3 to be amorphous, but WO_3 films prepared by rf sputtering are partially crystalline.[292] The spacegroup of crystalline $D_{0.52}WO_3$ is $Im3$.[287]

Annealing WO_3 results in enhanced response times,[271] caused by the increased proportion of crystalline WO_3. The temperature at which the (endothermic[12]) amorphous-to-crystalline transition occurs is $ca.$ 90 °C, as determined by thermal gravimetric analysis (TGA).[293] By contrast, for crystallinity Deepa *et al.*[56] and Bohnke and Bohnke[271] both annealed samples at 250 °C, and in the study by Deb and co-workers[294] of thermally evaporated WO_3 the crystallisation process is said to start at 390 °C and is complete at 450 °C, while Antonaia *et al.*[295] maintain that annealing commences at 400 °C.

As the physical (and optical) properties of WO_3, and its reduced forms, are highly preparation-sensitive, the apparent contradictions noted here and elsewhere in this text are almost certainly ascribable to intrinsic variability in (sometimes marked, sometimes minute) structural aspects of the solids.

Preparation of tungsten oxide electrochromes

Thermal evaporation Pure bulk tungsten trioxide is pale yellow. The colour of the WO_3 deposited depends on the preparative method, thin films sometimes showing a pale-blue aspect owing to oxygen deficiency in a sub-stoichiometric oxide $WO_{(3-y)}$, y lying between 0.03[270] and 0.3[271] (see p. 137).

The extent of oxygen deficiency depends principally on the temperature of the evaporation boat.[272] Sun and Holloway employ a modification of this method in which evaporation occurs in a relatively high partial pressure of oxygen. They call it 'oxygen backfilling',[296,297] which partly remedies the non-stoichiometry.

Chemical vapour deposition, CVD (see p. 131). The volatile carbonyl CVD precursor $W(CO)_6$ is the most widely used. Pyrolysis in a stream of gaseous oxygen generates finely divided tungsten, and then thin-film WO_3 after annealing in an oxygen-rich atmosphere.[62,64,65,66,67,68,69,70] Other organometallic precursors include tungsten(pentacarbonyl-1-methylbutylisonitrile)[298,299] and tungsten tetrakis(allyl), $W(\eta^3\text{-}C_3H_5)_4$.[300]

Sputtering (Section 6.1.4, p. 136). Many studies[221,292,301,302,303,304,305,306,307,308,309,310] involve sputtered WO_3 films which are chemically more robust than evaporated films. Pilkington plc employed rf sputtering, bombarding a tungsten target with reactive argon ions in a low-pressure oxygen to sputter WO_3 onto ITO.[27,311,312] Direct-current magnetron sputtering is less often employed.[221,222]

Electrodeposition WO_3 films electrodeposited onto ITO or Pt from a solution of the peroxotungstate anion,[56,88,89,94,95,96,97,99,198,313,314,315] (putatively $[(O_2)_2\text{-}(O)\text{-}W\text{-}(O)\text{-}(O_2)_2)]^{2-}$, formed by oxidative dissolution of powdered tungsten metal in hydrogen peroxide) sometimes appear gelatinous, and are essentially amorphous in XRD. The tungsten carboxylates represent a different class of precursor for electrodeposition, yielding products that are amorphous.[314]

Sol–gel The sol–gel technique is widely used,[46,55,99,118,123,127,129,130,131,175,176,180,181,196,197,239,316,317,318,319,320,321,322,323,324,325,326,327,328] applying the sol–gel precursor by spin coating,[129,190,194,195,196,197,198] dip-coating[29,129,130,131,175,176,177,178,179,180,181] and spray pyrolysis.[143,144,145,146] Livage *et al.*[129,176,180,329,330,331,332] often made their WO_3 films from a gel of colloidal hydrogen tungstate applied to an OTE and annealed. Other sol–gel precursors include $WOCl_4$ in *iso*-butanol,[176] ethanolic WCl_6,[197] tungsten alkoxides[333,334,335] and phosphotungstic acids.[140,336]

The sol–gel method is often deemed particularly suited to producing large-area ECDs, for example for fabricating electrochromic windows.[310] Response times of 40 s are reported,[329] together with an open-circuit memory in excess of six months.[331,337]

Redox properties of WO_3 electrochromes

On applying a reductive potential, electrons enter the WO_3 film via the conductive electrode substrate, while cationic counter charges enter concurrently through the other (electrolyte-facing) side of the WO_3 film, Eq. (6.8),

$$W^{VI}O_3 \text{(s)} + x(M^+\text{(soln)}) + xe^- \rightarrow M_x(W^{VI})_{(1-x)}(W^V)_xO_3 \text{(s)}, \quad (6.8)$$

very pale yellow   intense blue

(where M = Li usually). For convenience, we abbreviate $M_x(W^{VI})_{(1-x)}(W^V)_xO_3$ to M_xWO_3. The speed of ion insertion is slower for larger cations. Babinec,[338] studying the coloration reaction with an EQCM (see p. 88), found the insertion reaction to be complicated, depending strongly on the deposition rate employed in forming the electrochromic layer.

Cation diffusion through WO_3 has received particular study with the cations of hydrogen ions,[339,340,341,342,343] deuterium cation,[344,345,346] Li^+,[271,339,347,348] Na^+,[40,349,350,351] K^+,[352] or even Ag^+.[339,353] The overwhelming majority of these cations cannot be inserted reversibly into WO_3, as only H^+ and Li^+ can be expelled readily following electro-insertion. In a further EQCM study, the coloration usually attributable to Li^+ is suggested to result rather from proton insertion, the proton then swapping with Li^+ at longer times.[354]

Consequences of electron localisation/delocalisation The non-metal-to-metal transition in H_xWO_3 occurs at a critical composition $x_c = 0.32$, determined for an amorphous H_xWO_3 by conductimetry[355] (the precise value cited no doubt applies exactly only to that type of product). Below x_c, the bronze is a mixed-valence species[356] in the Robin–Day[348] Group II (involving moderate electron delocalisation of the 'extra' W^V electron acquired by injection, that conducts by the sitewise hopping mechanism, or 'polaron hopping'). H_xWO_3 with $x > x_c$ is metallic with completely delocalised transferable electrons (the Robin and Day[347] Group IIIB). It is this unbound electron plasma in metallic WO_3 bronzes that confers reflectivity, as in Drude-type delocalisation,[302,340,357,358,359,360] an essentially free-electron model (but dismissed by Schirmer et al.[361,362] for amorphous WO_3). Dickens et al. analysed the reflectance spectra of Na_xWO_3 in terms of modified Drude–Zener theory that includes lattice interactions.[363]

Kinetic dependences on x The rates of charge transport in electrochromic WO_3 films are reviewed by Monk[285] and Goldner,[364] and salient details from Chapter 5 are reiterated here.

6.2 Metal oxides: primary electrochromes

Considerable evidence now suggests that the value of the insertion coefficient x influences the rates of electrochromic coloration, because the electronic conductivity[362] σ follows x. At very low x, the charge mobility μ of the inserted electron is low,[362] hence rate-limiting, owing to the minimal delocalisation of conduction electrons which conduct by polaron-hopping. The electronic conductivity of evaporated WO_3, subsequently reduced, has been determined as a function of x.[362,365,366] Figure 5.4 shows H_xWO_3 to be effectively an insulator at $x=0$, but σ increases rapidly until at about $x \approx 0.3$ the electronic conductivity becomes metallic following the delocalisation at this and higher x values.

Most properties of the proton tungsten bronzes H_xWO_3 depend on the insertion coefficient x, such as the *emf*,[367] the reflectance spectra,[363] and the dielectric-[368] and ferroelectric properties.[369] (It is notable that the alignment of spins in the ferroelectric states differs in proton-containing bronzes compared with that in Na_xWO_3, owing to the occupation of different crystallographic sites by the minute protons.[35])

The ellipsometric studies by Ord and co-workers[370,371] of thin-film WO_3 (grown anodically) show little optical hysteresis associated with coloration, provided the reductive current is only applied for a limited duration: films then return to their original thicknesses and refractive indices. Colour cycles of longer duration, however, reach a point at which further coloration is accompanied by film dissolution (*cf.* comments in Section 1.4 and above, concerning cycle lives). The optical data for WO_3 grown anodically on W metal best fit a model in which the colouring process takes place by a progressive change throughout the film, rather than by the movement of a clear interface that separates coloured and uncoloured regions of the material. The former therefore represents a diffuse interface between regions of the film, the latter a 'colour front'. Furthermore, Ord *et al.* conclude that a 'substantial' fraction of the H^+ inserted during coloration cycles is still retained within the film when bleaching is complete.[371]

The different mechanisms of colouring and bleaching discussed in Chapter 5 may be sufficient to explain the significant extent of optical hysteresis observed.[7,372] Figure 6.2 demonstrates such hysteresis for coloration and bleaching.

Structural changes occurring during redox cycling In Whittingham's 1988 review 'The formation of tungsten bronzes and their electrochromic properties'[373] the structures and thermodynamics of phases formed during the electro-reduction of WO_3 are discussed. Other studies of structure changes during redox change are cited in references 37 and 374–376. The effects of structural change are discussed in greater depth in Section 5.2 on p. 86.

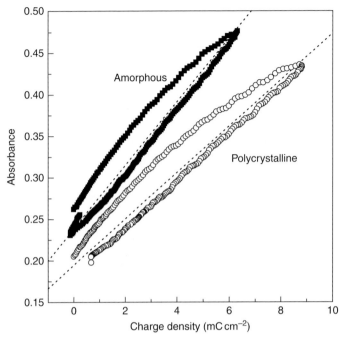

Figure 6.2 Optical density vs. intercalated charge density obtained for polycrystalline and amorphous WO₃ films during dynamic coloration and bleaching. (Figure reproduced from: Scarminio, J., Urbano, A. and Gardes, B. 'The Beer–Lambert law for electrochromic tungsten oxide thin films'. *Mater. Chem. Phys.*, **61**, 1999, 143–146, by permission of Elsevier Science.)

Some authors, such as Kitao *et al.*,[377] say that when the mobile ion in Eq. (6.8) is the proton, it forms a hydrogen bond with bridging oxygen atoms. However, the X-ray and neutron study by Wiseman and Dickens[287] of $D_{0.53}WO_3$ shows the O–D and O–D–O distances are almost certainly too large for hydrogen bonding to occur. Similarly, Georg *et al.*[378] suggest the proton resides at the centres of the hexagons created by WO_6 octahedra. Whatever its position, X-ray results[379] suggest that extensive write–erase (on–off) electrochromic cycling generates non-bridging oxygen, i.e. causes fragmentation of the lattice structure.

Optical properties of tungsten oxide electrochromes

Optical effects: absorption The intense blue colour of reduced films gives a UV-visible spectrum exhibiting a broad, structureless band peaking in the near infra-red. Figure 6.3 shows this (electronic) spectrum of H_xWO_3. In transmission, the electrochromic transition is effectively colourless-to-blue at low x (≤ 0.2). At higher values of x, insertion irreversibly forms a reflecting, metallic (now properly named) 'bronze', red or golden in colour.

6.2 Metal oxides: primary electrochromes

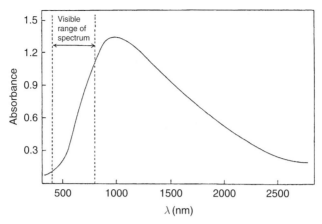

Figure 6.3 UV-visible spectrum of thin-film $H_{0.17}WO_3$ deposited by sputtering on ITO. The visible region of the spectrum is indicated. (Figure reproduced from Baucke, F. G. K., Bange, K. and Gambke, T. 'Reflecting electrochromic devices'. *Displays*, **9**, 1988, 179–187, by permission of Elsevier Science.)

The origin of the blue colour of low-x tungsten oxides is contentious. The absorption is often attributed to an F-centre-like phenomenon, localised at oxygen vacancies within the WO_3 sub-lattice.[270] Elsewhere the blue colour is attributed to the electrochemical extraction of oxygen, forming the coloured sub-stoichiometric product $WO_{(3-y)}$.[272,380] Faughnan et al.[381] and Krasnov et al.[382] proposed that injected electrons are predominantly localised on W^V ions, the electron localisation and the accompanying lattice distortion around the W^V being treated as a bound small polaron.[270,276,364,382,383,384,385] The colour was attributed[381] to the intervalence transition $W_A^V + W_B^{VI} \rightarrow W_A^{VI} + W_B^V$ (subscripts A and B being just site labels). While this is now widely accepted, among critics Pifer and Sichel,[386] studying the ESR spectrum of H_xWO_3 at low x, could find no evidence for unpaired electrons on the W^V sites. Could the ground-state electrons form paired rather than single spins, at adjacent loosely interacting W^V sites?[384,385]

Provisionally we assign the blue colour to an intervalence charge-transfer transition. While the wavelength maximum λ_{max} of a particular H_xWO_3 is essentially independent of the insertion coefficient x, the value of λ_{max} does vary considerably with the preparative method (see p. 146): λ_{max} depends crucially on morphology and occluded impurities such as water, electrolyte, and also the extent and nature of the electronic surface states (i.e. vacant electronic orbitals on the surface). Thus the value of λ_{max} shifts from 900 nm in amorphous and hydrated reduced films of H_xWO_3,[5,361,387,388,389] to longer wavelengths in polycrystalline[389] materials, where λ_{max} can reach 1300 nm for average grain sizes of 250 Å.[339,361] Intervalence optical transitions are known to be neighbour-sensitive.

As outlined in Chapter 4, a graph of absorbance *Abs* against the charge density consumed in forming a bronze M_xWO_3 is akin to a Beer's-law plot of absorbance versus concentration, since each electron acquired generates a colour centre. The gradient of such a graph is the coloration efficiency η (see Equations (4.5) and (4.6)). Most authors[5,355] believe the colour of the bronze is independent of the cation used during reduction, be it $M = H^+$, Li^+, Na^+, K^+, Cs^+, Ag^+ or Mg^{2+} (here $M = \frac{1}{2}Mg^{2+}$). However, Dini et al.[349] state that the coloration efficiency η *does* depend on the counter ion, and, for $\lambda = 700$ nm, give values of $\eta(H_xWO_3) = 63$, $\eta(Li_xWO_3) = 36$ and $\eta(Na_xWO_3) = 27 \text{ cm}^2 \text{ C}^{-1}$.

While sputtered films are more robust chemically than evaporated films, their electrochromic colour formed per unit charge density is generally weaker, i.e. η is smaller, although one sputtered film[390] had a contrast ratio *CR* of 1000:1, which is high enough to implicate reflection effects (as below, possibly even specular reflection). The higher absorbances of evaporated samples arise because the W species will be on average closer within (amorphous) grain boundaries, as discussed in ref. 285. Close proximities increase the probability of the optical intervalence transition in the electron-excitation colour-forming process, Eq. (6.8). This could explain why films sputtered from a target of W metal show different Beer's-law behaviour from sputtered films made from targets of WO_3.[312]

The role of defects, and their influence on electrochromic properties, turns out to be far from clear, but the amorphous material (of course) contains a very high proportion of (what are from a crystal viewpoint) defects. The forms of defect in polycrystalline and amorphous WO_3 influence the optical spectra of WO_3 and its coloured reduction products.[391] Chadwick and co-workers[392] analysed the interdependence of defects and electronic structure, using WO_3 as a case study. They show how structural defects exert a strong influence upon electronic structure and hence on chemical properties. For example, while little is known about how the chemical activity at the interface is affected by interaction of liquid, their results suggest that any liquid suppresses water dissociation at the surface and the formation of OH_3^+ structures near to it.

As expected, the overall absorbance *Abs* of any particular WO_3 film always increases as the insertion coefficient x increases, although *Abs* is never a simple function of the electrochemical charge Q passed over all (especially high) values of x. Beer's law is therefore not followed except over limited ranges of Q and hence of x; see Figure 5.12.[393]

Probably reflecting the preparation-dependence of film properties, there are considerable discrepancies in such graphs. At one extreme, the coloration efficiency for Li$^+$ insertion is asserted to be essentially independent of x, so a Beer's-law plot is linear until x is quite large.[394] Contrarily, for H$^+$ or Na$^+$, the gradient of a Beer's-law plot is claimed to decrease with increasing x, i.e. for coloration efficiency η decreasing as x increases. The non-linearity in such Beer's-law graphs seems not to be due to competing electrochemical side-reactions[5] but is, rather, attributed to either a decrease in the oscillator strength per electron,[393,a] or a broadening of the envelope of the absorption band owing to differing neighbour-interactions.

In the middle ground, workers such as Batchelor et al.,[311] who used sputtered WO$_3$ to form Li$_x$WO$_3$, found only two distinct regions, ε in the range $0 < x < 0.2$ being higher than when $x > 0.2$. At the other extreme, other workers suggest that Beer's-law plots for thin-film WO$_3$ are only linear for small x values $(0 < x \leq 0.03)$[5,381] or $(0 < x \leq 0.04)$.[393] This result applies both for the insertion of protons[5,381,393] and sodium ions[394] in evaporated (amorphous) WO$_3$ films. Beer's-law plots are linear to larger x values from data for the insertion of Li$^+$ into evaporated therefore amorphous WO$_3$. Such graphs have a smaller gradient, so η is smaller.[387]

The most intense coloration per electron (that is, the highest values of η) is seen when x is very small (<0.04).[393] The higher intensities follow since, at low x, the electron is localised within a very deep potential well described as a W^V polaron or, possibly, as a spin-paired (diamagnetic) W^V–W^V dimeric 'bipolaron', located at defect sites.[395] Only at higher values of x, as the extent of electronic delocalisation increases, will conduction bands start to form as polaron distortions extend and coalesce (as mentioned under *Kinetic dependences on x* on p. 142). The existence of polarons may explain the finding that oxygen deficiency improves the coloration efficiency.[396]

Duffy and co-workers[393] conducted extensive studies of such Beer's-law graphs on a range of H$_x$WO$_3$ films made by immersing evaporated (hence amorphous) WO$_3$ on ITO in dilute acid. Beer's-law plots showed four linear regions, each with a different apparent extinction coefficient ε. Structural changes accompanying electro-reduction were inferred, that resulted in stepwise alteration of oscillator strength or optical bandwidth. These accord somewhat with views of Tritthart et al.[397] who proposed three definite types of colour centre in H$_x$WO$_3$.

---

[a] The oscillator strength f_{ij} is defined by IUPAC as a measure for integrated intensity of electronic transitions and related to the Einstein transition probability coefficient A_{ij}:

$$f_{ij} = 1.4992 \times 10^{-14} (A_{ij}/s^{-1})(\lambda/nm)^2,$$

where λ is the transition wavelength.

Table 6.2. *Sample values of coloration efficiency η for WO_3 electrochromes.*

Preparative route	Morphology	η/cm^2 C^{-1} ($\lambda_{(obs)}$ in nm)	Ref.
Electrodeposition	Amorphous	118 (633)	400
Thermal evaporation	Amorphous	115 (633)	206
Thermal evaporation	Amorphous	115 (633)	401
Thermal evaporation	Amorphous	79 (800)	206
rf sputtering	Polycrystalline	21	307
Sputtering	Polycrystalline	42 (650)	401
Dip coating	Amorphous	52	402
Sol–gel[a]	PAA composite	38	403
Sol–gel	Crystalline	70 (685)	404
Sol–gel	Crystalline	167 (800)	405
Sol–gel	Crystalline	36 (630)	406
Spin-coated gel	Crystalline	64 (650)	197
Effect of counter cation – all samples prepared by thermal evaporation			
H$_x$WO$_3$	Amorphous	63 (700)	349
Li$_x$WO$_3$	Amorphous	36 (700)	349
Na$_x$WO$_3$	Amorphous	27 (700)	349

PAA = poly(acrylic acid); [a] alternate layers of PAA and WO$_3$.

A wholly different behaviour is exhibited by films of polycrystalline WO$_3$, prepared, e.g., by rf sputtering or by high-temperature annealing of amorphous WO$_3$. At low x, the Beer's-law plot is linear (but of low gradient) but η *increases* with an increase in x,[387,398] possibly due to specular reflection, clearly not a wholly absorptive phenomenon.

For thin films of WO$_3$ prepared by CVD,[62,66,67,68,69] Beer's-law plots are said to be linear for H$^+$ or Li$^+$ only when the insertion coefficient x is low. Coloration efficiency η decreases at higher x, but the x value at the onset of curvature was not reported.

Table 6.2 cites some coloration efficiencies η. Other Beer's-law plots appear in refs. 393 and 399. The wide variations in η are no doubt caused in part by monitoring the optical absorbance at different wavelengths, but also result from morphological and other differences arising from the preparative methods.

Optical effects by reflection As recorded in Table 6.3, the colour of crystalline M$_x$WO$_3$, when viewed by reflected light, shows a colour that depends on x, where x is proportional to charge injected.[363,373,407] For x values at and beyond the insulator/metal transition – i.e. those exceeding *ca.* $x = 0.2$ or 0.3

Table 6.3. *Colours of light reflected from tungsten oxides of varying insertion extents of reduction x.*

x	Colour
0.1	Grey
0.2–0.4	Blue
0.6	Purple
0.7	Brick red
0.8–1.0	Golden bronze

To repeat: $x > 0.3$ prevents electrochromic reversal.

depending on preparation – the reflections become ever more metallic in origin. In consequence, crystalline WO_3 is both optically absorbent and also partially reflective. Amorphous M_xWO_3 does not show the same clear changes in reflected colour, probably because its insulator–metal transition is much less distinct.

Devices containing tungsten trioxide electrochrome

Much device-led research into solid-state ECDs concentrates on the tungsten trioxide electrochrome in, for example, 'smart windows',[408,409] alphanumeric watch-display characters,[410] electrochromic mirrors[393,411,412,413,414,415,416,417,418,419] and display devices.[387,420,421,422,423,424,425,426] When the second electrochrome is a metal oxide, the WO_3 will be the primary electrochrome owing to the greater intensity of its optical absorption. Electrochromic devices of WO_3 have been fabricated with the oxides of iridium,[427] nickel,[428,429,430,431,432] niobium[433] and vanadium (as pentoxide)[242,277,434,435] as the secondary electrochrome. Thin-film WO_3 has also been used in ECDs in conjunction with the hexacyanoferrates of indium[436,437] or iron (i.e. Prussian blue),[438,439,440,441,442] and the organic polymers poly(aniline),[443,444,445,446,447,448,449,450,451,452,453] the thiophene-based polymer PEDOT[454] and poly(pyrrole).[455,456,457]

A response time of 40 s is reported for a WO_3 film prepared by a sol–gel technique,[329] together with an open-circuit memory in excess of six months.[331,337]

Following Deb's 1969 electrochromic experiments on solid WO_3 (p. 29) significant progress ensued in 1975 when Faughnan *et al.*[381] published the construction of a device with WO_3 in contact with liquid electrolyte (see Chapter 2). This ECD worked well at short times, but failed rapidly owing to film dissolution in the H_2SO_4 solution employed. The effect of steadily drying the electrolyte has been studied often.[7,12,13,14,15,354,458,459,460] To summarise, the rate and extent of film dissolution decreases as the water content decreases, but the rate of coloration also decreases.

Reichman and Bard[461] showed, for the electrochromic processes of WO_3 on samples prepared by either anodic oxidation of tungsten metal or by vacuum evaporation onto ITO, that the electrochromic response time τ was faster with the anodically grown material because it is microscopically porous. Furthermore, the value of τ was an incremental function of the water content and film porosity, both properties unfortunately producing films susceptible to dissolution, which is accelerated by aqueous Cl^-.[460] WO_3 films, in aqueous sulfuric acid as ECD electrolyte,[b] form crystalline hydrates such as $WO_3 \cdot m(SO_4) \cdot n(H_2O)$ which decrease the electrochromic efficiency considerably.[463]

Film dissolution can be prevented by two means; the use of non-aqueous acidic solutions, for example, anhydrous perchloric acid in DMSO (dimethyl sulfoxide),[464] or, rather than the use of acid, a non-protonic (alkali-metal) cation, usually lithium, is employed as insertion ion. Examples include films of WO_3 immersed in lithium-containing electrolytes such as $LiClO_4$, lithium triflate ($LiCF_3CH_2CO_2$), or occasionally $LiAlF_6$ or $LiAsF_5$, in dry propylene carbonate. Alternatively, WO_3 ECDs have been constructed which incorporate solid inorganic electrolytes such as Ta_2O_5, or organic polymers such as poly(acrylic acid), poly(AMPS) or poly(ethylene oxide) – PEO, each containing a suitable ionic electrolyte; see Section 14.2 for further detail. Such cells have slower response times and also a poorer open-circuit memory, although Tell[465,466] has made such a solid-state ECD from phosphotungstic acid, claiming a τ of 10 ms (but for an unspecified change in absorbance). Such liquid-free devices are preferred for their chemical and mechanical robustness.

Tungsten trioxide in aqueous acidic electrolytes is more durable if the electrochrome–electrolyte interface is protected with a very thin over-layer of Nafion^{TM},[467] Ta_2O_5,[468] or tungsten oxyfluoride,[469] although charge transport through such layers will be slower. Other layers used to protect WO_3 are described on p. 446.

Other over-layers can speed up the electrochromic response. For example, a layer of gold enhances the response time τ and also protects against chemical degradation.[470,471] Clearly, the layer needs to be ion-permeable, hence very thin or porous.

In solid-state WO_3 devices, the stability of the electrochromic colour is generally good, despite some loss of absorbance with time. This 'self bleaching' or 'spontaneous hydrogen deintercalation',[472] has been studied often:[15,295,473,474] in one study, CVD-prepared WO_3 returned to its initial transparency

---

[b] The reaction of acid with WO_3 prepared by anodising W metal is found to be kinetically first order with respect to acid,[460] and zeroth order with respect to film thickness.[462]

after only three minutes.[475] Deb and co-workers[474] have also investigated the chemistry underlying the self bleaching of evaporated WO_3 on ITO, suggesting that adsorbed water in the films reacts with the coloured Li_xWO_3 to form LiOH and molecular hydrogen.

6.2.2 Molybdenum oxide

Preparation of molybdenum oxide electrochromes

Molybdenum trioxide films may be formed with amorphous or polycrystalline morphologies. Amorphous material can be formed by vacuum evaporation of solid, powdered MoO_3,[21,23,273,274,275,476,477] by anodic oxidation of molybdenum metal immersed in e.g. acetic acid,[478] or deposited electrochemically; a widely used precursor is prepared by oxidative dissolution of molybdenum metal in hydrogen peroxide solution.[91,92,313,479]

Sputtering yields polycrystalline material. The product of dc magnetron sputtering is of good quality and colourless.[209] In rf sputtering, however, over-rapid rates of deposition can yield oxygen-deficient material, which is blue,[20,209,480] and clearly different from the desired 'bronze', M_xMoO_3, being in fact substoichiometric[23,275,481,482] with composition $Mo^{VI}_c Mo^{V}_{(1-c)} O_{(3-c/2)}$, where c can be as high as 0.3. Granqvist and co-workers[209] show that substoichiometric blue 'MoO_3' forms at deposition rates up to $1.5\,\mathrm{nm\,s^{-1}}$, whereas clear MoO_3 requires a deposition rate of about 0.85 and $0.1\,\mathrm{nm\,s^{-1}}$ for films made with dual-target and single-target sputtering, respectively. (Dual-target sputtering is twenty times faster than single-target deposition.[480]) Nevertheless, the electrochromic properties, particularly in bleaching, of sub-stoichiometric films improve after about five colour/bleach cycles in a $LiClO_4$/PC electrolyte.[209] Gorenstein and co-workers found[481] that blue sputtered 'MoO_3' forms particularly at *low* fluxes of ionised Ar^+, which could be a result of differing conditions such as the sputtering geometries.

In rf sputtering a target of metallic molybdenum and low-pressure $Ar + O_2$[20,483] are employed. Controlling the flow and composition of the atmosphere dictates the composition and structure of the final electrochrome.[484] The flow rate and hence the exact composition have a profound effect on the optical properties of the film.[20] The best films were made with a low rate of oxygen flow that gave a sub-stoichiometric oxide, although the relationship(s) between optical, electrochemical and mechanical properties and flow rate are complex.[20,481,485]

Chemical vapour deposition also yields polycrystalline material from an initial deposit of usually finely divided metal. This needs to be roasted in an oxidising atmosphere, that causes amorphous material to crystallise. Chemical vapour deposition precursors include gaseous molybdenum hexacarbonyl[62] or organometallics like the pentacarbonyl-1-methylbutylisonitrile compound.[486]

Molybdenum trioxide films derived from sol–gel precursors are also polycrystalline as a consequence of high-temperature annealing after deposition. The most common precursor is a spin-coated gel of peroxopolymolybdate[189,487] resulting from oxidative dissolution of metallic molybdenum in hydrogen peroxide. Such films are claimed to show a superior memory effect to sputtered films of MoO_3.[488] Other sol–gel precursors include alkoxide species such as[190] $MoO(OEt)_4$.

Films have also been made by spray pyrolysis, spraying aqueous lithium molybdate at low pH onto ITO, itself deposited on a copper substrate[489] by electron-beam evaporation.[232] Thermal oxidation of thin-film MoS_3 also yields electrochromic MoO_3.[490] Finally, solid phosphomolybdic acid is also found to be electrochromic.[466]

Redox chemistry of molybdenum oxide electrochromes

The electrochromism of molybdenum oxide is similar to that of WO_3, above, so little detail will be given here. There is a considerable literature on the electrochemistry of thin-film MoO_3, but smaller than for WO_3.

As with WO_3, annealing amorphous MoO_3 causes crystallisation. The electrochromic behaviour of the films depend on the extent of crystallinity, and therefore on the annealing. McEvoy et al.[491] suggest that electrodeposited films of MoO_3 on ITO are completely amorphous if not heated beyond about 100 °C. Films heated to 250 °C comprise a disordered mixture of orthorhombic α-MoO_3 and monoclinic β-MoO_3 phases, giving voltammetry which is 'complicated'. Crystallisation to form the thermodynamically stable α phase occurs at temperatures above 350 °C.

The dark-blue coloured form of the electrochrome is generated by simultaneous electron and proton injection into the MoO_3, in the electrochromic reaction Eq. (6.9):

$$Mo^{VI}O_3 + x(H^+ + e^-) \rightarrow H_xMo^{V,VI}O_3. \quad (6.9)$$
colourless $\qquad\qquad\qquad$ intense blue

Whittingham[492] considers H^+ mobility in layered $H_xMoO_3(H_2O)_n$ but many workers prefer to insert lithium ions Li^+, from anhydrous solutions of salts such as $LiAlF_6$ or $LiClO_4$ in PC,[20,51,209,480,488] while Sian and Reddy preferred Mg^{2+} as the mobile counter cation.[275,477]

Equation (6.9) is over-simplified because Mo^{IV} appears in the XPS of the coloured bronze, as well as the expected valence states of Mo^V and Mo^{VI}.[19,275]

6.2 Metal oxides: primary electrochromes

Some oxygen deficiency can complicate spectroscopic analyses:[275] evaporated MoO$_3$ films, colourless when deposited, nevertheless give an ESR signal characteristic of Mo^V at[23] $g = 1.924$.

Molybdenum bronzes H$_x$MoO$_3$ show an improved open-circuit memory compared with the tungsten bronzes H$_x$WO$_3$, since H$_x$MoO$_3$ films oxidise more slowly than do films of H$_x$WO$_3$ having the same value of x.[5] Also, protons enter the molybdenum films at potentials more cathodic than $+0.4$ V (against the SHE), leaving a coloration range of about 0.4 V prior to formation of molecular hydrogen; the gas possibly forms catalytically on the surface of the bronze, as in Eq. (6.10):

$$2H^+ (aq) + 2e^- \rightarrow H_2 (g). \quad (6.10)$$

The corresponding range for H$_x$WO$_3$ is larger, about 0.5 V.[5] Additionally beneficial, the chemical diffusion coefficients $\bar{D}$ of H$^+$ through MoO$_3$ are faster than through the otherwise similar WO$_3$, implying faster electrochromic operation.[5]

Ord and DeSmet[478,493] interpret their ellipsometric study of the proton injection into MoO$_3$ as showing two distinct insertion sites for the mobile hydrogen ion within the reduced film. There is a readily observed, well-defined boundary between the oxidised and reduced regions within the oxide, perhaps in contrast to WO$_3$, implying a somewhat different mechanism for electro-reduction. The XRD study by Crouch-Baker and Dickens[494] suggests that hydrogen insertion proceeds without the occurrence of major structural rearrangement in the bulk of the oxide film.

The electrochromism of molybdenum oxide is enhanced when coated with a thin, 20 nm, transparent film of Au or Pt,[495] presumably because the precious metal helps minimise the effects of *IR* drop caused by the poor electronic conductivity across the surface of the MoO$_3$. Coating the MoO$_3$ with precious metal also decreases the extent of oxide corrosion,[495] perhaps similarly to protecting WO$_3$ with a thin film of gold[470] or tungsten oxyfluoride.[288,496]

Optical properties of molybdenum oxide electrochromes

An XPS study[476] shows that the colour in the reduced state of the film arises from an intervalence transition between Mo^V and Mo^{VI} in the partially reduced oxide, *cf.* WO$_3$.

In appearance, the optical absorption spectrum of H$_x$MoO$_3$ is very similar to that of H$_x$WO$_3$ (e.g. see Figure 6.4) except that the wavelength maximum of H$_x$MoO$_3$ falls at shorter wavelengths than does λ_{max} for H$_x$WO$_3$. The wavelength maximum of the partly reduced oxide is centred at[23] 770 nm. This band

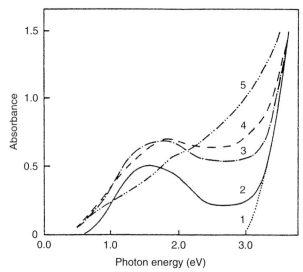

Figure 6.4 UV-visible spectrum of thin-film molybdenum oxide for various amounts of inserted charge: (1) 0; (2) 490; (3) 1600; (4) 2200 and (5) 3800 mC cm^{-3}. (Figure reproduced from Hiruta, Y., Kitao, M. and Yamada, M. 'Absorption bands of electrochemically-coloured films of WO$_3$, MoO$_3$ and Mo$_c$W$_{1-c}$O$_3$.' *Jpn. J. Appl. Phys.*, **23**, 1984, 1624–7, with permission of The Institute of Pure and Applied Physics.)

is clearly not of simple origin,[23] but comprises a collection of discrete bands having maxima at around 500 nm, 625 nm, and 770 nm. The absorption edge of MoO$_3$ occurs at[476] 385 nm, but shifts to ≈390 nm for the coloured reduced film.[476] The 'apparent coloration efficiency' for partly reduced molybdenum oxide is therefore slightly greater than for partly reduced tungsten trioxide since the absorption envelope coincides more closely with the visible region of the spectrum. The optical constants n and k of thermally annealed MoO$_3$ (i.e. amorphous MoO$_3$ that was formed by thermal evaporation but then roasted) depend quite strongly on the annealing temperature.[275]

Unlike H$_x$WO$_3$, the value of λ_{max} for H$_x$MoO$_3$ is *not* independent of x,[292] but moves to shorter wavelengths as x increases; see Figure 6.5.

Table 6.4 contains a few representative values of coloration efficiency η.

Devices containing molybdenum oxide electrochromes

Devices containing MoO$_3$ are comparatively rare. For example, Kuwabara *et al.*[497,498] made several cells of the form WO$_3$| tin phosphate |H$_x$MoO$_3$. The solid electrolyte layer is opaque: otherwise, no discernible change in absorbance would occur during device operation. The response times of ECDs may be enhanced by depositing an ultra-thin layer of platinum or gold on the

6.2 Metal oxides: primary electrochromes

Table 6.4. *Sample values of coloration efficiency η for molybdenum oxide electrochromes.*

Preparative route	$\eta/\mathrm{cm}^2\,\mathrm{C}^{-1}$ ($\lambda_{(\mathrm{obs})}$/nm)	Ref.
Thermal evaporation of MoO_3	77	7
Evaporation of Mo metal in vacuo	19.5 (700)	482
Oxidation of thin-film MoS_3	35 (634)	490

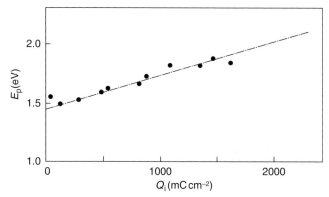

Figure 6.5 Plot of $\mathcal{E}$ (as $\mathcal{E} = h\nu$, where ν is the frequency maximum of the intervalence band) for the reduced oxides H_xMoO_3 as a function of the electrochemical charge inserted, Q_i, which is proportional to the hydrogen content, x. (Figure redrawn from Fig. 4 of Hurita, Y., Kitao, M. and Yamada, W. 'Absorption bands of electrochemically coloured films of WO_3, MoO_3 and $Mo_cW_{(1-c)}O_3$.' *Jpn. J. Appl. Phys.*, **23**, 1984, 1624–162, by permission of The Japanese Physics Society.)

electrolyte-facing side of the electrochrome.[495] As with WO_3, clearly the layer of precious metal must be permeable to ions.

6.2.3 Iridium oxide

Preparation of iridium oxide electrochromes

There are now two commonly employed methods of film preparation: firstly, electrochemical deposition to form an 'anodic iridium oxide film' ('AIROF' in a jargon abbreviation). The second major class are 'sputtered iridium oxide films' ('SIROFs').

The anodically grown films[464,499,500,501,502,503,504,505,506,507,508,509] are made by the potentiostatic cycling between -0.25 V and $+1.25$ V (against SCE) of an iridium electrode immersed in a suitable aqueous solution. Such AIROFs are largely amorphous.[510] They have a *CR* as high as 70:1 which forms within $\tau = 20$ to 40 ms;[504] such response times are considerably faster than for WO_3 or V_2O_5 films of similar thickness and morphology. Anodic iridium oxide films degrade badly under intense illumination,[16] sometimes a serious disadvantage.

Anodic iridium oxide films can also be generated by immersing a suitable electrode (e.g. ITO) into an aqueous solution of iridium trichloride.[499,511,512] The solution must also contain hydrogen peroxide and oxalic acid. (Following the usual desire for acronyms, such films are now designated as 'AEIROFs' i.e. anodically electrodeposited iridium oxide films.) Once formed and dried, the electrochromic activity of an AEIROF increases as the proportion of water in the electrolyte increases. Conversely, if annealed, the electrochromic activity *decreases* as the anneal temperature increases.

The second method of forming films is reactive sputtering in an oxygen–argon atmosphere (the respective partial pressures being 1:4).[506] Hydrogen may also be added.[513] Such films are grey–blue in the coloured state with $\lambda_{max} = 610$ nm. A denser SIROF, that forms a black electrochromic colour, can be made with oxygen alone as the flow-gas during the sputtering process. Sputtered iridium oxide films have a complicated structure which, unlike AIROFs, is not macroscopically porous, i.e. decreased response times are observed since ionic insertion is slowed. These black SIROFs are deposited as coloured films which can be decolorised by up to 85% on cycling, while blue SIROFs give superior films which may be transformed to a truly colourless state. In fact, blue SIROFs are very similar to AIROFs in being totally decolorisable. Furthermore, in terms of write–erase response times and absorbance spectra, blue SIROFs and AIROFs are again similar, cyclic voltammetry confirming the similarity.[506] Blue SIROFs have superior response times to black SIROFs, and a longer open-circuit memory. Beni and Shay[506] view the blue SIROFs as aesthetically the more pleasing. The reliability of AIROFs is apparently variable.[511]

Extremely porous films of iridium oxide can be prepared by thermally oxidising vacuum-deposited iridium–carbon composites.[514]

The electrochromically generated colour of a SIROF is only moderately stable, and decreases by about 8% per day.[509]

Sol–gel methods also yield polycrystalline iridium oxide, and start from a sol formed from iridium trichloride solution in an ethanol–acetic acid mixture,[118,153,239] and iridium oxide films have been prepared by sputtering metallic iridium onto an OTE in an oxygen atmosphere.[505]

Finally, electrochromic films are formed on ITO when γ-rays irradiate solutions of iridium chloride in ethanol.[118,515]

The redox chemistry of iridium oxide electrochromes

In aqueous solution, the mechanism of coloration is still uncertain,[234] so two different reactions are current. The first is described in terms of proton loss,[502,503] Eq. (6.11):

$$Ir(OH)_3 \rightarrow IrO_2 \cdot H_2O + H^+(soln) + e^-, \qquad (6.11)$$
colourless   blue–grey

which is confirmed by probe-beam deflection methods.[516] The second involves anion insertion,[507] Eq. (6.12):

$$Ir(OH)_3\,(s) + OH^-(soln) \rightarrow IrO_2 \cdot H_2O\,(s) + H_2O + e^-. \qquad (6.12)$$

While XPS measurements[500] seem to confirm Eq. (6.11), AIROFs do not colour in anhydrous acid solutions, e.g. $HClO_4$ in anhydrous DMSO,[464] so the reaction (6.11) probably applies only to aqueous electrolytes. While protons are ejected from AIROFs during oxidation,[516] their electrochromic behaviour is independent of the pH of the electrolyte solution,[501] suggesting that both protons and hydroxide ions are involved in the electrochromic process. Equation (6.12) is not without question: some workers[507] assert that AIROFs will colour when oxidised while immersed in solutions containing the counter ions of[507] F^- or CN^-; others disagree.[517] Regardless of whether the mechanism is hydroxide insertion or proton extraction, $Ir(OH)_3$ is the bleached form of the oxide and the coloured form is IrO_2.

Ellipsometric data[518] suggest little hysteresis during redox cycling, the optical constants during reduction retracing the path followed during oxidation. Unlike other metal oxides, neither the coloration nor bleaching reactions proceed by movement of an interface between oxidised and reduced material traversing a 'duplex' film. Nor does redox conversion proceed with a single-stage conversion of a homogeneous film. In fact, the optical and electrochemical data both suggest that conversion occurs in two distinct stages: Rice[519] suggests that a satisfactory model requires the recognition that AIROFs act as a conductor of both electrons and anions during the electrochromic reaction, which helps explain the relatively low faradaic efficiency in dilute acid.[520] The participation of the electrons, and the sudden change in electrochromic rate, may correlate with the occurrence of a non-metal-to-metal transition between 0 and 0.12 V (vs. SCE).[521] Phase changes in iridium oxide are discussed by Hackwood and Beni.[522]

Gutierrez et al.[523] have investigated AIROFs using potential-modulated reflectance tentatively to assign the peaks in the cyclic voltammetry of anodic films of iridium oxide to the various redox processes occurring.

Optical properties of iridium oxide electrochromes

Figure 6.6 shows an absorbance spectrum of thin-film iridium oxide sputtered onto quartz.[524] The change in transmittance of crystalline Ir_2O_3 films made by sol–gel techniques is larger than that of the amorphous Ir_2O_3 under the same experimental conditions.[118,153]

There are relatively few coloration efficiencies η in the literature: η for an oxide film made by thermal oxidation of an iridium–carbon composite[525,526] is quite low at $-(15 \text{ to } 20) \text{ cm}^2 \text{ C}^{-1}$ at a λ_{max} of 633 nm. The AIEROF film[511] is characterised by η of $-22 \text{ cm}^2 \text{ C}^{-1}$ at 400 nm, $-38 \text{ cm}^2 \text{ C}^{-1}$ at 500 nm and $-65.5 \text{ cm}^2 \text{ C}^{-1}$ at 600 nm; η for spray-deposited oxide depends strongly on the annealing temperature,[512] varying from $-10 \text{ cm}^2 \text{ C}^{-1}$ at 630 nm for films annealed at 400 K to $-26 \text{ cm}^2 \text{ C}^{-1}$ for films annealed at 250 K.

Optical study of the electrochromic transition of AIROFs is greatly complicated by anion adsorption at the electrochrome–solution interphase.[527]

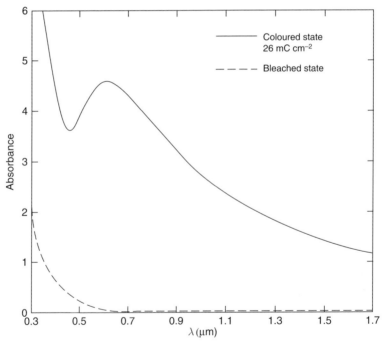

Figure 6.6 UV-visible spectrum of thin-film iridium oxide sputtered onto quartz. The broken line is the reduced (uncoloured) form of the film and the continuous line is the spectrum following oxidative electro-coloration with 26 mC cm^{-2}. (Figure reproduced from Kang, K. S. and Shay, J. L. 'Blue sputtered iridium oxide films (blue SIROF's)'. *J. Electrochem. Soc.*, **130**, 1983, 766–769, by permission of The Electrochemical Society, Inc.).

6.2 Metal oxides: primary electrochromes

Electrochromic devices containing iridium oxide electrochromes

Thin-film iridium oxide was one of the first metal-oxide electrochromes to be investigated for ECD use. Electrochromic cells containing iridium oxide generate colour rapidly: the cell SnO$_2$|AIROF|fluoride|Au develops colour in 0.1 second (where 'fluoride' represents PbF$_2$ on PbSnF$_4$).[528]

Another ECD was prepared with two iridium oxide films in different oxidation states, 'ox-AIROF' being one oxide film in its oxidised form while 'red-AIROF' is the second film in its reduced form.[508] The cell fabricated was 'ox-AIROF|Nafion®|red-AIROF', the Nafion® containing an opaque whitener against which the coloration was observed; otherwise, the electrochromic colour of the two AIROF layers would change in a complementary sense, with the overall result of almost negligible modulation. When a voltage of 1.5 V was applied across the cell, the maximum colour formed in about 1 second.[509] Clearly, the device can only operate when initially one iridium layer is oxidised and the other reduced. This cell is described in detail in ref. 508. Solid-state AIROFs have been made with polymer electrolyte, but these have slower response times.[509] Ishihara[529] used iridium oxide in a solid-state device in which reduced chromium oxyhydroxide was the source of protons migrating into the electrochrome layer.

Anodic iridium oxide films are superior to WO$_3$-based electrochromes since they do not degrade in water but retain a high cycle life (of about 10^5) even in solutions of low pH,[507] provided the temperature remains low:[507] the bleached form of iridium oxide decomposes thermally above about 100 °C.[530]

A composite device based on iridium oxide and poly(*p*-phenylene terephthalate) on ITO shows different electrochromic colours: blue–green when oxidised, but colourless when reduced.[531] The reaction at the counter electrode is unidentified.

Other ECDs have been made with sputtered IrO$_x$ as the secondary electrochrome and WO$_3$ as the primary layer. On fabrication, one layer contains ionic charge; both layers colour in a complemetary sense as charge is decanted from one electrochrome layer to the other.[427,532]

6.2.4 Nickel oxide

Much of the nickel oxide prepared in thin-film form is oxygen deficient. The extent of deficiency varies according to the choice of preparative route and deposition parameters. For this reason, 'nickel oxide' is often written as NiO$_x$ or NiO$_y$ where the symbols x or y indicate oxygen non-stoichiometry. We prefer an alternative notation and denote oxygen non-stoichiometry by $NiO_{(1+y)}\,H_z$ when hydroxyl is a ligand, otherwise for hydroxyl free species, by $NiO_{(1+y)}$.

Preparation of nickel oxide electrochromes

There is a large literature on making thick films of nickel oxide owing to its use in secondary batteries.[247,533]

One of the principal difficulties in making thin-film nickel oxide is its thermal instability: heating an oxide film can cause degradation or outright decomposition. The thermal stability of thin-film nickel oxide is the subject of several investigations: by Cerc Korošec and co-workers[148,534,535] on electrochromes made via sol–gel methods; by Jiang et al.[251] studying the effects of annealing rf-sputtered $NiO_{(1-y)}$; by Natarajan et al.[536] probing the stability of electrodeposited samples; and by Kamal et al.,[141] examining samples made by spray pyrolysis.

Thin films of nickel oxide electrochromes are usually made by sputtering in vacuo, by the dc-magnetron[211,212,213,214,215,216,537] or rf-beam techniques.[86,248,249,250,251,252,253,254,255,256,538,539] The target is usually a block of solid nickel oxide,[211,214,215,540] but a nickel target and a relatively high partial pressure of oxygen is also common.[211,214,215,253,254,255,256] Rutherford backscattering[c] suggests that rf-sputtered NiO is rich in oxygen, i.e. nickel oxide of composition $NiO_{(1+y)}$.[256] Excess oxygen at grain boundaries enhances the extent of electrochromic colour.[539] A target of solid $LiNiO_2$ generates a pre-lithiated film.[252,541] Addition of gaseous hydrogen to the sputtering chamber has profound effects on the optical properties of the resultant films.[542]

Other films of $NiO_{(1-y)}$ are reported via electron-beam sputtering,[543,544] or pulsed laser ablation,[545,546,547,548,549] e.g. from a target of compacted $LiNiO_2$ powder.[546,548] A cathodic-arc technique also yields $NiO_{(1-y)}$ if metallic nickel is sputtered in vacuo in an oxidising atmosphere.[550]

Thermal vacuum evaporation seems a poor way of making $NiO_{(1+y)}$ films since the electrochrome readily decomposes in vacuo to yield a material with little oxygen. Nevertheless, this technique is reported to generate $NiO_{(1+y)}$ films satisfactorily.[544,551,552]

Electrodeposition of thin-film nickel oxide is more widely used, e.g. from solutions of aqueous nickel nitrate.[82,212,553] Equations (6.3) and (6.4) describe

---

[c] In the backscattering experiment, alpha particles typically possessing energies of several MeV are fired at a thin sample. The majority of alpha particles remain embedded in the sample, but a small proportion scatter from the atomic nuclei in the near surface (1 to 2 μm) of the sample. The energy with which they backscatter relates to the mass of the target element. For heavy target atoms such as tungsten, the backscattered energy is high – almost as high as the incident energy, but for lighter target atoms such as oxygen, the backscattered energy is low. Analysis of the backscattering pattern enables Rutherford backscattering (RBS) to measure the stoichiometry of thin films.

the reactions that form the immediate oxyhydroxide product NiO(OH)$_z$, which can be dehydrated according to Eq. (6.5) by annealing. Other aqueous electrodeposition solutions include an alkaline nickel–urea complex,[554] nickel diammine[554,555,556,557] nickel diacetate,[558] [Ni(NH$_3$)$_2$]$^{2+}$ or nickel sulfate,[17,536,559,560] albeit by an unknown deposition mechanism. Electrodeposition from a part-colloidal slurry has also been achieved.[561]

Fewer sol–gel films of nickel oxide electrochrome have been made, in part because the necessary annealing can damage the films. Electrochromic films have been made via sol–gels derived from NiSO$_4$ with formamide and PVA,[152] or nickel diacetate dimethylaminoethanol, although the resulting solid film is not durable.[149] Precursors of nickel bis(2-ethylhexanoate)[562] or NiCl$_2$ in butanol and ethylene glycol[151] have been employed in spin coating prior to thermal treatment to effect dehydration and crystallisation.

Dip-coating has also been used: electrodes are immersed repeatedly into a nickel-containing solution, like buffered NiF$_2$,[563] NiSO$_4$ in water[564] or polyvinyl alcohol,[152] or NiCl$_2$ in butanol and ethylene glycol.[151] Again, sol–gels have been made by adding LiOH drop-wise to NiSO$_4$ solution until quite alkaline,[148,534,535] then peptising (i.e making colloidal) the resulting green precipitate with glacial acetic acid. Such precursors are often termed a 'xerogel', although the sols are not completely desiccated.^{d} Additional water is added to ensure an appropriate viscosity prior to dipping. Conversely, an (uncharged) conducting electrode may be dipped alternately in solutions of aqueous NiSO$_4$ and either NaOH[563] or NH$_4$OH.[560] In all cases, the precursor film on the electrode is heated to effect dehydration, chemical oxidation and crystallisation.

Electrochromes are also reported[141,565,566,567] to have been made by spray pyrolysis, e.g. from a precursor of aqueous nickel chloride solution.[567] Chemical vapour deposition is not a popular route to forming NiO$_{(1+y)}$, perhaps again owing to the need for annealing. Precursors include nickel acetylacetonate.[568] Finally, NiO films have been made by plasma oxidation of Ni–C composite films, previously deposited by co-evaporation of Ni and C from two different sources.[569]

Redox electrochemistry

'Hydrated nickel oxide' (also called nickel 'hydroxide') is an anodically colouring electrochrome, the redox now differing in direction from that with the

---

^{d} A xerogel is defined by IUPAC as, 'the dried out open structures which have passed a gel stage during preparation (e.g. silica gel).'

preceding metals. In acidic media, the electrode reaction for nickel oxide follows Eq. (6.13):

$$\text{Ni}^{II}\text{O}_{(1-y)}\text{H}_z \rightarrow [\text{Ni}^{II}_{(1-x)}\text{Ni}^{III}_x]\text{O}_{(1-y)}\text{H}_{(z-x)} + x(\text{H}^+ + \text{e}^-) \cdot \quad (6.13)$$

colourless   brown–black

Nakaoka *et al.*[17] believe the coloured form is blue.

Equation (6.13) is an amended form of the reaction in ref. 570. Furthermore, the sub-stoichiometric 'NiO$_{(1-y)}$H$_z$' is, in reality, Ni$^{II}_{(1-x)}$Ni^{III}O$_{(1+y)}$H$_z$. The values of y and z in Eq. (6.13) are unknown and likely to depend on the pH of the electrolyte solution. Proton egress from rf-sputtered NiO$_{(1-y)}$H$_z$ is more difficult than entry to the oxide.[571]

The mechanism is different in alkaline solution: Murphy and Hutchins[572] cite the simplified reaction in Eq. (6.14),

$$\text{Ni(OH)}_2(\text{s}) + \text{OH}^-(\text{aq}) \rightarrow \text{NiO} \cdot \text{OH}(\text{s}) + \text{e}^- + \text{H}_2\text{O}. \quad (6.14)$$

Granqvist and Svensson believe that ^{15}N nuclear reaction analysis (see page 110) shows that coloration is accompanied by proton extraction.[253] Furthermore, Murphy and Hutchins[572] suggest that the following nickel species: Ni$_3$O$_4$, Ni$_2$O$_4$, Ni$_2$O$_3$ and NiO$_2$ are all involved. In this analysis, the bleached state is Ni$_3$O$_4$ and the coloured form is Ni$_2$O$_3$. Additionally, anodic coloration occurs in two distinct stages.[572] Chigane *et al.*[555] cite the involvement of: α-Ni(OH)$_2$, γ$_2$-2NiO$_2$–NiO·OH, β-Ni(OH)$_2$ and β-NiO·OH; Bouessay *et al.*[573] suggest that conversion of NiO into Ni(OH)$_2$ is a major cause of device degradation. The complex structures and phase changes occurring during the redox cycling of 'nickel oxide' were reviewed by Oliva *et al.*[574] in 1992.

The problem of mass balance in thin-film 'nickel oxide' has been described in great detail by Bange and co-workers,[575] Córdoba-Torresi *et al.*,[11] Giron and Lampert,[39] Lampert,[576] Gorenstein and co-workers[577] and Granqvist and co-workers.[216,253] Svensson and Granqvist[253] conclude that the bleached state in a nickel oxide based display is β-NiO·OH, and the coloured state is β-Ni(OH)$_2$. Conell *et al.* concur in this assignment.[86] They also suggest that only a minority of the film participates in the electrochromic reaction. Furthermore, the reduced form of the oxide contains a small amount of Ni^{III}: a startling result. Some workers have detected Ni^{IV} in the oxidised form of electrochromic NiO$_{(1-y)}$ films.[85,572]

Gorenstein, Scrosati and co-workers[578] suggest the electronic conductivities of the coloured and bleached states (which are said to differ dramatically) play a major role in the electrochromic process, although the rate of *ion* movement dictates the overall kinetic behaviour of nickel oxide based films. The kinetic

6.2 Metal oxides: primary electrochromes

behaviour is described further by MacArthur[579] and by Arviá and co-workers[580] The mechanism is, not unusually, quite sensitive to the method of film preparation. As Granqvist et al.[216] say, incontrovertibly,

Electrochromic nickel-oxide based films produced by different types of sputtering, evaporation, anodic oxidation and cathodic deposition, [and] thermal conversion all can have different optical, electrochemical and durability related properties, and therefore be more or less well suited for technical applications.

Water trapped preferentially at defect and grain boundaries (which are numerous in $NiO_{(1+y)}$[250]) plays a crucial role in the electrochromic reaction. Water is formed as a product of $NiO_{(1+y)}H_z$ degradation, the amount of water in the solid film increasing with cycle life. Its role is not beneficial, though, for it promotes chemical degradation. The efficiency of this electrochromic oxide, as prepared by rf sputtering, has been analysed in terms of microstructure, morphology and stoichiometry by Gorenstein and co-workers;[581] and Cordóba-Torresi et al.[11] in support say that the presence of lattice defects is a prerequisite for electrochromic activity. Furthermore, they believe that neither $Ni(OH)_2$ nor $NiO \cdot OH$ are beneficial to device operation because of their solubility in water.[11]

The tendency for water to cause deterioration is such that many workers now avoid water and hydroxide ions altogether, and prefer non-aqueous electrolytes. The reaction cited for electrochromic activity is then Eq. (6.15):

$$NiO_{(1+y)} + x(Li^+ + e^-) \rightarrow Li_xNiO_{(1+y)}, \quad (6.15)$$

brown–black　　　　　　　　colourless

the mobile Li^+ ion most commonly coming from $LiClO_4$ dissolved in a polymeric electrolyte.[49] Even at quite low potentials, the rate of electrochromic coloration and bleaching is dictated by the rates of ionic movement.[578] Detailed measurements with the electrochemical EQCM suggest cation swapping, e.g. H^+ being the first counter cation to enter the lattice, with subsequent insertion of Li^+.[43]

Optical properties of nickel oxide electrochromes

The electrochromic colour in $NiO_{(1+y)}$ undoubtedly derives from an Ni^{III}/Ni^{II} intervalence transition. Figure 6.7 shows absorption spectra of nickel oxide.[18]

There are wide variations reported in the values of coloration efficiency η. For example, although η is said to be $-36 \, cm^2 \, C^{-1}$ at 640 nm for nickel oxide made by rf sputtering,[542] the value depends strongly on the sputtering

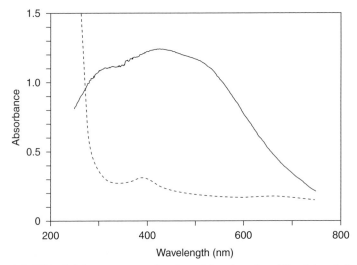

Figure 6.7 UV-visible spectrum of reduced (· · · ·) and oxidised (——) forms of thin-film nickel oxide on ITO. The film was electrodeposited onto ITO with a thickness of about 1 μm. Electro-coloration was performed with the film immersed in 0.1 mol dm^{-3} KOH solution. (Figure reproduced from Carpenter, M. K. and Corrigan, D. A. 'Photoelectrochemistry of nickel hydroxide thin films'. *J. Electrochem. Soc.*, **136**, 1989, 1022–6, by permission of The Electrochemical Society, Inc.)

conditions. This value of η was cited for a film obtained at a total pressure of 8 Pa, of which gaseous hydrogen accounted for 40%. Other values of η are cited in Table 6.5; a value of $-10\,\text{cm}^2\,\text{C}^{-1}$ is cited for thin-film lithium nickel oxide deposited by rf sputtering from a stoichiometric LiNiO$_2$ target.[252]

Electrochromic devices containing nickel oxide electrochromes

Films made by rf sputtering are significantly more durable than those made by electrodeposition: Conell cites 2500 and 500 write–erase cycles for the respective preparations.[86] Xu et al. suggest that 10^5 cycles are possible for dc magnetron sputtered samples.[211] Corrigan[82] reports that the durability can be improved to thousands of cycles by incorporating cobalt or lanthanum, but nevertheless, Ushio et al.[540] show that such sputtered NiO$_x$ degrades relatively easily.

Coloration/bleaching times of electrodeposited films range between 20 and 40 s, and depend on the applied potential.[584]

The speed of electrochromic operation often depends on so-called 'terminal effects' that arise because optically transparent conductive layers such as ITO have only modest electronic conductivities. Depositing an ultra-thin layer of metallic nickel between the ITO and NiO layers significantly improves the response time τ.[585]

Table 6.5. *Sample values of coloration efficiency η for nickel oxide electrochromes.*

Preparative route	$\eta/\text{cm}^2\,\text{C}^{-1}$ (λ/nm)	Ref.
CVD (from a nickel acetylacetonate precursor)	−44	568
dc sputtering	−25 to 41	537
Dipping technique	−35	564
Electrodeposition	−20	78
Electrodeposition	≈ −50 (450)	582
Electrodeposition	−24 (670)	560
rf sputtering	−36	542
Sol–gel (NiSO$_4$, PVA and formamide)	−35 to 40 (450)	152
Sol–gel (NiSO$_4$, glycerol, PVA and formamide)	−23.5 (450)	583
Sonicated solution	−80.3 (457)	107
Spray pyrolysis	−37	565
Spray pyrolysis	−30	566
Vacuum evaporation	−32 (670)	551

At present, much of the interest in nickel oxide electrochromes is focussed on their use as secondary electrochrome (i.e. not the main colourant) on the *counter* electrode, i.e. as redox reagent on the second electrode in an ECD cell where a primary electrochrome is redox reagent on the other electrode. Primary electrochromes so partnered could be WO$_3$,[42,49,149,248,254,310,431,544,548,586,587,588] or poly(pyrrole),[589] poly(thiophene)[590] or poly(methylthiophene).[590] However, in some prototype ECDs, NiO was the *primary* electrochrome on the one electrode while on the other, CuO,[591] MnO[592] or SnO$_2$[560] acted as secondary electrochrome.

6.3 Metal oxides: secondary electrochromes

6.3.1 Introduction

As outlined in Section 1.1, while for the usual two-electrode ECD it would generally be advantageous that both electrodes bear strongly colourant electrochromes, final conditions may dictate that one electrode provides the major colourant (hence, bears the *primary electrochrome*). The counter electrode would bear a feebly colouring *secondary electrochrome*, or even a non-colouring (*passive*) redox couple, either of the latter being chosen simply for superior electrochemical properties, stability and durability. This chapter covers the latter classes of 'electrochrome'. ("Electrochrome" here is not a misnomer because as has been established in Section 1.1, even invisible – IR and/or UV – changes, that attend all redox reactions, are nowadays being deemed 'electrochromic'.)

Secondary electrochromes

Bismuth oxide

An electrochromic bismuth oxide formed by sputtering or vacuum evaporation was studied by Shimanoe et al.[593] The best electrochromic performance was observed for a sputtered oxide annealed at 300–400 °C in air for 30 min. Films showed an electrochromic transition when immersed in LiClO$_4$–propylene carbonate electrolyte, Eq. (6.16):

$$Bi_2O_3 + x(Li^+ + e^-) \rightarrow Li_xBi_2O_3. \qquad (6.16)$$

transparent   dark brown

Bleaching ocurred at +1.2 V and coloration at −2.0 V vs. SCE. The response time either way was about 10 s, with coloration efficiency η of 3.7 cm^2 C^{-1}.

Bismuth oxide has also been co-deposited with other oxides.[594]

Cerium oxide

Preparation of cerium oxide Thin-film CeO$_2$ can be prepared by spray pyrolysis via spraying aqueous cerium chloride (CeCl$_3 \cdot$ 7H$_2$O) onto ITO.[137] Films prepared at temperatures below about 300 °C were amorphous, while those prepared at higher temperatures have a cubic ('cerianite') crystal structure.

Özer et al.[184,595] made cerium oxide films on fluoride-doped SnO$_2$ electrodes using a sol–gel procedure. The precursor derived from cerium ammonium nitrate in ethanol, with diethanolamine as a complexing agent. They recommend annealing at 450 °C or higher. Spectroelectrochemistry showed that these films were optically passive, and therefore ideal as counter electrodes in transmissive ECDs.

Porqueras et al.[596] deposited the oxide by electron-beam PVD (physical vapour deposition) on various substrates, such as glass, ITO-coated glass, Si wafers and fused silica. The substrate temperature was maintained at 125 °C. In contrast, ion-bombarded films show a denser structure and a different layer growth.[597]

The utility of cerium oxide derives from its near optical passivity. The redox reaction follows Eq. (6.17):

$$CeO_2 + x(Li^+ + e^-) \rightarrow Li_xCeO_2. \qquad (6.17)$$

Both redox states are essentially colourless in the visible region. Porqueras claims that films on ITO remain 'fully transparent after' Li^+ insertion and egress.[137] Cerium oxide is therefore not electrochromic, but is a widely-used choice of counter electrode material.[137,184,595,596,597] It is also widely used as a matrix in which other, electrochromic, oxides are dispersed. These mixed-metal oxide electrochromes are described in Section 6.4.

Chromium oxide

The electrochromism of chromium oxide has received little attention. The properties of a sputtered oxide are described as 'only slightly inferior to those of Ni oxide and with good stability in acidic electrolytes'.[598] The composition of the material is nowhere mentioned; the sputtered materials made by Cogan et al.[599] are said to be similar, and are called 'lithium chromate'.

In a fundamental study, Azens, Granqvist and co-workers[598] immersed films made by rf sputtering in aqueous H_3PO_4. The electrochromic colour did not vary by more than 10% during redox cycling, making it almost optically 'passive'.

Alternatively, thin films of chromium oxide, identified only as 'CrO_y', can be formed by electron-beam evaporation of Cr_2O_3.[231] The electrochromic operation was studied with films immersed in γ-butyrolactone containing $LiClO_4$.

Chromium oxide has been studied extensively for battery applications,[600,601,602] with the redox reaction Eq. (6.18):

$$Cr_2O_3 \text{ (s)} + x(Li^+ + e^-) \rightarrow Li_xCr_2O_3 \text{ (s)}. \qquad (6.18)$$

Chromium oxide allows device operation with a lower voltage than do most other electrochromic oxides.[603]

The only coloration efficiency available is that for vacuum-evaporated material, for which η is $-4 \text{ cm}^2 \text{ C}^{-1}$.[231]

Cobalt oxide

Preparation of cobalt oxide electrochromes Thin-film $LiCoO_2$ is made by rf sputtering from a target of $LiCoO_2$, and is polycrystalline. Because the as-deposited films are lithium deficient,[236,237,238] such nominal '$LiCoO_2$' shows significant absorption at $\lambda < 600$ nm; Goldner et al.[238] state that films can be coloured electrochemically, but will not decolour completely. Controlling the amount of lithium within films of rf-sputtered lithium cobalt oxide is, however, difficult.[238]

Other vacuum methods such as CVD generate thin films of metallic cobalt as initial layer, which is converted to CoO by being annealed in an oxidising atmosphere. Chemical vapour deposition precursors include Co(acetylacetonate)$_2$.[604]

Electrochemical studies of anodically generated layers of oxide on metallic cobalt,[605,606] for example, of pure cobalt metal anodised in a solution of aqueous 1 molar NaOH or a solution buffered to pH 7, show the films to be blue,[605] but the colour soon changes to brown on standing,[607] owing to atmospheric oxidation.

Electrodeposited oxyhydroxide, $CoO \cdot OH$,[84,608] may be electrodeposited on Pt or ITO from an aqueous solution of $Co(NO_3)_2$ via Eqs. (6.3) and (6.4). Subsequent thermal annealing converts most of the oxyhydroxide to oxide CoO, but some $CoO \cdot OH$ persists.[103,608] For this reason, such 'cobalt oxide' is sometimes written as CoO_x or, better, $CoO_{(1+y)}$. This $CoO_{(1+y)}$ has a pale green colour owing to a slight stoichiometric excess of oxide ion, causing a weak charge-transfer transition from O^{2-} to the Co^{2+} ion.[609] Gorenstein et al.[608] suggest the as-grown film may be $Co(OH)_3$, unlikely in our view owing to the strongly oxidising nature of Co^{III}.

As with W and Mo, Co metal can be dissolved oxidatively in H_2O_2,[80,81] to form the peroxo anion for use in sol–gel or electrodeposition procedures. Cobalt oxide can also be deposited from a Co^{II}–(tartrate) complex via $Co^{II}(OH)_2$ in aqueous sodium carbonate.[100]

Thin-film cobalt oxide can be made by spray pyrolysis in oxygen of aqueous $CoCl_2$ solutions[139] onto e.g. fluorine-doped tin oxide (FTO) coatings on glass substrates. These films change electrochemically from grey to pale yellow, with a response time of 2 to 4 s. Alternatively, sols of Co_3O_4 have been applied to an electrode substrate by both dipping and spraying.[610]

Redox chemistry of cobalt oxide electrochromes Equation (6.19) is the supposed electrochromic reaction of cobalt oxide grown anodically in aqueous electrolytes on cobalt metal:[605,606,607]

$$3Co^{II}O\,(s) + 2OH^-\,(soln.) \rightarrow Co_3^{II,III}O_4\,(s) + 2e^- + H_2O. \qquad (6.19)$$
pale yellow \qquad\qquad\qquad\qquad dark brown

The Co_3O_4 product would formally be $Co^{II}O + Co_2^{III}O_3$ (*cf.* magnetite, the iron equivalent). The colour of the brown form is probably due to a mixed-valence charge-transfer transition in the Co_3O_4, although the identity of the Co^{III}

6.3 *Metal oxides: secondary electrochromes* 169

oxide(s) formed by oxidation of Co(OH)$_2$ could not be assigned conclusively by FTIR.[611] Reference 611 cites IR data for all the known oxides of cobalt including those above, together with CoO and CoO·OH.

In non-aqueous solutions, e.g. LiClO$_4$ in propylene carbonate, oxidation of sputtered LiCoO$_2$ electrochrome results in an electrochromic colour change from effectively transparent to dark brown. The electrochromic reaction is Eq. (6.20):

$$\text{LiCoO}_2 + x(\text{M}^+ + \text{e}^-) \rightarrow \text{M}_x\text{LiCoO}_2, \qquad (6.20)$$
$$\text{pale yellow–brown} \qquad\qquad \text{dark brown}$$

where M$^+$ is generally Li$^+$, when the rate-limiting process during coloration and bleaching is the movement of the Li$^+$ counter ion.[612] The study by Pyun *et al.*[613] clearly demonstrates the complexity of the charge-transfer process(es) across the oxide–electrolyte interphase.

For the novel green product formed by reductive electrolysis of nitrate ion, the electrochromic transition is green → brown, in the electrochromic reaction[80,81] in Eq. (6.21):

$$3\text{CoO} + 2\text{OH}^- \rightarrow \text{Co}_3\text{O}_4 + 2\text{e}^- + \text{H}_2\text{O}. \qquad (6.21)$$
$$\text{pale green} \qquad\qquad \text{brown}$$

Optical properties of cobalt oxide electrochromes Figure 6.8 shows UV-visible spectra of electrodeposited CoO (pale green) and Co$_3$O$_4$ (dark brown),[81] and Figure 6.9 shows a coloration-efficiency plot of absorbance against charge passed[614] Q. This figure demonstrates how absorbance is generally not proportional to Q, since the graph is only linear for addition of small-to-medium amounts of inserted charge.

Table 6.6 cites representative values of coloration coefficient η.

Behl and Toni[618] find that many electrochromic colours may be achieved in films generated on metallic cobalt, presumably from varying oxide–hydroxide compositions, accompanied by composition-dependent CT or intervalence absorptions. Colours include white, pink, brown and black, confirming Benson *et al.*'s views.[619] Below 1.47 V vs. SCE, films are orange (or yellow–brown) but above this potential the films become dark brown (or even black if films are thick). The orange form of the oxide may also contain hydrated Co(OH)$_2$ following H$_2$O uptake; on Co metal anodised in NaOH (0.1 or 1.0 mol dm^{-3}) this oxide is predominantly the low-valence product, as demonstrated by FTIR.[611]

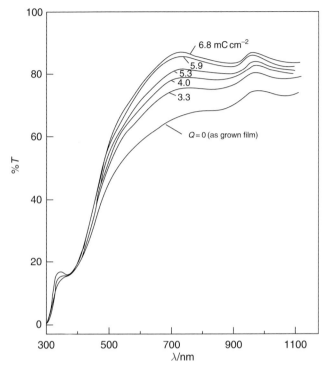

Figure 6.8 UV-visible spectra of thin-film cobalt oxide electrodeposited onto ITO. The figures above each trace represent the charge passed in mC cm^{-2}, beginning with the most coloured state at the bottom of the figure, and progressively bleaching. (Figure reproduced from Polo da Fontescu, C. N., De Paoli, M.-A. and Gorenstein, A. 'The electrochromic effect in cobalt oxide thin films'. *Adv. Mater.*, **3**, 1991, 553–5, with permission of Wiley–VCH.)

Films made by spray pyrolysis from CoCl$_2$ solution exhibited anodic electrochromism, changing colour from grey to pale yellow.[139]

Electrochromic devices containing cobalt oxide electrochromes Cobalt oxide is usually employed as a secondary electrochrome (on the counter electrode) against a more strongly colouring primary electrochrome on the major colourant electrode comprising e.g. WO$_3$.[248]

Copper oxide

Preparation of copper oxide electrochromes Özer and Tepehan[620,621] prepared a copper oxide electrochrome from sol–gel precursors, hydrolysing copper ethoxide, then annealing in an oxidising atmosphere. Ray[622] prepared a different sol–gel precursor via copper chloride in methanol, yielding films of either CuO or Cu$_2$O, the product depending on the annealing conditions.

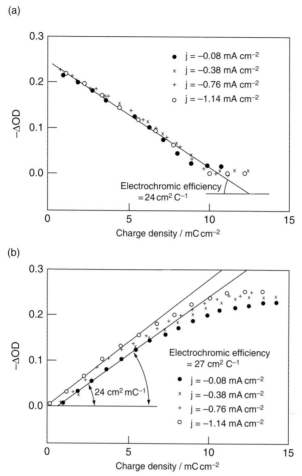

Figure 6.9 Coloration-efficiency plot of absorbance ($-\Delta OD$) against charge passed Q for thin-film CoO electrodeposited onto ITO, and immersed in NaOH solution (0.1 mol dm^{-3}): (a) during coloration and (b) during bleaching. The current density i during coloration was ● = 0.08 mA cm^{-2}; x = 0.38 mA cm^{-2}; + = 0.76 mA cm^{-2}; o = 1.14 mA cm^{-2}. The wavelength at which Abs was determined is not known. (Figure reproduced from Polo da Fontescu, C. N., De Paoli, M.-A. and Gorenstein, A. 'The electrochromic effect in cobalt oxide thin films'. *Adv. Mater.*, **3**, 1991, 553–5, with permission of Wiley–VCH.)

Richardson *et al.*[591,623] have made transparent films of Cu_2O on conductive SnO_2:F (FTO) substrates by anodic oxidation of sputtered copper films, or by electrodeposition.

The electrochromic transition is colourless to pale brown but, apparently, neither redox state has yet been identified. Özer and Tepehan[620] call their

Table 6.6. *Sample values of coloration efficiency η for cobalt oxide electrochromes*

Preparative route	$\eta/\text{cm}^2\,\text{C}^{-1}$ ($>\lambda_{(obs)}/\text{nm}$)	Ref.
CVD[a]	21.5	604
Electrodeposited	24	614, 615
Sol–gel	25	616
Sonicated solution	130	107
Spray pyrolysis	12 (633 nm)[b]	139
Thermal evaporation	20–27	617

[a] The precursor was Co(acetylacetonate)$_2$. [b] Figure in parenthesis is λ_{max}.

electrochrome 'Cu_wO'. The response time and optical properties of this electrochrome depend markedly on the temperature and duration of post-deposition annealing.

Redox chemistry of copper oxide electrochromes The Cu_2O films transform reversibly to black CuO at more anodic potentials.[622] In alkaline solution, a suggested redox reaction is Eq. (6.22):

$$2\text{CuO}\,(s) + 2e^- + 2\text{H}_2\text{O} \rightarrow \text{Cu}_2\text{O}\,(s) + 2\text{OH}^-. \tag{6.22}$$

    black   red-brown

In acidic electrolytes,[622] Cu_2O is transformed reversibly to opaque and highly reflective copper metal, according to Eq. (6.23):

$$\text{Cu}_2\text{O}\,(s) + 2e^- + 2\text{H}^+ \rightarrow 2\text{Cu}\,(s) + \text{H}_2\text{O}. \tag{6.23}$$

The cycle life of such electrochromic materials is said to be poor at *ca.* 20–100 cycles[591] owing to the large increase in molar volume of about 65% during conversion from Cu to Cu^{II}, Eq. (6.23). A large change in optical transmittance is claimed, from 85 to 10% transmittance. The coloration efficiency is about 32 cm^2 C^{-1}.[591] However, the usefulness in ECDs is virtually zero unless display applications are found, and this entry merely records an EC electrochemistry.

Iron oxide

Yellow–green films of iron oxide form on the surface of an iron electrode anodised in 0.1 M NaOH.[624,625,626] Such films display significant electrochromism. For successful film growth, the pH must exceed 9, and the temperature lower than 80 °C. This coloured material may be hydrated $Fe^{III}O \cdot OH$; the film becomes transparent at cathodic potentials as hydrated $Fe(OH)_2$ is formed, so the electrochromic reaction is Eq. (6.24):

$$Fe^{III}O \cdot OH\,(s) + e^- + H_2O \rightarrow Fe^{II}(OH)_2\,(s) + OH^-\,(soln). \quad (6.24)$$
yellow–green   transparent

Gutiérrez and Beden use differential reflectance spectroscopy to show that iron oxyhydroxide underlies the electrochromic effect.[624] These films are prone to slight electrochemical irreversibility owing to a surface layer of anhydrous FeO or $Fe(OH)_2$, which may preclude their use as ECD electrochromes.[625] The oxides γ-Fe_2O_3 (maghemite) and α-Fe_2O_3 (hematite) are also formed in the passivating layer.[624]

Thin films of Fe_2O_3 may be formed by electro-oxidation of $Fe(ClO_4)_2$ in MeCN solution.[104] This oxide is amorphous, the polycrystalline analogue being formed by annealing at high-temperature; polycrystalline Fe_2O_3 is essentially electro-inert. During electrochromic reactions, the first reduction product is Fe_3O_4, according to Eq. (6.25):

$$3Fe_2O_3\,(s) + 2H^+(soln) + e^- \rightarrow 2Fe_3O_4\,(s) + H_2O. \quad (6.25)$$
brown   black

This black Fe_3O_4 contains the mixed-valence oxide formally $FeO \cdot Fe_2O_3$, magnetite.

Such Fe_3O_4 can be further reduced to form a colourless oxide, FeO – Eq. (6.26):

$$Fe_3O_4\,(s) + 2H^+(soln) + 2e^- \rightarrow 3FeO\,(s) + H_2O. \quad (6.26)$$
black   colourless

Electrochromic Fe_2O_3 was made by Özer and Tepehan[627] from a sol of the iron alkoxide $Fe(O^iPr)_3$. After annealing, the Fe_2O_3 was immersed in $LiClO_4$–PC solution.[627] The electrochromic reaction, formally Eq. (6.27), showed good electro-reversibility.

$$Fe_2O_3\,(s) + x(Li^+ + e^-) \rightarrow Li_xFe_2O_3\,(s). \quad (6.27)$$
pale brown   black

The product may be thought of as mixed-valence Fe_3O_4, the lithium counter ion being incorporated for charge balancing during reaction. The source of the lithium ion is $LiClO_4$ in PC; the lithium insertion reaction here is wholly reversible.[627] Such $Li_xFe_2O_3$ is of good optical quality, although the coloured

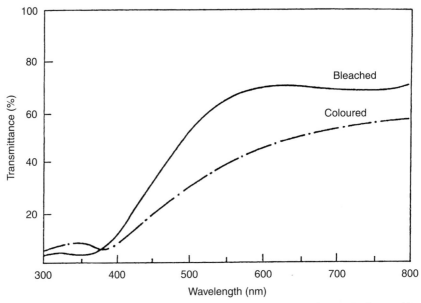

Figure 6.10 Transmittance spectrum of thin-film iron oxide Fe_2O_3 formed by spin-coated sol–gel onto an ITO electrode. The coloured form was generated at -2.0 V, and the bleached form at $+0.5$ V. (Figure reproduced from Özer, N. and Tepehan, F. 'Optical and electrochemical characteristics of sol–gel deposited iron oxide films'. *Sol. Energy Mater. Sol. Cells*, **56**, 1999, 141–52, by permission of Elsevier Science.)

films were insufficiently intense to consider their use as a primary electrochrome, but counter-electrode use is suggested.

Other sol–gel precursors have yielded electrochromic iron oxide films. Electrochromic films were made from a gel prepared by raising the pH of aqueous ferric chloride during addition of ammonium hydroxide, then homogenising the resultant precipitate with ethanoic acid to form a sol.[154] A dip-coating procedure, repeatedly immersing an electrode in the precursor solution and then annealing, yields Fe_2O_3 which bleaches cathodically and colours anodically in lithium-containing electrolytes of aqueous 10^{-3} mol dm^{-3} LiOH. Similar but inferior electrochromic activity was seen when the film was immersed in NaOH or KOH of the same concentration:[154] Na^+ and K^+ cations were presumably too large to enter the lattice readily.

Spin coating a further sol–gel film, based on iron pentoxide in propanol, yields Fe_2O_3 after firing at 180 °C;[628] Li^+ insertion into this oxide is fully reversible. Figure 6.10 shows the electronic spectra.

Iron(acetylacetonate)$_2$ is a suitable CVD precursor for iron oxide electrochromes. A thin film of metallic iron is formed first, which yields an

6.3 Metal oxides: secondary electrochromes

Table 6.7. *Sample values of coloration efficiency η for iron oxide electrochromes.*

Preparative route	$\eta/\mathrm{cm}^2\,\mathrm{C}^{-1}$	Ref.
CVD	−6.0 to 6.5	629
Sol–gel	−28	627
Electrodeposition	−30	104

electrochromic oxide after annealing.[629] The coloured form is Fe_2O_3, so the redox reaction is that given in Eq. (6.28):

$$2\,FeO\,(s) + H_2O \rightarrow Fe_2O_3\,(s) + 2e^- + 2\,H^+(soln). \qquad (6.28)$$

colourless brown

Optical properties of iron oxide electrochromes Values of η are relatively rare for this electrochrome; see Table 6.7.

Manganese oxide

Preparation of manganese oxide electrochromes Anodising metallic manganese in base (alkali) yields a thin surface film of electrochromic oxide.[605] Films of electrochromic MnO_2 can also be formed by reductive electrodeposition from aqueous $MnSO_4$,[630,631,632] the oxide originating from H_2O.

A sol–gel precursor, prepared by adding fumaric acid to sodium permanganate, can yield MnO_2 films. This electrochrome contains some immobile sodium ions, and has been formulated as $Na_\delta MnO_2 \cdot nH_2O$.[633]

Films can also be formed by rf sputtering,[246,247] while electron-beam evaporation yields an electron-deficient oxide, denoted here as $MnO_{(2-y)}$.[231]

Redox chemistry of manganese oxide electrochromes The electrochromic mechanism of MnO_2 grown on Mn metal is complicated. In aqueous solutions, electrochromic coloration involves hydroxide expulsion when solutions are alkaline,[634] according to Eq. (6.29):

$$2MnO_2\,(s) + H_2O + e^- \rightarrow Mn_2O_3\,(s) + 2OH^-(aq). \qquad (6.29)$$

dark brown pale yellow

The colours stated are for thin films; the electrochrome is black in thick films. The colourless form may comprise some hydrated hydroxide $Mn(OH)_3$ or oxyhydroxide, $MnO \cdot OH$. The couple responsible for the electrochromic transition is probably MnO_2–$MnO \cdot OH$,[634] which is confirmed by XPS spectroscopy.[635]

If the pH is low, coloration proceeds in accompaniment with proton uptake according to Eq. (6.30):

$$Mn^{IV}O_2 \,(s) + x(H^+ + e^-) \rightarrow Mn^{III,IV}O_{(2-x)}(OH)_x \,(s). \qquad (6.30)$$

The redox reactions of manganese dioxide in non-aqueous electrolytes are straightforward, and generally involve the insertion and extraction of Li^+, e.g. from $LiClO_4$ in PC via Eq. (6.31):

$$MnO_2 \,(s) + x(Li^+ + e^-) \rightarrow Li_xMnO_2 \,(s). \qquad (6.31)$$

brown   yellow

X-Ray photoelectron spectroscopy suggested that hydrated MnO_2 represents the composition in the oxidised state.[592] The redox process in Eq. (6.31) is better understood than for many other electrochromes, since MnO_2 is the vital component in many rechargeable and alkaline batteries.[247]

The electrochromic operation of MnO_2 films made from sol–gel precursors is said to perform best when immersed in aqueous base.[633] The films are very stable and are said to show high write–erase efficiencies in this electrolyte. Lithium ion can also be inserted from aqueous solution into sputtered MnO_2.[246]

Optical properties of manganese oxide electrochromes Figure 6.11 shows the spectrum of sputter-deposited MnO_2.

Sol–gel drived electrochromic MnO_2 follows Beer's law fairly closely[633] on electro-inserting Li^+ from $LiClO_4$–PC solution. A plot of *Abs* against x for Eq. (6.31) is linear, with a coloration efficiency of 12 to 14 $cm^2\,C^{-1}$, depending slightly on preparation conditions.[633] The value of η for thin-film $Li_xMnO_{(2-y)}$ made by electron-beam evaporation is 7.2 $cm^2\,C^{-1}$.[231] Electrochromic efficiencies as high as 130 $cm^2\,C^{-1}$ have been reported for MnO_y films in aqueous borate buffer solution.[631]

Electrochromic devices containing manganese oxide electrochromes Manganese oxide has been suggested as a counter electrode (or secondary electrochrome) since its coloration efficiency η is relatively low.[592] A device has been made by Ma *et al.*[636] in which the primary electrochrome was nickel oxide.

Niobium oxide

Preparation of niobium oxide electrochromes Sol–gel methods are now the most widely used procedure for forming electrochromic Nb_2O_5 films, for

6.3 Metal oxides: secondary electrochromes

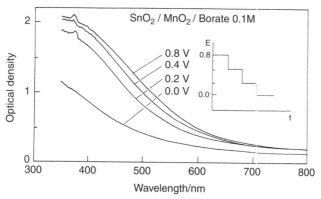

Figure 6.11 UV-visible spectrum of sputter-deposited thin-film manganese oxide at a variety of potentials (vs. SCE, as indicated on the figure). The oxide film was electrodeposited onto a SnO_2-coated optical electrode, and analysed while immersed in a borate electrolyte at pH = 9.2. (Figure reproduced from Córdoba de Torresi, S. I. and Gorenstein, A. 'Electrochromic behaviour of manganese dioxide electrodes in slightly alkaline solutions.' *Electrochim. Acta*, **37**, 1992, 2015–19, with permission of Elsevier Science.)

example by hydrolysing niobium alkoxides.[637,638] Precursors include ethoxide,[191] butoxide[155] or pentachloride[639,640,641] salts. Chloralkoxide sols of the type $NbCl_x(OEt)_{5-x}$, formed by mixing $NbCl_5$ and anhydrous ethanol,[123,170,642] are also used. Hydrolysis yields the solid oxide, Eq. (6.32):

$$2NbCl_x(OEt)_{5-x}\,(aq) + 5H_2O \rightarrow Nb_2O_5\,(s) + 2(5-x)EtOH + 2xHCl\,(aq). \quad (6.32)$$

The gel is then spin coated. Such films are 'slightly crystalline'[192] since they require high-temperature annealing, between 560 and 600 °C.[168] Niobium pentoxide films annealed at temperatures below 450 °C are said to be still amorphous.[643]

Films of Nb_2O_5 have also been prepared by anodising Nb metal, for example by redox cycling Nb metal in dilute aqueous acid.[644,645,646] An electrochromic layer of Nb_2O_5 can also be prepared on niobium metal by thermal oxidation.[647,648]

Direct-current (dc) magnetron sputtering is only occasionally used in preparations of Nb_2O_5.[192,217,218] Lampert and co-workers,[192] comparing the properties of films prepared by dc-magnetron sputtering and by the spin coating of gels subsequently annealed, found that the films were electrochromically essentially equivalent.

Redox electrochemistry of niobium pentoxide electrochromes The accepted redox reaction describing the process of Nb_2O_5 coloration is Eq. (6.33):

$$Nb_2O_5 \text{ (s)} + x(M^+ + e^-) \rightarrow M_xNb_2O_5 \text{ (s)}, \quad (6.33)$$

colourless　　　　　　　　　　　blue

where M^+ is generally Li^+. The response time of Nb_2O_5 grown on Nb metal in aqueous 1 M H_2SO_4 is said to be less than 1 s.[644] The cycle life of crystalline sol–gel-derived films is cited variously as 'up to 2000 voltammetry cycles between 2 and −1.8 V'[168] and 'beyond 1200 cycles without change in performance'.[191]

Films of sol–gel-derived Nb_2O_5 are superior if they are made to contain up to about 20 mole per cent of lithium oxide.[637] Firstly they can accommodate a larger charge (see the cyclic voltammograms in Figure 6.12); secondly, they do not degrade so fast, and thirdly, they can be decoloured completely, whereas sputtered Nb_2O_5 films retain some slight residual coloration.

Optical properties of niobium pentoxide electrochromes Thin films of niobium oxide are transparent and essentially colourless when fully oxidised, and present a deep blue colour on Li^+ ion insertion.[168] Some sol–gel-derived films of Nb_2O_5 also form a brown colour between the tonal extremes of colourless and blue.[170] Figure 6.13 depicts spectra of Nb_2O_5 and $Li_xNb_2O_5$.

The coloration efficiencies of niobium oxide electrochromes are listed in Table 6.8.

Use of niobium oxide electrochromes in devices Owing to its low coloration efficiency, Nb_2O_5 has been used as a 'passive' counter electrode, generally with WO_3[433] as primary electrochrome.

Table 6.8. *Coloration efficiencies η of niobium oxide electrochromes.*

Preparative procedure	$\eta/cm^2\ C^{-1}$ ($\lambda_{(obs)}$/nm)	Ref.
rf sputtering	5	258
rf sputtering	10	649
rf sputtering	100	401
Sol–gel	22 (600)	170, 172, 650
Sol–gel	28 (550)	171
Sol–gel	38 (700)	405
Spraying[a]	6 (800)	641

[a] $NbCl_5$ in ethanol

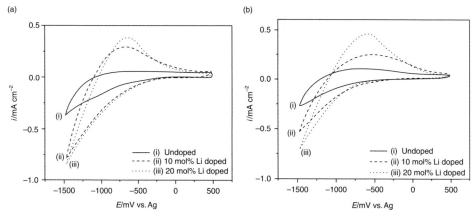

Figure 6.12 The effect of cycle number on the cyclic voltammogram of thin-film Nb_2O_5, deposited onto ITO by a sol–gel process. (a) The first cycle and (b) the twenty-first cycle. During redox cycling, the film was immersed in propylene carbonate solution itself comprising $LiClO_4$ (0.1 mol dm^{-3}). Note also the higher charge capacity of the lithium-containing films. (Figure reproduced from Bueno, P. R., Avellaneda, C. O., Faria, R. C. and Bulhões, L. O. S. 'Electrochromic properties of undoped and lithium doped Nb_2O_5 films prepared by the sol–gel method'. *Electrochimica Acta*, **46**, 2001, 2113–18, by permission of Elsevier Science.)

Palladium oxide

Amongst the few studies of electrochromic PdO_2, the most extensive, by Bolzán and Arvia,[651] concerns hydrated PdO_2 (prepared by anodising Pd metal in acidic solution), revealing some redox complexity. The coloured (black) form is hydrated PdO, hydrated PdO_2 is yellow, while anhydrous PdO_2 is reddish brown. This electrochemical complexity, coupled with high cost, means that palladium electrochromes are unlikely to be viable.

Praseodymium oxide

Electrochromic praseodymium oxide was studied by Granqvist and co-workers[219] who made thin-film PrO_2 by dc-magnetron sputtering, varying the ratio of O_2 to argon from 0.025 to 0.005. Thomas and Owen[652] used CVD from a metallo-organic precursor. The electrochromic reaction is[652] Eq. (6.34):

$$PrO_{(2-y)}\,(s) + x(Li^+ + e^-) \rightarrow Li_xPrO_{(2-y)}\,(s). \qquad (6.34)$$

dark orange   colourless

Films of electrochromic oxide switch in colour from dark orange (presumably PrO_2-like) to transparent. X-Ray diffraction of the CVD-derived samples suggest that the first lithium insertion cycle was accompanied by an irreversible

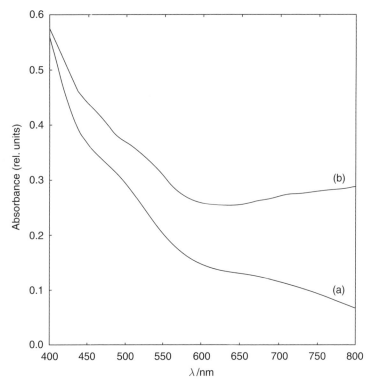

Figure 6.13 UV-visible spectrum of thin-film niobium pentoxide on ITO. The spectrum (a) of the reduced form at −0.875 V and (b) of the oxidised form was obtained at 0 V against SCE. The film was prepared by a sol–gel method and had a thickness of *ca.* 5 μm. The electro-coloration was performed in 1.0 mol dm^{-3} H$_2$SO$_4$ solution. (Figure reproduced from Lee, G. R. and Crayston, J. A. 'Electrochromic Nb$_2$O$_5$ and Nb$_2$O$_5$/silicone composite thin films prepared by sol–gel processing'. *J. Mater. Chem.*, **1**, 1991, 381–6, by permission of The Royal Society of Chemistry.)

phase change.[652] Thereafter, provided the switching was relatively fast and that the film was not left in the reduced state for long periods, the charge insertion was reversible over 500 cycles.

The charge capacity ranged from comparability with that of WO$_3$, for oxygen-rich films, to virtually zero for oxygen-depleted films.[652] The initially dark films made by sputtering showed strong anodic electrochromism. In a device incorporating WO$_3$ as the primary electrochrome, the use of PrO$_2$ as the secondary layer made the colour more 'neutral' (i.e. more grey).

Praseodymium films do not promise wide usage, but PrO$_2$ has been added to films of cerium oxide;[653] see p. 193.

6.3 Metal oxides: secondary electrochromes

Rhodium oxide

Electrochromic rhodium oxide has been little studied. Films may be formed on Rh metal by anodising metallic rhodium immersed in concentrated solution of alkali.[654,655] It can also be made from sol–gel precursors.[656] In an early study, Gottesfeld[657] cites the electrochromic reaction Eq. (6.35):

$$Rh_2O_3 \text{ (s)} + 2OH^-\text{(aq)} \rightarrow 2\,RhO_2 \text{ (s)} + H_2O + 2e^-. \qquad (6.35)$$

yellow   dark green

Both the rhodium oxides in Eq. (6.35) are hydrated, Rh_2O_3 probably more so than RhO_2.

Dark-green RhO_2 appears black if films are sufficiently thick. A fully colourless state is not attainable. The oxide RhO_2 is unusual in being green; the only other inorganic electrochromes evincing this colour are Prussian green (a mixed-valence species of partly oxidised Prussian blue), and electrodeposited cobalt oxide; see p. 168.

Rhodium oxide made by a sol–gel procedure switched from bright yellow to olive green.[656] Such films are polycrystalline, owing to annealing after deposition. The coloration efficiency at 700 nm was $29\,cm^2\,C^{-1}$.

Figure 6.14 shows a cyclic voltammogram of rhodium oxide;[657] reflectance and charge insertion are also shown as a function of potential.

Ruthenium oxide

Thin films of hydrous ruthenium oxide can be prepared by repeated cyclic voltammetry on ITO-coated glass substrates immersed in an aqueous solution of ruthenium chloride.[658] Films may also be generated by anodising metallic Ru in alkaline solution.[659]

The oxide changes colour electrochemically[659] according to Eq. (6.36):

$$RuO_2 \cdot 2H_2O \text{ (s)} + H_2O + e^- \rightarrow \tfrac{1}{2}(Ru_2O_3 \cdot 5H_2O) \text{ (s)} + OH^-. \qquad (6.36)$$

blue–brown   black

The electrogenerated colour is not intense. The ruthenium oxide electrode exhibits a 50% modulation of optical transmittance at 670 nm wavelength.[658]

Tantalum oxide

Preparation of tantalum oxide electrochromes Few electrochromism studies have been performed on tantalum oxide Ta_2O_5, but it has been used sometimes as a layer of ion-conductive electrolyte.[74,173,220,257,259,260,468,660,661,662,663,664,665,666]

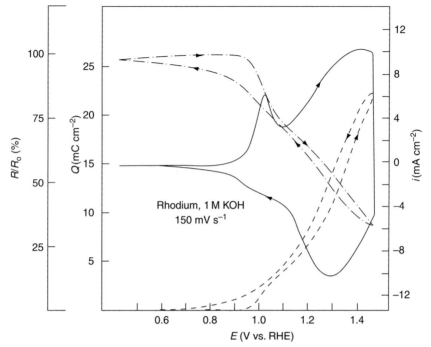

Figure 6.14 Cyclic voltammogram of rhodium oxide grown on an electrode of metallic rhodium, immersed in hydroxide solution (5 mol dm^{-3} KOH). Also included on the figure are the reflectance at 546 nm (–·–) and charge inserted (– – –) as a function of potential. The scan rate was 150 mV s^{-1}. (Figure reproduced from Gottesfeld, S. 'The anodic rhodium oxide film: a two-colour electrochromic system'. *J. Electrochem. Soc.*, **127**, 1980, 272–7, by permission of The Electrochemical Society, Inc.).

Thin films may be prepared by anodising Ta metal in sulfuric acid[662,667,668] or thermal oxidation of sputtered Ta metal.[666] Other films have been made by rf sputtering from a target of Ta_2O_5,[257,258,259,260] reactive dc sputtering[220] or thermal evaporation.[220]

The most widely used tantalum CVD precursors of Ta_2O_5 are $Ta(OEt)_5$,[72,74,173,663] $TaCl_5$[74] or TaI_5,[74] each volatilised in an oxygen-rich atmosphere. Carbon or halide impurities are however incorporated into the resultant films. Otherwise, solutions of the supposed peroxypolytantalate may be spin coated onto ITO, and then sintered; this solute is prepared by reactive dissolution in H_2O_2 of either Ta[93] or $Ta(OEt)_5$.[193]

Thin-film Ta_2O_5 can also be formed by dip coating using a liquor rich in $Ta(OEt)_5$ as the precursor. An electrode substrate is repeatedly dipped into the liquor, or slowly immersed and withdrawn at a predetermined rate.

Redox chemistry of tantalum oxide electrochromes The electrochromic reaction of thin-film Ta_2O_5 in aqueous alkali is Eq. (6.37):

$$Ta_2^V O_5 \text{ (s)} + H_2O + e^- \rightarrow 2\,Ta^{IV}O_2 \text{ (s)} + 2OH^- \text{(aq)}. \qquad (6.37)$$

colourless   very pale blue

While the kinetics here have been little studied, the kinetics of charge transport are dominated by movement of polaron species.[3] Garikepati and Xue,[665] studying the dynamics of charge movement (comprising proton conductance) across the Ta_2O_5–WO_3 interphase, found the rate of proton movement was dictated by water adsorbed within the interphase. While in the studies of Ahn *et al.*[664] on the interface comprising Ta_2O_5 and NiO or $Ni(OH)_2$, the authors do mention the effects of such adsorbed water on the rate of ionic movement across the interphase. However, they conclude that the rate is dictated by the extent to which the crystal structures of the oxides making the interface are complementary, i.e. how well structurally the oxides join.

The conductivity of protons through Ta_2O_5 is so fast that it is often classed as a 'fast ion conductor'.[669] Accordingly, workers are increasingly choosing to employ thin-film Ta_2O_5 as the ion-conductive electrolyte layer between the solid layers of primary and secondary electrochrome in all-solid-state devices.[220,257,259,260,660,663,670,671,672,673]

Optical properties of tantalum oxide electrochromes The value of λ_{max} for Ta_2O_5 made by anodised tantalum metal is 541 nm,[674] but the electrochromic effect is weak. For example, films made by rf sputtering have η values as low as $5\,cm^2\,C^{-1}$,[206] while material made by laser ablation has η of $10\,cm^2\,C^{-1}$.[675] The Ta_2O_5 films exhibit high transmittance except in the UV, where the films absorb strongly.

Tin oxide

In the few studies on the electrochromism of tin oxide, Eq. 6.38:

$$SnO_2 \text{ (s)} + x(Li^+ + e^-) \rightarrow Li_xSnO_2 \text{ (s)}, \qquad (6.38)$$

colourless   blue–grey

the tin(IV) oxide films were made by reactive rf-magnetron sputtering.[676] The films are conductive, by both electrons and ions. The wavelength maximum of Li_xSnO_2 lies in the infrared.[676] Granqvist and co-workers[677] assign the peak to intervalency transitions as in other cathodically colouring electrochromic oxides. The peak occurs in the near infrared.[676]

At low insertion coefficients ($0 < x < \sim 0.1$), the electro-inserted lithium ions appear to be located in internal double layers within the film.[677] Increasing the insertion coefficient x from ~ 0.1 to ~ 0.2 yielded significant transmittance drops, and Mössbauer spectra unambiguously show the conversion $Sn^{IV} \rightarrow Sn^{II}$. Electrocrystallisation appears to dominate the electrochemistry at $x > 0.2$.[677]

The electronic spectrum of tin-oxide films remains relatively unchanged following electro-insertion of lithium ion, but optical constants such as the refractive index increase with increasing insertion coefficient.

Titanium oxide

Thin-film TiO_2 can be made in vacuo by thermal evaporation of TiO_2,[678] reactive rf sputtering from a titanium target,[261] or pulsed laser ablation.[679] Alternatively, non-vacuum techniques involve alkoxides or the peroxo precursor made by dissolving a titanium alkoxide $Ti(OBu)_4$ in H_2O_2.[128,134] Methods involve sol–gel,[127,128,319,680] spin coating[128] and dip-coating procedures.[158,174]

The electrochromic reaction of TiO_2 is usually written as Eq. (6.39):

$$TiO_2(s) + x(Li^+ + e^-) \rightarrow Li_xTiO_2(s). \qquad (6.39)$$

colourless → blue–grey

Ord et al.[681] have studied the electrochromism of titanium oxide grown anodically on metallic titanium via in situ ellipsometry. Both reduction and oxidation processes occur via movement of a phase boundary which separates the reduced and oxidised regions within the TiO_2. The rate of TiO_2 reduction is controlled by the rate of counter-ion diffusion through the solid:[682,683] ionic insertion into the crystal form of anatase (the Li^+ deriving from a $LiClO_4$–propylene carbonate electrolyte) is characterised by a diffusion coefficient of[682] 10^{-10} cm^2 s^{-1}. To accelerate diffusion, Scrosati and co-workers[682] drove the electrochromic process with potentiostatic pulses.

Titanium oxide-based electrochromes show two optical bands at 420 and 650 nm.[679] The coloration efficiency is low, hence TiO_2 is used as a secondary electrochrome or even as an 'optically passive' counter electrode, with WO_3 as the primary electrochrome.[158,319,678,684]

Values of coloration efficiency η for thin-film TiO_2 are low; see Table 6.9. Nevertheless, Yoshimura et al.[685] claim to have modulated an incident beam by between 14% and 18%.

Thin-film titanium oxynitride is also electrochromic.[686]

Table 6.9. *Sample values of coloration efficiency η for titanium oxide electrochromes.*

Preparative procedure	$\eta/\text{cm}^2\,\text{C}^{-1}$ ($\lambda_{(\text{obs})}/\text{nm}$)	Ref.
Reactive thermal evaporation	7.6	678
Thermal evaporation	8 (646)	401
rf sputtering	14	261
Sol–gel	50	641

Vanadium oxide

Preparation of vanadium oxide electrochromes Thin-film V_2O_5 is commonly made by reactive rf sputtering,[206,263,264,265,266,267] with a high pressure of oxygen and a target of vanadium metal. Direct-current sputtering is also used.[50,223,224,225,226] Other vacuum methods employed include pulsed laser ablation,[687,688,689] cathodic arc deposition[550] and electron-beam sputtering.[233,265] Thermal evaporation in vacuo[476,526,690] affords a different class of preparative method, and includes flash evaporation.[691]

Films of V_2O_5 deposited by thermal evaporation in vacuo are amorphous,[476] sputtered samples are more crystalline,[264,267,692] although X-ray diffraction suggests the extent of crystallinity is low.[264] Annealing a sample of thermally evaporated V_2O_5 to above 180 °C improves the electrochemical performance,[693] presumably by increasing the extent of crystallinity in the amorphous material. Thin films of V_2O_5 from vanadium metal anodised in acetic acid[36,694] are essentially amorphous.

Electrochromic thin films have often been prepared using xerogels of V_2O_5, the precursor of choice generally being an alkoxide species such as $VO(O^iPr)_3$.[199,200] Subsequent annealing yields the desired electrochrome, which is always hydrated.[695] The preparation and use of such gels has been reviewed extensively by Livage[696,697] (in 1991 and 1996 respectively). A more general review was published in 2001.[120] Livage made VO_2 films by sol–gel methods, generally via alkoxide precursors.[698] Alkoxide precursors are also used in preparing films by CVD, like $VO(O^iPr)_3$ in 2-propanol;[71,699,700] bis(acetylacetonato)vanadyl has also been employed.[73] The deposition product is immediately annealed in an oxidising atmosphere, ensuring polycrystallinity.

Spin coating has also been used to prepare films of V_2O_5. Coating solutions include the liquor made by dissolving V_2O_5 powder in a mixed solution of benzyl alcohol and *iso*-butanol,[701,702] or that produced by oxidative dissolution of powdered vanadium in hydrogen peroxide.[132,133] The liquor made by dissolving metallic vanadium in H_2O_2 can also be spin coated e.g. onto ITO substrates.[132]

Deb and co-workers[703] prepared thin films of mesoporous vanadium oxide by electrochemical deposition from a water–ethanol solution of vanadyl sulfate and a non-ionic polymer surfactant. Aggregates of the polymer surfactant appeared to act as a form of template during deposition.

Electrochemical methods of making V_2O_5 electrochrome are rarely used, no doubt owing to their sensitivity to water. Nevertheless, thin-film V_2O_5 has been grown anodically on vanadium metal immersed in dilute acetic acid.[36,694,704]

Redox chemistry of vanadium oxide electrochromes The electrochromism of thin-film V_2O_5 was apparently first mentioned in 1977 by Gavrilyuk and Chudnovski,[705] who prepared samples by thermal evaporation in vacuo. Since thin-film V_2O_5 dissolves readily in dilute acid, alternative electrolytes have been used, for example, distilled water,[705] LiCl in anhydrous methanol[706] or $LiClO_4$ in propylene carbonate[263,264,266] or γ-butyrolactone.[707]

The electrochromic reaction in non-aqueous solution follows Eq. (6.40):

$$V_2^V O_5\,(s) + x(M^+ + e^-) \rightarrow M_x V_2^{IV,V} O_5\,(s), \qquad (6.40)$$

brown–yellow very pale blue

where M^+ is almost universally Li^+ owing to appreciable solubility of V_2O_5 in aqueous acid. The rates of ion insertion and egress are so much slower for Na^+ than for Li^+ that the sodium ions in $Na_{0.33}V_2O_5$ may be regarded as immobile.

In aqueous solution,[708] an alternative reaction is Eq. (6.41):

$$V_2^V O_5 + 2H^+ + 2e^- \rightarrow V_2^{IV} O_4 + H_2O. \qquad (6.41)$$

The relationships between the structure of V_2O_5 films (prepared by sol–gel) and their redox state has been described at length by Meulenkamp et al:[709] a transition occurs from α-V_2O_5 at $x=0.0$ to ε-$Li_xV_2O_5$ at $x=0.4$. These phases are nearly identical. For larger insertion coefficients, however, the structure undergoes significant changes: firstly, the phase for $x=0.8$ shows an elongated c-axis relative to ε-$Li_xV_2O_5$, which may represent a monoclinic structure. Secondly, at $x=1.0$ the structure distorts further and shows features in common with δ-LiV_2O_5. Thirdly, at $x=1.4$, the structure bears further resemblance to δ-LiV_2O_5. (Here, ε and δ are but phase labels.) Granqvist et al.[226] describe the structure of $Li_xV_2O_5$ as orthorhombic, later with additional details.[49]

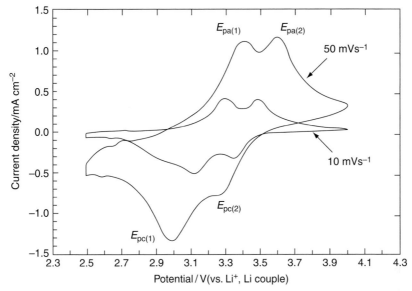

Figure 6.15 Cyclic voltammogram of thin film of V_2O_5 sputtered on an OTE, and immersed in propylene carbonate containing $LiClO_4$ (1.0 mol dm^{-3}). (Figure reproduced from Cogan, S. F., Nguyen, N. M., Perrotti, S. J. and Rauh, R. D. 'Electrochromism in sputtered vanadium pentoxide'. *Proc. SPIE*, **1016**, 1988, 57–62, with permission of the International Society for Optical Engineering.)

Cyclic voltammetry of sputtered V_2O_5, as a thin film supported on an OTE immersed in a lithium-containing PC electrolyte, shows two well-defined quasi-reversible redox couples[263] with anodic peaks at 3.26 and 3.45 V, and cathodic peaks at 3.14 and 3.36 V relative to the Li^+, Li couple in propylene carbonate; see Figure 6.15. Benmoussa *et al.*[710] produced V_2O_5 films by rf sputtering, obtaining 'excellent cyclic voltammograms', again with a two-step electrochromism: they cite yellow to green, and then green to blue during reduction. These two pairs of peaks may correspond to the two phases of $Li_xV_2O_5$ identified by Dickens and Reynolds.[711]

Ord *et al.*[36,694] grew thin anodic films of V_2O_5 on vanadium metal immersed in acetic acid, and studied the redox processes using the *in situ* technique of ellipsometry, in tandem with more traditional electrochemical methods such as cyclic voltammetry. As soon as the film is made cathodic, the outer surface is converted to $H_4V_2O_5$. Thereafter, their results clearly suggest how, in common with MoO_3 (but unlike WO_3), a well-defined boundary forms between the coloured and bleached phases during redox cycling: this boundary sweeps inward toward the substrate from the film–electrolyte interface during the bleaching and coloration processes. (Higher fields are required for bleaching than for coloration.) The rates of coloration and bleaching are both dictated by the rate of proton movement.[694] The bleaching process is complicated, and proceeds in three stages.[694]

A study by Scarminio et al.[225] monitored the stresses induced in V_2O_5 during redox cycling, this time with Li^+ as the mobile ion; their thin-film V_2O_5 was immersed in a solution of $LiClO_4$ in PC. Their results suggest the crystal structure within the film is determined by the sputter conditions employed during film fabrication. Deep charge–discharge cycles (performed under constant current density) allow correlations to be drawn between the stress changes in the crystalline film and the electrode potential steps. The authors say this behaviour is typical of the lithium insertion mechanisms in bulk V_2O_5 prepared as a cathode material for secondary lithium batteries. They also suggest the redox cycling is somewhat irreversible, implying a poor write–erase efficiency.

The crystal structure of vanadium pentoxide is complicated, with the nominally octahedral vanadium being almost tetragonal bipyramidal, with one distant oxygen.[712] Reductive injection of lithium ion into V_2O_5 forms $Li_xV_2O_5$. The $Li_xV_2O_5$ (of $x<0.2$) prepared by sputtering is the a-phase, which is not readily distinguishable from the starting pentoxide.[263] At higher injection levels ($0.3 < x < 0.7$), the crystalline form of the oxide is ε-$Li_xV_2O_5$,[263] as identified by the groups of Hub et al.[706] and Murphy et al.[713] The generation of the ε-phase of $Li_xV_2O_5$ in V_2O_5 thin films accompanies the electrochromic colour change. Also, a-$Li_xV_2O_5$ from the un-lithiated oxide is formed, and contributes an additional, slight change in absorbance.[263] Since several species participate in the spectrum of the partially reduced oxide, spectral regions following the Beer–Lambert law cannot be identified readily.[266]

Films of mesoporous V_2O_5 colour faster than evaporated films,[703] attributable to enhanced ion mobility. Such vanadium oxide also exhibits a higher lithium storage capacity and greatly enhanced charge-discharge rate.

$Na_{0.33}V_2O_5$ made by a sol–gel process is also electrochromic;[714] see Eq. (6.42):

$$Na_{0.33}V_2O_5\,(s) + x(Li^+ + e^-) \rightarrow Li_xNa_{0.33}V_2O_5\,(s). \quad (6.42)$$

The sodium ions are essentially immobile.

Optical properties of vanadium oxide electrochromes The absorption bands formed on reduction are generally considered to be too weak to imply the formation of any intervalence optical parameters (although there are other arguments formulated by Nabavi et al.[715]). Wu et al.[716] suggest the anodic electrochromism of V_2O_5 is due to a blue shift of the absorption edge, and the

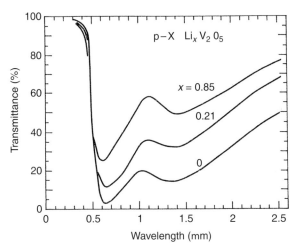

Figure 6.16 UV-visible spectrum of thin-film vanadium pentoxide on ITO. The polycrystalline V_2O_5 was sputter deposited to a thickness of 0.25 μm. The numbers refer to values of insertion coefficient x. (Figure reproduced in slightly altered form from Talledo, A., Andersson, A. M. and Granqvist, C. G. 'Structure and optical absorption of $Li_xV_2O_5$ thin films'. *J. Appl. Phys.*, **69**, 1991, 3261–5, by permission of Professor Granqvist and The American Institute of Physics.)

near-infrared electrochromism arises from absorption by small polarons in the V_2O_5. From X-ray photoelectron spectroscopy, Fujita *et al.*[690] assign the colour change in evaporated films incorporating lithium to the formation of VO_2 (which is blue) in the V_2O_5. Colten *et al.*,[476] using the same technique, did infer a weak charge-transfer transition between the oxygen 2p and vanadium 3d states, but only for an entirely V^{IV} solid.

Vanadium pentoxide films have a characteristic yellow–brown colour, attributable to the tail of an intense optical UV band appearing in the visible region;[224] see Figure 6.16. The electrogenerated colour is blue–green for evaporated films[717] at low insertion levels, going via dark blue to black at higher insertion levels.[705] The colour changes from purple to grey if films are sputtered.[5] Rauh and co-workers[263] state that certain film thicknesses of V_2O_5 yield colourless films between the brown and pale-blue conditions. The value of λ_{max} of the yellow–brown form lies in the range 1100–1250 nm. The loss of the yellow colour is attributed to the shift of the band edge from about 450 to 250 nm during reductive bleaching from V_2O_5 to $Li_{0.782}V_2O_5$.[266]

A few representative values of η are listed in Table 6.10.

Electrochromic devices containing vanadium pentoxide Since the electrochromic colours of V_2O_5 films are yellow and very pale blue, the *CR* values

Table 6.10. *Coloration efficiencies η of thin-film vanadium oxide electrochromes.*

Deposition method	$\eta/\text{cm}^2\,\text{C}^{-1}$ ($\lambda_{(obs)}$/nm)	Ref.
rf magnetron sputtering	−35	263
rf magnetron sputtering	−15 (600–1600)	264
Sol–gel	−50	641
CVD	−34	699

for such films are not great, hence the system is generally investigated for possible ECD use as a secondary electrochrome, i.e. in counter-electrode use.[263,266,526,718,719,720,721] For example, cells have often been constructed with V_2O_5 as the secondary material to WO_3 as the primary, e.g. ITO | Li_xWO_3 | electrolyte | V_2O_5 | ITO.[266,277,722] Gustaffson et al.[723] made similar cells but with the conducting polymer PEDOT as the primary electrochrome.

Thin-film vanadium dioxide VO_2 is electrochromic,[724,725] and lithium vanadate ($LiVO_2$) is not only electrochromic but also *thermo*-chromic,[726] and can be prepared by reactive sputtering;[723] $LiVO_2$ doped with titanium oxide is also thermochromic.[727]

Finally, composites of V_2O_5 in poly(aniline) and a 'melanin-like' polymer have been reported.[728]

6.4 Metal oxides: dual-metal electrochromes

6.4.1 Introduction

Preparing *mixtures* of metal oxide has been a major research goal during the past few years, for two reasons. Firstly, mixing these oxides can modify the solid-state structure through which the mobile ion moves, and thus increase the chemical diffusion coefficient $\bar{D}$ that results in superior response times τ.

Secondly, mixtures are capable of providing different colours. In particular, there is a desire for so-called 'neutral' electrochromic colours; see p. 399. Varying the energies of the optical bands by altering the mix allows the colour to be adjusted to that desired. Thus the choice of constituent oxides and their relative mole fractions allows a wide array of options. There are several models to correlate these variables with the electrochromic colour.

One of the most successful is the so-called 'site-saturation' model. Here, all inserted electrons are considered to be localised, and optical absorptions are considered to occur simply by photo-excitation of an electron to an empty redox site. To a good first approximation, the optical absorption intensity is proportional to the number of vacant redox sites surrounding the reductant site. As the insertion coefficient x increases, so the proportion of vacant sites neighbouring a given electron on a reductant's site decreases, with the effect of decreasing the oscillator strength. The treatment by Denesuk and Uhlman[729] has been tested with data for Li_xWO_3 on ITO – a system displaying a curious dependence of λ_{max} and η on the insertion coefficient x. Their model only applies to situations in which a dominant fraction of the electrons associated with the intercalating species are appreciably localised. The computed and experimental data correlated well, with published traces showing only slight deviation between respective values.

The earlier work of Hurita et al.[730] relates to mixtures of MoO_3 and WO_3. Again, the computed and experimental data correlate well, although published traces show somewhat more scatter. Several other reports[24,283,731] have discussed electrochromic colours in terms of this model. van Driel et al.[731] again studied λ_{max} and η for the Li_xWO_3 system. Published traces show significant divergences between calculated results and experiment, which are explained in terms of partial irreversibility during coloration.

In the discussion below, tungsten-based systems are considered first, as exemplar systems, since they were among the first mixed-metal oxides to receive attention for electrochromic applications. The chemistry of tungsten–molybdenum oxides has been reviewed briefly by Gérand and Seguin[732] (1996). Other host oxides are listed alphabetically.

6.4.2 Electrochromic mixtures of metal oxide

A full, systematic evaluation of the data below is not yet possible because the electrochromic properties of films depend so strongly on the modes of preparation, as has been copiously illustrated above, and such a wide range of preparative techniques has been employed.

Clearly, oxide mixtures of the type X–Y can be incorporated into either a section on oxides of X or of Y, so some slight duplication is inevitable.

Tungsten oxide as electrochromic host

Tungsten trioxide has been employed as a host or 'matrix' for a series of electrochromic oxides, containing the following oxides: (in alphabetical order) Ba,[733] Ce,[734] Co,[5,80,94,99,735,736] Mo,[61,62,91,292,361,475,730,737,738,739,740,741,742,743,744,745,746,747] Nb,[29,317,748,749,750] Ni,[53,80,81,89,94,99,735,736,751,752] Re,[753] Ta,[29,661,754] Si,[316,333] Ti.[127,129,134,203,316,333,335,404,755,756,757,758,759,760]

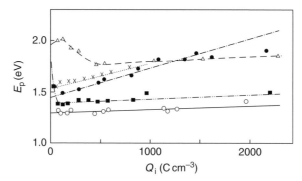

Figure 6.17 Photon energy of the absorption peak E_p as a function of the inserted charge: thin-film samples of partially reduced oxides of composition $Mo_cW_{(1-c)}O_3$. o = WO_3, ● = MoO_3, ■ = mixed film with $c = 0.008$, $\Delta = c = 0.13$, and x = 0.80. (Figure reproduced from Hiruta, Y., Kitao, M. and Yamada, M. Absorption bands of electrochemically-colored films of WO_3, MoO_3 and $Mo_cW_{1-c}O_3$. Jpn. J. Appl. Phys., **23**, 1984, 1624–7, with permission of The Institute of Pure and Applied Physics.)

or V.[204,254,325,327,687,761] Thin-film WO_3 can also be co-electrodeposited with phosphomolybdic acid to yield an electrochrome having a colour change described as 'light yellow → bluish brown', although the transition is reported not to be particularly intense.[745,762]

Kitao et al.[292] prepared a range of films of molybdenum–tungsten oxide of the formula $Mo_cW_{(1-c)}O_3$, and analysed the shift in wavelength maxima as a function of the mole fractions of either constituent oxide (see Figure 6.17) and found a complicated relationship, sometimes described within the 'site saturation model'; see p. 190. Here, it is recognised that electrons are captured (that is, they effect reduction) first at the sites of lowest energy. In practice, it is found that Mo sites are of lower energy than W, thereby explaining why Mo–Mo and Mo–W intervalence bands are formed at lower insertion coefficients x than are any W–W bands.

The value of λ_{max} for mixed films of WO_3–MoO_3 shifts to higher energy (lower λ) relative to the pure oxides. Since the wavelength of the shifted λ_{max} corresponds more closely to the sensitive range of the human eye, mixing the oxides effectively enhances the coloration efficiency η in the visible region. Faughnan and Crandall found the highest value of η occurs with a mole fraction of 0.05 of MoO_3.[738] Deb and Witzke[763] say the range of η is 30–40%. Additionally, the films of W–Mo oxide become darker because they can accommodate more charge, i.e. have a larger maximal insertion coefficient.[61] Disadvantageously, the electron mobility is decreased in thin-film WO_3–MoO_3 relative to the pure oxides.[738]

In the study by Hiruta et al.,[746] the optical band for W–Mo oxide is said to comprise two bands. The first is the intervalence band, and the second is a new band at higher energies, which is thought to relate to the Mo ions. Furthermore, the energy of the absorption band depends on the concentration of Mo in the film and the insertion coefficient x.[747]

Gérand and Seguin[732] suggest that ion insertion into W–Mo oxide occurs readily, but ion removal is usually somewhat difficult, thus precluding all but the slowest of electrochromic applications. The slowness is ascribed to induced 'amorphisation' of the mixed-metal oxide at high insertion coefficients. Such a result, if confirmed, would contradict the usual assumption that values of $\bar{D}$ for ion movement through amorphous material are higher than through polycrystalline material. The cause of such 'amorphisation' is as yet not clear.

The temperature dependence of the electrochromic response of sol–gel deposited titanium–tungsten mixed oxide was shown by Bell and Matthews[335] to be highly complicated, implicating multiple competing processes.

The W–Ce film consisted of a self-assembly structure based on the poly(oxotungsceriumate) cluster $K_{17}[Ce^{III}(P_2W_{17}O_{61})_2] \cdot 30H_2O[Ce(P_2W_{17})_2]$ and poly(allylamine) hydrochloride.[734] Comparatively long response times of 108 and 350 s were found for coloration and bleaching respectively.

Adding about 5% of nickel oxide significantly improves the cyle life of WO_3.[53]

Finally, a hybrid of WO_3 and Perspex (polymethylmethacrylate) has a relatively low η of 38 cm^2 C^{-1}.[403]

Antimony oxide as an electrochromic host

Thin-film antimony–tin oxide (ATO), grown by pulsed laser deposition, colours cathodically.[764] Its electrochromic properties are 'poor'; the electrochemical and optical properties were found to be extremely sensitive to their morphology. Naghavi et al.[765] suggested that the best electrochromic films were obtained by depositing at 200 °C in an oxygen atmosphere at a pressure of 10^{-2} mbar, followed by annealing at 550 °C. This last condition is described as 'critical'.

Cerium oxide as electrochromic host

Electrochromic mixtures have been prepared of cerium oxide together with the oxides of Co,[766] Hf,[767] Mo,[768] Nb,[769] Pr,[653] Si,[768,770] Sn,[755,771] Ti,[122,205,323,324,755, 767,771,772,773,774,775,776,777,778,779,780,781,782,783,784,785,786,787] V,[788,789,790,791,792,793] W,[734] and Zr.[122,783,786,794,795,796] The relative amount of the second oxide varies from a trace to a molar majority.

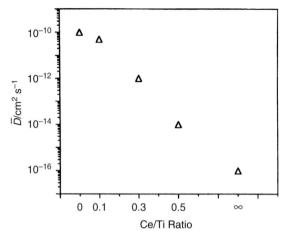

Figure 6.18 Graph of chemical diffusion coefficient $\bar{D}$ of Li^+ ion moving through films of CeO_2–TiO_2: the effect of varying the composition. (Figure reproduced in slightly altered form from Kullman, L., Azens, A. and Granqvist, C. G. 'Decreased electrochromism in Li-intercalated Ti oxide films containing La, Ce, and Pr'. *J. Appl. Phys.*, **81**, 1997, 8002–10, by permission of Professor Granqvist and The American Institute of Physics.)

Thin-film Ce–Ti oxide is more stable than CeO_2 alone[797] although, interestingly, evidence from EXAFS suggests that the electrons inserted into Ce–Ti oxide reside preferentially at cerium sites;[774] the oxide layer was prepared by dc magnetron sputtering. The charge movement necessary for electrochromic operation involves insertion and/or extraction of electrons via the Ce 4f states,[779,780] which are located in the gap between the valence and conduction bands of the CeO_2.

The chemical diffusion coefficient $\bar{D}$ of mobile Li^+ ions through thin-film Ce–Ti oxide increases as the mole fraction of cerium oxide decreases:[779,780] a plot of $\ln \bar{D}$ against mole fraction of CeO_2 (see Figure 6.18) is almost linear: $\bar{D}_{(Li^+)}$ increases from 10^{-16} cm^2 s^{-1} for pure CeO_2 to 10^{-10} cm^2 s^{-1} for pure TiO_2. These values of $\bar{D}$ suggest the extent of electron trapping is slighter (or at least the depths of such traps are shallower) for TiO_2 than for CeO_2. Clearly, then, electrochromes having as high a proportion of TiO_2 as possible are desirable to achieve rapid ECD operation. Conversely, adding CeO_2 to TiO_2 increases the cycle life, the cycle life of pure TiO_2 (as prepared by sol–gel techniques) being relatively low.[319]

Addition of cerium oxide to TiO_2 also decreases[779,780] the coloration efficiency η until, as the ratio Ce:Ti (call it γ) exceeds 0.3, the electrochromic 'absorbance' is essentially independent of the insertion coefficient, i.e. films of Ce–Ti oxide (with Li^+ as the counter ion) are optically passive and can function as ECD counter electrodes. Films prepared by magnetron sputtering with $\gamma > 0.6$ are not chemically stable.[781]

Clearly, optimising the electrochromic response of Ce–Ti oxide will require that all three of the parameters τ, η and cycle life are considered.

Cobalt oxide as electrochromic host

Many electrochromic mixtures of cobalt oxide have been prepared, e.g. with oxides of Al,[617,798] Ce,[766] Cr,[80] Fe,[80,81] Ir,[799] Mo,[80,81] Ni,[79,80,83,140,158,799,800,801] W[80,81,736] or Zn.[80,81]

Diffusion through films of cobalt oxide mixed with other d-block oxides can be considerably faster than through CoO alone: the value of $\bar{D}$ for the OH^- ion is 2.3×10^{-8} cm^2 s^{-1} through CoO, 5.5×10^{-8} cm^2 s^{-1} through WO_3, but 48.7×10^{-8} cm^2 s^{-1} through Co–W oxide.[80] All these films were electrodeposited. The value of $\bar{D}$ relates to H^+ as the mobile ion through WO_3, and to OH^- ions for CoO and Co–W oxide. The larger value of $\bar{D}$ probably reflects a more open, porous structure. These values of $\bar{D}$ are summarised in Table 6.11.

Thin-film Co–Al oxide[617] prepared by dip coating has a coloration efficiency of 22 cm^2 C^{-1}, which compares with η for CoO alone of 21.5 cm^2 C^{-1} (as prepared by CVD[604]) or 25 cm^2 C^{-1} (the CoO having been prepared by a sol–gel method[616]). Cobalt–aluminium oxide has a coloration efficiency of 25 cm^2 C^{-1}, and Co–Al–Si oxide has η of 22 cm^2 C^{-1}.[401] Since thin-film Al_2O_3 is rarely electroactive let alone electrochromic, the similarity between these η values probably indicates that the alumina component acts simply as a kind of matrix or 'filler', allowing of a more open structure; but any increase in the rate of electro-coloration follows from enhancements of $\bar{D}$ rather than from increases in η.

Since effective intervalence relies on juxtaposition of Co sites, and admixture would inevitably increase the mean Co–Co distance within this solid-state

Table 6.11. *Comparative speeds of hydroxide-ion movement through electrodeposited cobalt, tungsten and Co–W mixed oxides. The $\bar{D}$ data come from ref. 80.*

Oxide film	$\bar{D}$/cm^2 s^{-1}
CoO	2.3×10^{-8}
Co–WO$_3$	48.7×10^{-8}
WO$_3$	5.5×10^{-8}

mixture, possibly the 'Co–Al oxide' here in reality comprises aggregated clusters of the two constituent oxides, each as a pure oxide.

Indium oxide as an electrochromic host

The most commonly encountered mixed-metal oxide is indium–tin oxide (ITO), which is widely used in the construction of ECDs, and typically comprises about 9 mol% SnO_2.[802] Some of the tin oxide dopant has the composition of Sn_2O_3.[803] The most common alternative to ITO as an optically transparent electrode is tin oxide doped with fluoride (abbreviated to FTO), although the oxides of Ni[804] and Sb[764,765] have also been incorporated into In_2O_3.

While old, a review in 1983 by Chopra et al.[805] still contains information of interest, although the majority concerns ITO acting as a *conductive* electrode rather than a redox-active *insertion* electrode. The more recent review (2001) by Nagai[806] discusses the electrochemical properties of ITO films; however, the most recent review was in 2002 by Granqvist and Hultåker.[807]

Preparation of ITO electrochromes Electrochromic ITO is generally made by rf sputtering,[239,240,241,242,243,244,245] or reactive dc sputtering.[208] Room-temperature pulsed-laser deposition can also yield ITO.[808] Reactive electron-beam deposition onto heated glass also yields good-quality ITO,[229] but is not employed often since the resultant film is oxygen deficient and has a poorer transparency than material of complete stoichiometry. When preparing ITO films by electron-beam evaporation,[229,809] the precursor is $In_2O_3 + 9$ mol% of SnO_2, evaporated directly onto a glass substrate in an oxygen atmosphere of pressure of $\sim 5 \times 10^{-4}$ Torr. The ITO made by these routes is largely amorphous.

Other electrochromic ITO layers have been made via sol–gel,[183] and spin coating a dispersion of tin-doped indium oxide 'nanoparticle'.[186,187]

Redox electrochemistry of ITO When a thin-film ITO immersed in a solution of electroactive reactants has a negative potential applied, it will conduct charge to and/or from the redox species in solution. It behaves as a typical electrode substrate (see, for example, Section 14.3). By contrast, if the surrounding electrolyte solution contains no redox couple, then some of the metal centres within the film are themselves electroreduced.[810] Curiously, doubt persists whether it is the tin or the indium species of ITO which are reduced: the majority view is that all redox chemistry in such ITO occurs at the tin sites, the product being a solid solution; Eq. (6.43):

$$\text{ITO (s)} + x(\text{M}^+ + \text{e}^-) \rightarrow \text{M}_x\text{ITO (s)}, \qquad (6.43)$$

colourless   pale brown

where M is usually Li$^+$, e.g. from LiClO$_4$ electrolyte in PC, but it may be H$^+$. The resultant partially reduced oxide M$_x$ITO may be symbolised as M$_x$Sn^{IV,II}O$_2$(In$_2$O$_3$), where the indium is inert.

The reduced form of ITO is chemically unstable, as outlined in Section 16.2.

Ion insertion into ITO is extremely slow, with most of the cited values of chemical diffusion coefficient $\bar{D}$ lying in the range 10^{-13} to 10^{-16} cm^2 s^{-1},[809] although Yu et al.[811] cite 1 × 10^{-11} cm^2 s^{-1}. These low values may also be the cause of hysteresis in coulometric titration curves.[811] Electroreversibility is problematic if Li$^+$ rather than H$^+$ is the mobile ion inserted, so redox cycles ought to be shallow (i.e. with x in Eq. (6.43) kept relatively small). Contrarily, reductive incorporation of Li$^+$ *increases* the electronic conductivity of the ITO.[241,243]

Few cycle lives are cited in the literature: Golden and Steele[244] and Corradini et al.[812] are probably the only authors to cite a high write–erase efficiency (of 10^4 and 2 × 10^4 cycles, respectively).

Optical properties of ITO Some ITO has no visible electrochromism,[809] and is therefore a perfect choice for a 'passive' counter electrode. The colour of reduced ITO of different origin is pale brown (possibly owing to Sn^{II}); see Figure 6.19. The coloration efficiency η is 2.8 cm^2 C^{-1} at 600 nm;[244] M$_x$ITO is too pale to adopt as a primary electrochrome since its maximal *CR* is only 1:1.2.[241,242,243,802,809,812,813,814,815,816]

A recent report suggesting a yellow–blue colour was formed during electroreduction of ITO is intriguing since the source of the blue is, as yet, quite unknown.[817] Perhaps similar is the mixed-valent behaviour recently inferred for[677] Li$_x$SnO$_2$, as determined by Mössbauer measurements. The Li$_x$SnO$_2$ in that study was made by Li$^+$ insertion into sputtered SnO$_2$.[676]

Devices containing ITO counter electrodes When considered for use as an electrochrome, ITO is always the secondary 'optically passive' ion-insertion layer, e.g. with WO$_3$[243,277,811,818,819] or poly(3-methylthiophene)[812] as the primary electrochrome.

Bressers and Meulenkamp[820] consider that ITO 'probably cannot be used as a combined ion-storage layer and transparent conductor for all-solid-state ... switching device in view of its [poor] long-term stability'. X-Ray photoelectron spectroscopy studies seem to support this conclusion.[821]

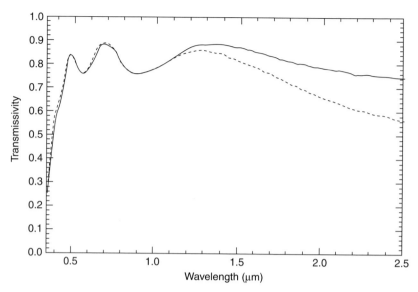

Figure 6.19 UV-visible spectrum of thin-film ITO in its oxidised (— clear) and partially reduced (· · · · pale brown) forms. (Figure reproduced from Goldner, R. B. et al. 'Electrochromic behaviour in ITO and related oxides'. *Appl. Opt.*, **24**, 1985, 2283–4, by permission of The Optical Society of America.)

Iridium oxide as electrochromic host

Iridium oxide has been doped with the oxides of magnesium[822] and with tantalum.[823] Films of composition $IrMg_yO_z$ ($2.5 < y < 3$) are superior to iridium oxide alone, for the electrochromic modulation is wider, and the bleached state is more transparent.[822] Such a high proportion of magnesium is surprising, considering the electro-inactive nature of MgO.

Addition of Ta_2O_5 decreases the coloration efficiency η but increases chemical diffusion coefficient $\overline{D}$. The changes are thought to be the result of diluting the colouring IrO_2 with Ta_2O_5, which supports a superior ionic conductivity.

Iridium oxide has also been incorporated into aramid resin, poly(*p*-phenylene terephthalamide).[531]

Iron oxide as electrochromic host

Iron oxide has been host to the oxides of Si and Ti, as prepared by sol–gel methods.[824] The films investigated are able reversibly to take up Li^+, Na^+ and K^+ ions. The coloration efficiencies η of this mixed oxide lie in the range[824] 6–14 cm^2 C^{-1} at λ_{max} of 450 nm (the authors do not say which compositions

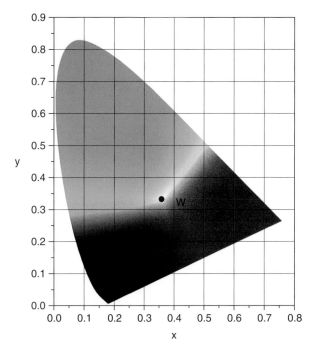

Plate 1 Colour CIE 1931 *xy* chromaticity diagram with labelled white point (**W**).

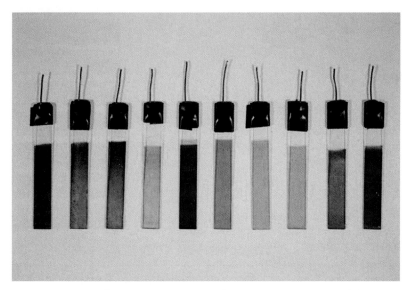

Plate 2 A series of neutral EDOT and BEDOT-arylene variable colour electrochromic polymer films on ITO–glass illustrating range of colours available. (Original figure as used for published black and white photo from Sapp, S., Sotzing, G. A. and Reynolds, J. R. 'High contrast ratio and fast-switching dual polymer electrochromic devices'. *Chem. Mater.*, **10**, 1998, 2101–8, by permission of The American Chemical Society.)

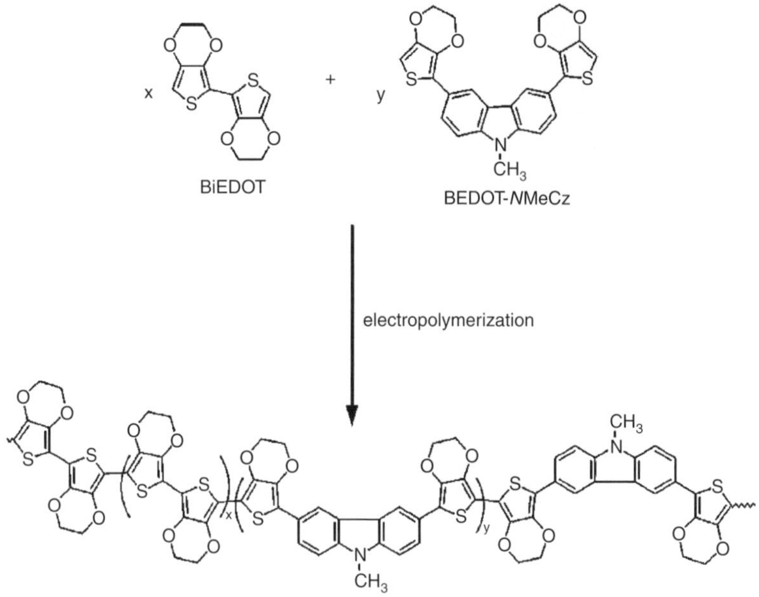

Comonomer Solution Composition	Neutral Polymer λ_{max} (nm)	Neutral Electrochromic Response (Photograph)
100% BiEDOT	577	
90:10	559	
80:20	530	
70:30	464	
50:50	434	
30:70	431	
20:80	429	
10:90	420	
100% BEDOT-NMeCz	420	

Plate 3 Representative structures and electrochromic properties of electrochemically prepared copolymers of varied compositions. (Figure reproduced from Gaupp, C. L. and Reynolds, J. R. 'Multichromic copolymers based on 3,6-bis[2-(3,4-ethylenedioxythiophene)]-N-alkylcarbazole derivatives'. *Macromolecules*, **36**, 2003, 6305–15, by permission of The American Chemical Society.)

Plate 4 Gentex window of area $1 \times 2\,\text{m}^2$. The top right pane has been electro-coloured. The other three panes are bleached. (Reproduced with permission from Rosseinsky, D. R. and Mortimer, R. J. 'Electrochromic systems and the prospects for devices'. *Adv. Mater.*, **13**, 2001, 783–93, with permission of VCH–Wiley.)

Plate 5 All-solid-state electrochromic motorcycle helmet manufactured in Sweden by Chromogenics AB. The primary electrochrome layer is WO_3, and the secondary layer is NiO_x. (Reproduced with permission of Professor C. G. Granqvist, of Uppsala University.)

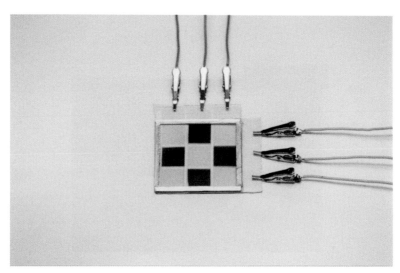

Plate 6 Pixel array showing no cross-talk between close picture elements ('pixels'), with solution-phase electrochromes. The unconnected pixels experience insufficient potential for coloration spread to ensue, even though the electrochromes (TMPD and heptyl viologen) are always in solution. The pixels can be made virtually microscopic in size. (Reproduced with permission from Leventis, N., Chen, M., Liapis, A. I., Johnson, J. W. and Jain, A. 'Characterization of 3×3 matrix arrays of solution-phase electrochromic cells'. *J. Electrochem. Soc.*, **145**, 1998, L55–8, with permission of The Electrochemical Society.)

Plate 7 Gentex windows being tested in Florida. A man is just visible beneath the nearest. (Reproduced with permission from Rosseinsky, D. R. and Mortimer, R. J. 'Electrochromic systems and the prospects for devices'. *Adv. Mater.*, **13**, 2001, 783–93, with permission of VCH–Wiley.)

6.4 Metal oxides: dual-metal electrochromes

relate to these values of η except 'the largest extent of colouring and bleaching was for pure iron oxide').

Molybdenum oxide as electrochromic host

Thin-film molybdenum oxide has also been made as a mixture with the oxides of Co,[80,81,91] Cr,[91] Fe,[91] Nb,[165,171,750,825] Ni,[91] Sn,[826,827] Ti,[22,826,828] V[124,202,829] or W.[61,62,91,292,361,475,730,737,738,739,740,741,742,743,744,745,746,747]

Most thin films of Mo–W oxide were prepared from reactive sputtering, but others have been prepared by sol–gel techniques, e.g. from a solution of peroxopolymolybdotungstate,[125] itself made by oxidative dissolution of both metallic molybdenum and tungsten in hydrogen peroxide (see p. 133 ff.). Molybdenum–vanadium oxide is also made by dissolving the respective metals in H_2O_2.[124,202] The electrochromic transition for the resultant film is 'green–yellow → violet' when cycled in $LiClO_4$–PC solution as the ion-providing electrolyte.

An additional benefit of incorporating molybdenum into an electrochromic mixture is its ability to extend the overpotential for hydrogen evolution (a nuisance if occurring at lower potentials) when in contact with a protonic acid. As an example, H_2 is first formed at the surface of the MoO_3 layer at -0.85 V (vs. SCE), cf. -0.75 V for electrodeposited Mo–W oxide (electrodeposited together on gold). More impressive still, no gas whatsoever forms when a gold electrode is coated with similarly formed Mo–Cr and Mo–Fe oxides.[830]

The coloration efficiency η for MoO_3–SnO_2 films[826] is low, being in the range 2–10 cm^2 C^{-1}, cf. 77 cm^2 C^{-1} for MoO_3 alone[7] and 3 cm^2 C^{-1} for ITO alone[244] (although some ITO is completely passive optically[809]). These data are summarised in Table 6.12, which clearly shows how the optical behaviour of Mo–Sn oxide is more akin to SnO_2 than to MoO_3. Possibly the tin sites are electroactive while the Mo sites are not.

The value of η for Mo–Ti oxide lies in the range 10–50 cm^2 C^{-1}, the value increasing as the mole fraction of molybdenum increases.[22]

Table 6.12. *Effect on the coloration efficiency η of mixing molybdenum and tin oxides.*

Films	η/cm^2 C^{-1}	Ref.
MoO_3	77	7
ITO	3	244
ITO	0	809
MoO_3–SnO_2	2–10	826

Nickel oxide as electrochromic host

Several electrochromic mixtures have been prepared of nickel oxide, e.g. with oxides of Ag,[77,831] Al,[831,832,833,834] Cd,[77,83] Ce,[77] Co,[77,79,80,81,82,83,140,801] Cr,[77,835] Cu,[77] Fe,[77,836] La,[77,82,84,837,838] Mg,[77,831,832,833,839] Mn,[636,833,840] Nb,[636,831] Pb,[77] Si,[167,831] Sn,[841] Ta,[831] V,[761,831,832,833,834] W,[53,80,81,89,94,99,735,736,751,752] Y[77] and Zn.[83] Nickel oxide has also been mixed with particles of various alloys, such as Ni–Au alloy,[842] to yield films with markedly different spectra. Traces of ferrocyanide have been incorporated,[82] and films containing gold are also made readily.[161] Nickel tungstate is also electrochromic.[843]

Nickel oxide often shows a residual absorption, an unwanted brown tint, but incorporating Al or Mg in the film virtually eliminates this colour.[832] Thus for applications requiring a highly bleached transmittance, such as architectural windows, the Al- and Mg-containing oxides are superior to conventional nickel oxide, for their greatly enhanced transparency.[832,834] Such films also show superior charge capacity.[832]

Incorporation of Ce, Cr or La into NiO improves the rates of electrocoloration, while adding Ce, Cr or Pb retards the rates of bleaching.[77] Addition of yttrium oxide severely impedes the rate of NiO electrocoloration, for reasons not yet clear.[77] An important observation for ECD construction is that electrodeposited Ni–La and Ni–Ce oxides are significantly more durable than NiO alone,[77] as evidenced by longer cycle life.

Tungsten trioxide is cathodically colouring while NiO is anodically colouring, so it is interesting that electrodeposited Ni–W oxide has a rather low coloration efficiency of[99] $4.4\,cm^2\,C^{-1}$ while η for sol–gel-derived NiO is[152] $-(35\ to\ 40)\,cm^2\,C^{-1}$.

The complex $[Ru_3O(acetate)_6\text{-}\mu\text{-}\{pyrazine\}_3\text{-}[Fe(CN)_5]_3]^{n-}$ has also been incorporated into NiO_x.[844]

Niobium oxide as electrochromic host

Electrochromic films have been prepared that are doped with the oxides of Ce,[769] Fe,[845] Mo,[171,750,825] Ni,[831] Sn,[171] Ti,[171,846] W[749,750] and Zn.[165,171] Lee and Crayston have also made a Nb–silicone composite.[642]

In a recent study of sol–gel deposited Nb_2O_5, Schmitt and Aegerter[171] prepared a variety of films that were doped with a variety of d-block oxides. The coloration efficiencies of such films were not particularly sensitive to the other metals, the highest being for Nb_2O_5 containing 20% TiO_2, which has a coloration efficiency of $27\,cm^2\,C^{-1}$. The maximum change in transmittance was observed for films comprising 20% Mo.

Table 6.13. *Effect on the coloration efficiency η of mixing niobium oxides with iron or titanium oxide: the effect of mixing and preparation method.*

Components	Preparation route	$\eta/\text{cm}^2\,\text{C}^{-1}$	Ref.
Nb_2O_5	rf sputtering	22	170
Nb_2O_5	rf sputtering	<12	258
Nb_2O_5	Sol–gel	16	171
Nb_2O_5	Sol–gel	25–30	748
FeO	CVD	−6 to −6.5	629
Fe_2O_3	Electrodeposition	−30	102
Nb_2O_5–FeO	CVD	20	845
$Nb_2O_5 + 20\%\ TiO_2$	Sol–gel	27	171
$HNbWO_6$ (hydrated)	Sol–gel	54	748
WO_3	Sol–gel	48	748

It is clear that Nb–W oxide behaves more like WO_3 than Nb_2O_5;[749] and Nb–Fe oxide behaves more like Nb_2O_5 than either FeO or Fe_2O_3.[845] Hydrated $HNbWO_6$ also has a superior chemical stability to that of WO_3 alone,[748] and doped niobium oxide is also more electrochemically stable.[171]

The coloration efficiencies η for such mixed Nb–metal oxide films are all low. Representative values are summarised in Table 6.13. Hydrated $HNbWO_6$ has a similar coloration efficiency ($54\,\text{cm}^2\,\text{C}^{-1}$)[748,847] to that of WO_3; *cf.* $48\,\text{cm}^2\,\text{C}^{-1}$ for WO_3 prepared by the same procedures.

Tin oxide as electrochromic host

Electrochromic mixtures have been prepared of tin oxide, with the oxides of Ce,[755,771] Mo,[826,827] Ni,[841] Sb[827] or V.[848] The film of Ce–Sn oxide was wholly optically inactive, with a transparency higher than 90%.

Titanium oxide as electrochromic host

Electrochromic mixtures of titanium are at present much used. Electrochromic mixtures have been prepared of TiO_2 with oxides of Ce,[122,205,323,324,755,771,772,773,774,775,776,777,778,779,780,781,782,783,784,785,786,787] Fe,[849,850] La,[779,780] Mo,[22,826,828] Nb,[165,171,825] Ni,[150,841] Pr,[779,780] Ta,[754] V,[721,851,852,853,854] W,[127,129,134,203,316,335,404,755,756,757,758,759,760,855] and Zn.[563] A mixture of TiO_2 and phosphotungstic acid has been made via sol–gel techniques,[768] and TiO_2 containing hexacyanoferrate has also been produced.[856]

Most of these electrochrome mixtures were made by sol–gel or sputtering techniques. For example, Ni–Ti oxide is made from $NiCl_2$ and Ti alkoxide,[150] and Ti–Fe oxide was prepared by a dip-coating procedure[850] via a liquor comprising alcoholic ferric nitrate and $Ti(O^iPr)_4$, followed by annealing in air. In ref. 680, however, the layer of W–Ti oxide was made by pulsed cathodic electrodeposition.

The value of λ_{max} for the Ni–Ti oxide[150] is 633 nm, and η lies in the range $-(10–42)\,cm^2\,C^{-1}$. The optical charge-transfer transition in the Ti–Fe system is responsible for the blue colour of naturally occurring sapphire;[857] but thin-film Ti–Fe oxide (prepared in this case by a dip-coating procedure[850]) did not possess the same colour as sapphire, probably having a different structure.

Vanadium oxide as electrochromic host

Electrochromic mixtures have been prepared of vanadium oxide, with the oxides of Bi,[594] Ce,[788,789,790,791,792,793] Dy,[858] Fe,[159] In,[859] Mo,[124,202,829] Nd,[858] Ni,[761,831,832] Pa,[703] Pr,[858,860] Sm,[858] Sn,[848] Ti[166,687,721,851,852,853,854] or W.[204,254,325,327,687,761]

Thin films of composition $(V_2O_5)_3$–$(TiO_2)_7$ oxide form a reddish brown colour at anodic potentials which Nagase et al.[854] attribute to the vanadium component, implying the majority TiO_2 component is optically passive.

When doped with the rare-earth oxides of Nd, Sm, Dy,[858] films of V_2O_5 show a considerably enhanced cycle life. X-Ray diffraction results suggest the formation of the respective orthovanadate species $SmVO_4$ and $DyVO_4$. The V–Sm oxide film showed a very small coloration efficiency η of only $0.6\,cm^2\,C^{-1}$, so the authors suggest counter-electrode use. Similarly, a film of Ni–V oxide is 'virtually [optically] passive', although no values of η are cited.[761] Other electrochromic vanadates include $FeVO_4$[159] and $CeVO_4$.[861]

The electrochromic behaviour of V–Ti oxide films is complicated:[853] in the best explanatory model, the inserted electrons are supposed to be localised, residing preferentially at vanadium sites. The V–Ti films have a larger charge capacity if the mole fraction of vanadium is relatively high.[862]

Oxide electrochromes having a grey hue, rather than blue, are said to be 'neutral' in colour; see p. 399. Such neutral colours have been made with V–Ti oxide (with brown–blue electrochromism);[687] and for V–W oxide which has a coloration efficiency in the range 7 to $30\,cm^2\,C^{-1}$, the value depending on the composition, with η decreasing as the mole % of vanadium increases.[863]

Composites of vanadium oxide have been formed by reacting a xerogel (see p. 161) with organic materials such as the nanocomposite [poly(aniline N-propanesulfonic acid)$_{0.3}V_2O_5$].[864] This material has a superior electronic conductivity to the precursor V_2O_5 xerogel alone and exhibits shorter ionic diffusion pathways, both properties implying a fast electrochromic transition.[864] The second V_2O_5–organic composite is a 'melanine like' material formed by reacting 3,4-dihydroxyphenylalanine with a V_2O_5 xerogel. This latter material generates a dark blue metallic electrochromic colour.[728,865]

Zirconium oxide as electrochromic host

Pure zirconium oxide is not electrochromic and has practically zero charge capacity,[796] but has been host to a large number of other oxides. It is now a popular choice of optically passive electrochromic layer when mixed with cerium oxide.[122,767,783,786,794,795,796] For example, in Granqvist et al.'s 1998 review of devices,[797] they cite Zr–Ce as the optically passive secondary layer, referring to material in the compositional range $Zr_{0.4}Ce_{0.6}O_2$ to $Zr_{0.25}Ce_{0.75}O_2$. The charge capacity of Zr–Ce oxide increases with increasing cerium content.[796]

Miscellaneous electrochromic hosts

Tantalum–zirconium oxide is electrochromic.[866] Its electrochromic qualities are said to be superior to either constituent oxide, suggesting a new phase rather than a mixture. Its coloration efficiency η is estimated to be 47 cm^2 C^{-1} at 650 nm.

Electrochromic iridium–ruthenium oxide in the molar ratio 40:50% is said to be 300 times more stable than either constituent oxide.[867]

Ternary and higher oxides

A few multiple-metal oxides have been made: for example, electrodeposition can be employed to produce mixtures of tungsten oxide together with three or even four additional metal oxides.[96] A notable mixture is W–Cr–Mo–Ni oxide,[96] which forms a green electrochromic colour – a colour not often seen in the field of inorganic electrochromism, and, though insufficiently analysed, possibly not caused here by charge transfer.

Most of these mixtures were prepared to 'tweak' the optical properties of a host oxide. For example, thin films of oxides based on Ni–V–Mg (made by reactive dc magnetron sputtering) show pronounced anodic electrochromism. The addition of magnesium significantly enhances the optical transparency of the films in their bleached state,[839] over the wavelength range $400 < \lambda < 500$ nm.

With counter-electrode use in mind, Orel and co-workers[166] made V–Ti–Zr and V–Ti–Ce oxides, and Avandano et al. made CeO_2–TiO_2–ZrO_2,[156] $NiV_{0.08}Mg_{0.5}$ oxides,[832] and CeO_2–TiO_2–ZrO_2.[156]

Samples of $NiO\cdot WO_xP_y$ were obtained from a polytungsten gel in which H_3PO_4 was added. The electrochromism was optimised when the P:W ratio was 100:8.3.[140]

Several other ternary oxides comprising three transition-metal oxides have received attention: the oxides of Co–Ni–Ir[868] and Cr–Fe–Ni (this latter oxide being grown anodically on the metallic alloy *Inconel-600*)[869] and W–V–Ti oxide.[727] Ternary oxides comprising p-block metals include Co–Al–Si[617,798] and Ce–Mo–Si.[768] The electrochromic behaviour of the materials $(WO_3)_x(Li_2O)_y(MO)_z$ where M = Ce, Fe, Mn, Nb, Sb or V has also been studied.[870]

Finally, Lian and Birss[871] have studied the electrochromism of the hydrous oxide layer formed on the alloy $Ni_{51}Co_{23}Cr_{10}Mo_7Fe_{5.5}B_{3.5}$. Its electrochromic behaviour is, apparently, similar to that of NiO_x.

6.4.3 Electrochromic oxides incorporating precious metals

Several workers have incorporated particulate precious metal in an oxide host. Table 6.14 lists a few such studies.

Such composites can be made in various ways: dual-target sputtering, mixed sputtering and sol–gel, or all sol–gel.[161] In the study of Au–NiO films by Fantini et al.,[874] the Au mole fraction of gold varied between from 0.0 to 0.05. The films reflected the different colours blue, green, yellow and orange–red, depending on mole fraction.

The electrochromic ceramic metal ('cermet') $Au–WO_3$ prepared by Sichel and Gittleman[879] comprised a matrix of amorphous WO_3 containing grains of Au of approximate diameter 20–120 Å. The cermet is blue as prepared, but is red or pink when electrochemically coloured – a relatively rare colour for an electrochromic oxide. The matrix must be amorphous in order for the red colour to develop.

In the study in ref. 877, Yano et al. also incorporated particulate gold (and V_2O_5) in an aramid resin.

Table 6.14. *Electrochromic mixtures of metal oxide incorporating precious metal.*

Precious metal	Host	Ref.
Ag	ITO	800, 872
Ag	V_2O_5	873
Ag	WO_3	830
Au	CoO	874
Au	IrO_2	531
Au	NiO	161, 874, 875
Au	MoO_3	495, 876
Au	V_2O_5	531, 877, 878
Au	WO_3	201, 830, 879, 880, 881
Pt	MoO_3	495
Pt	RuO_2	882
Pt	Ta_2O_5	883
Pt	WO_3	879, 884

6.4.4 Metal oxyfluorides

Many thin-film metal oxyfluorides are electrochromic. In the literature, the exact stoichiometry is often indefinite or unknown. In effect, they represent fluorinated analogues of the respective metal oxide. For this reason, we term the oxides, 'F:MO$_x$'.

Tin Films of F:SnO$_2$ were made by reactive rf sputtering in Ar + O$_2$ + CF$_4$ atmosphere. Rutherford backscattering (RBS) suggests the film composition is SnO$_{2.1}$F$_{0.6}$C$_{0.3}$.[885]

When such films are immersed in PC containing LiClO$_4$, the electrochromic effect is weak. The redox reaction causing the colour is:

$$\text{F:SnO}_2 + x(\text{Li}^+ + \text{e}^-) \rightarrow \text{Li}_x\text{F:SnO}_2. \tag{6.44}$$

It is easier to electro-insert Li$^+$ into SnO$_2$ electrodes than into fluorinated F:SnO$_2$.[885] For this reason, fluorinated tin oxide is superior as an optically transparent electrode, but is a poor electrochromic oxide.

Titanium Thin-film titanium oxyfluoride is made by reactive dc sputtering in an Ar + O$_2$ + CF$_4$ atmosphere. The amount of fluorine incorporated in the film is quite small: results from RBS suggest a composition of TiO$_{1.95}$F$_{0.1}$.[886]

When such films are immersed in PC containing LiClO$_4$, the electrochromic effect is 'pronounced'. The redox reaction causing the colour is:

$$\text{F:TiO}_2 + x(\text{Li}^+ + \text{e}^-) \rightarrow \text{Li}_x\text{F:TiO}_2. \tag{6.45}$$

The coloration efficiency is 37 cm^2 C^{-1} at 700 nm, the colour said to derive from photo-effected polaron interaction. The cycle life is as high as 2 × 10^4 cycles.[887] As expected, the diffusion of Na$^+$ or K$^+$ through F:TiO$_2$ is too slow to countenance inclusion within devices. In fact, structural changes accompany the incorporation of K$^+$.[888]

Tungsten Granqvist and co-workers[889] made thin-film tungsten oxyfluoride by reactive dc magnetron sputtering in plasmas containing O$_2$ + CF$_4$. Elevated target temperatures yielded strongly enhanced rates of electrochromic coloration. The coloration efficiency η is 60 cm^2 C^{-1}, and the wavelength maximum occurs at ∼780 nm.[890] The redox reaction causing the colour is:

$$\text{F:WO}_3 + x(\text{Li}^+ + \text{e}^-) \rightarrow \text{Li}_x\text{F:WO}_3. \tag{6.46}$$

The durability of such films with extensive Li$^+$ intercalation and egress was said to be poor, but the electrochromic colour–bleach dynamics are faster than for films of WO$_3$. Covering the film with a thin, protective layer of electron-bombarded WO$_3$ yields an electrochrome with rapid dynamics and good durability. The exact rôle of the oxide coating is uncertain, but it is conceivable that it may prevent dissolved oxyfluoride species from leaving the film.[496,891]

References

1. Rest, A. Polyene and linearly conjugated dyes. In Coyle, J. D., Hill, R. R. and Roberts, D. R. (eds.), *Light, Chemical Change and Life: A Source Book in Photochemistry*, Milton Keynes, Open University, 1982.
2. Granqvist, C. G. Electrochromic oxides: a unified view. *Solid State Ionics*, **70–1**, 1994, 678–85.
3. Granqvist, G. C. *Handbook of Inorganic Electrochromic Materials*, Amsterdam, Elsevier, 1995.
4. Monk, P. M. S., Mortimer, R. J. and Rosseinsky, D. R. *Electrochromism: Fundamentals and Applications*, Weinheim, VCH, 1995.
5. Dautremont-Smith, W. C. Transition metal oxide electrochromic materials and displays: a review. Part 1: oxides with cathodic coloration. *Displays*, **3**, 1982, 3–22.
6. Dautremont-Smith, W. C. Transition metal oxide electrochromic materials and displays, a review. Part 2: oxides with anodic coloration. *Displays*, **3**, 1982, 67–80.
7. Faughnan, B. W. and Crandall, R. S. Electrochromic devices based on WO$_3$. In Pankove J. L. (ed.), *Display Devices*, Berlin, Springer-Verlag, 1980, pp. 181–211.
8. Hagenmuller, P. Tungsten bronzes, vanadium bronzes and related compounds. In *Comprehensive Inorganic Chemistry*, New York, Pergamon, 1973, vol. 4, pp. 541–605.
9. van den Meerakker, J. E. A. M., Baarslag, P. C. and Scholten, M. On the mechanism of ITO etching in halogen acids: the influence of oxidizing agents. *J. Electrochem. Soc.*, **142**, 1995, 2321–6.
10. Monk, P. M. S. and Man, C. M. Reductive ion insertion into thin-film indium tin oxide (ITO) in aqueous acidic solutions: the effect of leaching of indium from the ITO *J. Mater. Sci., Electron. Mater.*, **10**, 1999, 101–7.
11. Córdoba-Torresi, S. I., Gabrielli, C., Hugot-Le Goff, A. and Torresi, R. Electrochromic behaviour of nickel oxide electrodes, I: identification of the colored state using quartz-crystal microbalance. *J. Electrochem. Soc.*, **138**, 1991, 1548–1553.
12. Randin, J.-P. Chemical and electrochemical stability of WO$_3$ electrochromic films in liquid electrolytes. *J. Electron. Mater.*, **7**, 1978, 47–63.
13. Randin, J.-P. Ion-containing polymers as semisolid electrolytes in WO$_3$-based electrochromic devices. *J. Electrochem. Soc.*, **129**, 1982, 1215–1220.
14. Arnoldussen, T. C. A model for electrochromic tungsten oxide microstructure and degradation *J. Electrochem. Soc.*, **128**, 1981, 117–23.
15. Duffy, J. A., Ingram, M. D. and Monk, P. M. S. The effect of moisture on tungsten oxide electrochromism in polymer electrolyte devices. *Solid State Ionics*, **58**, 1992, 109–14.
16. Burke, L. D. and Scannell, R. A. The effect of UV light on the hydrous oxides of iridium. *J. Electroanal. Chem.*, **257**, 1988, 101–21.

17. Nakaoka, K., Ueyama, J. and Ogura, K. Semiconductor and electrochromic properties of electrochemically deposited nickel oxide films. *J. Electroanal. Chem.*, **571**, 2004, 93–9.
18. Carpenter, M. K. and Corrigan, D. A. Photoelectrochemistry of nickel hydroxide thin films. *J. Electrochem. Soc.*, **136**, 1989, 1022–6.
19. Fleisch, T. H. and Mains, G. J. An XPS study of the UV reduction and photochromism of MoO_3 and WO_3. *J. Chem. Phys.*, **76**, 1982, 780–6.
20. Scarminio, J., Lourenco, A. and Gorenstein, A. Electrochromism and photochromism in amorphous molybdenum oxide films. *Thin Solid Films*, **302**, 1997, 66–70.
21. Mondragon, M. N., Zelaya-Angel, O., Ramirez-Bon, R., Herrera, J. L. and Reyes-Betanzo, C. Refraction index and oscillator strength in MoO_3 photocolored films. *Physica B: Condens. Matter*, **271**, 1999, 369–73.
22. Kullman, L., Azens, A. and Granqvist, C. G. Electrochromism and photochromism of reactively DC magnetron sputtered Mo–Ti oxide films. *Sol. Energy Mater. Sol. Cells*, **61**, 2000, 189–96.
23. Deb, S. K. and Chopoorian, J. A. Optical properties and color-formation in thin films of molybdenum trioxide. *J. Appl. Phys.*, **37**, 1966, 4818–25.
24. Özer, N. Reproducibility of the coloration processes in TiO_2 films. *Thin Solid Films*, **214**, 1992, 17–24.
25. Gomez, M., Rodriguez, J., Lindquist, S.-E. and Granqvist, C. G. Photoelectrochemical studies of dye-sensitized polycrystalline titanium oxide thin films prepared by sputtering. *Thin Solid Films*, **342**, 1999, 148–52.
26. Gomez, M. M., Beermann, N., Lu, J. *et al.* Dye-sensitized sputtered titanium oxide films for photovoltaic applications: influence of the O_2/Ar gas flow ratio during the deposition. *Sol. Energy Mater. Sol. Cells*, **76**, 2003, 37–56.
27. Bechinger, C., Burdis, M. S. and Zhang, J.-G. Comparison between electrochromic and photochromic coloration efficiency of tungsten oxide thin films. *Solid State Commun.*, **101**, 1997, 753–6.
28. Gavrilyuk, A. I. Photochromism in WO_3 thin films. *Electrochim. Acta*, **44**, 1999, 3027–37.
29. Avellaneda, C. O. and Bulhões, L. O. S. Photochromic properties of WO_3 and WO_3:X (X = Ti, Nb, Ta and Zr) thin films. *Solid State Ionics*, **165**, 2003, 117–121.
30. Scarminio, J. Stress in photochromic and electrochromic effects on tungsten oxide film. *Sol. Energy Mater. Sol. Cells*, **79**, 2003, 357–68.
31. Argazzi, R., Murakami Iha, N. Y., Zabri, H., Odobel, F. and Bignozzi, C. A. Design of molecular dyes for application in photoelectrochemical and electrochromic devices based on nanocrystalline metal oxide semiconductors. *Coord. Chem. Rev.*, **248**, 2004, 1299–316.
32. Bedja, I., Hotchandani, S., Carpentier, R., Vinodgopal, K. and Kamat, P. V. Electrochromic and photoelectrochemical behavior of thin WO_3 films prepared from quantized colloidal particles. *Thin Solid Films*, **247**, 1994, 195–200.
33. Su, L., Zhang, L., Fang, J., Xu, M. and Lu, Z. Electrochromic and photoelectrochemical behavior of electrodeposited tungsten trioxide films. *Sol. Energy Mater. Sol. Cells*, **58**, 1999, 133–40.
34. Loo, B. H., Yao, J. N., Dwain Coble, H., Hashimoto, K. and Fujishima, A. A Raman microprobe study of the electrochromic and photochromic thin films of molybdenum trioxide and tungsten trioxide. *Appl. Surf. Sci.*, **81**, 1994, 175–81.
35. Green, M. Atom motion in tungsten bronze thin films. *Thin Solid Films*, **50**, 1978, 148–50.

36. Ord, J. L., Bishop, S. D. and DeSmet, D. J. Hydrogen insertion into anodic oxide films on vanadium. *Proc. Electrochem. Soc.*, **90–2**, 1990, 116–24.
37. Dini, D., Passerini, S., Scrosati, B. and Decker, F. Stress changes in electrochromic thin film electrodes: laser beam deflection method (LBDM) as a tool for the analysis of intercalation processes. *Sol. Energy Mater. Sol. Cells*, **56**, 1999, 213–21.
38. Barbero, C., Miras, M. C. and Kotz, R. Electrochemical mass transport studied by probe beam deflection: potential step experiments. *Electrochim. Acta*, **37**, 1992, 429–37.
39. Giron, J.-C. and Lampert, C. M. Study by laser probe deflection of the ionic mechanisms of nickel oxide thin films. *Proc. Electrochem. Soc.*, **94–2**, 1994, 82–99.
40. Dini, D. and Decker, F. Stress in thin films of metal oxide electrodes for intercalation reactions. *Electrochim. Acta*, **43**, 1998, 2919–23.
41. Plinchon, V., Giron, J.-C., Deloulbe, J. P. and Lerbet, F. Detection by mirage effect of the counter-ion flux between an electrochrome and a liquid electrolyte: application to WO_3, Prussian blue and lutetium diphthalocyanine film. *Proc. SPIE*, **1536**, 1991, 37–47.
42. Nagai, J. Characterization of evaporated nickel oxide and its application to electrochromic glazing. *Sol. Energy Mater. Sol. Cells*, **31**, 1993, 291–9.
43. Faria, I. C., Torresi, R. and Gorenstein, A. Electrochemical intercalation in NiO_x thin films. *Electrochim. Acta*, **38**, 1993, 2765–71.
44. Krtil, P., Fattakhova, D., Kavan, L., Burnside, S. and Grätzel, M. Lithium insertion into self-organized mesoscopic TiO_2 (anatase) electrodes. *Solid State Ionics*, **135**, 2000, 101–6.
45. Bohnke, O., Vuillemin, B., Gabrielli, C., Keddam, M. and Perrot, H. An electrochemical quartz crystal microbalance study of lithium insertion into thin films of tungsten trioxide, II: experimental results and comparison with model calculations. *Electrochim. Acta*, **40**, 1995, 2765–73.
46. Avellaneda, C. O., Bueno, P. R., Faria, R. C. and Bulhões, L. O. S. Electrochromic properties of lithium doped WO_3 films prepared by the sol–gel process. *Electrochim. Acta*, **46**, 2001, 1977–81.
47. de Torresi, S. I. C., Gorenstein, A., Torresi, R. M. and Vazquez, M. V. Electrochromism of WO_3 in acid solutions: an electrochemical, optical and electrogravimetric study. *J. Electroanal. Chem.*, **318**, 1991, 131–44.
48. Vergé, M.-G., Olsson, C.-O. A. and Landolt, D. Anodic oxide growth on tungsten studied by EQCM, EIS and AES. *Corros. Sci.*, **46**, 2004, 2583–600.
49. Decker, F., Passerini, S., Pileggi, R. and Scrosati, B. The electrochromic process in non-stoichiometric nickel oxide thin film electrodes. *Electrochim. Acta*, **37**, 1992, 1033–8.
50. Talledo, A. and Granqvist, C. G. Electrochromic vanadium pentoxide based films: structural, electrochemical, and optical properties. *J. Appl. Phys.*, **77**, 1995, 4655–66.
51. Lee, S.-H., Seong, M. J., Tracy, C. E., Mascarenhas, A., Pitts, J. R. and Deb, S. K. Raman spectroscopic studies of electrochromic a-MoO_3 thin films. *Solid State Ionics*, **147**, 2002, 129–33.
52. Goldner, R. B., Arntz, F. O., Dickson, K., Goldner, M. A., Haas, T. E., Liv, T. Y., Slaven, S., Wei, G., Wong, K. K. and Zerigian, P. Some lessons learned from research on a thin film electrochromic window. *Solid State Ionics*, **70–1**, 1994, 613–18.
53. Penin, N., Rougier, A., Laffont, L., Poizot, P. and Tarascon, J.-M. Improved cyclability by tungsten addition in electrochromic NiO thin films. *Sol. Energy Mater. Sol. Cells*, **90**, 2005, 422–33.

54. Yishiike, N. and Kondo, S. Electrochemical properties of $WO_3.x(H_2O)$, II: the influence of crystallization as hydration. *J. Electrochem. Soc.*, **131**, 1984, 809–13.
55. Bell, J. M. and Skryabin, I. L. Failure modes of sol–gel deposited electrochromic devices. *Sol. Energy Mater. Sol. Cells*, **56**, 1999, 437–48.
56. Deepa, M., Kar, M. and Agnihotry, S. A. Electrodeposited tungsten oxide films: annealing effects on structure and electrochromic performance. *Thin Solid Films*, **468**, 2004, 32–42.
57. Granqvist, C. G. Electrochromic tungsten oxide films: review of progress 1993–1998. *Sol. Energy Mater. Sol. Cells*, **60**, 2000, 201–62.
58. Kullman, L. *Components of Smart Windows: Investigations of Electrochromic Films, Transparent Counter Electrodes and Sputtering Techniques*, Philadelphia, Coronet, 1999.
59. Venables, J. A. *Introduction to Surface and Thin Film Processes*, Cambridge, Cambridge University Press, 2000.
60. Abdellaoui, A., Bouchikhi, B., Leveque, G., Donnadieu, A. and Bath, A. Iteratively derived optical constants of MoO_3 polycrystalline thin films prepared by CVD. *Thin Solid Films*, **304**, 1997, 39–44.
61. Gesheva, K., Szekeres, A. and Ivanova, T. Optical properties of chemical vapor deposited thin films of molybdenum and tungsten based metal oxides. *Sol. Energy Mater. Sol. Cells*, **76**, 2003, 563–76.
62. Donnadieu, A., Davazoglou, D. and Abdellaoui, A. Structure, optical and electro-optical properties of polycrystalline WO_3 and MoO_3 thin films prepared by chemical vapour deposition. *Thin Solid Films*, **164**, 1988, 333–8.
63. Tracy, C. E. and Benson, D. K. Preparation of amorphous electrochromic tungsten oxide and molybdenum oxide by plasma enhanced chemical vapour deposition, *J. Vac. Sci. Technol., A*, **4**, 1986, 2377–83.
64. Gogova, D., Iossifova, A., Ivanova, T., Dimitrova, Z. and Gesheva, K. Electrochromic behavior in CVD grown tungsten oxide films. *J. Cryst. Growth*, **198–9**, 1999, 1230–4.
65. Gogova, D., Stoyanov, G. and Gesheva, K. A. Optimization of the growth rate of electrochromic WO_3 coatings, in-situ grown by chemical vapor deposition at atmospheric pressure. *Renewable Energy*, **8**, 1996, 546–50.
66. Davazoglou, D. and Donnadieu, A. Electrochromism in polycrystalline WO_3 thin films prepared by chemical vapour deposition at high temperature. *Thin Solid Films*, **164**, 1988, 369–74.
67. Davazoglou, D. and Donnadieu, A. Structure and optical properties of WO_3 thin films prepared by chemical vapour deposition. *Thin Solid Films*, **147**, 1987, 131–42.
68. Bohnke, O., Bohnke, C., Donnadieu, A. and Davazoglou, D. Electrochromic properties of polycrystalline thin films of tungsten trioxide prepared by chemical vapour deposition. *J. Appl. Electrochem.*, **18**, 1988, 447–53.
69. Davazoglou, D., Donnadieu, A. and Bohnke, O. Electrochromic effect in WO_3 thin films prepared by CVD. *Sol. Energy Mater.*, **16**, 1987, 55–65.
70. Donnadieu, A., Regragui, M., Abdellaoui, A. and Davazoglou, D. Optical and electrical properties of coloured and transparent states of polycrystalline WO_3 thin films prepared by CVD. *Proc. SPIE*, **1272**, 1990, 197–206.
71. Kuypers, A. D., Spee, C. I. M. A., Linden, J. L., Kirchner, G., Forsyth, J. F. and Mackor, A. Plasma-enhanced CVD of electrochromic materials, *Surf. Coat. Technol.*, **74–5**, 1995, 1033–7.
72. Kajiwara, K., Isobe, C. and Saitoh, M. An AES study of LPCVD Ta_2O_5 films on Si. *Surf. Interface Anal.*, **19**, 1992, 331–5.

73. Watanabe, H., Itoh, K.-I. and Matsumoto, O. Properties of V_2O_5 thin films deposited by means of plasma MOCVD. *Thin Solid Films*, **386**, 2001, 281–5.
74. Forsgren, K. and Harsta, A. Halide chemical vapour deposition of Ta_2O_5. *Thin Solid Films*, **343–4**, 1999, 111–14.
75. Meulenkamp, E. A. Mechanism of WO_3 electrodeposition from peroxy-tungstate solution. *J. Electrochem. Soc.*, **144**, 1997, 1664–72.
76. Falk, U. and Salkind, A. J. *Alkaline Storage Batteries*, New York, Wiley, 1969.
77. Corrigan, D. A. and Carpenter, M. K. Electrochromic nickel hydroxide films and the effect of foreign metal ions. *SPIE Institute Series*, **IS4**, 1990, 298–312.
78. Carpenter, M. K., Conell, R. S. and Corrigan, D. A. The electrochromic properties of hydrous nickel oxide. *Sol. Energy Mater.*, **16**, 1987, 333–46.
79. Monk, P. M. S. and Ayub, S. Solid-state properties of thin film electrochromic cobalt–nickel oxide. *Solid State Ionics*, **99**, 1997, 115–24.
80. Monk, P. M. S., Chester, S. L. and Higham, D. S. Electrodeposition of cobalt oxide doped with additional metal oxides: a new electrochromic counter-electrode material. *Proc. Electrochem. Soc.*, **94–2**, 1994, 100–12.
81. Monk, P. M. S., Chester, S. L., Higham, D. S. and Partridge, R. D. Electrodeposition of cobalt oxide doped with additional metal oxides. *Electrochim. Acta*, **39**, 1994, 2277–84.
82. Corrigan, D. A. Durable electrochromic films of nickel hydroxide via chemical modifications. *Sol. Energy Mater. Sol. Cells*, **25**, 1992, 293–300.
83. Provazi, K., Giz, M. J., Dall'Antonia, L. H. and Córdoba de Torresi, S. I. The effect of Cd, Co, and Zn as additives on nickel hydroxide opto-electrochemical behavior. *J. Power Sources*, **102**, 2001, 224–32.
84. Bendert, M. and Corrigan, C. A. Effect of co-precipitated metal ions on the electrochromic properties of nickel hydroxide. *J. Electrochem. Soc.*, **136**, 1989, 1369–74.
85. Corrigan, D. A. and Knight, S. L. Electrochemical and spectroscopic evidence on the participation of quadrivalent nickel in the nickel hydroxide redox reaction. *J. Electrochem. Soc.*, **136**, 1989, 613–19.
86. Conell, R. S., Corrigan, D. A. and Powell, B. R. The electrochromic properties of sputtered nickel oxide films. *Sol. Energy Mater. Sol. Cells*, **25**, 1992, 301–13.
87. Bendert, R. M. and Corrigan, D. A. Effect of co-precipitated metal ions on the electrochromic properties of nickel hydroxide. *J. Electrochem. Soc.*, **136**, 1989, 1369–74.
88. Yamanaka, K., Oakamoto, H., Kidou, H. and Kudo, T. Peroxotungstic acid coated films for electrochromic display devices. *Jpn. J. Appl. Phys.*, **25**, 1986, 1420–6.
89. Shen, P. K. and Tseung, A. C. C. Study of electrodeposited tungsten trioxide thin films. *J. Mater. Chem.*, **2**, 1992, 1141–7.
90. Streinz, C. C., Motupally, S. and Widner, J. W. The effect of temperature and ethanol on the deposition of nickel hydroxide films. *J. Electrochem. Soc.*, **142**, 1995, 4051–6.
91. Monk, P. M. S., Ali, T. and Partridge, R. D. The effect of doping electrochromic molybdenum oxide with other metal oxides: correlation of optical and kinetic properties. *Solid State Ionics*, **80**, 1995, 75–85.
92. Guerfi, A. and Dao, L. H. Electrochromic molybdenum oxide thin films by electrodeposition. *J. Electrochem. Soc.*, **136**, 1989, L2435–6.
93. Kishimoto, A., Nanba, T. and Kudo, T. Spin-coated $Ta_2O_5.nH_2O$ films derived from peroxo poly-tantalate solution. Seventh International Conference on Solid State Ionics, Japan, 1989, abs. 8pb–24.

94. Syed-Bokhari, J. K. and Tseung, A. C. C. The performance of electrochromic tungsten trioxide films doped with cobalt or nickel. *J. Electrochem. Soc.*, **138**, 1991, 2778–83.
95. Pei, K. S. and Tseung, A. C. C. *In situ* monitoring of electrode polarisation during the operation of an electrochromic device based on WO_3. *J. Electroanal. Chem.*, **389**, 1995, 219–22.
96. Monk, P. M. S., Partridge, R. D., Janes, R. and Parker, M. Electrochromic tungsten oxide: doping with two or three other metal oxides. *J. Mater. Chem.*, **4**, 1994, 1071–4.
97. Monk, P. M. S., Akhtar, S. P., Boutevin, J. and Duffield, J. R. Toward the tailoring of electrochromic bands of metal-oxide mixtures. *Electrochim. Acta*, **46**, 2001, 2091–6.
98. Andrukaitis, E. and Hill, I. Diffusion of lithium in electrodeposited vanadium oxides. *J. Power Sources*, **136**, 2004, 290–5.
99. Monk, P. M. S. and Chester, S. L. Electro-deposition of films of electrochromic tungsten oxide containing additional metal oxides. *Electrochim. Acta*, **38**, 1993, 1521–6.
100. Casella, I. G. Electrodeposition of cobalt oxide films from carbonate solutions containing Co(II)-tartrate complexes. *J. Electroanal. Chem.*, **520**, 2002, 119–225.
101. Pauporté, T. A simplified method for WO_3 electrodeposition. *J. Electochem. Soc.*, **149**, 2002, C539–45.
102. Sotti, G., Schiavon, G., Zecchin, S. and Castellato, U. Electrodeposition of amorphous Fe_2O_3 films by reduction of iron perchlorate in acetonitrile. *J. Electrochem. Soc.*, **145**, 1998, 385–9.
103. Yoshino, T. and Baba, N. Characterization and properties of electrochromic cobalt oxide thin film prepared by electrodeposition. *Sol. Energy Mater. Sol. Cells*, **39**, 1995, 391–7.
104. Zotti, G., Schiavon, G., Zecchin, S. and Casellato, U. Electrodeposition of amorphous Fe_2O_3 films by reduction of iron perchlorate in acetonitrile. *J. Electrochem. Soc.*, **145**, 1998, 385–9.
105. Monk, P. M. S., Janes, R. and Partridge, R. D. Speciation analysis applied to the electrodeposition of precursors of neodymium cuprate and related phases: the first application of speciation modelling to a solution not at equilibrium. *J. Chem. Soc., Faraday Trans.*, **93**, 1997, 3985–90.
106. Monk, P. M. S., Janes, R. and Partridge, R. D. Speciation modelling of the electroprecipitation of rare-earth cuprate and nickelate materials: speciation of aqueous solutions not at equilibrium. *J. Chem. Soc., Faraday Trans.*, **93**, 1997, 3991–7.
107. Vidotti, M., van Greco, C., Ponzio, E. A. and Córdoba de Torresi, S. Sonochemically synthesized $Ni(OH)_2$ and $Co(OH)_2$ nanoparticles and their application in electrochromic electrodes. *Electrochem. Commun.*, **8**, 2006, 554–60.
108. Gedanken, A. Using sonochemistry for the fabrication of nanomaterials. *Ultrason. Sonochem.*, **11**, 2004, 47–55.
109. Jeevanandam, P., Koltypin, Y. and Gedanken, A. Synthesis of nanosized α-nickel hydroxide by a sonochemical method. *Nano Lett.*, **1**, 2001, 263–6.
110. Jeevanandam, P., Koltypin, Y., Gedanken, A. and Mastai, Y. Synthesis of α-cobalt(II) hydroxide using ultrasound radiation. *J. Mater. Chem.*, **10**, 2000, 511–14.
111. Evans, D. F. and Wennerström, H. *The Colloidal Domain: Where Physics, Chemistry, Biology and Technology Meet*, 2nd edn, New York, Wiley, 1999, p. 497.

112. Bell, J. M., Skryabin, I. L. and Koplick, A. J. Large area electrochromic films – preparation and performance. *Sol. Energy Mater. Sol. Cells*, **68**, 2001, 239–47.
113. Lakeman, C. D. E. and Payne, D. A. Sol–gel processing of electrical and magnetic ceramics. *Mater. Chem. Phys.*, **38**, 1994, 305–24.
114. Whittingham, M. S., Guo, J.-D., Chen, R., Chirayil, T., Janaver, G. and Zavilij, P. The hydrothermal synthesis of new oxide materials. *Solid State Ionics*, **75**, 1995, 257–68.
115. Alber, K. S. and Cox, J. A. Electrochemistry in solids prepared by sol–gel processes. *Mikrochim. Acta*, **127**, 1997, 131–47.
116. Lev, O., Wu, Z., Bharathi, S., Glezer, V., Modestov., Gun, J., Rabinovich, L. and Sampath, S. Sol–gel materials in electrochemistry. *Chem. Mater.*, **9**, 1997, 2354–75.
117. Therese, G. H. A. and Kamath, P. V. Electrochemical synthesis of metal oxides and hydroxides. *Chem. Mater.*, **12**, 2000, 1195–204.
118. Nishio, K. and Tsuchiya, T. Electrochromic thin films prepared by sol–gel process. *Sol. Energy Mater. Sol. Cells*, **68**, 2001, 279–93.
119. Chen, D. Anti-reflection (AR) coatings made by sol–gel processes: a review. *Sol. Energy Mater. Sol. Cells*, **68**, 2001, 313–36.
120. Livage, J. and Ganguli, D. Sol–gel electrochromic coatings and devices: a review. *Sol. Energy Mater. Sol. Cells*, **68**, 2001, 365–81.
121. Klein, L. C. Electrochromic sol–gel coatings. In Schwartz M. (ed.), *Encyclopedia of Smart Materials*, New York, Wiley, 2002, pp. 356–62.
122. Valla, B., Tonazzi, J. C. L., Macêdo, M. A. *et al.* Transparent storage layers for H^+ and Li^+ ions prepared by sol–gel technique. *Proc. SPIE*, **1536**, 1991, 48–62.
123. Alquier, C., Vandenborre, M. T. and Henry, M. Synthesis of niobium pentoxide gels. *J. Non-Cryst. Solids*, **79**, 1986, 383–95.
124. Li, Y.-M. and Kudo, T. Properties of mixed-oxide MoO_3/V_2O_5 electrochromic films coated from peroxo-polymolybdovanadate solutions. *Sol. Energy Mater. Sol. Cells*, **39**, 1995, 179–90.
125. Takano, S., Kishimoto, A., Hinokuma, K. and Kudo, T. Electrochromic thin films coated from peroxo-polymolybdotungstate solutions. *Solid State Ionics*, **70–1**, 1994, 636–41.
126. Svegl, F., Orel, B. and Kaucic, V. Electrochromic properties of lithiated Co-oxide (Li_xCoO_2) and Ni-oxide (Li_xNiO_2) thin films prepared by the sol–gel route. *Sol. Energy*, **68**, 2000, 523–40.
127. Pecquenard, B., Le Cacheux, H., Livage, J. and Julien, C. Orthorhombic WO_3 formed via a Ti-stabilized $WO_3 | H_2O$ phase. *J. Solid State Chem.*, **135**, 1998, 159–68.
128. Wang, Z. and Hu, X. Fabrication and electrochromic properties of spin-coated TiO_2 thin films from peroxo-polytitanic acid. *Thin Solid Films*, **352**, 1999, 62–5.
129. Livage, J. and Guzman, G. Aqueous precursors for electrochromic tungsten oxide hydrates. *Solid State Ionics*, **84**, 1996, 205–11.
130. Wang, H., Zhang, M., Yang, S., Zhao, L. and Ding, L. Preparation and properties of electrochromic tungsten oxide film. *Sol. Energy Mater. Sol. Cells*, **43**, 1996, 345–52.
131. Biswas, P. K., Pramanik, N. C., Mahapatra, M. K., Ganguli, D. and Livage, J. Optical and electrochromic properties of sol–gel WO_3 films on conducting glass. *Mater. Lett.*, **57**, 2003, 4429–32.
132. Hibino, M., Ugaji, M., Kishimoto, A. and Kudo, T. Preparation and lithium intercalation of a new vanadium oxide with a two-dimensional structure. *Solid State Ionics*, **79**, 1995, 239–44.

133. Wang, Z., Chen, J. and Hu, X. Electrochromic properties of aqueous sol–gel derived vanadium oxide films with different thickness. *Thin Solid Films*, **375**, 2000, 238–41.
134. Wang, Z. and Hu, X. Electrochromic properties of TiO_2-doped WO_3 films spin-coated from Ti-stabilized peroxotungstic acid. *Electrochim. Acta*, **46**, 2001, 1951–6.
135. Patil, P. R., Pawar, S. H. and Patil, P. S. The electrochromic properties of tungsten oxide thin films deposited by solution thermolysis. *Solid State Ionics*, **136–137**, 2000, 505–11.
136. Patil, P. S., Patil, P. R., Kamble, S. S. and Pawar, S. H. Thickness-dependent electrochromic properties of solution thermolyzed tungsten oxide thin films. *Sol. Energy Mater. Sol. Cells*, **60**, 2000, 143–53.
137. El Idrissi, B., Addou, M., Outzourhit, A., Regragui, M., Bougrine, A. and Kachouane, A. Sprayed CeO_2 thin films for electrochromic applications. *Sol. Energy Mater. Sol. Cells*, **69**, 2000, 1–8.
138. Patil, P. S., Kadam, L. D. and Lokhande, C. D. Studies on electrochromism of spray pyrolyzed cobalt oxide thin films. *Sol. Energy Mater. Sol. Cells*, **53**, 1998, 229–34.
139. Kadam, L. D., Pawar, S. H. and Patil, P. S. Studies on ionic intercalation properties of cobalt oxide thin films prepared by spray pyrolysis technique. *Mater. Chem. Phys.*, **68**, 2001, 280–2.
140. Gomez, M., Medina, A. and Estrada, W. Improved electrochromic films of NiO_x and WO_xP_y obtained by spray pyrolysis. *Sol. Energy Mater. Sol. Cells*, **64**, 2000, 297–309.
141. Kamal, H., Elmaghraby, E. K., Ali, S. A. and Abdel-Hady, K. Characterization of nickel oxide films deposited at different substrate temperatures using spray pyrolysis. *J. Crystal Growth*, **262**, 2004, 424–34.
142. Wang, S.-Y., Wang, W., Wang, W.-Z. and Du, Y.-W. Preparation and characterization of highly oriented $NiO_{(200)}$ films by a pulse ultrasonic spray pyrolysis method. *Mater. Sci. Eng. B*, **90**, 2002, 133–7.
143. Regragui, M., Addou, M., Outzourhit, A., Bernede, J. C., El Idrissi, E., Bensseddik, E. and Kachouane, A. Preparation and characterization of pyrolytic spray deposited electrochromic tungsten trioxide films. *Thin Solid Films*, **358**, 2000, 40–5.
144. Regragui, M., Addou, M., Outzourhit, A. El Idrissi, E., Kachouane, A. and Bougrine, A. Electrochromic effect in WO_3 thin films prepared by spray pyrolysis. *Sol. Energy Mater. Sol. Cells*, **77**, 2003, 341–50.
145. Sivakumar, R., Moses Ezhil Raj, A., Subramanian, B., Jayachandran, M., Trivedi, D. C. and Sanjeeviraja, C. Preparation and characterization of spray deposited n-type WO_3 thin films for electrochromic devices. *Mater. Res. Bull.*, **39**, 2004, 1479–89.
146. Zhang, J., Wessel, S. A. and Colbow, K. Spray pyrolysis elecrochromic WO_3 films: electrical and X-ray diffraction measurements. *Thin Solid Films*, **185**, 1990, 265–77.
147. Štangar, L. U., Orel, B., Grabec, I., Ogorevc, B. and Kalcher, K. Optical and electrochemical properties of CeO_2 and CeO_2–TiO_2 coatings. *Sol. Energy Mater. Sol. Cells*, **31**, 1993, 171–85.
148. Cerc Korošec, R. and Bukovec, P. The role of thermal analysis in optimization of the electrochromic effect of nickel oxide thin films, prepared by the sol–gel method: part II. *Thermochim. Acta*, **410**, 2004, 65–71.
149. Garcia-Miquel, J. L., Zhang, Q., Allen, S. J., Rougier, A., Blyr, A., Davies, H. O., Jones, A. C., Leedhan, T. J., Williams, P. A. and Impey, S. A. Nickel oxide sol–gel films from nickel diacetate for electrochromic applications. *Thin Solid Films*, **424**, 2003, 165–70.

150. Martini, M., Brito, G. E. S., Fantini, M. C. A., Craievich, A. F. and Gorenstein, A. Electrochromic properties of NiO-based thin films prepared by sol–gel and dip coating. *Electrochim. Acta*, **46**, 2001, 2275–9.
151. Sharma, P. K., Fantini, M. C. A. and Gorenstein, A. Synthesis, characterization and electrochromic properties of NiO_xH_y thin film prepared by a sol–gel method. *Solid State Ionics*, **113–15**, 1998, 457–63.
152. Šurca, A., Orel, B., Pihlar, B. and Bukovec, P. Optical, spectroelectrochemical and structural properties of sol–gel derived Ni-oxide electrochromic film. *J. Electroanal. Chem.*, **408**, 1996, 83–100.
153. Nishio, K., Watanabe, Y. and Tsuchiya, T. Preparation and properties of electrochromic iridium oxide thin film by sol–gel process. *Thin Solid Films*, **350**, 1999, 96–100.
154. Orel, B., Maček, M., Švegl, F. and Kalcher, K. Electrochromism of iron oxide films prepared via the sol–gel route by the dip-coating technique. *Thin Solid Films*, **246**, 1994, 131–42.
155. Avellaneda, C. O., Aegerter, M. A. and Pawlicka, A. Caracterização de filmes finos de Nb_2O_5 com propriedades eletrocrômicas [Characterisation and electrochromic properties of films of Nb_2O_5], *Quim. Nova*, **21**, 1998, 365–7.
156. Avellaneda, C. O., Bulhões, L. O. S. and Pawlicka, A. The CeO_2–TiO_2–ZrO_2 sol–gel film: a counter-electrode for electrochromic devices. *Thin Solid Films*, **84**, 2004, 337–50.
157. Avellaneda, C. O. and Pawlicka, A. Preparation of transparent CeO_2–TiO_2 coatings for electrochromic devices. *Thin Solid Films*, **335**, 1998, 245–8.
158. Baudry, P., Rodrigues, A. C. M., Aegerter, M. A. and Bulhões, L. O. Dip-coated TiO_2–CeO_2 films as transparent counter-electrode for transmissive electrochromic devices. *J. Non-Cryst. Solids*, **121**, 1990, 319–22.
159. Benčič, S., Orel, B., Surca, A. and Stangar, U. L. Structural and electrochromic properties of nanosized Fe/V-oxide films with $FeVO_4$ and $Fe_2V_4O_{13}$ grains: comparative studies with crystalline V_2O_5. *Sol. Energy*, **68**, 2000, 499–515.
160. Berton, M. A. C., Avellaneda, C. O. and Bulhões, L. O. S. Thin film of CeO_2–SiO_2: a new ion storage layer for smart windows. *Sol. Energy Mater. Sol. Cells*, **80**, 2003, 443–9.
161. Ferreira, F. F., Haddad, P. S., Fantini, M. C. A. and Brito, G. E. S. Composite Au–NiO films. *Solid State Ionics*, **165**, 2003, 161–8.
162. Ghodsi, F. E., Tepehan, F. Z. and Tepehan, G. G. Optical and electrochromic properties of sol–gel made CeO_2–TiO_2 thin films. *Electrochim. Acta*, **44**, 1999, 3127–36.
163. Orel, B., Lavrencic-Štangar, U., Hutchins, M. G. and Kalcher, K. Mixed phosphotungstic acid/titanium oxide gels and thin solid xerogel films with electrochromic-ionic conductive properties. *J. Non-Cryst. Solids*, **175**, 1994, 251–62.
164. Özer, N., Sabuncu, S. and Cronin, J. Electrochromic properties of sol–gel deposited Ti-doped vanadium oxide film. *Thin Solid Films*, **338**, 1999, 201–6.
165. Schmitt, M. and Aegerter, M. A. Electrochromic properties of Nb_2O_5 and Nb_2O_5:X sol–gel coatings (X = Sn, Zr, Li, Ti, Mo). *Proc. SPIE*, **3788**, 1999, 93–102.
166. Šurca, A., Benčič, S., Orel, B. and Pihlar, B. Spectroelectrochemical studies of V/Ti-, V/Ti/Zr- and V/Ti/Ce-oxide counter-electrode films. *Electrochim. Acta*, **44**, 1999, 3075–84.

167. Šurca, A., Orel, B., Cerc-Korosec, R., Bukovec, P. and Pihlar, B. Structural and electrochromic properties of sol–gel derived Ni(Si)-oxide films, *J. Electroanal. Chem.*, **433**, 1997, 57–72.
168. Avellaneda, C. O., Pawlicka, A. and Aegerter, M. A. Two methods of obtaining sol–gel Nb_2O_5 thin films for electrochromic devices. *J. Mater. Sci.*, **33**, 1998, 2181–5.
169. Özer, N., Rubin, M. D. and Lampert, C. M. Optical and electrochemical characteristics of niobium oxide films prepared by sol–gel process and magnetron sputtering: a comparison. *Sol. Energy Mater. Sol. Cells*, **40**, 1996, 285–96.
170. Pawlicka, A., Atik, M. and Aegerter, M. A. Synthesis of multicolor Nb_2O_5 coatings for electrochromic devices. *Thin Solid Films*, **301**, 1997, 236–241.
171. Schmitt, M. and Aegerter, M. A. Electrochromic properties of pure and doped Nb_2O_5 coatings and devices. *Electrochim. Acta*, **46**, 2001, 2105–11.
172. Schmitt, M., Heusing, S., Aegerter, M. A., Pawlicka, A. and Avellaneda, C. Electrochromic properties of Nb_2O_5 sol–gel coatings. *Sol. Energy Mater. Sol. Cells*, **54**, 1998, 9–17.
173. Tepehan, Z., Ghodsi, F., Ferhad, E. F., Ozer, N. and Tepehan, G. G. Optical properties of sol–gel dip-coated Ta_2O_5 films for electrochromic applications. *Sol. Energy Mater. Sol. Cells*, **59**, 1999, 265–75.
174. Bell, J. M., Barczynska, J., Evans, L. A., MacDonald, K. A., Wang, J., Green, D. C. and Smith, G. B. Electrochromism in sol–gel deposited TiO_2 films. *Proc. SPIE*, **2255**, 1994, 324–31.
175. Badilescu, S. and Ashrit, P. V. Study of sol–gel prepared nanostructured WO_3 thin films and composites for electrochromic applications. *Solid State Ionics*, **158**, 2003, 187–97.
176. Bessière, A., Badot, J.-C., Certiat, M.-C., Livage, J., Lucas, V. and Baffier, N. Sol–gel deposition of electrochromic WO_3 thin film on flexible ITO/PET substrate. *Electrochim. Acta*, **46**, 2001, 2251–6.
177. Dong Lee, K. Preparation and electrochromic properties of WO_3 coating deposited by the sol–gel method. *Sol. Energy Mater. Sol. Cells*, **57**, 1999, 21–30.
178. Opara Krašovec, U., Orel, B., Georg, A. and Wittwer, V. The gasochromic properties of sol–gel WO_3 films with sputtered Pt catalyst. *Sol. Energy*, **68**, 2000, 541–51.
179. Opara Krašovec, U., Surca, A. V. and Orel, B. IR spectroscopic studies of charged–discharged crystalline WO_3 films. *Electrochim. Acta*, **46**, 2001, 1921–9.
180. Patra, A., Auddy, K., Ganguli, D., Livage, J. and Biswas, P. K. Sol–gel electrochromic WO_3 coatings on glass. *Mater. Lett.*, **58**, 2004, 1059–63.
181. Wang, J., Bell, J. M. and Skryabin, I. L. Kinetics of charge injection in sol–gel deposited WO_3. *Sol. Energy Mater. Sol. Cells*, **56**, 1999, 465–75.
182. Šurca, A. and Orel, B. IR spectroscopy of crystalline V_2O_5 films in different stages of lithiation. *Electrochim. Acta*, **44**, 1999, 3051–7.
183. Djaoued, Y., Phong, V. H., Badilescu, S., Ashrit, P. V., Girouard, F. E. and Truong, V.-V. Sol–gel-prepared ITO films for electrochromic systems. *Thin Solid Films*, **293**, 1997, 108–12.
184. Özer, N. Optical properties and electrochromic characterization of sol–gel deposited ceria films. *Sol. Energy Mater. Sol. Cells*, **68**, 2001, 391–400.
185. Özer, N., Chen, D.-G. and Buyuklimanli, T. Electrochromic characterization of $Co(OH)_2$ thin film prepared by sol–gel process. *Sol. Energy Mater. Sol. Cells*, **52**, 1998, 223–30.
186. Ederth, J., Heszler, P., Hultåker, A., Niklasson, G. A. and Granqvist, C. G. Indium tin oxide films made from nanoparticles: models for the optical and electrical properties. *Thin Solid Films*, **445**, 2003, 199–206.

187. Ederth, J., Hultåker, A., Heszler, P., Niklasson, G. A., Granqvist, C. G., van Doorn, A. K., van Haag, C., Jongerius, M. J. and Burgard, D. Electrical and optical properties of thin films prepared by spin coating a dispersion of nano-sized tin-doped indium-oxide particles. *Proc. SPIE*, **4590**, 2001, 280–5.
188. Özer, N., Tepehan, F. and Tepehan, G. Preparation and optical properties of sol gel deposited electrochromic iron oxide films. *Proc. SPIE*, **3138**, 1997, 31–9.
189. Hinokuma, K., Ogasawara, K., Kishimoto, A., Takano, S. and Kudo, T. Electrochromism of spin-coated $MoO_3.nH_2O$ thin films from peroxo-polymolybdate. *Solid State Ionics*, **53–6**, 1992, 507–12.
190. Yanovskaya, M. I., Obvintseva, I. E., Kessler, V. G., Galyamov, B. S., Kucheiko, S. I., Shifrina, R. R. and Turova, N. Y. Hydrolysis of molybdenum and tungsten alkoxides: sols, powders and films. *J. Non-Cryst. Solids*, **124**, 1990, 155–66.
191. Özer, N., Chen, D.-G. and Lampert, C. M. Preparation and properties of spin-coated Nb_2O_5 films by the sol–gel process for electrochromic applications. *Thin Solid Films*, **277**, 1996, 162–8.
192. Özer, N., Rubin, M. D. and Lampert, C. M. Optical and electrochemical characteristics of niobium oxide films prepared by sol–gel process and magnetron sputtering: a comparison. *Sol. Energy Mater. Sol. Cells*, **40**, 1996, 285–96.
193. Sone, Y., Kishimoto, A. and Kudo, T. Proton conductivity of spin-coated $Ta_2O_5.nH_2O$ amorphous thin films from peroxo-polytantalate solution. *Solid State Ionics*, **66**, 1993, 53–9.
194. Krings, L. H. M. and Talen, W. Wet chemical preparation and characterization of electrochromic WO_3. *Sol. Energy Mater. Sol. Cells*, **54**, 1998, 27–37.
195. Li, Y., Aikawa, Y., Kishimoto, A. and Kudo, T. Coloration dynamics of tungsten oxide based all solid state electrochromic device. *Electrochim. Acta*, **39**, 1994, 807–12.
196. Özer, N. Optical and electrochemical characteristics of sol–gel deposited tungsten oxide films: a comparison. *Thin Solid Films*, **304**, 1997, 310–14.
197. Özkan, E., Lee, S.-H., Liu, P., Tracy, C. E., Tepehan, F. Z., Pitts, J. R. and Deb, S. K. Electrochromic and optical properties of mesoporous tungsten oxide films. *Solid State Ionics*, **149**, 2002, 139–46.
198. Sharma, N., Deepa, M., Varshney, P. and Agnihotry, S. A. FTIR investigations of tungsten oxide electrochromic films derived from organically modified peroxotungstic acid precursors. *Thin Solid Films*, **401**, 2001, 45–51.
199. Özer, N. Electrochemical properties of sol–gel deposited vanadium pentoxide films. *Thin Solid Films*, **305**, 1997, 80–7.
200. Passerini, S., Tipton, A. L. and Smyrl, W. H. Spin coated V_2O_5 XRG [xerogel] as optically passive electrode in laminated electrochromic devices. *Sol. Energy Mater. Sol. Cells*, **39**, 1995, 167–77.
201. He, T., Ma, Y., Cao, Y., Yang, W. and Yao, J. Enhanced electrochromism of WO_3 thin film by gold nanoparticles. *J. Electroanal. Chem.*, **514**, 2001, 129–32.
202. Li, Y.-M. and Kudo, T. Lithium intercalation dynamics of spin-coated amorphous $Mo_{0.5}V_{0.5}O_{2.75}$ thin film. *Solid State Ionics*, **86–8**, 1996, 1295–9.
203. Özer, N. and Dogan, N. Study of electrochromism in $Ti:WO_3$ films by sol–gel process. *Proc. SPIE*, **3424**, 1998, 106–14.
204. Özer, N. and Lampert, C. M. Electrochromic performance of sol–gel deposited WO_3–V_2O_5 films. *Thin Solid Films*, **349**, 1999, 205–11.
205. Verma, A., Samanta, S. B., Bakhshi, A. K. and Agnihotry, S. A. Optimization of CeO_2–TiO_2 composition for fast switching kinetics and improved Li ion storage capacity. *Solid State Ionics*, **171**, 2004, 81–90.

206. Cogan, S. F., Plante, T. D., Anderson, E. J. and Rauh, R. D. Materials and devices in electrochromic window development, *Proc. SPIE*, **562**, 1985, 23–31.
207. Azens, A., Kullman, L., Ragan, D. D. and Granqvist, C. G. Optically passive counter electrodes for electrochromic devices: transition metal–cerium oxide thin films. *Sol. Energy Mater. Sol. Cells*, **54**, 1998, 85–91.
208. Teixeira, V., Cui, H. N., Meng, L. J., Fortunato, E. and Martins, R. Amorphous ITO thin films prepared by DC sputtering for electrochromic applications. *Thin Solid Films*, **420–1**, 2002, 70–5.
209. Kharrazi, M., Kullman, L. and Granqvist, C. G. High-rate dual-target DC magnetron sputter deposition of 'blue' electrochromic Mo oxide films. *Sol. Energy Mater. Sol. Cells*, **53**, 1998, 349–56.
210. Estrada, W., Andersson, A. M. and Granqvist, C. G. Electrochromic nickel-oxide-based coatings made by reactive dc magnetron sputtering: preparation and optical properties. *J. Appl. Phys.*, **64**, 1988, 3678–83.
211. Xu, Y. Z., Qiu, M. Q., Qiu, S. C., Dai, J., Cao, G. J., He, H. H. and Wang, J. Y. Electrochromism of NiO_xH_y films grown by DC sputtering. *Sol. Energy Mater. Sol. Cells*, **45**, 1997, 105–13.
212. Ragan, D. D., Svedlindh, P. and Granqvist, C. G. Electrochromic Ni oxide films studied by magnetic measurements. *Sol. Energy Mater. Sol. Cells*, **54**, 1998, 247–54.
213. Yueyan, S., Zhiyang, Z. and Xiaoji, Y. Electrochromic properties of NiO_xH_y thin films. *Sol. Energy Mater. Sol. Cells*, **71**, 2002, 51–9.
214. Xuping, Z. and Guoping, C. The microstructure and electrochromic properties of nickel oxide films deposited with different substrate temperatures. *Thin Solid Films*, **298**, 1997, 53–6.
215. Yoshimura, K., Miki, T. and Tanemura, S. Cross-sectional observations by HRTEM of the structure of nickel oxide electrochromic thin films in the as-deposited state and the bleached state. *Mater. Res. Bull.*, **32**, 1997, 839–45.
216. Estrada, W., Andersson, A. M., Granqvist, C. G., Gorenstein, A. and Decker, F. Infrared spectroscopy of electrochromic NiO_xH_y films made by reactive dc sputtering. *J. Mater. Res.*, **6**, 1991, 1715–19.
217. Yoshimura, K., Miki, T., Tanemura, S. and Iwama, S. Characterization of niobium oxide electrochromic thin films prepared by reactive d.c. magnetron sputtering. *Thin Solid Films*, **281–2**, 1996, 235–8.
218. Huang, Y., Zhang, Y. and Hu, X. Structural, morphological and electrochromic properties of Nb_2O_5 films deposited by reactive sputtering. *Sol. Energy Mater. Sol. Cells*, **77**, 2003, 155–62.
219. Kullman, L., Azens, A. and Granqvist, C. G. Electrochromic praseodymium oxide films. *Proc. SPIE*, **3138**, 1997, 2–8.
220. Al-Jumaily, G. A. and Edlou, S. M. Optical properties of tantalum pentoxide coatings deposited using ion beam processes. *Thin Solid Films*, **209**, 1992, 223–9.
221. Berggren, L., Ederth, J. and Niklasson, G. A. Electrical conductivity as a function of temperature in amorphous lithium tungsten oxide. *Sol. Energy Mater. Sol. Cells*, **84**, 2005, 329–36.
222. Salinga, C., Weis, H. and Wuttig, M. Gasochromic switching of tungsten oxide films: a correlation between film properties and coloration kinetics. *Thin Solid Films*, **414**, 2002, 288–95.
223. Wruck, D., Ramamurthi, S. and Rubin, M. Sputtered electrochromic V_2O_5 films. *Thin Solid Films*, **182**, 1989, 79–86.
224. Talledo, A., Andersson, A. M. and Granqvist, C. G. Structure and optical absorption of $Li_yV_2O_5$ thin films. *J. Appl. Phys.*, **69**, 1991, 3261–5.

225. Scarminio, J., Talledo, A., Andersson, A. A., Passerini, S. and Decker, F. Stress and electrochromism induced by Li insertion in crystalline and amorphous V_2O_5 thin film electrodes. *Electrochim. Acta*, **38**, 1993, 1637–42.
226. Talledo, A., Andersson, A. M. and Granqvist, C. G. Electrochemically lithiated V_2O_5 films: an optically passive ion storage layer for transparent electrochromic devices. *J. Mater. Res.*, **5**, 1990, 1253–6.
227. Hamberg, I. and Granqvist, C. G. Dielectric function of 'undoped' In_2O_3. *Thin Solid Films*, **105**, 1983, L83–6.
228. Hamberg, I. and Granqvist, C. G. Optical properties of transparent and heat-reflecting indium-tin-oxide films: experimental data and theoretical analysis. *Sol. Energy Mater.*, **11**, 1984, 239–48.
229. Hamberg, I., Granqvist, C., Berggren, K.-F., Sernelius, B. and Engstrom, L. Optical properties of transparent and infra-red-reflecting ITO films in the 0.2–50 µm range. *Vacuum*, **35**, 1985, 207–9.
230. Hjortsberg, A., Hamberg, I. and Granqvist, C. G. Transparent and heat-reflecting indium tin oxide films prepared by reactive electron beam evaporation. *Thin Solid Films*, **90**, 1982, 323–6.
231. Seike, T. and Nagai, J. Electrochromism of 3d transition metal oxides. *Sol. Energy Mater.*, **22**, 1991, 107–17.
232. Yahaya, M., Salleh, M. M. and Talib, I. A. Optical properties of MoO_3 thin films for electrochromic windows. *Solid State Ionics*, **113–15**, 1998, 421–3.
233. Ramana, C. V., Hussain, O. M., Naidu, B. S., Julien, C. and Balkanski, M. Physical investigations on electron-beam evaporated vanadium pentoxide films. *Mater. Sci. Eng. B.*, **52**, 1998, 32–9.
234. Pauporté, T., Aberdam, D., Hazemann, J.-L., Faure, R. and Durand, R. X-Ray absorption in relation to valency of iridium in sputtered iridium oxide films. *J. Electroanal. Chem.*, **465**, 1999, 88–95.
235. Pauporté, T. and Durand, R. Impedance spectroscopy study of electrochromism in sputtered iridium oxide films. *J. Appl. Electrochem.*, **30**, 2000, 35–41.
236. Wei, G., Haas, T. E. and Goldner, R. B. Thin films of lithium cobalt oxide. *Solid State Ionics*, **58**, 1992, 115–22.
237. Wei, G., Goldner, R. B. and Haas, T. E. Lithium cobalt oxide and its electrochromism. *Proc. Electrochem. Soc.*, **90–2**, 1990, 80–9.
238. Goldner, R. B., Arntz, F. O., Berera, G., Haas, T. E., Wei, G., Wong, K. K. and Yu, P. C. A monolithic thin-film electrochromic window. *Solid State Ionics*, **53–6**, 1992, 617–27.
239. Nishio, K., Sei, T. and Tsuchiya, T. Preparation and properties of fully solid state electrochromic-display thin film from a sol–gel process. *Proc. SPIE*, **3136**, 1997, 419–25.
240. Wang, Z. and Hu, X. Structural and electrochemical characterization of 'open-structured' ITO films. *Thin Solid Films*, **392**, 2001, 22–8.
241. Goldner, R. B., Foley, G., Goldner, E. L., Norton, P., Wong, K., Haas, T., Seward, G. and Chapman, R. Electrochromic behaviour in ITO and related oxides. *Appl. Opt.* **24**, 1985, 2283–4.
242. Goldner, R. B., Seward, G., Wong, K., Haas, T., Foley, G. H., Chapman, R. and Schulz, S. Completely solid lithiated smart windows. *Sol. Energy Mater.*, **19**, 1989, 17–26.
243. Benkhelifa, F., Ashrit, P. V., Bader, G., Girouard, F. E. and Truong, V.-V. Near room temperature deposited indium tin oxide films as transparent conductors and counterelectrodes in electrochromic systems. *Thin Solid Films*, **232**, 1993, 83–6.

244. Golden, S. J. and Steele, B. C. H. Characterisation of I. T. O. thin film electrodes in Li-based systems and their use in electrochromic windows. *Mater. Res. Soc. Symp. Proc.*, **293**, 1993, 395–400.
245. Golden, S. J. and Steele, B. C. H. Thin-film tin-doped indium oxide counter electrode for electrochromic applications. *Solid State Ionics*, **28–30**, 1988, 1733–7.
246. Kanoh, H., Hirotsu, T. and Ooh, K. Electrochromic behavior of a λ-MnO_2 electrode accompanying Li^+-insertion in an aqueous phase. *J. Electrochem. Soc.*, **143**, 1996, 905–8.
247. Besenhard, J. O. *Handbook of Battery Materials*, Chichester, Wiley, 1998.
248. Cantao, M. P., Laurenco, A., Gorenstein, A., Córdoba de Torresi, S. I. and Torresi, R. M. Inorganic oxide solid state electrochromic devices. *Mater. Sci. Eng. B*, **26**, 1994, 157–61.
249. Song, X. Y., He, Y. X., Lampert, C. M., Hu, X. F. and Chen, X. F. Cross-sectional high-resolution transmission electron microscopy of the microstructure of electrochromic nickel oxide. *Sol. Energy Mater. Sol. Cells*, **63**, 2000, 227–35.
250. Chen, X., Hu, X. and Feng, J. Nanostructured nickel oxide films and their electrochromic properties. *Nanostruct. Mater.*, **6**, 1995, 309–12.
251. Jiang, S. R., Feng, B. X., Yan, P. X., Cai, X. M. and Lu, S. Y. The effect of annealing on the electrochromic properties of microcrystalline NiO_x films prepared by reactive magnetron rf sputtering. *Appl. Surf. Sci.*, **174**, 2001, 125–31.
252. Michalak, F., von Rottkay, K., Richardson, T., Slack, J. and Rubin, M. Electrochromic lithium nickel oxide thin films by RF-sputtering from a $LiNiO_2$ target. *Electrochim. Acta*, **44**, 1999, 3085–92.
253. Svensson, J. S. E. M. and Granqvist, C. G. Electrochromism of nickel-based sputtered coatings. *Sol. Energy Mater.*, **16**, 1987, 19–26.
254. Lechner, R. and Thomas, L. K. All solid state electrochromic devices on glass and polymeric foils. *Sol. Energy Mater. Sol. Cells*, **54**, 1998, 139–46.
255. Jiang, S. R., Yan, P. X., Feng, B. X., Cai, X. M. and Wang, J. The response of a NiO_x thin film to a step potential and its electrochromic mechanism. *Mater. Chem. Phys.*, **77**, 2003, 384–9.
256. Ferreira, F. F., Tabacniks, M. H., Fantini, M. C. A., Faria, I. C. and Gorenstein, A. Electrochromic nickel oxide thin films deposited under different sputtering conditions. *Solid State Ionics*, **86–8**, 1996, 971–6.
257. Corbella, C., Vives, M., Pinyol, A., Porqueras, I., Person, C. and Bertran, E. Influence of the porosity of RF sputtered Ta_2O_5 thin films on their optical properties for electrochromic applications. *Solid State Ionics*, **165**, 2003, 15–22.
258. Cogan, S. F., Anderson, E. J., Plante, T. D. and Rauh, R. D. Materials and devices in electrochromic window development. *Proc. SPIE*, **562**, 1985, 23–31.
259. Kitao, M., Akram, H., Machida, H. and Urabe, K. Ta_2O_5 electrolyte films and solid-state EC cells. *Proc. SPIE*, **1728**, 1992, 165–72.
260. Kitao, M., Akram, H., Urabe, K. and Yamada, S. Properties of solid-state electrochromic cells using Ta_2O_5 electrolyte. *J. Electron. Mater.*, **21**, 1992, 419–22.
261. Kitao, M., Oshima, Y. and Urabe, K. Preparation and electrochromism of RF-sputtered TiO_2 films. *Jpn. J. Appl. Phys.*, **36**, 1997, 4423–6.
262. Paul, J.-L. and Lassegues, J.-C. Infrared spectroscopic study of sputtered tungsten oxide films. *J. Solid State Chem.*, **106**, 1993, 357–71.
263. Cogan, S. F., Nguyen, N. M., Perrotti, S. J. and Rauh, R. D. Optical properties of electrochromic vanadium pentoxide. *J. Appl. Phys.*, **66**, 1989, 1333–7.
264. Cogan, S. F., Nguyen, N. M., Perrotti, S. J. and Rauh, R. D. Electrochromism in sputtered vanadium pentoxide. *Proc. SPIE*, **1016**, 1988, 57–62.

265. Cogan, S. F., Rauh, R. D., Plante, T. D., Nguyen, N. M. and Westwood, J. D. Morphology and electrochromic properties of V_2O_5 films. *Proc. Electrochem. Soc.*, **90–2**, 1990, 99–111.
266. Rauh, R. D. and Cogan, S. F. Counter electrodes in transmissive electrochromic light modulators. *Solid State Ionics*, **28–30**, 1988, 1707–14.
267. Hansen, S. D. and Aita, C. R. Low temperature reactive sputter deposition of vanadium oxide. *J. Vac. Sci. Technol., A*, **3**, 1985, 660–3.
268. de Wijs, G. A. and de Groot, R. A. Amorphous WO_3: a first-principles approach. *Electrochim. Acta*, **46**, 2001, 1989–93.
269. Ai-Kuhaili, M. F., Khawaja, E. E., Ingram, D. C. and Durrani, S. M. A. A study of thin films of V_2O_5 containing molybdenum from an evaporation boat. *Thin Solid Films*, **460**, 2004, 30–5.
270. Deb, S. K. Optical and photoelectric properties and colour centres in thin films of tungsten oxide. *Philos. Mag.*, **27**, 1973, 801–22.
271. Bohnke, C. and Bohnke, O. Heat treatment of amorphous electrochromic WO_3 thin films deposited onto indium-tin oxide substrates. *J. Appl. Electrochem.*, **18**, 1988, 715–23.
272. Gérard, P., Deneuville, A., Hollinger, G. and Duc, T. M. Color in 'tungsten trioxide' thin films. *J. Appl. Phys.*, **48**, 1977, 4252–5.
273. Arnoldussen, T. C. Electrochromism and photochromism in MoO_3 films. *J. Electrochem. Soc.*, **123**, 1976, 527–31.
274. Miyata, N., Suzuki, T. and Ohyama, R. Physical properties of evaporated molybdenum oxide films. *Thin Solid Films*, **281–2**, 1996, 218–22.
275. Sian, T. S. and Reddy, G. B. Optical structural and photoelectron spectroscopic studies on amorphous and crystalline molybdenum oxide thin films. *Sol. Energy Mater. Sol. Cells*, **82**, 2004, 375–86.
276. Colton, R. J., Guzman, A. M. and Rabalais, J. W. Electrochromism in some thin-film transition-metal oxides characterised by X-ray electron spectroscopy. *J. Appl. Phys.*, **49**, 1978, 409–16.
277. Monk, P. M. S., Duffy, J. A. and Ingram, M. D. Electrochromic display devices of tungstic oxide containing vanadium oxide or cadmium sulphide as a light-sensitive layer. *Electrochim. Acta*, **38**, 1993, 2759–64.
278. Monk, P. M. S., Duffy, J. A. and Ingram, M. D. Pulsed enhancement of the rate of coloration for tungsten trioxide based electrochromic devices. *Electrochim. Acta*, **43**, 1998, 2349–57.
279. Schlotter, P. and Pickelmann, L. Xerogel structure of thermally evaporated tungsten oxide layers. *J. Electron. Mater.*, **11**, 1982, 207–36.
280. Holland, L. *Vacuum Deposition of Thin Films*, London, Chapman Hall, 1956.
281. Goldenberg, L. M. Electrochemical properties of Langmuir–Blodgett films. *J. Electroanal. Chem.*, **379**, 1994, 3–19.
282. Granqvist, C. G. Electrochromic tungsten-oxide based thin films: physics, chemistry and technology. In Francombe M. H. and Vossen, J. L. (eds.), *Physics of Thin Films*, New York, Academic, 1993, pp. 301–70.
283. Granqvist, C. G. Progress in electrochromics: tungsten oxide revisited. *Electrochim. Acta*, **44**, 1999, 3005–15.
284. Azens, A., Le Bellac, D., Granqvist, C. G., Barczynska, J., Pentjuss, E., Gabrusenoks, J. and Wills, J. M. Electrochromism of W-oxide-based thin films: recent advances. *Proc. Electrochem. Soc.*, **95–22**, 1995, 102–24.
285. Monk, P. M. S. Charge movement through electrochromic thin-film tungsten trioxide. *Crit. Rev. Solid State Mater. Sci.*, **24**, 1999, 193–226.

286. Bange, K. Colouration of tungsten oxide films: a model for optically active coatings. *Sol. Energy Mater. Sol. Cells*, **58**, 1999, 1–131.
287. Wiseman, P. J. and Dickens, P. G. The crystal structure of cubic hydrogen tungsten bronze. *J. Solid State Chem.*, **6**, 1973, 374–7.
288. Azens, A., Hjelm, A., Le Bellac, D., Granqvist, C. G., Barczynska, J., Pentuss, E., Gabrusenoks, J. and Wills, J. M. Electrochromism of W-oxide-based thin films: recent advances. *Solid State Ionics*, **86–8**, 1996, 943–8.
289. Granqvist, C. G. Electrochromic oxides: a bandstructure approach. *Sol. Energy Mater. Sol. Cells*, **32**, 1994, 369–82.
290. Granqvist, C. G. Electrochromic materials: metal oxide nanocomposites with variable optical properties. *Mater. Sci. Eng. A*, **168**, 1993, 209–15.
291. Khatko, V., Guirado, F., Hubalek, J., Llobet, E. and Correig, Z. X-Ray investigation of nanopowder WO_3 thick films. *Physica Status Solidi*, **202**, 2005, 1973–9.
292. Kitao, M. and Yamada, S. Electrochromic properties of transition metal oxides and their complementary cells. In Chowdari B. V. R. and Radharkrishna, S. (eds.), *Proceedings of the International Seminar on Solid State Ionic Devices*, Singapore, World Scientific Publishing Co., 1988, pp. 359–78.
293. Yoshiike, N. and Kondo, S. Electrochemical properties of $WO_3.x(H_2O)$, II: the influence of crystallization as hydration. *J. Electrochem. Soc.*, **131**, 1984, 809–13.
294. Özkan, E., Lee, S. H., Tracy, C. E., Pitts, J. R. and Deb, S. K. Comparison of electrochromic amorphous and crystalline tungsten oxide films. *Sol. Energy Mater. Sol. Cells*, **79**, 2003, 439–48.
295. Antonaia, A., Polichetti, T., Addonizio, M. L., Aprea, S., Minarini, C. and Rubino, A. Structural and optical characterization of amorphous and crystalline evaporated WO_3 layers. *Thin Solid Films*, **354**, 1999, 73–81.
296. Sun, S.-S. and Holloway, P. H. Modification of vapor-deposited WO_3 electrochromic films by oxygen backfilling. *J. Vac. Sci. Technol., A*, **1**, 1983, 529–33.
297. Sun, S.-S. and Holloway, P. H. Modification of the electrochromic response of WO_3 thin films by oxygen backfilling. *J. Vac. Sci. Technol., A*, **2**, 1984, 336–40.
298. Gordon, R. G., Barry, S., Barton, J. T. and Broomhall-Dillard, R. N. R. Atmospheric pressure chemical vapor deposition of electrochromic tungsten oxide films. *Thin Solid Films*, **392**, 2001, 231–5.
299. Gordon, R. G., Barry, S., Broomhill-Dillard, R. N. R., Wagner, V. A. and Wang, Y. Volatile liquid precursors for the chemical vapor deposition (CVD) of thin films containing tungsten. *Mater. Res. Soc. Symp. Proc.*, **612**, 2000, D9121–6.
300. Meda, L., Breitkopf, R. C., Haas, T. E. and Kirss, R. U. Investigation of electrochromic properties of nanocrystalline tungsten oxide thin film. *Thin Solid Films*, **402**, 2002, 126–30.
301. Green, M., Smith, D. C. and Weiner, J. A. A thin film electrochromic display based on the tungsten bronzes. *Thin Solid Films*, **38**, 1976, 89–100.
302. Goldner, R. B., Norton, P., Wong, G., Foley, E. L. and Seward, G. R. C. Further evidence for free electrons as dominating the behaviour of electrochromic polycrystalline WO_3 films. *Appl. Phys. Lett.*, **47**, 1985, 536–8.
303. Barna, G. G. Material and device properties of a solid state electrochromic device. *J. Electron. Mater.*, **8**, 1979, 153–73.
304. Akl, A. A., Kamal, H. and Abdel-Hady, K. Characterization of tungsten oxide films of different crystallinity prepared by RF sputtering. *Physica B: Condensed Matter*, **325**, 2003, 65–75.
305. He, J. L. and Chiu, M. C. Effect of oxygen on the electrochromism of RF reactive magnetron sputter deposited tungsten oxide. *Surf. Coat. Technol.*, **127**, 2000, 43–51.

306. Huang, Y.-S., Zhang, Y.-Z., Zeng, X.-T. and Hu, X.-F. Study on Raman spectra of electrochromic c-WO_3 films and their infrared emittance modulation characteristics. *Appl. Surf. Sci.*, **202**, 2002, 104–9.
307. Hutchins, M., Kamel, N. and Abdel-Hady, K. Effect of oxygen content on the electrochromic properties of sputtered tungsten oxide films with Li^+ insertion. *Vacuum*, **51**, 1998, 433–9.
308. Kitao, M., Yamada, S., Yoshida, S., Akram, H. and Urabe, K. Preparation conditions of sputtered electrochromic WO_3 films and their infrared absorption spectra. *Sol. Energy Mater. Sol. Cells*, **25**, 1992, 241–55.
309. Masetti, E., Grilli, M. L., Dautzenberg, G., Macrelli, G. and Adamik, M. Analysis of the influence of the gas pressure during the deposition of electrochromic WO_3 films by reactive r.f. sputtering of W and WO_3 target. *Sol. Energy Mater. Sol. Cells*, **56**, 1999, 259–69.
310. Pennisi, A., Simone, F., Barletta, G., Di Marco, G. and Lanza, M. Preliminary test of a large electrochromic window. *Electrochim. Acta*, **44**, 1999, 3237–43.
311. Batchelor, R. A., Burdis, M. S. and Siddle, J. R. Electrochromism in sputtered WO_3 thin films. *J. Electrochem. Soc.*, **143**, 1996, 1050–5.
312. Burdis, M. S. and Siddle, J. R. Observation of non-ideal lithium insertion into sputtered thin films of tungsten oxide. *Thin Solid Films*, **237**, 1994, 320–5.
313. Dao, L. H. and Nguyen, M. T. Prototype solid-state electrochromic window device. *Proc. Electrochem. Soc.*, **90–2**, 1990, 246–60.
314. Shimizu, Y., Noda, K., Nagase, K., Miura, N. and Yamazoe, N. Sogo Rikogaku Kenkyuka Hokoku, *Kyusha Daigaku Diagakuin*, **12**, 1991, 367, as cited in *Chem. Abs.* **115**: 102,676k.
315. Sharma, N., Deepa, M., Varshney, P. and Agnihotry, S. A. FTIR and absorption edge studies on tungsten oxide based precursor materials synthesized by sol–gel technique. *J. Non-Cryst. Solids*, **306**, 2002, 129–37.
316. Aliev, A. E. and Shin, H. W. Nanostructured materials for electrochromic devices. *Solid State Ionics*, **154–5**, 2002, 425–31.
317. Avellaneda, C. O. and Bulhões, L. O. Electrochemical and optical properties of WO_3:X sol–gel coatings (X = Li, Ti, Nb, Ta). *Proc. SPIE*, **4104**, 2000, 57–63.
318. Avellaneda, C. O. and Bulhões, L. O. S. Intercalation in WO_3 and WO_3:Li films. *Solid State Ionics*, **165**, 2003, 59–64.
319. Baker, A. T., Bosi, S. G., Bell, J. M., MacFarlane, D. R., Monsma, B. G., Skryabin, I. and Wang, J. Degradation mechanisms in electrochromic devices based on sol–gel deposited thin films. *Sol. Energy Mater. Sol. Cells*, **39**, 1995, 133–43.
320. Bechinger, C., Muffler, H., Schafle, C., Sundberg, O. and Leiderer, P. Submicron metal oxide structures by a sol–gel process on patterned substrates. *Thin Solid Films*, **366**, 2000, 135–8.
321. Cronin, J. P., Tarico, D. J., Tonazzi, J. C. L., Agrawal, A. and Kennedy, S. R. Microstructure and properties of sol–gel deposited WO_3 coatings for large area electrochromic windows. *Sol. Energy Mater. Sol. Cells*, **29**, 1993, 371–86.
322. Leftheriotis, G., Papaefthimiou, S. and Yianoulis, P. The effect of water on the electrochromic properties of WO_3 films prepared by vacuum and chemical methods. *Sol. Energy Mater. Sol. Cells*, **83**, 2004, 115–24.
323. Munro, B., Conrad, P., Kramer, S., Schmidt, H. and Zapp, P. Development of electrochromic cells by the sol–gel process. *Sol. Energy Mater. Sol. Cells*, **54**, 1998, 131–7.
324. Munro, B., Kramer, S., Zapp, P., Krug, H. and Schmidt, H. All sol–gel electrochromic system for plate glass. *J. Non-Cryst. Solids*, **218**, 1997, 185–8.

325. Özkan, E. and Tepehan, F. Z. Optical and structural characteristics of sol–gel-deposited tungsten oxide and vanadium-doped tungsten oxide films. *Sol. Energy Mater. Sol. Cells*, **68**, 2001, 265–77.
326. Reisfeld, R., Zayat, M., Minti, H. and Zastrow, A. Electrochromic glasses prepared by the sol–gel method. *Sol. Energy Mater. Sol. Cells*, **54**, 1998, 109–20.
327. von Rottkay, K., Ozer, N., Rubin, M. and Richardson, T. Analysis of binary electrochromic tungsten oxides with effective medium theory. *Thin Solid Films*, **308–9**, 1997, 50–5.
328. Wang, J. and Bell, J. M. The kinetic behaviour of ion injection in WO_3 based films produced by sputter and sol–gel deposition, part II: diffusion coefficients. *Sol. Energy Mater. Sol. Cells*, **58**, 1999, 411–29.
329. Livage, J., Zarudiansky, A., Rose, R. and Judenstein, P. An 'all gel' electrochromic device. *Solid State Ionics*, **28–30**, 1988, 1722–8.
330. Chemseddine, A., Morineau, R. and Livage, J. Electrochromism of colloidal tungsten oxide. *Solid State Ionics*, **9–10**, 1983, 357–61.
331. Judeinstein, P. and Livage, J. Electrochemical degradation of WO_3 thin films. *J. Mater. Chem.*, **1**, 1991, 621–7.
332. Judeinstein, P., Morineau, R. and Livage, J. Electrochemical degradation of $WO_3.nH_2O$ thin films. *Solid State Ionics*, **51**, 1992, 239–47.
333. Yarovskaya, M. I., Obvintseva, I. E., Kessler, V. G., Galyamov, B. S., Kucheiko, S. I., Shifrina, R. R. and Turova, N. Y. Hydrolysis of molybdenum and tungsten alkoxides: sols, powders and films. *J. Non-Cryst. Solids*, **124**, 1990, 155–66.
334. Bell, J. M., Green, D. C., Patterson, A., Smith, G. B., MacDonald, K. A., Lec, K., Kirkup, L., Cullen, J. D., West, B. O., Spiccia, L., Kenny, M. J. and Wielunski, L. S. Structure and properties of electrochromic WO_3 produced by sol–gel methods. *Proc. SPIE*, **1536**, 1991, 29–36.
335. Bell, J. M. and Matthews, J. P. Temperature dependence of kinetic behaviour of sol–gel deposited electrochromics. *Sol. Energy Mater. Sol. Cells*, **68**, 2001, 249–63.
336. Medina, A., Solis, J. L., Rodriguez, J. and Estrada, W. Synthesis and characterization of rough electrochromic phosphotungstic acid films obtained by spray-gel process. *Sol. Energy Mater. Sol. Cells*, **80**, 2003, 473–81.
337. Judeinstein, P. and Livage, J. Synthesis and multispectroscopic characterization of organically modified polyoxometallates. *Proc. SPIE*, **1328**, 1990, 344–51.
338. Babinec, S. J. A quartz crystal microbalance analysis of ion insertion into WO_3. *Sol. Energy Mater. Sol. Cells*, **25**, 1992, 269–91.
339. Dautremont-Smith, W. C., Green, M. and Kang, K. S. Optical and electrical properties of thin films of WO_3 electrochemically coloured. *Electrochim. Acta*, **22**, 1977, 751–9.
340. Cogan, S. F., Plante, T. D., Parker, M. A. and Rauh, R. D. Free-electron electrochromic modulation in crystalline Li_xWO_3. *J. Appl. Phys.*, **60**, 1986, 2735–8.
341. Faughnan, B. W., Crandall, R. S. and Lampert, M. A. Model for the bleaching of WO_3 electrochromic films by an electric field. *Appl. Phys. Lett.*, **27**, 1975, 275–7.
342. Dickens, P. G., Murphy, D. J. and Holstead, T. K. Pulsed NMR study of proton mobility in a hydrogen tungsten bronze. *J. Solid State Chem.*, **6**, 1973, 370–3.
343. Vanice, M. A., Boudart, M. and Fripiat, J. J. Mobility of hydrogen in hydrogen tungsten bronze. *J. Catal.*, **17**, 1970, 359–65.
344. Kurita, S., Nishimura, T. and Taira, K. Proton injection phenomena in WO_3-electrolyte electrochromic cells. *Appl. Phys. Lett.*, **36**, 1980, 585–7.

345. Shiyanovskaya, I., Ratajczak, H., Baran, J. and Marchewka, M. Fourier transform Raman study of electrochromic crystalline hydrate films $WO_3 \cdot \frac{1}{3}(H_2O)$, *J. Mol. Struct.*, **348**, 1995, 99–102.
346. Shiyanovskaya, I. Isotopic effect in evolution of structure and optical gap during electrochromic coloration of $WO_3 \cdot \frac{1}{3}(H_2O)$ films. *Mikrochim. Acta*, **S14**, 1997, 819–22.
347. Ho, C.-K., Raistrick, I. D. and Huggins, R. A. Application of AC-techniques to the study of lithium diffusion in tungsten trioxide thin-films. *J. Electrochem. Soc.*, **127**, 1980, 343–50.
348. Bohnke, O. and Vuillermin, B. Proton insertion into thin films of amorphous WO_3: kinetics study. In Balkanski, M., Takahashi, T. and Tuller, H. L. (eds.), *Solid State Ionics*, Amsterdam, Elsevier, 1992, pp. 593–8.
349. Dini, D., Decker, F. and Masetti, E. A comparison of the electrochromic properties of WO_3 films intercalated with H^+, Li^+ and Na^+. *J. Appl. Electrochem.*, **26**, 1996, 647–53.
350. Masetti, E., Dini, D. and Decker, F. The electrochromic response of tungsten bronzes M_xWO_3 with different ions and insertion rates. *Sol. Energy Mater. Sol. Cells*, **39**, 1995, 301–7.
351. Kang, K. and Green, M. Solid state electrochromic cells: optical properties of the sodium tungsten bronze system. *Thin Solid Films*, **113**, 1984, L29–32.
352. Ho, K.-C. Cycling and at-rest stabilities of a complementary electrochromic device based on tungsten oxide and Prussian blue thin films. *Electrochim. Acta*, **44**, 1999, 3227–35.
353. Green, M. and Richman, D. A solid state electrochromic cell: the $RbAg_4I_5 | WO_3$ system. *Thin Solid Films*, **24**, 1974, S45–6.
354. Bohnke, O., Bohnke, C., Robert, G. and Carquille, B. Electrochromism in WO_3 thin films, I: $LiClO_4$–propylene carbonate–water electrolytes. *Solid State Ionics*, **6**, 1982, 121–8.
355. Crandall, R. S. and Faughnan, B. W. Electronic transport in amorphous H_xWO_3. *Phys. Rev. Lett.*, **39**, 1977, 232–5.
356. Goodenough, J. B. Metallic oxides. *Prog. Solid. State Chem.*, **5**, 1971, 145–399.
357. Goldner, R. B., Mendelsohn, D. H., Alexander, J., Henderson, W. R., Fitzpatrick, D., Haas, T. E., Sample, H. H., Rauh, R. D., Parker, M. A. and Rose, T. L. High near-infrared reflectivity modulation with polycrystalline electrochromic WO_3 films. *Appl. Phys. Lett.*, **43**, 1983, 1093–5.
358. Goldner, R. B. and Mendelsohn, D. H. Ellipsometry measurements as direct evidence of the Drude model for polycrystalline electrochromic WO_3 films. *J. Electrochem. Soc.*, **131**, 1984, 857–60.
359. Arntz, F. O., Goldner, R. B., Morel, T. E., Haas, T. E. and Wong, G. Near-infrared reflectance modulation with electrochromic crystalline WO_3 films deposited on ambient temperature glass substrates by an oxygen ion-assisted technique. *J. Appl. Phys.*, **67**, 1990, 3177–9.
360. Goldner, R. B., Seward, G., Wong, G., Berera, G., Haas, T. and Norton, P. Improved colored state reflectivity in lithiated WO_3 films. *Proc. SPIE*, **823**, 1987, 101–4.
361. Schirmer, O. F., Wittner, V., Baur, G. and Brandt, G. Dependence of WO_3 electrochromic absorption on crystallinity. *J. Electrochem. Soc.*, **124**, 1977, 749–53.
362. Wittwer, V., Schirmer, O. F. and Schlotter, P. Disorder dependence and optical detection of the Anderson transition in amorphous H_xWO_3 bronzes. *Solid State Commun.*, **25**, 1978, 977–80.
363. Dickens, P. G., Quilliam, R. M. P. and Whittingham, M. S. The reflectance spectra of the tungsten bronzes. *Mater. Res. Bull.*, **3**, 1968, 941–9.

364. Goldner, R. B. Some aspects of charge transport in electrochromic films. In Chowdari, B. V. R. and Radhakrishna, S. (eds.), *Proceedings of the International Seminar on Solid State Ionic Devices*, Singapore, World Publishing Company, 1988, pp. 351–8.
365. Bohnke, O., Gire, A. and Theobald, J. G. *In situ* detection of electrical conductivity variation of an a-WO_3 thin film during electrochemical reduction and oxidation in $LiClO_4(M)$–PC electrolyte. *Thin Solid Films*, **247**, 1994, 51–5.
366. Crandall, R. S. and Faughnan, B. W. Measurement of the diffusion coefficient of electrons in WO_3 films. *Appl. Phys. Lett*, **26**, 1975, 120–1.
367. Crandall, R. S., Wojtowicz, P. J. and Faughnan, B. W. Theory and measurement of the change in chemical potential of hydrogen in amorphous H_xWO_3 as a function of the stoichiometric parameter x. *Solid State Commun.*, **18**, 1976, 1409–11.
368. Kirlashkina, Z. I., Popov, F. M., Bilenko, D. L. and Kirlashkin, V. I. *Sov. Phys.-Tech. Phys. (Engl. Edn.)*, **2**, 1957, 69, as cited in ref. 361.
369. Matthias, B. T. Ferro-electric properties of WO_3. *Phys. Rev.*, **76**, 1949, 430–1.
370. Ord, J. L. An ellipsometric study of electrochromism in tungsten oxide. *J. Electrochem. Soc.*, **129**, 1982, 767–72.
371. Ord, J. L., Pepin, G. M. and Beckstead, D. J. An optical study of hydrogen insertion in the anodic oxide of tungsten. *J. Electrochem. Soc.*, **136**, 1989, 362–8.
372. Denesuk, M., Cronin, J. P., Kennedy, S. R. and Uhlmann, D. R. Relation between coloring and bleaching with lithium in tungsten oxide based electrochromic device. *J. Electrochem. Soc.*, **144**, 1997, 1971–9.
373. Whittingham, M. S. The formation of tungsten bronzes and their electrochromic properties. In Chowdari B. V. R. and Radhakrishna, S. (eds.), *Proceedings of the International Seminar on Solid State Ionic Devices*, Singapore, World Publishing Company, 1988, pp. 325–40.
374. Cheng, K. H. and Whittingham, M. S. Lithium incorporation in tungsten oxides. *Solid State Ionics*, **1**, 1980, 151–61.
375. Berezin, L. Y. and Malinenko, V. P. Electrochromic coloration and bleaching of polycrystalline tungsten trioxide. *Pis'ma. Zh. Tekh. Fiz*, **13**, 1987, 401–4 [in Russian], as cited in *Chem. Abs.* **107**: 449,382t.
376. Berezin, L. Y., Aleshina, L. A., Inyushin, N. B., Malinenko, V. P. and Fofanov, A. D. Phase transitions during electrochromic processes in tungsten trioxide. *Fiz. Tverd Tela (Leningrad)*, **31**, 1989, 41–9 [in Russian], as cited in *Chem. Abs.* **112**: 225,739.
377. Kitao, M., Makifuchi, M. and Urabe, K. Residual charges and infrared absorption in electrochromic WO_3 films prepared by hydrogen-introduced sputtering. *Sol. Energy Mater. Sol. Cells*, **70**, 2001, 219–30.
378. Georg, A., Schweiger, D., Graf, W. and Wittwer, V. The dependence of the chemical potential of WO_3 films on hydrogen insertion. *Sol. Energy Mater. Sol. Cells*, **70**, 2002, 437–46.
379. Nanba, T., Ishikawa, M., Sakai, Y. and Miura, Y. Changes in atomic and electronic structures of amorphous WO_3 films due to electrochemical ion insertion. *Thin Solid Films*, **445**, 2003, 175–81.
380. Chang, I. F., Gilbert, B. L. and Sun, T. I. Electrochemichromic systems for display applications. *J. Electrochem. Soc.*, **122**, 1975, 955–62.
381. Faughnan, B. W., Crandall, R. S. and Heyman, P. M. Electrochromism in WO_3 amorphous films. *RCA Rev.*, **36**, 1975, 177–97.
382. Krasnov, Y. S., Sych, O. A., Patsyuk, F. N. and Vas'ko, A. T. Electrochromism and diffusion of charge carriers in amorphous tungsten trioxide, taking into

account the electron capture on localized sites. *Electrokhimiya*, **24**, 1988, 1468–74, [in Russian], as cited in *Chem Abs*. **1110**: 1447,1513z.
383. Mott, N. F. *Conduction in Non-Crystalline Materials*, 2nd edn, Oxford, Clarendon Press, 1993.
384. Cox, P. A. *The Electronic Structure and Chemistry of Solids*, Oxford, Oxford University Press, 1987.
385. Cox, P. A. *Transition Metal Oxides: An Introduction to their Electronic Structure and Properties*, Oxford, Clarendon Press, 1992.
386. Pifer, J. H. and Sichel, E. K. Electron resonance study of hydrogen-containing WO_3 films. *J. Electron. Mater.*, **9**, 1980, 129–40.
387. Matsuhiro, K. and Masuda, Y. Transmissive electrochromic display using a porous crystalline WO_3 counter electrode. *Proc. SID*, **21**, 1980, 101–5.
388. Owen, J. F., Teegarden, K. J. and Shanks, H. R. Optical properties of the sodium-tungsten bronzes and tungsten trioxide. *Phys. Rev. B*, **18**, 1978, 3827–37.
389. Deneuville, A. and Gérard, P. Influence of non-stoichiometry, hydrogen content and crystallinity on the optical properties and electrical properties of H_xWO_y thin films. *J. Electron. Mater.*, **7**, 1978, 559–88.
390. Schlotter, P. High contrast electrochromic tungsten oxide layers. *Sol. Energy Mater. Sol. Cells*, **16**, 1987, 39–46.
391. Niklasson, G. A., Berggren, L. and Larsson, A.-L. Electrochromic tungsten oxide: the role of defects. *Sol. Energy Mater. Sol. Cells*, **84**, 2004, 315–28.
392. Lindan, P., Duplock, E., Zhang, C., Thomas, M., Chatten, R. and Chadwick, A. The interdependence of defects, electronic structure and surface chemistry. *J. Chem. Soc., Dalton Trans.*, 2004, 3076–84.
393. Baucke, F. G. K., Duffy, J. A. and Smith, R. I. Optical absorption of tungsten bronze thin films for electrochromic applications. *Thin Solid Films*, **186**, 1990, 47–51.
394. Green, M., Dautremont-Smith, W. C. and Kang, K. S. Second International Conference on Solid Electrolytes (St. Andrews, Scotland, UK), 1978. (Ref. (30) of our ref. 493).
395. Dixon, R. A., Williams, J. J., Morris, D., Rebane, J., Jones, F. H., Edgell, R. G. and Downes, S. W. Electronic states at oxygen deficient $WO_{3(001)}$ surfaces: a study by resonant photoemission. *Surf. Sci.*, **399**, 1998, 199–211.
396. Ho, J.-J., Chen, C. Y. and Lee, W. J. Improvement of electrochromic coloration efficiency by oxygen deficiency in sputtering $a\text{-}WO_x$ films. *Electron. Lett.*, **40**, 2004, 510–11.
397. Tritthart, U., Gey, W. and Gavrilyuk, A. Nature of the optical absorption band in amorphous H_xWO_3 thin films. *Electrochim. Acta*, **44**, 1999, 3039–49.
398. Yamada, S., Yoshida, S. and Kikao, M. Infrared absorption of colored and bleached films of tungsten oxide. Seventh International Conference on Solid State Ionics, Japan, 1989, abs. 6pB–34.
399. Scarminio, J., Urbano, A. and Gardes, B. The Beer–Lambert law for electrochromic tungsten oxide thin films. *Mater. Chem. Phys.*, **61**, 1999, 143–6.
400. Deepa, M., Srivastava, A. K., Singh, S. and Agnihotry, S. A. Structure–property correlation of nanostructured WO_3 thin films produced by electrodeposition. *J. Mater Res.*, **19**, 2004, 2576–85.
401. Bange, K. and Gambke, T. Electrochromic materials for optical switching devices. *Adv. Mater.*, **2**, 1992, 10–16.
402. Park, N.-G., Kim, M. W., Poquet, A., Campet, G., Portier, J., Choy, J.-H. and Kim, Y.-I. New and simple method for manufacturing electrochromic tungsten oxide films. *Active Passive Electron. Components*, **20**, 1998, 125–33.

403. Choy, J.-H., Kim, Y.-I., Park, N.-G., Campet, G. and Grenier, J.-C. New solution route to poly(acrylic acid)/WO$_3$ hybrid film. *Chem. Mater.*, **12**, 2000, 2950–6.
404. Göttsche, J., Hinsch, A. and Wittwer, V. Electrochromic mixed WO$_3$–TiO$_2$ thin films produced by sputtering and the sol–gel technique: a comparison. *Sol. Energy Mater. Sol. Cells*, **31**, 1993, 415–28.
405. Ohtani, B., Masuoka, M., Atsui, T., Nishimoto, S. and Kagiya, N. Electrochromism of tungsten oxide film prepared from tungstic acid *Chem. Express*, **3**, 1988, 319–22.
406. Bessiere, A., Badot, J.-C., Certiat, M.-C., Livage, J., Lucas, V. and Baffier, N. Sol–gel deposition of electrochromic WO$_3$ thin film on flexible ITO/PET substrate. *Electrochim. Acta*, **46**, 2001, 2251–6.
407. Dickens, P. G. and Whittingham, M. S. The tungsten bronzes and related compounds. *Quart. Rev. Chem. Soc*, **22**, 1968, 30–44.
408. Goldner, R. B. Electrochromic smart window™ glass. In Chowdari B. V. R. and Radhakrishna, S. (eds.), *Proceedings of the International Seminar on Solid State Ionic Devices*, Singapore, World Publishing Co., 1988, pp. 379–89.
409. Varjian, R. D., Shabrand, M. and Babinec, S. Application of a solid polymer electrolyte in one square foot electrochromic devices. *Proc. Electrochem. Soc.*, **94-2**, 1994, 278–89.
410. Kaneko, N., Tabata, J. and Miyoshi, T. Electrochromic device watch display. *SID Int. Symp. Digest*, **12**, 1981, 74–5.
411. Baucke, F. G. K. Electrochromic applications. *Mater. Sci. Eng. B*, **10**, 1991, 285–92.
412. Baucke, F. G. K. Electrochromic mirrors with variable reflectance. *Rivista della Staz. Sper. Vetro*, **6**, 1986, 119–22.
413. Baucke, F. G. K. Electrochromic mirrors with variable reflectance. *Sol. Energy Mater*, **16**, 1987, 67–77.
414. Baucke, F. G. K. Reflecting electrochromic devices – construction, operation and application. *Proc. Electrochem. Soc.*, **20-4**, 1990, 298–311.
415. Baucke, F. G. K., Bange, K. and Gambke, T. Reflecting electrochromic devices. *Displays*, **9**, 1988, 179–87.
416. Baucke, F. G. K. Beat the dazzlers, *Schott Information*, **1**, 1983, 11–13.
417. Baucke, F. G. K. Reflectance control of automotive mirrors. *Proc. SPIE*, **IS4**, 1990, 518–38.
418. Baucke, F. G. K. and Duffy, J. A. Darkening glass by electricity. *Chem. Br.*, **21**, 1985, 643–46 and 653.
419. Baucke, F. G. K. and Gambke, T. Electrochromic materials for optical switching devices. *Adv. Mater.*, **2**, 1990, 10–16.
420. Hersch, H. N., Kramer, W. E. and McGee, J. K. Mechanism of electrochromism in WO$_3$. *Appl. Phys. Lett.*, **27**, 1975, 646–8.
421. Mohapatra, S. K., Boyd, G. D., Storz, F. G., Wagner, S. and Wudl, F. Application of solid proton conductors to WO$_3$ electrochromic displays. *J. Electrochem. Soc.*, **126**, 1979, 805–8.
422. Howe, A. T., Sheffield, S. H., Childs, P. E. and Shilton, M. G. Fabrication of films of hydrogen uranyl phosphate tetrahydrate and their use as solid electrolytes in electrochromic displays. *Thin Solid Films*, **67**, 1980, 365–70.
423. Giglia, R. D. and Haacke, G. Performance improvements in WO$_3$-based electrochromic displays. *Proc. SID*, **12**, 1981, 41–5.
424. Cohen, C. Electrochromic display rivals liquid crystals for low-power needs. *Electronics*, **11**, 1981, 65–6.
425. Schlotter, P. and Pickelmann, L. The xerogel structure of thermally evaporated tungsten oxide layers. *J. Electron. Mater.*, **11**, 1982, 207–36.

426. Kamimori, T., Nagai, J. and Mizuhashi, M. Electrochromic devices for transmissive and reflective light control. *Sol. Energy Mater.*, **16**, 1987, 27–38.
427. Taunier, S., Guery, C. and Tarascon, J.-M. Design and characterization of a three-electrode electrochromic device, based on the system WO_3/IrO_2. *Electrochim. Acta*, **44**, 1999, 3219–25.
428. Larsson, A.-L. and Niklasson, G. A. Infrared emittance modulation of all-thin-film electrochromic devices. *Mater. Lett.*, **58**, 2004, 2517–20.
429. Jonsson, A., Furlani, M. and Niklasson, G. A. G. A. Isothermal transient ionic current study of laminated electrochromic devices for smart window applications. *Sol. Energy Mater. Sol. Cells*, **84**, 2004, 361–7.
430. Passerini, S., Scrosati, B., Hermann, V., Holmblad, C. and Bartlett, T. Laminated electrochromic windows based on nickel oxide, tungsten oxide, and gel electrolytes. *J. Electrochem. Soc.*, **141**, 1994, 1025–8.
431. Andrei, M., Roggero, A., Marchese, L. and Passerini, S. Highly conductive solid polymer electrolyte for smart windows. *Polymer*, **35**, 1994, 3592–7.
432. Scrosati, B. Ion conducting polymers and related electrochromic devices. *Mol. Cryst. Liq. Cryst.*, **190**, 1990, 161–70.
433. Orel, B., Opara Krašovec, U., Macek, M., Svegl, F. and Lavrencic Štangar, U. Comparative studies of 'all sol–gel' electrochromic devices with optically passive counter-electrode films, ormolyte Li^+ ion-conductor and WO_3 or Nb_2O_5 electrochromic films. *Sol. Energy Mater. Sol. Cells*, **56**, 1999, 343–73.
434. Papaefthimiou, S., Leftheriotis, G. and Yianoulis, P. Advanced electrochromic devices based on WO_3 thin films. *Electrochim. Acta*, **46**, 2001, 2145–50.
435. Yoo, S. J., Lim, J. W. and Sung, Y.-E. Improved electrochromic devices with an inorganic solid electrolyte protective layer. *Sol. Energy Mater. Sol. Cells*, **90**, 2006, 477–84.
436. Tung, T.-S., Chen, L.-C. and Ho, K.-C. An indium hexacyanoferrate–tungsten oxide electrochromic battery with a hybrid K^+/H^+-conducting polymer electrolyte. *Solid State Ionics*, **165**, 2003, 257–67.
437. Chen, L., Tseng, K., Huang, Y. and Ho, K. Novel electrochromic batteries, II: an InHCF–WO_3 cell with a high visual contrast. *J. New Mater. Electrochem. Syst.*, **5**, 2002, 213–21.
438. Su, L., Fang, J., Xiao, Z. and Lu, Z. An all-solid-state electrochromic display device of Prussian blue and WO_3 particulate film with a PMMA gel electrolyte. *Thin Solid Films*, **306**, 1997, 133–6.
439. Su, L., Xiao, Z. and Lu, Z. All solid-state electrochromic window of electrodeposited WO_3 and Prussian blue film with PVC gel electrolyte. *Thin Solid Films*, **320**, 1998, 285–9.
440. Su, L., Wang, H. and Lu, Z. All-solid-state electrochromic window of Prussian blue and electrodeposited WO_3 film with poly(ethylene oxide) gel electrolyte. *Mater. Chem. Phys.*, **56**, 1998, 266–70.
441. Chen, L.-C. and Ho, K.-C. Design equations for complementary electrochromic devices: application to the tungsten oxide–Prussian blue system. *Electrochim. Acta*, **46**, 2001, 2151–8.
442. Chen, L., Huang, Y., Tseng, K. and Ho, K. Novel electrochromic batteries, I: a PB–WO_3 cell with a theoretical voltage of 1.35 V. *J. New Mater. Electrochem. Syst.*, **5**, 2002, 203–12.
443. Bernard, M.-C., Hugot-Le Goff, A. and Zeng, W. Elaboration and study of a PANI/PAMPS/WO_3 all solid-state electrochromic device. *Electrochim. Acta*, **44**, 1998, 781–96.

444. Jelle, B. P. and Hagen, G. Performance of an electrochromic window based on polyaniline, Prussian blue and tungsten oxide. *Sol. Energy Mater. Sol. Cells*, **58**, 1999, 277–86.
445. Jelle, B. P., Hagen, G. and Nodland, S. Transmission spectra of an electrochromic window consisting of polyaniline, Prussian blue and tungsten oxide. *Electrochim. Acta*, **38**, 1993, 1497–500.
446. Marcel, C. and Tarascon, J.-M. An all-plastic $WO_3 \cdot H_2O$/polyaniline electrochromic device. *Solid State Ionics*, **143**, 2001, 89–101.
447. Michalak, F. and Aldebert, P. A flexible electrochromic device based on colloidal tungsten oxide and polyaniline. *Solid State Ionics*, **85**, 1996, 265–72.
448. Tassi, E. L. and De Paoli, M.-A. An electrochromic device based on association of the graft copolymer of polyaniline and nitrilic rubber with WO_3. *Electrochim. Acta*, **39**, 1994, 2481–4.
449. Bich, V. T., Bernard, M. C. and Hugot-Le Goff, A. Resonant Raman identification of the polaronic organization in PANI. *Synth. Met.*, **101**, 1999, 811–12.
450. Jelle, B. P., Hagen, G., Sunde, S. and Ødegård, R. Dynamic light modulation in an electrochromic window consisting of polyaniline, tungsten oxide and a solid polymer electrolyte. *Synth. Met.*, **54**, 1993, 315–20.
451. Topart, P. and Hourquebie, P. Infrared switching electroemissive devices based on highly conducting polymers. *Thin Solid Films*, **352**, 1999, 243–8.
452. Bernard, M. C. and Hugot-Le Goff, A. Reactions at the two sides of an ECD device studied by Raman spectroscopy. *Synth. Met.*, **102**, 1999, 1342–5.
453. Bernard, M.-C., Hugot-Le Goff, A. and Zeng, W. Characterization and stability tests of an all solid state electrochromic cell using polyaniline. *Synth. Met.*, **85**, 1997, 1347–8.
454. Rauh, R. D., Wang, F., Reynolds, J. R. and Meeker, D. L. High coloration efficiency electrochromics and their application to multi-color devices. *Electrochim. Acta*, **46**, 2001, 2023–9.
455. De Paoli, M.-A., Zanelli, A., Mastragostino, M. and Rocco, A. M. An electrochromic device combining polypyrrole and WO_3, II: solid-state device with polymeric electrolyte. *J. Electroanal. Chem.*, **435**, 1997, 217–24.
456. Lee, D. S., Lee, D. D., Hwang, H. R., Paik, J. H., Huh, J. S., Lim, J. O. and Lee, J. J. Characteristics of electrochromic device with polypyrrole and WO_3. *J. Mater. Sci.: Mater. Electron.*, **12**, 2001, 41–4.
457. Rocco, A. M., De Paoli, M.-A., Zanelli, A. and Mastragostino, M. An electrochromic device combining polypyrrole and WO_3, I: liquid electrolyte. *Electrochim. Acta*, **41**, 1996, 2805–16.
458. Hurditch, R. Electrochromism in hydrated tungsten-oxide films. *Electron. Lett.*, **11**, 1975, 142–4.
459. Stocker, R. J., Singh, S., van Uitert, L. G. and Zydzik, G. J. Efficiency and humidity dependence of WO_3–insulator electrochromic display structures. *J. Appl. Phys.*, **50**, 1979, 2993–4.
460. Hefny, M. M., Gadallah, A. G. and Mogoda, A. S. Some electrochemical properties of the anodic oxide film on tungsten. *Bull. Electrochem.*, **3**, 1987, 11–14.
461. Reichman, B. and Bard, A. J. The electrochromic process at WO_3 electrodes prepared by vacuum evaporation and anodic oxidation of W. *J. Electrochem. Soc.*, **126**, 1979, 583–91.
462. Perez, M. A. and Teijelo, M. L. Ellipsometric study of WO_3 films dissolution in aqueous solutions. *Thin Solid Films*, **449**, 2004, 138–46.

463. Gavrilko, T. A., Stepkin, V. I. and Shiyanovskaya, I. V. IR and optical spectroscopy of structural changes of WO_3 electrochromic thin films. *J. Mol. Struct.*, **218**, 1990, 411–16.
464. Rice, C. E. A comparison of the behaviours of tungsten trioxide and anodic iridium oxide film electrochromics in non-aqueous acidic medium. *Appl. Phys. Lett.*, **35**, 1979, 563–5.
465. Tell, B. Electrochromism in solid phosphotungstic acid. *J. Electrochem. Soc.*, **127**, 1980, 2451–4.
466. Tell, B. and Wudl, F. Electrochromic effects in solid phosphotungstic acid and phosphomolybdic acid. *J. Appl. Phys.*, **50**, 1979, 5944–6.
467. Shen, P. K., Huang, H. and Tseung, A. C. C. Improvements in the life of WO_3 electrochromic films. *J. Mater. Chem.*, **2**, 1992, 497–9.
468. Nah, Y.-C., Ahn, K.-S. and Sung, Y.-E. Effects of tantalum oxide films on stability and optical memory in electrochromic tungsten oxide films. *Solid State Ionics*, **165**, 2003, 229–33.
469. Azens, A., Hjelm, A., Le Bellac, D., Granqvist, C. G., Barczynska, J., Pentuss, E., Gabrusenoks, J. and Wills, J. M. Electrochromism of W-oxide-based films: some theoretical and experimental results. *Proc. SPIE*, **2531**, 1995, 92–104.
470. Haranahalli, A. R. and Dove, D. B. Influence of a thin gold surface layer on the electrochromic behavior of WO_3 films. *Appl. Phys. Lett.*, **36**, 1980, 791–3.
471. Haranahalli, A. R. and Holloway, P. H. The influence of metal overlayers on electrochromic behavior of tungsten trioxide films. *J. Electron. Mater.*, **10**, 1981, 141–72.
472. Denesuk, M., Cronin, J. P., Kennedy, S. R. and Uhlmann, D. R. Step-current analysis of the built-in potential of tungsten oxide-based electrochromic devices and the effects of spontaneous hydrogen deintercalation. *J. Electochem. Soc.*, **144**, 1997, 888–97.
473. Kamal, H., Akl, A. A. and Abdel-Hady, K. Influence of proton insertion on the conductivity, structural and optical properties of amorphous and crystalline electrochromic WO_3 films. *Physica B: Condensed Matter*, **349**, 2004, 192–205.
474. Zhang, J.-G., Benson, D. K., Tracy, C. E., Webb, J. and Deb, S. K. Self-bleaching mechanism of electrochromic WO_3 films. *Proc. SPIE*, **2017**, 1993, 104–12.
475. Ivanova, T., Gesheva, K. A., Popkirov, M., Ganchev, M. and Tzvetkova, E. Electrochromic properties of atmospheric CVD MoO_3 and MoO_3–WO_3 films and their application in electrochromic devices. *Mater. Sci. Eng. B*, **119**, 2005, 232–9.
476. Colten, R. J., Guzman, A. M. and Rabalais, J. W. Electrochromism in some thin-film transition-metal oxides characterised by X-ray electron spectroscopy. *J. Appl. Phys.*, **49**, 1978, 409–16.
477. Sian, T. S. and Reddy, G. B. Infrared and electrochemical studies on Mg intercalated a-MoO_3 thin films. *Solid State Ionics*, **167**, 2004, 399–405.
478. Ord, J. L. and DeSmet, D. J. Optical anisotropy and electrostriction in the anodic oxide of molybdenum. *J. Electrochem. Soc.*, **130**, 1983, 280–4.
479. Yao, J. N., Loo, B. H., Hashimoto, K. and Fujishima, A. Photochromic and electrochromic behavior of electrodeposited MoO_3 thin films. *J. Electroanal. Chem.*, **290**, 1990, 263–7.
480. Kharrazi, M., Azens, A., Kullman, L. and Granqvist, C. G. High-rate dual-target d.c. magnetron sputter deposition of electrochromic MoO_3 films. *Thin Solid Films*, **295**, 1997, 117–21.
481. Gorenstein, A., Scarminio, J. and Lourenço, A. Lithium insertion in sputtered amorphous molybdenum thin films. *Solid State Ionics*, **86–8**, 1996, 977–81.

482. Bica De Moraes, M. A., Transferetti, B. C., Rouxinol, F. P., Landers, R., Durant, S. F., Scarminio, J. and Urbano, B. Molybdenum oxide thin films obtained by hot-filament metal oxide deposition technique. *Chem. Mater.*, **163**, 2004, 513–20.
483. Cruz, T. G. S., Gorenstein, A., Landers, R., Kleiman, G. G. and deCastro, S. C. Electrochromism in MoO_x films characterized by X-ray electron spectroscopy. *J. Electron. Spectrosc. Relat. Phenom.*, **101–3**, 1999, 397–400.
484. Ferreira, F. F., Souza Cruz, T. G., Fantini, M. C. A., Tabacniks, M. H., de Castro, S. C., Morais, J., de Siervo, A., Landers, R. and Gorenstein, A. Lithium insertion and electrochromism in polycrystalline molybdenum oxide films. *Solid State Ionics*, **136–7**, 2000, 357–63.
485. Abe, Y., Imamura, H., Washizu, E. and Sasaki, K. Formation process of reactively sputtered MoO_3 thin films and their optical properties. *Proc. Electrochem. Soc.*, **2002–22**, 2003, 62–7.
486. Hamelmann, F., Brechling, A., Aschentrup, A., Heinzmann, U., Jutzi, P., Sandrock, J., Siemeling, U., Ivanova, T., Szekeres, A. and Gesheva, K. Thin molybdenum oxide films produced by molybdenum pentacarbonyl 1-methylbutylisonitrile with plasma-assisted chemical vapor deposition. *Thin Solid Films*, **446**, 2004, 167–71.
487. Hinokuma, K., Kishimoto, A. and Kudo, T. Coloration dynamics of spin-coated $MoO_3 \cdot nH_2O$ electrochromic films fabricated from peroxopolymolybdate solution. *J. Electrochem. Soc.*, **141**, 1994, 876–9.
488. Zhang, Y., Kuai, S., Wang, Z. and Hu, X. Preparation and electrochromic properties of Li-doped MoO_3 films fabricated by the peroxo sol–gel process. *Appl. Surf. Sci.*, **165**, 2000, 56–9.
489. Tolgyesi, M. and Novak, M. New method of preparation and some properties of electrochromic MoO_3 thin layer. *Jpn. J. Appl. Phys.*, **32**, 1993, 93–6.
490. Laperriere, G., Lavoie, M. A. and Belenger, D. Electrochromic behavior of molybdenum trioxide thin films, prepared by thermal oxidation of electrodeposited molybdenum trisulfide, in mixtures of nonaqueous and aqueous electrolytes. *J. Electrochem. Soc.*, **143**, 1996, 3109–17.
491. McEvoy, T. M., Stevenson, K. J., Hupp, J. T. and Dang, X. Electrochemical preparation of molybdenum trioxide thin films: effect of sintering on electrochromic and electroinsertion properties. *Langmuir*, **19**, 2003, 4316–26.
492. Whittingham, M. S. Hydrogen motion in oxides: from insulators to bronzes. *Solid State Ionics*, **168**, 2004, 255–63.
493. DeSmet, D. J. and Ord, J. L. An optical study of hydrogen insertion in the anodic oxide of molybdenum. *J. Electrochem. Soc.*, **134**, 1987, 1734–40.
494. Crouch-Baker, S. and Dickens, P. G. Hydrogen insertion compounds of the molybdic acids, $MoO_3 \cdot nH_2O$ ($n = 1, 2$). *Mater. Res. Bull.*, **19**, 1984, 1457–62.
495. Yao, J. N., Yang, Y. A. and Loo, B. H. Enhancement of photochromism and electrochromism in MoO_3/Au and MoO_3/Pt thin films. *J. Phys. Chem. B*, **102**, 1998, 1856–60.
496. Azens, A. and Granqvist, C. G. Electrochromic films of tungsten oxyfluoride and electron bombarded tungsten oxide. *Sol. Energy Mater. Sol. Cells*, **44**, 1996, 333–40.
497. Kuwabara, K., Sugiyama, K. and Ohno, M. All-solid-state electrochromic device, 1: electrophoretic deposition film of proton conductive solid electrolyte. *Solid State Ionics*, **44**, 1991, 313–18.
498. Kuwabara, K., Ohno, M. and Sugiyama, K. All-solid-state electrochromic device, 2: characterization of transition-metal oxide thin films for counter electrode. *Solid State Ionics*, **44**, 1991, 319–23.

499. Petit, M. A. and Plichon, V. Anodic electrodeposition of iridium oxide films. *J. Electroanal. Chem.*, **444**, 1998, 247–52.
500. Kötz, E. R. and Neff, H. Anodic iridium oxide films: an UPS study of emersed electrodes. *Surf. Sci.*, **160**, 1985, 517–30.
501. Gottesfeld, S. and Schiavone, L. M. Electrochemical and optical studies of thick oxide layers on iridium and their electrocatalytic activities for the oxygen evolution reaction. *J. Electroanal. Chem.*, **86**, 1978, 89–104.
502. Gottesfeld, S. and McIntyre, J. D. E. Electrochromism in anodic iridium oxide. *J. Electrochem. Soc.*, **126**, 1979, 742–50.
503. Gottesfeld, S., McIntyre, J. D. E., Beni, G. and Shay, J. L. Electrochromism in anodic iridium oxide films. *Appl. Phys. Lett.*, **33**, 1978, 208–10.
504. Beni, G. and Shay, J. L. Electrochromism of anodic iridium oxide films. In Vashishta, P., Mundy, J. N. and Shenoy, G. K. (eds.), *Fast Ion Transport in Solids*, Amsterdam, Elsevier, 1979, pp. 75–8.
505. Shay, J. L., Beni, G. and Schiavone, L. M. Electrochromism of anodic iridium oxide films on transparent substrates. *Appl. Phys. Lett.*, **33**, 1978, 942–4.
506. Beni, G. and Shay, J. L. Electrochromism of heat-treated anodic iridium oxide films in acidic, neutral, and alkaline solutions. *Appl. Phys. Lett.*, **33**, 1978, 567–8.
507. Beni, G., Rice, C. E. and Shay, J. L. Electrochromism of iridium oxide films, III: anion mechanism. *J. Electrochem. Soc.*, **127**, 1980, 1342–8.
508. Dautremont-Smith, W. C., Beni, G., Schiavone, L. M. and Shay, J. L. Solid-state electrochromic cell with anodic iridium oxide film electrodes. In Vashishta, P., Mundy, J. N. and Shenoy, G. K. (eds.), *Fast Ion Transport in Solids*, Amsterdam, Elsevier, 1979, pp. 99–101.
509. Dautremont-Smith, W. C., Beni, G., Schiavone, L. M. and Shay, J. L. Solid-state electrochromic cell with anodic iridium oxide film electrodes. *Appl. Phys. Lett.*, **35**, 1979, 565–7.
510. Mo, Y., Stefan, I. C., Cai, W.-B. *et al. In situ* L_{III}-edge X-ray absorption and surface enhanced Raman spectroscopy of electrodeposited iridium oxide films in aqueous electrolytes. *J. Phys. Chem. B*, **106**, 2002, 3681–6.
511. Yamanaka, K. The electrochemical behaviour of anodically deposited iridium oxide films and the reliability of transmittance variable cells. *Jpn. J. Appl. Phys.*, **30**, 1991, 1295–8.
512. Patil, P. S., Kawar, R. K. and Sadale, S. B. Effect of substrate temperature on electrochromic properties of spray-deposited Ir-oxide thin films. *Appl. Surf. Sci.*, **249**, 2005, 367–74.
513. Klein, J. D. and Clauson, S. L. Chemistry of electrochromic IrO_x films deposited under variable redox conditions. *Mater. Res. Soc. Symp. Proc.*, **369**, 1995, 149–54.
514. Sato, Y. Characterization of thermally oxidized iridium oxide films. *Vacuum*, **41**, 1990, 1198–200.
515. Michalak, F., Rault, L. and Aldebert, P. Electrochromism with colloidal WO_3 and IrO_2. *Proc. SPIE*, **1728**, 1992, 278–88.
516. Kötz, R., Barbero, C. and Haas, O. Probe beam deflection investigation of the charge storage reaction in anodic iridium and tungsten oxide films. *J. Electroanal. Chem.*, **296**, 1990, 37–49.
517. McIntyre, J. D. E., Basu, S., Peck, W. F., Brown, W. L. and Augustyniak, W. M. Cation insertion reactions of electrochromic iridium oxide films. *Solid State Ionics*, **5**, 1981, 359–62.
518. Ord, J. L. An ellipsometric study of electrochromism in iridium oxide. *J. Electrochem. Soc.*, **129**, 1982, 335–9.

519. Rice, C. E. Ionic conduction in electrochromic anodic iridium oxide films. In Vashishta, P., Mundy, J. N. and Shenoy, G. K. (eds.), *Fast Ion Transport in Solids*, Amsterdam, Elsevier, 1979, p. 103–4.
520. Sziráki, L. and Bóbics, L. Impedance study of electrochromism in anodic Ir oxide films. *Electrochim. Acta.*, **47**, 2002, 2189–97.
521. Sanjinés, R., Aruchamy, A. and Lévy, F. Metal–non metal transition in electrochromic sputtered iridium oxide films. *Solid State Commun.*, **64**, 1987, 645–50.
522. Hackwood, S. and Beni, G. Phase transitions in iridium oxide films. *Solid State Ionics*, **2**, 1981, 297–9.
523. Gutiérrez, C., Sanchez, M., Pena, J. I., Martinez, C. and Martinez, M. A. Potential-modulated reflectance study of the oxidation state of iridium in anodic iridium oxide films. *J. Electrochem. Soc.*, **134**, 1987, 2119–26.
524. Kang, K. S. and Shay, J. L. Blue sputtered iridium oxide films (blue SIROF's). *J. Electrochem. Soc.*, **130**, 1983, 766–9.
525. Sato, Y., Ono, K., Kobayashi, T., Watanabe, H. and Yamanoka, H. Electrochromism in iridium oxide films prepared by thermal oxidation of iridium–carbon composite films. *J. Electrochem. Soc.*, **134**, 1987, 570–5.
526. Baudry, P., Aegerter, M. A., Deroo, D. and Valla, B. Electrochromic window with lithium conductive polymer electrolyte. *Proc. Electrochem. Soc.*, **90–2**, 1990, 274–87.
527. Shamritskaya, I. G., Lazorenko-Manevich, R. M. and Sokolova, L. A. Effects of anions on the electroreflectance spectra of anodically oxidized iridium in aqueous solutions. *Russ. J. Electrochem.*, **33**, 1997, 645–52.
528. Rice, C. E. and Bridenbaugh, P. M. Observation of electrochromism in solid-state anodic iridium oxide film cells using fluoride electrolytes. *Appl. Phys. Lett.*, **38**, 1981, 59–61.
529. Ishihara, S. Erasable optical memory device, Jpn. Kokai Tokkyo Koho JP 63,119,035, as cited in *Chem. Abs.* **110**: P48,553z, 1989.
530. Sanjinés, R., Aruchamy, A. and Lévy, F. Thermal stability of sputtered iridium oxide films. *J. Electrochem. Soc.*, **136**, 1989, 1740–4.
531. Yano, J., Noguchi, K., Yamasaki, S. and Yamazaki, S. Novel color change of electrochromic iridium oxide in a matrix aramid resin film. *Electrochem. Commun.*, **6**, 2004, 110–14.
532. Saito, T., Ushio, Y., Yamada, M. and Niwa, T. Properties of all solid-state thin film electrochromic device. Seventh International Conference on Solid State Ionics, Japan, 1989, p. abs. 6pB–40.
533. Heckner, K.-H. and Kraft, A. Similarities between electrochromic windows and thin film batteries. *Solid State Ionics*, **152–3**, 2002, 899–905.
534. Cerc Korošec, R., Bukovec, P., Pihlar, B. and Gomilšek, J. P. The role of thermal analysis in optimization of the electrochromic effect of nickel oxide thin films, prepared by the sol–gel method: part I. *Thermochim. Acta*, **402**, 2003, 57–67.
535. Cerc Korošec, R., Bukovec, P., Pihlar, B., Šurca Vuk, A., Orel, B. and Drazic, G. Preparation and structural investigations of electrochromic nanosized NiO_x films made via the sol–gel route. *Solid State Ionics*, **165**, 2003, 191–200.
536. Natarajan, C., Ohkubo, S. and Nogami, G. Influence of film processing temperature on the electrochromic properties of electrodeposited nickel hydroxide. *Solid State Ionics*, **86–8**, 1996, 949–53.
537. Scarminio, J., Gorenstein, A., Decker, F., Passerini, S., Pileggi, R. and Scrosati, B. Cation insertion in electrochromic NiO_x films. *Proc. SPIE*, **1536**, 1991, 70–80.

538. Hutchins, M. G., McMeeking, G. and Xingfang, H. Rf diode sputtered nickel oxide films. *Proc. SPIE*, **1272**, 1990, 139–50.
539. Wruck, D. A. and Rubin, M. Structure and electronic properties of electrochromic NiO films. *J. Electrochem. Soc.*, **140**, 1993, 1097–104.
540. Ushio, Y., Ishikawa, A. and Niwa, T. Degradation of the electrochromic nickel oxide film upon redox cycling. *Thin Solid Films*, **280**, 1996, 233–7.
541. Urbano, A., Ferreira, F. F., deCastro, S. C., Landers, R., Fantini, M. C. A. and Gorenstein, A. Electrochromism in lithiated nickel oxide films deposited by rf sputtering. *Electrochim. Acta*, **46**, 2001, 2269–73.
542. Kitao, M., Izawa, K. and Yamada, S. Electrochromic properties of nickel oxide films prepared by introduction of hydrogen into sputtering atmosphere. *Sol. Energy Mater. Sol. Cells*, **39**, 1995, 115–22.
543. Agrawal, A., Habibi, H. R., Agrawal, R. K., Cronin, J. P., Roberts, D. M., Caron-Papowich, R. and Lampert, C. M. Effect of deposition pressure on the microstructure and electrochromic properties of electron-beam-evaporated nickel oxide films. *Thin Solid Films*, **221**, 1992, 239–53.
544. Porqueras, I. and Bertran, E. Electrochromic behaviour of nickel oxide thin films deposited by thermal evaporation. *Thin Solid Films*, **398–9**, 2001, 41–4.
545. Bouessay, I., Rougier, A., Beaudoin, B. and Leriche, J. B. Pulsed laser-deposited nickel oxide thin films as electrochromic anodic materials. *Appl. Surf. Sci.*, **186**, 2002, 490–5.
546. Wen, S.-J., von Rottkay, K. and Rubin, M. Electrochromic lithium nickel oxide thin film by pulsed laser deposition. *Proc. Electrochem. Soc.*, **96–24**, 1996, 54–63.
547. Rubin, M., Wen, S.-J., Richardson, T., Kerr, J., von Rottkay, K. and Slack, J. Electrochromic lithium nickel oxide by pulsed laser deposition and sputtering. *Sol. Energy Mater. Sol. Cells*, **54**, 1998, 59–66.
548. Wen, S.-J., Kerr, J., Rubin, M., Slack, J. and von Rottkay, K. Analysis of durability in lithium nickel oxide electrochromic materials and devices. *Sol. Energy Mater. Sol. Cells*, **56**, 1999, 299–307.
549. Bouessay, I., Rougier, A. and Tarascon, J.-M. Electrochromic mechanism in nickel oxide thin films grown by pulsed laser deposition, *Proc. Electrochem. Soc.*, **2003–22**, 2003, 91–102.
550. Anders, S., Anders, A., Rubin, M., Wang, Z., Raoux, S., Kong, F. and Brown, I. G. Formation of metal oxides by cathodic arc deposition. *Surf. Coat. Technol.*, **76–7**, 1995, 167–73.
551. Velevska, J. and Ristova, M. Electrochromic properties of NiO_x prepared by low vacuum evaporation. *Sol. Energy Mater. Sol. Cells*, **73**, 2002, 131–9.
552. Scarminio, J., Urbano, B., Gardes, J. and Gorenstein, A. Electrochromism in nickel oxide films obtained by thermal decomposition. *J. Mater. Sci. Lett.*, **11**, 1992, 562–3.
553. Córdoba de Terresi, S. I., Hugot le-Goff, A. and Takenouti, H. Electrochromism in metal oxide films studied by Raman spectroscopy and A.C. techniques: charge insertion mechanism. *Proc. SPIE*, **1272**, 1990, 152–61.
554. Torresi, R. M., Vazquez, M. V., Gorenstein, A. and de Torresi, S. I. C. Infrared characterization of electrochromic nickel hydroxide prepared by homogeneous chemical precipitation. *Thin Solid Films*, **229**, 1993, 180–6.
555. Chigane, M., Ishikawa, M. and Inoue, H. Further XRD characterization of electrochromic nickel oxide thin films prepared by anodic deposition. *Sol. Energy Mater. Sol. Cells*, **64**, 2000, 65–72.
556. Chigane, M. and Ishikawa, M. Electrochromic properties of nickel oxide thin films prepared by electrolysis followed by chemical deposition. *Electrochim. Acta*, **42**, 1997, 1515–19.

557. Chigane, M. and Ishikawa, M. XRD and XPS characterization of electrochromic nickel oxide thin films prepared by electrolysis–chemical deposition. *J. Chem. Soc., Faraday Trans.*, **94**, 1998, 3665–70.
558. Jiménez-González, A. E. and Cambray, J. G. Deposition of NiO_x thin films by sol–gel technique. *Surf. Eng.*, **16**, 2000, 73–6.
559. Mahmoud, S. A., Aly, S. A., Abdel-Rahman, M. and Abdel-Hady, K. Electrochromic characterisation of electrochemically deposited nickel oxide films, *Physica B: Condens. Matter*, **293**, 2000, 125–31.
560. Ristova, M., Velevska, J. and Ristov, M. Chemical bath deposition and electrochromic properties of NiO_x films. *Sol. Energy Mater. Sol. Cells*, **71**, 2002, 219–30.
561. Crnjak Orel, Z., Hutchins, M. G. and McMeeking, G. The electrochromic properties of hydrated nickel oxide films formed by colloidal and anodic deposition. *Sol. Energy Mater. Sol. Cells*, **30**, 1993, 327–37.
562. Sato, Y., Ando, M. and Murai, K. Electrochromic properties of spin-coated nickel oxide films. *Solid State Ionics*, **113–15**, 1998, 443–7.
563. Richardson, T. J. and Rubin, M. D. Liquid phase deposition of electrochromic thin films. *Electrochim. Acta*, **46**, 2001, 2119–23.
564. Fantini, M. C. A., Bezerra, G. H., Carvalho, C. R. C. and Gorenstein, A. Electrochromic properties and temperature dependence of chemically deposited $Ni(OH)_x$ thin films. *Proc. SPIE*, **1536**, 1991, 81–92.
565. Kadam, L. D. and Patil, P. S. Studies on electrochromic properties of nickel oxide thin films prepared by spray pyrolysis technique. *Sol. Energy Mater. Sol. Cells*, **69**, 2001, 361–9.
566. Arakaki, J., Reyes, R., Horn, M. and Estrada, W. Electrochromism in NiO_x and WO_x obtained by spray pyrolysis. *Sol. Energy Mater. Sol. Cells*, **37**, 1995, 33–41.
567. Mahmoud, S. A., Akl, A. A., Kamal, H. and Abdel-Hady, K. Opto-structural, electrical and electrochromic properties of crystalline nickel oxide thin films prepared by spray pyrolysis. *Physica B*, **311**, 2002, 366–75.
568. Maruyama, T. and Arai, S. The electrochromic properties of nickel oxide thin films prepared by chemical vapor deposition. *Sol. Energy Mater. Sol. Cells*, **30**, 1993, 257–62.
569. Murai, K., Mihara, T., Mochizuki, S., Tamura, S. and Sato, Y. Electrochromism in nickel oxide films prepared by plasma oxidation of nickel–carbon composite films. *Solid State Ionics*, **86–8**, 1996, 955–8.
570. Suiyang, H., Fengbo, C. and Jicai, Z. Electrochromism in hydrated nickel oxide films made by RF sputtering. In Chowdari B. V. R. and Radharkrishna, S. (eds.), *Proceedings of the International Seminar on Solid State Ionic Devices*, Singapore, World Publishing Co., 1988, pp. 521–6.
571. Ahn, K.-S., Nah, Y.-C. and Sung, Y.-E. Surface morphological, microstructural, and electrochromic properties of short-range ordered and crystalline nickel oxide thin films. *Appl. Surf. Sci.*, **199**, 2002, 259–69.
572. Murphy, T. P. and Hutchins, M. G. Oxidation states in nickel oxide electrochromism. *Sol. Energy Mater. Sol. Cells*, **39**, 1995, 377–89.
573. Bouessay, I., Rougier, A., Poizat, O., Moscovici, J., Michalowicz, A. and Tarascon, J. M. Electrochromic degradation in nickel oxide thin film: a self-discharge and dissolution phenomenon. *Electrochim. Acta*, **50**, 2005, 3737–45.
574. Oliva, P. J. L., Laurent, J. F., Delmas, C., Braconnier, J. J., Figlarz, M., Fievet, F. and de Guibert, A. Review of the structure and the electrochemistry of nickel hydroxides and oxy-hydroxides. *J. Power Sources*, **8**, 1992, 229–55.
575. Nemetz, A., Temmink, A., Bange, K., Córdoba de Torresi, S., Gabrielli, C., Torresi, R. and Hugot le-Goff., A. Investigations and modeling of e^--beam evaporated $NiO(OH)_x$ films. *Sol. Energy Mater. Sol. Cells*, **25**, 1992, 93–103.

576. Lampert, C. M. *In situ* spectroscopic studies of electrochromic hydrated nickel oxide films. *Sol. Energy Mater.*, **19**, 1989, 1–16.
577. Rosolen, J. M., Decker, F., Fracastoro-Decker, M., Gorenstein, A., Torresi, R. M. and Córdoba de Torresi, S. I. A mirage effect analysis of the electrochemical processes in nickel hydroxide electrodes. *J. Electroanal. Chem.*, **354**, 1993, 273–9.
578. Gorenstein, A., Decker, F., Estrada, W., Esteves, C., Andersson, A., Passerini, S., Pantaloni, S. and Scrosati, B. Electrochromic NiO_xH_y hydrated films: cyclic voltammetry and ac impedance spectroscopy in aqueous electrolyte. *J. Electroanal. Chem.*, **277**, 1990, 277–90.
579. MacArthur, D. M. The proton diffusion coefficient for the nickel hydroxide electrode. *J. Electrochem. Soc.*, **117**, 1970, 729–32. MacArthur consistently in this paper talks of 'ΔH for diffusion', but in fact the data from his Arrhenius-type graphs yield E_A.
580. Schrebler Guzmán, R. S., Vilche, J. R. and Arviá, A. J. Rate processes related to the hydrated nickel hydroxide electrode in alkaline solutions. *J. Electrochem. Soc.*, **125**, 1978, 1578–87.
581. Faria, I. C., Kleinke, M., Gorenstein, A., Fantini, M. C. A. and Tabacniks, M. H. Toward efficient electrochromic NiO_x films: a study of microstructure, morphology, and stoichiometry of radio-frequency sputtered films. *J. Electrochem. Soc.*, **145**, 1998, 235–41.
582. Liquan, C., Ming, D., Yunfa, C., Chunxiang, S. and Rungjian, X. Study on EC Ni-O thin film and new EC device. Seventh International Conference on Solid State Ionics, Japan, 1989, abs. 6pB–38.
583. Šurca, A., Orel, B. and Pihlar, B. Sol–gel derived hydrated nickel oxide electrochromic films: optical, spectroelectrochemical and structural properties. *J. Sol Gel Sci. Technol.*, **8**, 1997, 743–8.
584. Fantini, M. and Gorenstein, A. Electrochromic nickel hydroxide films on transparent/conducting substrates. *Sol. Energy Mater.*, **16**, 1987, 487–500.
585. Jeong, D. J., Kim, W.-S. and Sung, Y. E. Improved electrochromic response time of nickel hydroxide thin films by ultra-thin nickel metal underlayer. *Jpn. J. Appl. Phys.*, **40**, 2001, L708–10.
586. Decker, F., Pileggi, R., Passerini, S. and Scrosati, B. A comparison of the electrochromic behaviour and the mechanical properties of WO_3 and NiO_x thin-film electrodes. *J. Electrochem. Soc.*, **138**, 1991, 3182–6.
587. Azens, A., Kullman, L., Vaivars, G., Nordborg, H. and Granqvist, C. G. Sputter-deposited nickel oxide for electrochromic applications. *Solid State Ionics*, **113–15**, 1998, 449–56.
588. Trimble, C., DeVries, M., Hale, J. S., Thompson, D. W., Tiwald, T. E. and Woolan, J. A. Infrared emittance modulation devices using electrochromic crystalline tungsten oxide, polymer conductor, and nickel oxide. *Thin Solid Films*, **355–6**, 1999, 26–34.
589. Avino, C., Panero, S. and Scrosati, B. An electrochromic window based on a modified polypyrrole/nickel oxide combination. *J. Mater. Chem.*, **2**, 1993, 1259–61.
590. Arbizzani, C., Mastragostino, M., Passerini, S., Pillegi, R. and Scrosati, B. An electrochromic window based on polymethyl thiophene and nickel oxide electrodes. *Electrochim. Acta*, **36**, 1991, 837–40.
591. Richardson, T. J., Slack, J. L. and Rubin, M. D. Electrochromism in copper oxide thin films. *Electrochim. Acta*, **46**, 2001, 2281–4.
592. Garnich, F., Yu, P. C. and Lampert, C. M. Hydrated manganese oxide as a counter-electrode material for an electrochromic optical switching device. *Sol. Energy Mater.*, **20**, 1990, 265–75.

593. Shimanoe, K., Suetsugu, M., Miura, N. and Yamazoe, N. Bismuth oxide thin film as new electrochromic material. *Solid State Ionics*, **113–15**, 1998, 415–19.
594. Cazzanelli, E., Marino, S., Bruno, V., Castriosta, M., Scaramuzza, N., Strangi, G., Versace, C., Ceccato, R. and Carturan, G. Characterizations of mixed Bi/V oxide films, deposited via sol–gel route, used as electrodes in asymmetric liquid crystal cells. *Solid State Ionics*, **165**, 2003, 201–8.
595. Özer, N., Cronin, J. P. and Akyuz, S. Electrochromic performance of sol–gel-deposited CeO_2 films. *Proc. SPIE*, **3788**, 1999, 103–10.
596. Porqueras, I., Person, C., Corbella, C., Vives, M., Pinyol, A. and Bertran, E. Characteristics of e-beam deposited electrochromic CeO_2 thin films. *Solid State Ionics*, **165**, 2003, 131–7.
597. Porqueras, I., Person, C. and Bertran, E. Influence of the film structure on the properties of electrochromic CeO_2 thin films deposited by e-beam PVD. *Thin Solid Films*, **447–8**, 2004, 119–24.
598. Azens, A., Vaivars, G., Kullman, L. and Granqvist, C. G. Electrochromism of Cr oxide films. *Electrochim. Acta*, **44**, 1999, 3059–61.
599. Cogan, S. F., Rauh, R. D., Klein, J. D., Nguyen, N. M., Jones, R. B. and Plante, T. D. Variable transmittance coatings using electrochromic lithium chromate and amorphous WO_3 thin films. *J. Electrochem. Soc.*, **144**, 1997, 956–60.
600. Besenhard, J. O. and Schöllhörn, R. Chromium oxides as cathodes for secondary high energy density lithium batteries. *J. Electrochem. Soc.*, **124**, 1977, 968–71.
601. Arora, P., Zhang, D., Popov, B. N. and White, R. E. Chromium oxides and lithiated chromium oxides: promising cathode materials for secondary lithium batteries. *Electrochem. Solid State Lett.*, **1**, 198, 249–51.
602. Takeda, Y., Tsuji, Y. and Yashamoto, O. Rechargeable lithium/chromium oxide cells. *J. Electrochem. Soc.*, **131**, 1984, 2006–9.
603. Kullman, L., Azens, A., Vaivars, G. and Granqvist, C. G. Electrochromic devices incorporating Cr oxide and Ni oxide films: a comparison. *Sol. Energy*, **68**, 2000, 517–22.
604. Maruyama, T. and Arai, S. Electrochromic properties of cobalt oxide thin films prepared by chemical vapour deposition. *J. Electrochem. Soc.*, **143**, 1996, 1383–6.
605. Burke, L. D. and Murphy, O. J. Electrochromic behaviour of oxide films grown on cobalt and manganese in base. *J. Electroanal. Chem.*, **109**, 1980, 373–7.
606. Burke, L. D., Lyons, M. E. and Murphy, O. J. Formation of hydrous oxide films on cobalt under potential cycling conditions. *J. Electroanal. Chem.*, **132**, 1982, 247–61.
607. Burke, L. D. and Murphy, O. J. Electrochromic behaviour of electrodeposited cobalt oxide films. *J. Electroanal. Chem.*, **112**, 1980, 379–82.
608. Gorenstein, A., Polo Da Fonseca, C. N. and Torresi, R. Electrochromism in cobalt oxyhydroxide thin films. *Proc. SPIE*, **1536**, 1991, 104–15.
609. Cotton, F. A. and Wilkinson, G. *Advanced Inorganic Chemistry*, 4th edn, New York, Wiley, 1980, p. 767.
610. Unuma, H., Saito, Y., Watanabe, K. and Sugawara, M. Preparation of Co_3O_4 thin films by a modified chemical-bath method. *Thin Solid Films*, **468**, 2004, 4–7.
611. Bewick, A., Gutiérrez, C. and Larramona, G. An *in-situ* IR spectroscopic study of the anodic oxide film on cobalt in alkaline solutions. *J. Electroanal. Chem.*, **333**, 1992, 165–75.
612. Wei, G. *Diss. Abstr. Int. B.*, 52 (1991) 2247, as cited in *Chem. Abs.* **116**: 116, 951d. Ph.D. thesis, Tufts University, MA, 1991.
613. Go, J.-Y., Pyun, S.-I. and Shin, H.-C. Lithium transport through the $Li_{1-\delta}CoO_2$ film electrode prepared by RF magnetron sputtering. *J. Electroanal. Chem.*, **527**, 2002, 93–102.

614. Polo da Fonseca, C. N., De Paoli, M.-A. and Gorenstein, A. The electrochromic effect in cobalt oxide thin films. *Adv. Mater.*, **3**, 1991, 553–5.
615. Polo da Fonseca, C. N., De Paoli, M.-A. and Gorenstein, A. Electrochromism in cobalt oxide thin films grown by anodic electroprecipitation. *Sol. Energy Mater. Sol. Cells*, **33**, 1994, 73–81.
616. Svegl, F., Orel, B., Hutchins, M. G. and Kalcher, K. Structural and spectroelectrochemical investigations of sol–gel derived electrochromic spinel Co_3O_4 films. *J. Electrochem. Soc.*, **143**, 1996, 1532–9.
617. Svegl, F., Orel, B., Bukovec, P., Kalcher, K. and Hutchins, M. G. Spectroelectrochemical and structural properties of electrochromic Co(Al)-oxide and Co(Al, Si)-oxide films prepared by the sol–gel route. *J. Electroanal. Chem.*, **418**, 1996, 53–66.
618. Behl, W. K. and Toni, J. E. Anodic oxidation of cobalt in potassium hydroxide electrolytes. *J. Electroanal. Chem.*, **31**, 1971, 63–75.
619. Benson, P., Briggs, G. W. D. and Wynne-Jones, W. F. K. The cobalt hydroxide electrode, I: structure and phase transitions of the hydroxides. *Electrochim. Acta*, **9**, 1964, 275–80.
620. Özer, N. and Tepehan, F. Structure and optical properties of electrochromic copper oxide films prepared by reactive and conventional evaporation techniques. *Sol. Energy Mater. Sol. Cells*, **30**, 1993, 13–26.
621. Özer, N. and Tepehan, F. Sol–gel deposition of electrochromic copper oxide films. *Proc. SPIE*, **2017**, 1993, 113–31.
622. Ray, S. C. Preparation of copper oxide thin film by the sol–gel dip technique and study of their structural and optical properties. *Sol. Energy Mater. Sol. Cells*, **68**, 2001, 307–12.
623. Richardson, T. J. New electrochromic mirror systems. *Solid State Ionics*, **165**, 2003, 305–8.
624. Gutiérrez, C. and Beden, B. UV-Visible differential reflectance spectroscopy of the electrochromic oxide layer on iron in 0.1 M NaOH. *J. Electroanal. Chem.*, **293**, 1990, 253–9.
625. Burke, L. D. and Lyons, M. E. G. The formation and stability of hydrous oxide films of iron under potential cycling conditions in aqueous solution at high pH. *J. Electroanal. Chem.*, **198**, 1986, 247–68.
626. Burke, L. D. and Murphy, O. J. Growth of an electrochromic film on iron in base under potential cycling conditions. *J. Electroanal. Chem.*, **109**, 1980, 379–83.
627. Özer, N. and Tepehan, F. Optical and electrochemical characterisation of sol–gel deposited iron oxide films. *Sol. Energy Mater. Sol. Cells*, **56**, 1999, 141–52.
628. Özer, N., Tepehan, F. and Tepehan, G. Preparation and optical properties of sol gel deposited electrochromic iron oxide films. *Proc. SPIE*, **3138**, 1997, 31–9.
629. Maruyama, T. and Kanagawa, T. Electrochromic properties of iron oxide thin films prepared by chemical vapor deposition. *J. Electrochem. Soc.*, **143**, 1996, 1675–8.
630. Baba, N., Yoshino, T. and Watanabe, S. Preparation of electrochromic MnO_2 thin film by electrodeposition. Seventh International Conference on Solid State Ionics, Japan, 1989, abs. 6pB–39.
631. Córdoba De Torresi, S. I. and Gorenstein, A. Electrochromic behaviour of manganese dioxide electrodes in slightly alkaline solutions. *Electrochim. Acta*, **37**, 1992, 2015–19.
632. López de Mishima, B. A., Ohtsuka, T. and Sata, N. *In-situ* Raman spectroscopy of manganese dioxide during the discharge process. *J. Electroanal. Chem.*, **243**, 1988, 219–23.

633. Long, J. W., Qadir, L. R., Stroud, R. M. and Rolinson, D. R. Spectroelectrochemical investigations of cation-insertion reactions at sol–gel derived nanostructured, mesoporous thin films of manganese oxide. *J. Phys. Chem. B*, **105**, 2001, 8712–17.
634. Naghash, A. R. and Lee, J. Y. Preparation of spinel lithium manganese oxide by aqueous co-precipitation. *J. Power Sources*, **85**, 2000, 284–93.
635. Demishima, B., Ohtsuka, T., Konno, H. and Sata, N. XPS study of the MnO_2 electrode in borate solution during the discharge process. *Electrochim. Acta*, **36**, 1991, 1485–9.
636. Ma, Y.-P., Yu, P. C. and Lampert, C. M. Development of laminated nickel/manganese and nickel/niobium oxide electrochromic devices. *Proc. SPIE*, **1536**, 1991, 93–103.
637. Bueno, P. R., Avellaneda, C. O., Faria, R. C. and Bulhões, L. O. S. Electrochromic properties of undoped and lithium doped Nb_2O_5 films prepared by the sol–gel method. *Electrochim. Acta*, **46**, 2001, 2113–18.
638. Aegerter, M. A. Sol–gel niobium pentoxide: a promising material for electrochromic coatings, batteries, nanocrystalline solar cells and catalysis. *Sol. Energy Mater. Sol. Cells*, **68**, 2001, 401–22.
639. Maček, M. and Orel, B. Electrochromism of sol–gel derived niobium oxide films. *Sol. Energy Mater. Sol. Cells*, **54**, 1998, 121–30.
640. Maček, M. and Orel, B. Electrochromism of sol–gel derived niobium oxide films. *Turk. J. Chem*, **22**, 1998, 67–72.
641. Lee, G. R. and Crayston, J. A. Sol–gel processing of transition-metal alkoxides for electronics. *Adv. Mater.*, **5**, 1993, 434–42.
642. Lee, G. R. and Crayston, J. A. Electrochromic Nb_2O_5 and Nb_2O_5/silicone composite thin films prepared by sol–gel processing. *J. Mater. Chem.*, **1**, 1991, 381–6.
643. Özer, N., Barreto, T., Buyuklimanli, T. and Lampert, C. M. Characterization of sol–gel deposited niobium pentoxide films for electrochromic devices. *Sol. Energy Mater. Sol. Cells*, **36**, 1995, 433–43.
644. Reichman, B. and Bard, A. J. Electrochromism at niobium pentoxide electrodes in aqueous and acetonitrile solution. *J. Electrochem. Soc.*, **127**, 1980, 241–2.
645. Maranhão, S. L. D. A. and Torresi, R. M. Electrochemical and chromogenics kinetics of lithium intercalation in anodic niobium oxide films. *Electrochim. Acta*, **43**, 1998, 257–64.
646. Maranhão, S. L. D. A. and Torresi, R. M. Filmes de óxidos anódicos de nióbio: efeito eletrocrômico e cinética da reação de eletro-intercalação. *Quim. Nova*, **21**, 1998, 284–8.
647. Gomes, M. A. B., Bulhões, L. O. S., de Castro, S. C. and Damiao, A. J. The electrochromic process at Nb_2O_5 electrode prepared by thermal oxidation of niobium. *J. Electrochem. Soc.*, **137**, 1990, 3067–71.
648. Gomes, M. A. B. and Bulhões, L. O. S. Diffusion coefficient of H^+ at Nb_2O_5 layers prepared by thermal oxidation of niobium. *Electrochim. Acta*, **35**, 1990, 765–8.
649. Maruyama, T. and Arai, K. Electrochromic properties of niobium oxide thin films prepared by radio-frequency magnetron sputtering method. *Appl. Phys. Lett.*, **63**, 1993, 869–70.
650. Rosario, A. V. and Pereira, E. C. Optimisation of the electrochromic properties of Nb_2O_5 thin films produced by sol–gel route using factorial design. *Sol. Energy Mater. Sol. Cells*, **71**, 2002, 41–50.
651. Bolzán, A. Z. and Arvia, A. J. The electrochemical behaviour of hydrous palladium oxide layers formed at high positive potentials in different electrolyte solutions. *J. Electroanal. Chem.*, **322**, 1992, 247–65.

652. Thomas, G. R. and Owen, J. R. Rare earth oxides in electrochromic windows. *Solid State Ionics*, **53–6**, 1992, 513–19.
653. Hartridge, A., Ghanashyam Krishna, M. and Bhattacharya, A. K. A study of nanocrystalline CeO_2/PrO_x optoionic thin films: temperature and oxygen vacancy dependence. *Mater. Sci. Eng. B*, **57**, 1999, 173–8.
654. Burke, L. D. and O'Sullivan, E. J. M. Reactivity of hydrous rhodium oxide films in base. *J. Electroanal. Chem.*, **129**, 1981, 133–48.
655. Burke, L. D. and O'Sullivan, E. J. M. Enhanced oxide growth at a rhodium surface in base under potential cycling conditions. *J. Electroanal. Chem.*, **93**, 1978, 11–18.
656. Wang, H., Yan, M. and Jiang, Z. Electrochromic properties of rhodium oxide films prepared by a sol–gel method. *Thin Solid Films*, **401**, 2001, 211–15.
657. Gottesfeld, S. The anodic rhodium oxide film: a two-colour electrochromic system. *J. Electrochem. Soc.*, **127**, 1980, 272–7.
658. Lee, S.-H., Liu, P., Cheong, H. M., Tracy, C. E. and Deb, S. K. Electrochromism of amorphous ruthenium oxide thin films. *Solid State Ionics*, **165**, 2003, 217–21.
659. Burke, L. D. and Whelan, D. P. The behaviour of ruthenium anodes in base. *J. Electroanal. Chem.*, **103**, 1979, 179–87.
660. Hutchins, M. G., Butt, N. S., Topping, A. J., Gallego, J. M., Milne, P. E., Jeffrey, D. and Brotherton, I. D. Tantalum oxide thin film ionic conductors for monolithic electrochromic devices. *Proc. SPIE*, **4458**, 2001, 120–7.
661. Klingler, M., Chu, W. F. and Weppner, W. Three-layer electrochromic system. *Sol. Energy Mater. Sol. Cells*, **39**, 1995, 247–55.
662. Masing, L., Orme, J. E. and Young, L. Optical properties of anodic film oxide films on tantalum, niobium, and tantalum + niobium alloys, and the optical constants of tantalum. *J. Electrochem. Soc.*, **108**, 1961, 428–38.
663. Özer, N., He, Y. and Lampert, C. M. Ionic conductivity of tantalum oxide films prepared by sol–gel process for electrochromic devices. *Proc. SPIE*, **2255**, 1994, 456–66.
664. Ahn, K.-S., Nah, Y.-C. and Sung, Y.-E. Effect of interfacial property on electrochromic response speed of Ta_2O_5/NiO and $Ta_2O_5/Ni(OH)_2$. *Solid State Ionics*, **165**, 2003, 155–60.
665. Garikepati, P. and Xue, T. Study of the electrochromic film-solid electrolyte film interface (WO_3/Ta_2O_5) by impedance measurements. *Sol. Energy Mater. Sol. Cells*, **25**, 1992, 105–11.
666. Hensler, D. H., Cuthbert, J. D., Martin, R. J. and Tien, P. K. Optical propagation in sheet and pattern generated films of Ta_2O_5. *Appl. Opt.*, **10**, 1971, 1037–42.
667. Ord, J. L. and Wang, W. P. Optical anisotropy and electrostriction in the anodic oxide of tantalum. *J. Electrochem. Soc.*, **130**, 1983, 1809–14.
668. Ord, J. L., Hopper, M. A. and Wang, W. P. Field-dependence of the dielectric constant during anodic oxidation of tantalum, niobium, and tungsten. *J. Electrochem. Soc.*, **119**, 1972, 439–45.
669. Tuller, H. L. and Moon, P. K. Fast ion conductors: future trends. *Mater. Sci. Eng. B*, **1**, 1988, 171–91.
670. Sone, Y., Kishimoto, A. and Kudo, T. Amorphous tantalum oxide proton conductor derived from peroxo-polyacid and its application for EC device. *Solid State Ionics*, **70–1**, 1994, 316–20.
671. Nagai, J., McMeeking, G. D. and Saitoh, Y. Durability of electrochromic glazing. *Sol. Energy Mater. Sol. Cells*, **56**, 1999, 309–19.
672. Hale, J. S., DeVries, M., Dworak, B. and Woollam, J. A. Visible and infrared optical constants of electrochromic materials for emissivity modulation applications. *Thin Solid Films*, **313–14**, 1998, 205–9.

673. Hale, J. S. and Woollam, J. A. Prospects for IR emissivity control using electrochromic structures. *Thin Solid Films*, **339**, 1999, 174–80.
674. Matsuda, S. and Sugimoto, K. Ellipsometric analysis of changes in surface oxide films on tantalum during anodic and cathodic polarization. *J. Jpn. Inst. Met.*, **49**, 1985, 224–30 [in Japanese].
675. Fu, Z.-W. and Qui, Q.-Z. Pulsed laser deposite Ta_2O_5 thin films as an electrochromic material. *Electrochem. Solid-State Lett.*, **2**, 1999, 600–1.
676. Isidorsson, J. and Granqvist, C. G. Electrochromism of Li–intercalated Sn oxide films made by sputtering. *Sol. Energy Mater. Sol. Cells*, **44**, 1996, 375–81.
677. Isidorsson, J., Granqvist, C. G., Häggström, L. and Nordström, E. Electrochromism in lithiated Sn oxide: Mössbauer spectroscopy data on valence state changes. *J. Appl. Phys.*, **80**, 1996, 2367–71.
678. Yonghong, Y., Jiayu, Z., Peifu, G., Xu, L. and Jinfa, T. Electrochromism of titanium oxide thin film. *Thin Solid Films*, **298**, 1997, 197–9.
679. Fu, Z., Kong, J., Qin, Q. and Tian, Z. *In situ* spectroelectrochemical behaviour of nanocrystalline TiO_2 thin film electrode fabricated by pulsed laser ablation. *Chem. China*, **42**, 1999, 493–500.
680. de Tacconi, N. R., Chenthamarakshan, C. R., Wouters, K. L., MacDonnell, F. M., and Rajeshwar, K. Composite WO_3–TiO_2 films prepared by pulsed electrodeposition: morphological aspects and electrochromic behavior. *J. Electroanal. Chem.*, **566**, 2004, 249–56.
681. Ord, J. L., DeSmet, D. J. and Beckstead, D. J. Electrochemical and optical properties of anodic oxide films on titanium. *J. Electrochem. Soc.*, **136**, 1989, 2178–84.
682. Ottaviani, M., Panero, S., Morizilli, S., Scrosati, B. and Lazzari, M. The electrochromic characteristics of titanium oxide thin film. *Solid State Ionics*, **20**, 1986, 197–202.
683. Ohzuki, T. and Hirai, T. An electrochromic display based on titanium. *Electrochim. Acta*, **27**, 1982, 1263–6.
684. Bonhôte, P., Gogniat, E., Grätzel, M. and Ashrit, P. V. Novel electrochromic devices based on complementary nanocrystalline TiO_2 and WO_3 thin films. *Thin Solid Films*, **350**, 1999, 269–75.
685. Yoshimura, T., Miki, T. and Tanemura, S. TiO_2 electrochromic thins films by reactive direct current magnetron sputtering. *J. Vac. Sci. Technol. A*, **15**, 1997, 2673–6.
686. Rousselot, C., Chappé, J.-M., Martin, N. and Terwange, G. Properties and electrochromic performance of titanium oxynitride thin films prepared by reactive sputtering. *Proc. Electrochem. Soc.*, **2003–22**, 2003, 68–79.
687. Fang, G. J., Yao, K.-L. and Liu, Z.-L. Fabrication and electrochromic properties of double layer $WO_3(V)/V_2O_5(Ti)$ thin films prepared by pulsed laser ablation technique. *Thin Solid Films*, **394**, 2001, 63–70.
688. Rougier, A. and Blyr, A. Electrochromic properties of vanadium tungsten oxide thin films grown by pulsed laser deposition. *Electrochim. Acta*, **46**, 2001, 1945–50.
689. Fang, J. G., Liu, Y. H. and Yao, K. L. Synthesis and structural, electrochromic characterization of pulsed laser deposition of vanadium oxide thin films. *J. Vac. Sci. Technol. A*, **19**, 2001, 887–92.
690. Fujita, Y., Miyazaki, K. and Tatsuyama, C. On the electrochromism of evaporated V_2O_5 films. *Jpn. J. Appl. Phys.*, **24**, 1985, 1082–6.
691. Julien, C., Guesdon, J. P., Gorenstein, A., Khelfa, A. and Ivanova, T. Growth of V_2O_5 flash-evaporated films. *J. Mater. Sci. Lett.*, **14**, 1995, 934–6.
692. Aita, C. R., Liu, Y., Kao, M. L. and Hansen, S. D. Optical behaviour of sputter-deposited vanadium pentoxide. *J. Appl. Phys.*, **60**, 1986, 749–53.
693. Guan, Z. S., Yao, J. N., Yang, Y. A. and Loo, B. H. Electrochromism of annealed vacuum-evaporated V_2O_5 films. *J. Electroanal. Chem.*, **443**, 1998, 175–9.

694. Ord, J. L., Bishop, S. D. and DeSmet, D. J. An optical study of hydrogen insertion in the anodic oxide of vanadium. *J. Electrochem. Soc.*, **138**, 1991, 208–14.
695. Znaidi, Z., Baffier, N. and Lemordant, D. Kinetics of the H^+/M^+ ion exchange in V_2O_5 xerogel. *Solid State Ionics*, **28–30**, 1988, 1750–5.
696. Livage, J. Vanadium pentoxide gels. *Chem. Mater.*, **3**, 1991, 578–93.
697. Livage, J. Sol–gel chemistry and electrochemical properties of vanadium oxide gels. *Solid State Ionics*, **86–8**, 1996, 935–42.
698. Livage, J. Optical and electrical properties of vanadium oxides synthesized from alkoxides. *Coord. Chem. Rev.*, **190–2**, 1999, 391–403.
699. Vroon, Z. A. E. P. and Spee, C. I. M. A. Sol–gel coatings on large area glass sheets for electrochromic devices. *J. Non-Cryst. Solids*, **218**, 1997, 189–95.
700. Stewart, O., Rodriguez, J., Williams, K. B., Reck, G. P., Malani, N. and Proscia, J. W. Chemical vapor deposition of vanadium oxide thin films. *Mater. Res. Soc. Symp. Proc.*, **335**, 1994, 329–33.
701. Shimizu, Y., Nagase, K., Miura, N. and Yamazoe, N. Electrochromic properties of spin-coated V_2O_5 thin films. *Solid State Ionics*, **53–6**, 1992, 490–5.
702. Shimizu, Y., Nagase, K., Muira, N. and Yamazoe, N. Electrochromic properties of vanadium pentoxide thin films prepared by new wet process. *Appl. Phys. Lett.*, **60**, 1992, 802–4.
703. Liu, P., Lee, S.-H., Tracy, C. E., Turner, J. A., Pitt, J. R. and Deb, S. K. Electrochromic and chemochromic performance of mesoporous thin-film vanadium oxide. *Solid State Ionics*, **165**, 2003, 223–8.
704. Burke, L. D. and O'Sullivan, E. J. M. Electrochromism in electrodeposited vanadium oxide films. *J. Electroanal. Chem.*, **111**, 1980, 383–4.
705. Gavrilyuk, A. I. and Chudnovskii, F. A. Electrochromism in vanadium pentoxide films. *Pis'ma. Zh. Tekh. Fiz*, **3**, 1977, 174–7; also available as: *Sov. Tech. Phys. Lett.*, **3**, 1977, 69–70.
706. Hub, S., Trenchant, A. and Messina, R. X-Ray investigations on electroformed $Li_xV_2O_5$ bronzes. *Electrochim. Acta*, **33**, 1988, 997–1002.
707. Gavrilyuk, V. I. and Plakhotnik, V. N. Electrochromism of thin-films of vanadium(V) and tungsten(III) oxides in the system $LiBF_4$–γ-butyrolactone. *Vopr. Khim. Khim. Technol.*, **89**, 1989, 23–26 [in Russian], as cited in *Chem. Abs.*, **113**: 122,883.
708. Dickens, P. G., Hibble, S. J. and Jarman, R. H. Ion insertion at a vanadium pentoxide cathode. *J. Electrochem. Soc.*, **130**, 1983, 1787–8.
709. Meulenkamp, E. A., van Klinken, W. and Schlatmann, A. R. *In-situ* X-ray diffraction of Li intercalation in sol–gel V_2O_5 films. *Solid State Ionics*, **126**, 1999, 235–44.
710. Benmoussa, M., Outzourhit, A., Bennouna, A. and Ameziane, E. L. Electrochromism in sputtered V_2O_5 thin films: structural and optical studies. *Thin Solid Films*, **405**, 2002, 11–16.
711. Dickens, P. G. and Reynolds, G. J. Transport and equilibrium properties of some oxide insertion compounds. *Solid State Ionics*, **5**, 1981, 331–4.
712. Bachmann, H. G., Ahmed, F. R. and Barnes, W. H. The crystal structure of vanadium pentoxide. *Z. Kristall. Bd.*, **115**, 1961, 110–31.
713. Murphy, D. W., Christian, P. A., Disalvo, R. J. and Waszczak, J. V. Lithium incorporation by vanadium pentoxide. *Inorg. Chem.*, **18**, 1979, 2800–3.
714. Bach, S., Pereira-Ramos, J. P., Baffier, N. and Messina, R. A thermodynamic and kinetic study of electrochemical lithium intercalation in $Na_{0.33}V_2O_5$ bronze prepared by a sol–gel process. *J. Electrochem. Soc.*, **137**, 1990, 1042–8.

715. Nabavi, M., Sanchez, C., Taulelle, F. and Livage, J. Electrochemical properties of amorphous V_2O_5. *Solid State Ionics*, **28–30**, 1988, 1183–6.
716. Wu, G., Du, K., Xia, C., Kun, X., Shen, J., Zhou, B. and Wang, J. Optical absorption edge evolution of vanadium pentoxide films during lithium intercalation. *Thin Solid Films*, **485**, 2005, 284–9.
717. Murphy, D. W. and Christian, P. A. Solid state electrodes for high energy batteries. *Science*, **205**, 1979, 651–6.
718. Ashrit, P. V., Benaissa, K., Bader, G., Girouard, F. E. and Truong, V.-V. Lithiation studies on some transition metal oxides for an all-solid thin film electrochromic system. *Solid State Ionics*, **59**, 1993, 47–57.
719. Ashrit, P. V., Girouard, F. E. and Truong, V.-V. Fabrication and testing of an all-solid state system for smart window application. *Solid State Ionics*, **89**, 1996, 65–73.
720. Liu, G. and Richardson, T. J. Sb–Cu–Li electrochromic mirrors. *Sol. Energy Mater. Sol. Cells*, **86**, 2005, 113–21.
721. Zhang, Q., Wu, G., Zhou, B., Shen, J. and Wang, J. Electrochromic properties of sol–gel deposited V_2O_5 and TiO_2–V_2O_5 binary thin films. *J. Mater. Sci. Technol.*, **17**, 2001, 417–20.
722. Andersson, A. M., Granqvist, C. G. and Stevens, J. R. Towards an all-solid-state smart window: electrochromic coatings and polymer ion conductors. *Proc. SPIE*, **1016**, 1988, 41–9.
723. Gustafsson, J. C., Inganas, O. and Andersson, A. M. Conductive polyheterocycles as electrode materials in solid state electrochromic devices. *Synth. Met.*, **62**, 1994, 17–21.
724. Babulanam, S. M., Eriksson, T. S., Niklasson, G. A. and Granqvist, C. G. Thermochromic VO_2 films for energy-efficient windows. *Sol. Energy Mater.*, **16**, 1987, 347–63.
725. Hakim, M. O., Babulanam, S. M. and Granqvist, C. G. Electrochemical properties of thin VO_2 films on polyimide substrates. *Thin Solid Films*, **158**, 1988, L49–52.
726. Khan, M. S. R., Khan, K. A., Estrada, W. and Granqvist, C. G. Electrochromism and thermochromism of Li_xVO_2 thin films. *J. Appl. Phys.*, **69**, 1991, 3231–4.
727. Takahashi, I., Hibino, M. and Kudo, T. Thermochromic properties of double-doped VO_2 thin films fabricated from polyvanadate-based solutions. *Proc. SPIE*, **3788**, 1999, 26–33.
728. Oliveira, H. P., Graeff, C. F. O., Brunello, C. A. and Guerra, E. M. Electrochromic and conductivity properties: a comparative study between melanin-like/$V_2O_5.nH_2O$ and polyaniline/$V_2O_5.nH_2O$ hybrid materials. *J. Non-Cryst. Solids*, **273**, 2000, 193–7.
729. Denesuk, M. and Uhlmann, D. R. Site-saturation model for the optical efficiency of tungsten oxide-based devices. *J. Electrochem. Soc.*, **143**, 1996, L186–8.
730. Hurita, Y., Kitao, M. and Yamada, W. Absorption bands of electrochemically coloured films of WO_3, MoO_3 and $Mo_cW_{(1-c)}O_3$. *Jpn. J. Appl. Phys.*, **23**, 1984, 1624–7.
731. van Driel, F., Decker, F., Simone, F. and Pennisi, A. Charge and colour diffusivity from PITT in electrochromic Li_xWO_3 sputtered films. *J. Electroanal. Chem.*, **537**, 2002, 125–34.
732. Gérand, B. and Seguin, L. The soft chemistry of molybdenum and tungsten oxides: a review. *Solid State Ionics*, **84**, 1996, 199–204.
733. Molnar, B. J., Haranahalli, A. R. and Dove, B. D. Electrochromism in WO_3 films with BaO additions. *J. Vac. Sci. Technol. A*, **15**, 1978, 161–3.
734. Gao, G., Xu, L., Wang, W., An, W. and Qiu, Y. Electrochromic ultra-thin films based on cerium polyoxometalate. *J. Mater. Chem.*, **14**, 2004, 2024–9.

735. Lee, S.-H. and Joo, S.-K. Electrochromic behavior of Ni–W oxide electrodes. *Sol. Energy Mater. Sol. Cells*, **39**, 1995, 155–66.
736. Shen, P. K., Syed-Bokhari, J. K. and Tseung, A. C. C. Performance of electrochromic tungsten trioxide films doped with cobalt or nickel. *J. Electrochem. Soc.*, **138**, 1991, 2778–83.
737. Pennisi, A. and Simone, F. An electrochromic device working in absence of ion storage counter-electrode. *Sol. Energy Mater. Sol. Cells*, **39**, 1995, 333–40.
738. Faughnan, B. W. and Crandall, R. S. Optical properties of mixed-oxide WO_3/MoO_3 electrochromic films. *Appl. Phys. Lett.*, **31**, 1977, 834–6.
739. Yamada, S. and Kitao, M. Modulation of absorption spectra by the use of mixed films of $Mo_cW_{1-c}O_3$. *Proc. SPIE*, **IS4**, 1990, 246–59.
740. Pennisi, A., Simone, F. and Lampert, C. M. Electrochromic properties of tungsten–molybdenum oxide electrodes. *Sol. Energy Mater. Sol. Cells*, **28**, 1992, 233–47.
741. Patil, P. R. and Patil, P. S. Preparation of mixed oxide MoO_3–WO_3 thin films by spray pyrolysis technique and their characterisation. *Thin Solid Films*, **382**, 2001, 13–22.
742. Genin, C., Driouiche, A., Gerand, B. and Figlarz, M. Hydrogen bronzes of new oxides of the WO_3–MoO_3 system with hexagonal, pyrochlore and ReO_3-type structures. *Solid State Ionics*, **53–6**, 1992, 315–23.
743. Baeck, S.-H., Jaramillo, T. F., Jeong, D. H. and McFarland, E. W. Parallel synthesis and characterization of photoelectrochemically and electrochromically active tungsten–molybdenum oxides. *J. Chem. Soc., Chem. Commun.*, 2004, 390–1.
744. Ivanova, T., Gesheva, K. A., Ganchev, M. and Tzvetkova, E. Electrochromic behavior of CVD molybdenum oxide and Mo–W mixed-oxide thin films. *J. Mater. Sci.: Mater. Electron.*, **14**, 2003, 755–6.
745. Visco, S. J., Liu, M., Doeff, M. M., Ma, Y. P., Lampert, C. and Da Jonghe, L. C. Polyorganodisulfide electrodes for solid-state batteries and electrochromic devices. *Solid State Ionics*, **60**, 1993, 175–87.
746. Hiruta, Y., Kitao, M. and Yamada, M. Absorption bands of electrochemically-colored films of WO_3, MoO_3 and $Mo_cW_{1-c}O_3$. *Jpn. J. Appl. Phys.*, **23**, 1984, 1624–7.
747. Kitao, M., Yamada, M., Hiruta, Y., Suzuki, N. and Urabe, K. Electrochromic absorption spectra modulated by the composition of WO_3/MoO_3 mixed films. *Appl. Surf. Sci.*, **33–4**, 1985, 812–17.
748. Gillet, P. A., Fourquet, J. L. and Bohnke, O. Niobium tungsten titanium oxides: from 'soft chemistry' precursors to electrochromic thin layer materials. *Mater. Res. Bull.*, **27**, 1992, 1145–52.
749. Pehlivan, E., Tepehan, F. Z. and Tepehan, G. G. Comparison of optical, structural and electrochromic properties of undoped and WO_3-doped Nb_2O_5 thin films. *Solid State Ionics*, **165**, 2003, 105–10.
750. Sun, D. L., Heusing, S. and Aegerter, M. A. Electronic properties of Nb_2O_5:Mo, WO_3 and $(CeO_2)_x(TiO_2)_{1-x}$ sol–gel coatings and devices using dry and wet electrolytes. *Proc. Electrochem. Soc.*, **2003–22**, 2003, 119–29.
751. Lee, S.-H., Cheong, H. M., Park, N.-G., Tracy, C. E., Mascarenhas, A., Benson, D. K. and Deb, S. K. Raman spectroscopic studies of Ni–W oxide thin films. *Solid State Ionics*, **140**, 2001, 135–9.
752. Gao, W., Lee, S.-H., Benson, D. K. and Branz, H. M. Novel electrochromic projection and writing device incorporating an amorphous silicon carbide photodiode. *J. Non-Cryst. Solids*, **266–9**, 2000, 1233–7.

753. Cazzanelli, E., Vinegoni, C., Mariotti, G., Kuzmin, A. and Purans, J. Changes of structural, optical and vibrational properties of WO_3 powders after milling with ReO_3. *Proc. Electrochem. Soc.*, **96–24**, 1996, 260–274.
754. Özkan Zayim, E., Türham, I. and Tepehan, F. Z. Sol–gel made tantalum oxide doped tungsten oxide films. *Proc. Electrochem. Soc.*, **2003–22**, 2003, 40–8.
755. Aegerter, M. A., Avellaneda, C. O., Pawlicka, A. and Atik, M. Electrochromism in materials prepared by the sol–gel process. *J. Sol–Gel Sci. Technol.*, **8**, 1997, 689–96.
756. Yebka, B., Pecquenard, B., Julien, C. and Livage, J. Electrochemical Li^+ insertion in $WO_{3-x}TiO_2$ mixed oxides. *Solid State Ionics*, **104**, 1997, 169–75.
757. Macêdo, M. A. and Aegerter, M. A. Sol–gel electrochromic device. *J. Sol–Gel Sci. Technol.*, **2**, 1994, 667–71.
758. de Tacconi, N. R., Chenthamarakshan, C. R. and Rajeshwar, K. Electrochromic behaviour of WO_3, TiO_2 and WO_3–TiO_2 composite films prepared by pulsed electrodeposition. *Proc. Electrochem. Soc.*, **2003–22**, 2003, 28–39.
759. Göttsche, J. F., Hinsch, A. and Wittwer, V. Electrochromic and optical properties of mixed WO_3–TiO_2 thin films produced by sputtering and sol–gel technique. *Proc. SPIE*, **1728**, 1992, 13–25.
760. Patil, P. S., Mujawar, S. H., Inamdar, A. I. and Sadale, S. B. Structural, electrical and optical properties of TiO_2 doped WO_3 thin films. *Appl. Surf. Sci.*, **250**, 2005, 117–23.
761. Lourenco, A., Masetti, E. and Decker, F. Electrochemical and optical characterization of RF-sputtered thin films of vanadium–nickel mixed oxides. *Electrochim. Acta*, **46**, 2001, 2257–62.
762. Pan, B. H. and Lee, J. Y. Electrochromism of electrochemically codeposited composites of phosphomolybdic acid and tungsten trioxide. *J. Electrochem. Soc.*, **143**, 1996, 2784–9.
763. Deb, S. K. and Witzke, H. Abstract G7, Nineteenth Electronics Materials Conference, Cornell, New York 1977; as cited in Dautremont-Smith, W. C. Transition metal oxide electrochromic materials and displays: a review; part 1: oxides with cathodic coloration. *Displays*, **3**, 1982, 3–22.
764. Marcel, C., Hegde, M. S., Rougier, A., Maugy, C., Guery, C. and Tarascon, J.-M. Electrochromic properties of antimony tin oxide (ATO) thin films synthesized by pulsed laser deposition. *Electrochim. Acta*, **46**, 2001, 2097–104.
765. Naghavi, N., Marcel, C., Dupont, L., Leriche, J.-B. and Tarascon, J.-M. On the electrochromic properties of antimony–tin oxide thin films deposited by pulsed laser deposition. *Solid State Ionics*, **156**, 2003, 463–74.
766. Yoshino, T. and Masuda, H. Characterization of nano-structured thin films of electrodeposited Ce–Co mixed oxides for EC devices. *Solid State Ionics*, **165**, 2003, 123–9.
767. Veszelei, M., Kullman, L., Strømme Mattsson, M., Azens, A. and Granqvist, C. G. Optical and electrochemical properties of Li^+ intercalated Zr–Ce oxide and Hf–Ce oxide. *J. Appl. Phys.*, **833**, 1998, 1670–6.
768. Štangar, U. L., Opara, U. and Orel, B. Structural and electrochemical properties of sol–gel derived $Mo:CeO_2$, $Si:Mo:CeO_2$ and $Si:CeO_2$ nanocrystalline films for electrochromic devices. *J. Sol–Gel Sci. Technol.*, **8**, 1997, 751–8.
769. Oliveira, S., Faria, R. C., Terezo, A. J., Pereira, E. C. and Bulhões, L. O. S. The cerium addition effect on the electrochemical properties of niobium pentoxide electrochromic thin films. *Proc. Electrochem. Soc.*, **96–24**, 1996, 106–18.

770. Zhu, B., Luo, Z. and Xia, C. Transparent conducting CeO_2–SiO_2 thin films. *Mater. Res. Bull.*, **34**, 1999, 1507–12.
771. Rosario, A. V. and Pereira, E. C. Comparison of the electrochemical behavior of CeO_2–SnO_2 and CeO_2–TiO_2 electrodes produced by the Pechini method. *Thin Solid Films*, **410**, 2002, 1–7.
772. Keomany, D., Petit, J.-P. and Deroo, D. Electrochemical insertion in sol–gel made CeO_2–TiO_2 from lithium conducting polymer electrolyte: relation with the material structure. *Sol. Energy Mater. Sol. Cells*, **36**, 1995, 397–408.
773. Purans, J., Azens, A. and Granqvist, C. G. X-Ray absorption study of Ce–Ti oxide films. *Electrochim. Acta*, **46**, 2001, 2055–8.
774. Mattson, M. S., Azens, A., Niklasson, G. A., Granqvist, C. G. and Purans, J. Li intercalation in transparent Ti–Ce oxide films: energetics and ion dynamics. *J. Appl. Phys.*, **81**, 1997, 6432–7.
775. Kim, Y. I., Yoon, J. B., Choy, J. H., Campet, G., Camino, D., Portier, J. and Salardenne, J. RF sputtered SnO_2, Sn-doped In_2O_3 and Ce-doped TiO_2 films as transparent counter electrodes for electrochromic window. *Bull. Korean Chem. Soc.*, **19**, 1998, 107–9.
776. Tavcar, G., Kalcher, K. and Ogorvec, B. Applicability of a sol–gel derived CeO_2–TiO_2 thin film electrode as an amperometric sensor in flow injection. *Analyst*, **122**, 1997, 371–6.
777. Camino, D., Deroo, D., Salardenne, J. and Treuil, N. $(CeO_2)_x$–$(TiO_2)_{1-x}$: counter electrode materials for lithium electrochromic devices. *Sol. Energy Mater. Sol. Cells*, **39**, 1995, 349–66.
778. von Rottkay, K., Richardson, T., Rubin, M., Slack, J. and Kullman, L. Influence of stoichiometry on electrochromic cerium–titanium oxide compounds. *Solid State Ionics*, **113–15**, 1998, 425–30.
779. Kullman, L., Azens, A. and Granqvist, C. G. Decreased electrochromism in Li-intercalated Ti oxide films containing La, Ce, and Pr. *J. Appl. Phys.*, **81**, 1997, 8002–10.
780. Kullman, L., Veszelei, M., Ragan, D. D., Isidorsson, J., Vaivars, G., Kanders, U., Azens, A., Schelle, S., Hjorvarsson, B. and Granqvist, C. G. Cerium-containing counter electrodes for transparent electrochromic devices. *Proc. SPIE*, **2968**, 1997, 219–24.
781. Azens, A., Kullman, L., Ragan, D. D., Granqvist, C. G., Hjovarsson, B. and Vaivars, G. Optical and electrochemical properties of dc magnetron sputtered Ti–Ce oxide films. *Appl. Phys. Lett.*, **68**, 1996, 3701–3.
782. Granqvist, C. G., Azens, A., Kullman, L. and Rönnow, D. Progress in smart windows research: improved electrochromic W oxide films and transparent Ti–Ce oxide counter electrodes. *Renewable Energy*, **8**, 1996, 97–106.
783. Macrelli, G. and Poli, E. Mixed cerium/titanium and cerium/zirconium oxides as thin film counter electrodes for all solid state electrochromic transmissive devices. *Electrochim. Acta*, **44**, 1999, 3137–47.
784. Janke, N., Bieberle, A. and Weißmann, R. Characterization of sputter-deposited WO_3 and CeO_{2-x}–TiO_2 thin films for electrochromic applications. *Thin Solid Films*, **392**, 2001, 134–41.
785. Verma, A., Samanta, S. B., Mehra, N. C., Bakhshi, A. K. and Agnihotry, S. A. Sol–gel derived nanocrystalline CeO_2–TiO_2 coatings for electrochromic windows. *Sol. Energy Mater. Sol. Cells*, **86**, 2005, 85–103.

786. Veszelei, M., Kullman, L., Granqvist, C. G., von Rottkay, K. and Rubin, M. Optical constants of sputter-deposited Ti–Ce and Zr–Ce oxide films. *Appl. Opt.*, **37**, 1998, 5993–6001.
787. Veszelei, M., Kullman, L., Azens, A., Granqvist, C. G. and Hjörvarsson, B. Transparent ion intercalation films of Zr–Ce. *J. Appl. Phys.*, **81**, 1997, 2024–6.
788. Masetti, E., Varsano, F. and Decker, F. Sputter-deposited cerium vanadium mixed oxide as counter-electrode for electrochromic devices. *Electrochim. Acta*, **44**, 1999, 3117–19.
789. Flamini, C., Ciccioli, A., Traverso, P., Gnecco, F., Giardini Guidoni, A. and Mele, A. Laser-induced evaporation, reactivity and deposition of ZrO_2, CeO_2, V_2O_5 and mixed Ce-V oxides. *Appl. Surf. Sci.*, **168**, 2000, 104–7.
790. Kaneko, Y. and Chen, W. Electrochemical synthesis of electrochromic Ce–V oxide films in NH_4HSO_4 melts. *J. Electroanal. Chem.*, **559**, 2003, 87–90.
791. Crnjak Orel, Z., Gaberšček, M. and Turković, A. Electrical and spectroscopic characterisation of nanocrystalline V/Ce oxides. *Sol. Energy Mater. Sol. Cells*, **86**, 2005, 19–32.
792. Opara Krašovec, U., Orel, B., Surca, A., Bukovec, N. and Reisfeld, R. Structural and spectroelectrochemical investigations of tetragonal $CeVO_4$ and Ce/V-oxide sol–gel derived ion-storage films. *Solid State Ionics*, **118**, 1999, 195–214.
793. Opara Krašovec, U., Orel, B. and Reisfeld, R. Electrochromism of $CeVO_4$ and Ce/V-oxide ion-storage films prepared by the sol–gel route. *Electrochem. Solid-State Lett.*, **1**, 1998, 104–6.
794. Varsano, F., Decker, F., Masetti, E., Cardellini, F. and Licciulli, A. Optical and electrochemical properties of cerium–zirconium mixed oxide thin films deposited by sol–gel and r.f. sputtering. *Electrochim. Acta*, **44**, 1999, 3149–56.
795. Luo, X., Zhu, B., Xia, C., Niklasson, G. A. and Granqvist, C. G. Transparent ion-conducting ceria–zirconia films made by sol–gel technology. *Sol. Energy Mater. Sol. Cells*, **53**, 1998, 341–7.
796. Veszelei, M., Strømme Mattsson, M., Kullman, L., Azens, A. and Granqvist, C. G. Zr–Ce oxides as candidates for optically passive counter electrodes. *Sol. Energy Mater. Sol. Cells*, **56**, 1999, 223–30.
797. Granqvist, C. G., Azens, A., Hjelm, A., Kullman, L., Niklasson, G. A., Rönnow, D., Strømme Mattson, M., Veszelei, M. and Vaivars, G. Recent advances in electrochromics for smart windows applications. *Sol. Energy*, **63**, 1998, 199–216.
798. Svegl, F., Orel, B. and Hutchins, M. G. Structural and electrochromic properties of Co-oxide and Co/Al/Si-oxide films prepared by the sol–gel dip coating technique. *J. Sol-Gel Sci. Technol.*, **8**, 1997, 765–9.
799. K. K. Canon. Electrochromic device, Jpn. Kokai Tokkyo Koho, Japanese Patent JP 59,232,316; as cited in *Chem. Abs.* **102**: P212,795, 1985.
800. Bertran, E., Corbella, C., Vives, M., Pinyol, A., Person, C. and Porqueras, I. RF sputtering deposition of Ag/ITO coatings at room temperature. *Solid State Ionics*, **165**, 2003, 139–48.
801. Serebrennikova, I. and Birss, V. I. Electrochemical behavior of sol–gel produced Ni and Ni–Co oxide films. *J. Electrochem. Soc.*, **144**, 1997, 566–73.
802. Cogan, S. F., Anderson, E. J., Plante, T. D. and Rauh, R. D. Electrochemical investigation of electrochromism in transparent conductive oxides. *Appl. Opt.*, **24**, 1984, 2282–3.

803. Rauf, I. A. and Walls, M. G. A comparative study of microstructure (in ITO films) and techniques (CTEM and STM). *Ultramicroscopy*, **35**, 1991, 19–26.
804. Ahn, K.-S., Nah, Y.-C. and Sung, Y.-E. Electrochromic properties of SnO_2-incorporated Ni oxide films grown using a cosputtering system. *J. Appl. Phys.*, **92**, 2002, 7128–32.
805. Chopra, K. L., Major, S. and Pandya, D. K. Transparent conductors: a status review. *Thin Solid Films*, **102**, 1983, 1–46.
806. Nagai, J. Electrochemical properties of ITO electrodes. *Proc. SPIE*, **3788**, 1999, 22–5.
807. Granqvist, C. G. and Hultåker, A. Transparent and conducting ITO films: new developments and applications. *Thin Solid Films*, **411**, 2002, 1–5.
808. Adurodija, F. O., Izumi, H., Ishihara, T., Yoshioka, H. and Motoyama, M. The electro-optical properties of amorphous indium tin oxide prepared at room temperature by pulsed laser deposition. *Sol. Energy Mater. Sol. Cells*, **71**, 2002, 1–8.
809. Svensson, J. S. E. M. and Granqvist, C. G. No visible electrochromism in high-quality e-beam evaporated In_2O_3:Sn films. *Appl. Opt.*, **24**, 1984, 2284–5.
810. Armstrong, N. R., Liu, A. W. C., Fujihira, M. and Kuwana, T. Electrochemical and surface characterics of tin oxide and indium oxide electrodes. *Anal. Chem.*, **48**, 1976, 741–50.
811. Yu, P. C., Haas, T., Goldner, R. B. and Cogan, S. F. Characterization of indium oxide for the use as a counter-electrode in an electrochromic device. *Mater. Res. Soc. Symp. Proc.*, **210**, 1991, 63–8.
812. Corradini, A., Marinangeli, A. M. and Mastragostino, M. ITO as counter-electrode in a polymer based electrochromic device. *Electrochim. Acta*, **35**, 1990, 1757–60.
813. Hamberg, I. and Granqvist, C. G. Theoretical model for the optical properties of In_2O_3:Sn films in the 0.3–50 μm range. *Proc. SPIE*, **562**, 1985, 137–46.
814. Golden, S. J. and Steele, B. C. H. Variable transmission electrochromic windows utilizing tin-doped indium oxide counter electrodes. *Appl. Phys. Lett.*, **59**, 1991, 2357–9.
815. Kaneko, H. and Miyake, K. Effects of transparent electrode resistance on the performance characteristics of electrochemichromic cells. *Appl. Phys. Lett.*, **49**, 1986, 112–14.
816. Yu, P. C., Haas, T. E., Goldner, R. B. and Cogan, S. F. Characterisation of indium-tin oxide for the use of counter electrode in an electrochromic device. *Mater. Res. Soc. Symp. Proc.*, **210**, 1991, 63–8.
817. Coleman, J. P., Freeman, J. J., Lynch, A. T., Madhukar, P. and Wagenknecht, J. H. Unexpected yellow–blue electrochromism of ITO powders at modest potentials in aqueous electrolytes. *Acta Chem. Scand.*, **52**, 1998, 86–94.
818. Ingram, M. D., Duffy, J. A. and Monk, P. M. S. Chronoamperometric response of the cell ITO | H_xWO_3 | PEO–H_3PO_4 (MeCN) | ITO. *J. Electroanal. Chem.*, **380**, 1995, 77–82.
819. Steele, B. C. H. and Golden, S. J., Variable transmission electrochromic windows utilizing tin-doped indium oxide counterelectrodes. *Appl. Phys. Lett.*, **59**, 1991, 2357–9.
820. Bressers, P. M. M. C. and Meulenkamp, E. A. Electrochromic behavior of indium tin oxide in propylene carbonate. *J. Electrochem. Soc.*, **145**, 1998, 2225–30.
821. Radhakrisnan, S., Unde, S. and Mandale, A. B. Source of instability in solid state polymeric electrochromic cells: the deterioration of indium tin oxide electrodes. *Mater. Chem. Phys.*, **48**, 1997, 268–71.

822. Azens, A. and Granqvist, C. G. Electrochromism in Ir–Mg oxide films. *Appl. Phys. Lett.*, **81**, 2002, 928–9.
823. Backholm, J., Azens, A. and Niklasson, G. A. Electrochemical and optical properties of sputter deposited Ir–Ta and Ir oxide thin films. *Sol. Energy Mater. Sol. Cells*, **90**, 2006, 414–20.
824. Orel, B., Maček, M. and Surca, A. Electrochromism of dip-coated Fe-oxide, Fe/Ti-oxide and Fe/Si-oxide films prepared by the sol–gel route. *Proc. SPIE*, **2255**, 1994, 273–84.
825. Schmitt, M. and Aegerter, M. A. Properties of electrochromic devices made with Nb_2O_5 and Nb_2O_5:X (X = Li, Ti, or Mo) as coloring electrode. *Proc. SPIE*, **3788**, 1999, 75–83.
826. Opara Krašovec, U., Orel, B., Hocevar, S. and Musevic, I. Electrochemical and spectro-electrochemical properties of SnO_2 and SnO_2/Mo transparent electrodes with high ion-storage capacity. *J. Electrochem. Soc.*, **144**, 1997, 3398–409.
827. Orel, B., Opara Krašovec, U., Štangar, U. L. and Judenstein, P. All sol–gel electrochromic devices with Li^+ ionic conductor, WO_3 electrochromic films and SnO_2 counter-electrode films. *J. Sol–Gel Sci. Technol.*, **11**, 1998, 87–104.
828. Wang, Z., Hu, X. and Helmersson, U. Peroxo sol–gel preparation: photochromic/electrochromic properties of Mo–Ti oxide gels and thin films. *J. Mater. Chem.*, **10**, 2000, 2396–400.
829. Acharya, B. S., Pradhan, L. D., Nayak, B. B. and Mishar, P. Vacancy-induced electronic states in substoichiometric $V_{2-x}Mo_xO_{3\pm y}$ thin films and powders: a soft X-ray emission study. *Bull. Mater. Sci.*, **22**, 1999, 981–6.
830. Ashrit, P. V., Bader, G., Girouard, F. E., Truong, V.-V. and Yamaguchi, T. Optical properties of cermets consisting of metal in a WO_3 matrix. *Physica A*, **157**, 1989, 333–8.
831. Avendaño, E., Azens, A., Niklasson, G. A. and Granqvist, C. G. Electrochromism in nickel oxide films containing Mg, Al, Si, V, Zr, Nb, Ag, or Ta. *Sol. Energy Mater. Sol. Cells*, **84**, 2004, 337–50.
832. Avendaño, E., Azens, A., Isidorsson, J., Harmhag, R., Niklasson, G. A. and Granqvist, C. G. Optimized nickel-oxide-based electrochromic thin films. *Solid State Ionics*, **165**, 2003, 169–73.
833. Granqvist, C. G., Avendaño, E. and Azens, A. Electrochromic coatings and devices: survey of some recent advances. *Thin Solid Films*, **442**, 2003, 201–11.
834. Avendaño, E., Azens, A., Niklasson, G. A. and Granqvist, C. G. Nickel-oxide based electrochromic films with optimized optical properties. *J. Solid State Electrochem.*, **8**, 2003, 37–9.
835. Azens, A. and Granqvist, C. G. Electrochromism of sputter deposited Ni–Cr oxide. *J. Appl. Phys.*, **84**, 1998, 6454–6.
836. Miller, E. L. and Rocheleau, R. E. Electrochemical behavior of reactively sputtered iron-doped nickel oxide. *J. Electrochem. Soc.*, **144**, 1997, 3072–7.
837. Campet, G., Morel, B., Bourrel, M., Chabagno, J. M., Ferry, D., Garie, R., Quet, C., Geoffrey, C., Videau, J. J., Portier, J., Delmas, C. and Salardenne, J. Electrochemistry of nickel oxide films in aqueous and Li^+ containing non-aqueous solutions: an application for a new lithium-based nickel oxide electrode exhibiting electrochromism by a reversible Li^+ ion insertion mechanism. *Mater. Sci. Eng. B*, **8**, 1991, 303–8.
838. Šurca, A., Orel, B. and Pihlar, B. Characterisation of redox states of Ni(La)-hydroxide films prepared via the sol–gel route by *ex situ* IR spectroscopy. *J. Solid State Electron.*, **2**, 1998, 38–49.

839. Azens, A., Isidorsson, J., Karmhag, R. and Granqvist, C. G. Highly transparent Ni–Mg and Ni–V–Mg oxide films for electrochromic applications. *Thin Solid Films*, **422**, 2002, 1–3.
840. de Torresi, S. I. C. The effect of manganese addition on nickel hydroxide electrodes with emphasis on its electrochromic properties. *Electrochim. Acta*, **40**, 1995, 1101–7.
841. Hutchins, M. G. and Murphy, T. P. The electrochromic behaviour of tin–nickel oxide. *Sol. Energy Mater. Sol. Cells*, **54**, 1998, 75–84.
842. Ferreira, F. F. and Fantini, M. C. A. Theoretical optical properties of composite metal–NiO films. *J. Phys. D: Appl. Phys.*, **36**, 2003, 2386–92.
843. Kuzmin, A., Purans, J., Kalendarev, R., Pailharey, D. and Mathey, Y. XAS, XRD, AFM and Raman studies of nickel tungstate electrochromic thin films. *Electrochim. Acta*, **46**, 2001, 2233–6.
844. Toma, H. E., Matsumoto, F. M. and Cipriano, C. Spectroelectrochemistry of the hexanuclear cluster $[Ru_3O(acetate)_6\text{-}\mu\text{-}(pyrazine)_3\text{-}\{Fe(CN)_5\}_3]^{n-}$ and of its modified nickel electrode in aqueous solution. *J. Electroanal. Chem.*, **346**, 1993, 261–70.
845. Orel, B., Macek, M., Lavrencic-Štanger, U. and Pihlar, B. Amorphous Nb/Fe-oxide ion-storage films for counter electrode applications in electrochromic devices. *J. Electrochem. Soc.*, **145**, 1998, 1607–14.
846. Rosario, A. V. and Pereira, E. C. Lithium insertion in TiO_2 doped Nb_2O_5 electrochromic thin films. *Electrochim. Acta*, **46**, 2001, 1905–10.
847. Gillet, P. A., Fourquet, J. L. and Bohnke, O. New electrochromic thin-film materials. *Proc. SPIE*, **1728**, 1992, 82–91.
848. Manno, D., Serra, A., Micocci, G., Siciliano, T., Filippo, E. and Tepore, A. Morphological, structural and electronic characterization of nanostructured vanadium–tin mixed oxide thin films. *Sol. Energy Mater. Sol. Cells*, **341**, 2004, 68–76.
849. Wu, Y., Hu, L. L., Jiang, Z. H. and Ke, Q. Study on the electrochemical properties of Fe_2O_3–TiO_2 films prepared by sol–gel. *J. Electrochem. Soc.*, **144**, 1997, 1728–34.
850. Maček, M., Orel, B. and Meden, T. Electrochemical and structural characterisation of dip-coated Fe/Ti oxide films prepared by the sol–gel route. *J. Sol–Gel. Sci. Technol.*, **8**, 1997, 771–9.
851. Bellenger, F., Chemarin, C., Deroo, D., Maximovitch, S., Šurca Vuk, A. and Orel, B. Insertion of lithium in vanadium and mixed vanadium–titanium oxide films. *Electrochim. Acta*, **46**, 2001, 2263–8.
852. Burdis, M. S. Properties of sputtered thin films of vanadium–titanium oxide for use in electrochromic windows. *Thin Solid Films*, **311**, 1997, 286–98.
853. Burdis, M. S., Siddle, J. R., Batchelor, R. A. and Gallego, J. M. $V_{0.50}Ti_{0.50}O_x$ thin films as counter-electrodes for electrochromic devices. *Sol. Energy Mater. Sol. Cells*, **54**, 1998, 93–8.
854. Nagase, K., Shimizu, S., Miura, N. and Yamazoe, N. Electrochromism of vanadium–titanium oxide thin films prepared by spin-coating method. *Appl. Phys. Lett.*, **61**, 1992, 243–5.
855. Özkan Zayim, E. Optical and electrochromic properties of sol–gel made anti-reflective WO_3–TiO_2 films. *Sol. Energy Mater. Sol. Cells*, **87**, 2005, 695–703.
856. de Tacconi, N. R., Rajeshwar, K. and Lezna, R. O. Preparation, photoelectrochemical characterization, and photoelectrochromic behavior of metal hexacyanoferrate–titanium dioxide composite films. *Electrochim. Acta*, **45**, 2000, 3403–11.
857. Duffy, J. A. *Bonding, Energy Levels and Inorganic Solids*. London, 1990, Longmans.

858. Chen, W. and Kaneko, Y. Electrochromism of vanadium oxide films doped by rare-earth (Pr, Nd, Sm, Dy) oxides. *J. Electroanal. Chem.*, **559**, 2003, 83–6.
859. Coluzza, C., Cimino, N., Decker, F., Santo, G. D., Liberatore, M., Zanoni, R., Bertolo, M. and Rosa, S. L. Surface analyses of In V oxide films aged electrochemically by Li insertion reactions. *Phys. Chem. Chem. Phys.*, **5**, 2003, 5489–98.
860. Kaneko, Y., Mori, S. and Yamanaka, J. Synthesis of electrochromic praseodymium-doped vanadium oxide films by molten salt electrolysis. *Solid State Ionics*, **151**, 2002, 35–9.
861. Artuso, F., Picardi, G., Bonino, F., Decker, F., Benčič, S., Šurca, Vuk, A., Opara Krašovec, U. and Orel, B. Fe-containing $CeVO_4$ films as Li intercalation transparent counter-electrodes. *Electrochim. Acta*, **46**, 2001, 2077–84.
862. Štanger, U. L., Orel, B., Regis, A. and Colomban, P. Chromogenic WPA/TiO_2 hybrid gels and films. *J. Sol–Gel Sci. Technol.*, **8**, 1997, 965–71.
863. Rougier, A., Blyr, A., Garcia, J., Zhang, Q. and Impey, S. A. Electrochromic W–M–O (M = V, Nb) sol–gel thin films: a way to neutral colour. *Sol. Energy Mater. Sol. Cells*, **71**, 2002, 343–57.
864. Huguenin, F., Torresi, R. M., Buttry, D. A., da Silva, J. E. P. and de Torresi, S. I. C. Electrochemical and Raman studies on a hybrid organic–inorganic nanocomposite of vanadium oxide and a sulfonated polyaniline. *Electrochim. Acta*, **46**, 2001, 3555–62.
865. Oliveira, H. P., Graeff, C. F. O., Zanta, C. L. P. S., Galina, A. C. and Gonçalves, P. J. Synthesis, characterization and properties of a melanin-like/vanadium pentoxide hybrid compound. *J. Mater. Chem.*, **10**, 2000, 371–5.
866. NuLi, Y.-N., Fu, Z.-W., Chu, Y.-Q. and Qin, Q.-Z. Electrochemical and electrochromic characteristics of Ta_2O_5–ZnO composite films. *Solid State Ionics*, **160**, 2003, 197–207.
867. Vukovic, M., Cukman, D., Milun, M., Atanasoska, L. D. and Atanasoski, R. T. Anodic stability and electrochromism of electrodeposited ruthenium–iridium coatings on titanium. *J. Electroanal. Chem.*, **330**, 1992, 663–73.
868. K. K. Canon. Electrochromic device, Jpn. Kokai Tokkyo Koho, Japanese Patent JP 6,004,925; as cited in *Chem. Abs.* **102**: P212,797, 1985.
869. Marijan, D., Vukovic, M., Parvan, P. and Milun, M. Surface modification of *Inconel-600* by growth of a hydrous oxide film. *J. Appl. Electrochem.*, **28**, 1998, 96–106.
870. Chu, W. F., Hartmann, R., Leonhard, V. and Ganson, G. Investigations on counter electrode materials for solid state electrochromic systems. *Mater. Sci. Eng. B*, **13**, 1992, 235–7.
871. Lian, K. K. and Birss, V. I. Hydrous oxide growth on amorphous Ni–Co alloys. *J. Electrochem. Soc.*, **1991**, 1991, 2877–84.
872. Hultåker, A., Jarrendahl, K., Lu, J., Granqvist, C. G. and Niklasson, G. A. Electrical and optical properties of sputter deposited tin doped indium oxide thin films with silver additive. *Thin Solid Films*, **392**, 2001, 305–10.
873. Coustier, F., Passerini, S. and Smyrl, W. H. Dip-coated silver-doped V_2O_5 xerogels as host materials for lithium intercalation. *Solid State Ionics*, **100**, 1997, 247–58.
874. Fantini, M. C. A., Ferreira, F. F. and Gorenstein, A. Theoretical and experimental results on Au–NiO and Au–CoO electrochromic composite films. *Solid State Ionics*, **152–3**, 2002, 867–72.
875. Ferreira, F. F. and Fantini, M. C. A. Multilayered composite Au–NiO_x electrochromic films. *Solid State Ionics*, **175**, 2004, 517–20.

876. He, T., Ma, Y., Cao, Y., Yin, Y., Yang, W. and Yao, J. Enhanced visible-light coloration and its mechanism of MoO_3 thin films by Au nanoparticles. *Appl. Surf. Sci.*, **180**, 2001, 336–40.
877. Yano, J., Hirayama, T., Yamasaki, S., Yamazaki, S. and Kanno, Y. Stable freestanding aramid resin film containing vanadium pentoxide and new colour electrochromism of the film by electrodeposition of gold. *Electrochem. Commun.*, **3**, 2001, 263–6.
878. Nagase, K., Shimizu, Y., Miura, N. and Yamazoe, N. Electrochromism of gold–vanadium pentoxide composite thin films prepared by alternating thermal deposition. *Appl. Phys. Lett.*, **9**, 1994, 1059–61.
879. Sichel, E. K. and Gittleman, G. I. Characteristics of the electrochromic materials $Au-WO_3$ and $Pt-WO_3$. *J. Electron. Mater.*, **8**, 1979, 1–9.
880. Heszler, P., Reyes, L. F., Hoel, A., Landstrom, L., Lantto, V. and Granqvist, C. G. Nanoparticle films made by gas phase synthesis: comparison of various techniques and sensor applications. *Proc. SPIE*, **5055**, 2003, 106–19.
881. Park, K.-W. Electrochromic properties of $Au-WO_3$ nanocomposite thin-film electrode. *Electrochim. Acta*, **50**, 2005, 4690–3.
882. Park, K.-W. and Sung, Y. E. Modulation of electrochromic performance and *in situ* observation of proton transport in $Pt-RuO_2$ nanocomposite thin-film electrodes. *J. Appl. Phys.*, **94**, 2003, 7276–80.
883. Park, K.-W. and Toney, M. F. Electrochemical and electrochromic properties of nanoworm-shaped Ta_2O_5–Pt thin-films. *Electrochem. Commun.*, **7**, 2005, 151–5.
884. Chen, K. Y. and Tseung, A. C. C. Effect of Nafion dispersion on the stability of Pt/WO_3 electrodes. *J. Electrochem. Soc.*, **143**, 1996, 2703–8.
885. Strømme, M., Isidorsson, J., Niklasson, G. A. and Granqvist, C. G. Impedance studies on Li insertion electrodes of Sn oxide and oxyfluoride. *J. Appl. Phys.*, **80**, 1996, 233–41.
886. Strømme, M., Gutarra, A., Niklasson, G. A. and Granqvist, C. G. Impedance spectroscopy on lithiated Ti oxide and Ti oxyfluoride thin films. *J. Appl. Phys.*, **79**, 1996, 3749–57.
887. Gutarra, A., Azens, A., Stjerna, B. and Granqvist, C. G. Electrochromism of sputtered fluorinated titanium oxide thin films. *Appl. Phys. Lett.*, **64**, 1994, 1604–6.
888. Strømme Mattson, M., Niklasson, G. A. and Granqvist, C. G. Diffusion of Li, Na, and K in fluorinated Ti dioxide films: applicability of the Anderson–Stuart model. *J. Appl. Phys.*, **81**, 1997, 2167–72.
889. Azens, A., Stjerna, B. and Granqvist, C. G. Chemically enhanced sputtering in fluorine-containing plasmas: application to tungsten oxyfluoride. *Thin Solid Films*, **254**, 1995, 1–2.
890. Azens, A., Stjerna, B., Granqvist, C. G., Gabrusenoks, J. and Lusis, A. Electrochromism in tungsten oxyfluoride films made by chemically enhanced d. c. sputtering. *Appl. Phys. Lett.*, **65**, 1994, 1998–2000.
891. Azens, A., Granqvist, C. G., Pentjuss, E., Gabrusenoks, J. and Barczynska, J. Electrochromism of fluorinated and electron-bombarded tungsten oxide. *J. Appl. Phys.*, **78**, 1995, 1968–74.

7

Electrochromism within metal coordination complexes

7.1 Redox coloration and the underlying electronic transitions

Metal coordination complexes show promise as electrochromic materials because of their intense coloration and redox reactivity.[1] Chromophore properties arise from low-energy metal-to-ligand charge-transfer (MLCT), intervalence charge-transfer (IVCT), intra-ligand excitation, and related visible-region electronic transitions. Because these transitions involve valence electrons, chromophoric characteristics are altered or eliminated upon oxidation or reduction of the complex, as touched on in Chapter 1. A familiar example used in titrations is the redox indicator ferroin, $[Fe^{II}(phen)_3]^{2+}$ (phen = 1,10-phenanthroline), which has been employed in a solid-state ECD, the deep red colour of which is transformed to pale blue on oxidation to the iron(III) form.[2] Often more markedly than other chemical groups, a coloured metal coordination complex susceptible to a redox change will in general undergo an accompanying colour change, and will therefore be electrochromic to some extent. The redox change – electron loss or gain – can be assigned to either the central coordinating cation or the bound ligand(s); often it is clear which, but not always. If it is the central cation that undergoes redox change, then its initial and final oxidation states are shown in superscript roman numerals, while the less clear convention for ligands is usually to indicate the extra charge lost or gained by a superscripted + or −. As mentioned in Chapter 1, whilst the term 'coloured' generally implies absorption in the visible region, metal coordination complexes that switch between a colourless state and a state with strong absorption in the near infra red (NIR) region are now being intensively studied.[3]

While these spectroscopic and redox properties alone would be sufficient for direct use of metal coordination complexes in solution-phase ECDs, in addition, polymeric systems based on metal coordination-complex monomer units, which have prospective use in all-solid-state systems, have also been investigated.

Following usage in the field, in this chapter an arrow between two species can indicate the direction of transfer of an electron.

7.2 Electrochromism of polypyridyl complexes

7.2.1 Polypyridyl complexes in solution

The complexes [M^{II}(bipy)$_3$]$^{2+}$ (M = Fe, Ru, Os; bipy = 2,2'-bipyridine) are respectively red, orange and green, due to the presence of an intense MLCT absorption band.[4] Electrochromism results from *loss* of the MLCT absorption band on switching to the M^{III} redox state. Such complexes also exhibit a series of ligand-based redox processes, the first three of which are accessible in solvents such as acetonitrile and dimethylformamide (DMF).[4] Attachment of electron-withdrawing substituents to the 2,2'-bipyridine ligands allows additional ligand-based redox processes to be observed, due to the anodic shift of the redox potentials induced by these substituents. Thus Elliott and co-workers have shown that a series of colours is available with [M(bipy)$_3$]$^{2+}$ derivatives when the 2,2'-bipyridine ligands have electron-withdrawing substituents at the 5,5' positions (see below).[5] The electrochromic colours established by bulk electrochemical reactions in acetonitrile are given in Table 7.1.

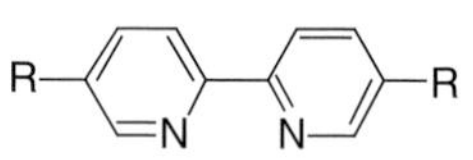

L^1 R = CO$_2$Et
L^2 R = CONEt$_2$
L^3 R = CON(Me)Ph
L^4 R = CN
L^5 R = C(O)^{n}Bu

A surface-modified polymeric system can be obtained by spin coating or heating [Ru(L^6)$_3$]$^{2+}$ as its *p*-tosylate salt.[6] The resulting film shows seven-colour electrochromism with colours covering the full visible region spectral range, which can be scanned in 250 ms.

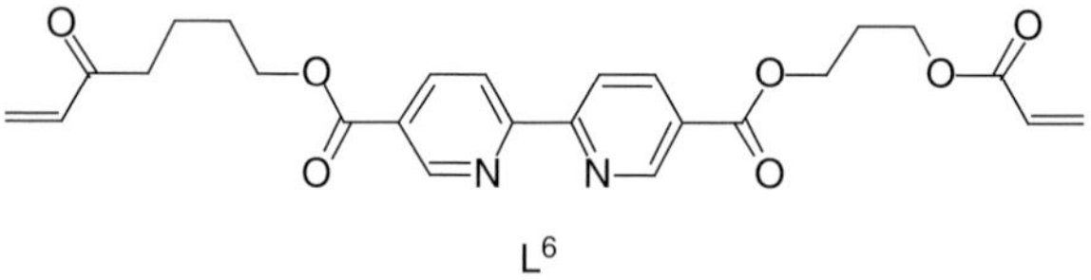

L^6

Spectral modulation in the NIR region has been reported for the related complex [Ru(L^7)$_3$]$^{2+}$ which undergoes six ligand-centred reductions, two per ligand.[7] The complex initially shows no absorption between 700 and 2100 nm; however, upon reduction by one electron a very broad pair of overlapping peaks appear with maxima at 1210 nm (ε = 2600 dm^3 mol^{-1} cm^{-1}) and 1460 nm (ε = 3400 dm^3 mol^{-1} cm^{-1}). Following the second one-electron reduction, the peaks shift to slightly lower energy (1290 and 1510 nm) and increase in

Table 7.1. *Colours (established by bulk electrolysis in acetonitrile) of the ruthenium(II) tris-bipyridyl complexes of the ligands L^1-L^5, in all accessible oxidation states (from ref. 5).*

Charge on RuL$_3$ unit	L^1	L^2	L^3	L^4	L^5
+2	Orange	Orange	Orange	Red–orange	Red–orange
+1	Purple	Wine red	Grey–blue	Purple	Red–brown
0	Blue	Purple	Turquoise	Blue	Purple–brown
−1	Green	Blue	Green	Turquoise	Grey–blue
−2	Brown			Aquamarine	Green
−3	Red			Brown–green	Purple

intensity ($\varepsilon = 6000$ and $7300\,\text{dm}^3\,\text{mol}^{-1}\,\text{cm}^{-1}$ respectively). Following the third one-electron reduction, the two peaks coalesce into a broad absorption at 1560 nm, which is again enhanced in intensity ($\varepsilon = 12\,000\,\text{dm}^3\,\text{mol}^{-1}\,\text{cm}^{-1}$). Upon reduction by the fourth and subsequent electrons the peak intensity diminishes continuously to approximately zero for the six-electron reduction product. These NIR transitions are almost exclusively ligand-based.

An optically transparent thin-layer electrode (OTTLE) study[8] revealed that the visible spectra of the reduced forms of [Ru(bipy)$_3$]$^{2+}$ derivatives can be separated into two classes. Type-A complexes, such as [Ru(bipy)$_3$]$^{2+}$, [Ru(L^7)$_3$]$^{2+}$ and [Ru(L^1)$_3$]$^{2+}$ show spectra on reduction which contain low-intensity ($\varepsilon < 2500\,\text{dm}^3\,\text{mol}^{-1}\,\text{cm}^{-1}$) bands; these spectra are similar to those of the reduced free ligand and are clearly associated with ligand radical anions. In contrast, type-B complexes such as [Ru(L^8)$_3$]$^{2+}$ and [Ru(L^9)$_3$]$^{2+}$ on reduction exhibit spectra containing broad bands of greater intensity ($1000 < \varepsilon < 15\,000\,\text{dm}^3\,\text{mol}^{-1}\,\text{cm}^{-1}$).

L^7 R = Me
L^8 R = COOEt
L^9 R = CONEt$_2$

7.2.2 Reductive electropolymerisation of polypyridyl complexes

The reductive electropolymerisation technique relies on the ligand-centred nature of the three sequential reductions of complexes such as [Ru(L^{10})$_3$]$^{2+}$ (L^{10} = 4-vinyl-4′-methyl-2,2′-bipyridine), combined with the anionic polymerisability of suitable ligands.[9] Vinyl-substituted pyridyl ligands such as L^{10}–L^{12} are generally employed, although metallopolymers have also been formed from chloro-substituted pyridyl ligands, via electrochemically initiated carbon–halide bond cleavage. In either case, electrochemical reduction of their metal complexes generates radicals leading to carbon–carbon bond formation and oligomerisation. Oligomers above a critical size are insoluble and thus thin films of the electroactive metallopolymer are deposited on the electrode surface.

7.2.3 Oxidative electropolymerisation of polypyridyl complexes

Oxidative electropolymerisation has been described for iron(II) and ruthenium(II) complexes containing amino-[10] and pendant aniline-substituted[11] 2,2'-bipyridyl ligands, and amino- and hydroxy- substituted 2,2':6',2''-terpyridinyl ligands.[12] Analysis of IR spectra suggests that the electropolymerisation of $[Ru(L^{13})_2]^{2+}$, via the pendant aminophenyl substituent, proceeds by a reaction mechanism similar to that of aniline.[12] The resulting modified electrode reversibly switched from purple to pale pink on oxidation of Fe^{II} to Fe^{III}. For polymeric films formed from $[Ru(L^{14})_2]^{2+}$, via polymerisation of the pendant hydroxyphenyl group, the colour switch was from brown to dark yellow. The dark yellow was attributed to an absorption band at 455 nm, probably due to quinone moieties in the polymer formed during electropolymerisation. Infrared spectra confirmed the absence of hydroxyl groups in the initially deposited brown films.

Metallopolymer films have also been prepared by oxidative polymerisation of complexes of the type $[M(phen)_2(4,4'-bipy)_2]^{2+}$ (M = Fe, Ru or Os; 4,4'-bipy = 4,4'-bipyridine).[13] Such films are both oxidatively and reductively electrochromic; reversible film-based reduction at potentials below −1 V results in dark purple films,[13] the colour and potential region being consistent with the viologen-dication/radical-cation electrochromic response. A purple state at high negative potentials has also been observed for polymeric films prepared from $[Ru(L^{15})_3]^{2+}$.[14] Electropolymerised films prepared from the complexes $[Ru(L^{16})(bipy)_2][PF_6]_2$[15] and $[Ru(L^{17})_3][PF_6]_2$[16,17] exhibit reversible orange–transparent electrochromic behaviour associated with the Ru^{II}/Ru^{III} interconversion.

7.2 Electrochromism of polypyridyl complexes

[Structures of ligands L¹³, L¹⁴, L¹⁵, L¹⁶, L¹⁷]

7.2.4 Spatial electrochromism of polymeric polypyridyl complexes

Spatial electrochromism has been demonstrated in metallopolymeric films.[18] Photolysis of poly[Ru^{II}(L^{10})$_2$(py)$_2$]Cl$_2$ thin films on tin-doped indium oxide-coated (ITO) glass in the presence of chloride ions leads to photochemical loss of the photolabile pyridine ligands, and sequential formation of poly[Ru^{II}(L^{10})$_2$(py)Cl]Cl and poly[Ru^{II}(L^{10})$_2$Cl$_2$] (see Scheme 7.1).

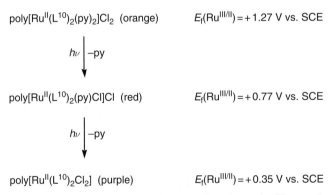

poly[Ru^{II}(L^{10})$_2$(py)$_2$]Cl$_2$ (orange) E_f(Ru$^{III/II}$) = +1.27 V vs. SCE

$h\nu$ ↓ −py

poly[Ru^{II}(L^{10})$_2$(py)Cl]Cl (red) E_f(Ru$^{III/II}$) = +0.77 V vs. SCE

$h\nu$ ↓ −py

poly[Ru^{II}(L^{10})$_2$Cl$_2$] (purple) E_f(Ru$^{III/II}$) = +0.35 V vs. SCE

Scheme 7.1 Spatial electrochromism in metallopolymeric films using photolabile pyridine ligands. (Scheme reproduced from Leasure, R. M., Ou, W., Moss, J. A., Linton, R. W. and Meyer, T. J. 'Spatial electrochromism in metallo-polymeric films of ruthenium polypyridyl complexes.' *Chem. Mater.*, **8**, 1996, 264–73, with permission of The American Chemical Society.)

Contact lithography can be used to spatially control the photosubstitution process to form laterally resolved bicomponent films with image resolution below 10 μm. Dramatic changes occur in the colours and redox potentials of such ruthenium(II) complexes upon substitution of chloride for the pyridine ligands (Scheme 7.1). Striped patterns of variable colours are observed on addressing such films with a sequence of potentials.

7.3 Electrochromism in metallophthalocyanines and porphyrins

7.3.1 Introduction to metal phthalocyanines and porphyrins

The porphyrins are a group of highly coloured, naturally occurring pigments containing a tetrapyrrole porphine nucleus (see below) with substituents at the eight β-positions of the pyrroles, and/or the four *meso*-positions between the pyrrole rings.[19] The natural pigments themselves are metal chelate complexes of the porphyrins. Phthalocyanines are tetraazatetrabenzo derivatives of porphyrins with highly delocalised π-electron systems. Metallophthalocyanines are

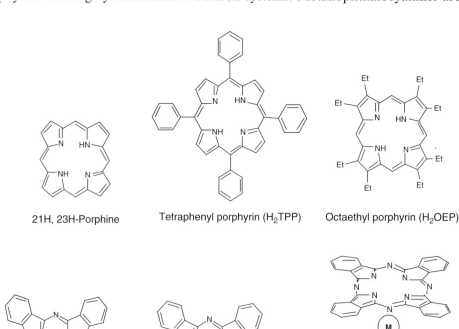

21H, 23H-Porphine   Tetraphenyl porphyrin (H₂TPP)   Octaethyl porphyrin (H₂OEP)

29H, 31H-Phthalocyanine   1:1 Metallophthalocyanine complex   A 'sandwich'-type metallophthalocyanine complex

important industrial pigments, blue to green in colour, used primarily in inks and for colouring plastics and metal surfaces.[19,20,21] The water-soluble sulfonate derivatives are used as dyestuffs for clothing. In addition to these uses, the metallophthalocyanines have been extensively investigated in many fields including catalysis, liquid crystals, gas sensors, electronic conductivity, photosensitisers, non-linear optics and electrochromism.[20] The purity and depth of the colour of metallophthalocyanines arise from the unique property of having an isolated, single band located in the far-red end of the visible spectrum (near 670 nm), with ε often exceeding $10^5\,\mathrm{dm}^3\,\mathrm{mol}^{-1}\,\mathrm{cm}^{-1}$. The next, more energetic, set of transitions is generally much less intense, near 340 nm. Charge transfer transitions between a chosen metal and the phthalocyanine ring introduce additional bands around 500 nm that allow tuning of the hue.[20]

The metal ion in metallophthalocyanines lies either at the centre of a single phthalocyanine (Pc = dianion of phthalocyanine), or between two rings in a sandwich-type complex.[20] Phthalocyanine complexes of transition metals usually contain only a single Pc ring while lanthanide-containing species usually form bis(phthalocyanines), where the π-systems interact strongly with each other, resulting in characteristic features such as the semiconducting ($\kappa = 5 \times 10^{-5}\,\Omega^{-1}\,\mathrm{cm}^{-1}$) properties of thin films of bis-(phthalocyaninato)lutetium(III) [Lu(Pc)$_2$].[22]

7.3.2 Sublimed bis(phthalocyaninato)lutetium(III) films

The electrochromism of the phthalocyanine ring-based redox processes of vacuum-sublimed thin films of [Lu(Pc)$_2$] was first reported in 1970,[23] and since that time this complex has received most attention, although many other (mainly lanthanide) metallophthalocyanines have been investigated for their electrochromic properties. The complex Lu(Pc)$_2$ has been studied extensively by Collins and Schiffrin[24,25] and by Nicholson and Pizzarello.[26,27,28,29,30,31] It was initially studied as a film immersed in aqueous electrolyte, but hydroxide ion from water causes gradual film destruction, attacking nitrogens of the Pc ring.[24] Acidic solution allows a greater number of stable write–erase cycles, up to 5×10^6 cycles in sulfuric acid,[24] approaching exploitable device requirements. Films of [Lu(Pc)$_2$] in ethylene glycol solution were found to be even more stable.[25]

Fresh [Lu(Pc)$_2$] films (likely to be singly protonated,[31] although this issue is contentious[24,32]), which are brilliant green in colour ($\lambda_{\max} = 605$ nm), are electro-oxidised to a yellow–tan form, Eq. (7.1):[26,29,32]

$$[\mathrm{Pc}_2\mathrm{LuH}]^+ \,(s) \rightarrow [\mathrm{Pc}_2\mathrm{Lu}]^+(s) + \mathrm{H}^+ + \mathrm{e}^-. \qquad (7.1)$$
$$\text{green} \qquad\qquad \text{yellow-tan}$$

A further oxidation product is red,[26,29,32] yet of unknown composition. Electroreduction of [Lu(Pc)$_2$] films gives a blue-coloured film, Eq. (7.2):[33]

$$[Pc_2LuH]^+ \text{ (s)} + e^- \rightarrow [Pc_2LuH] \text{ (s)}, \quad (7.2)$$
$$\text{green} \qquad\qquad\qquad \text{blue}$$

with further reduction yielding a violet–blue product, Eqs. (7.3) and (7.4):[29]

$$[Pc_2LuH] \text{ (s)} + e^- \rightarrow [Pc_2LuH]^- \text{(s)}; \quad (7.3)$$
$$\text{blue} \qquad\qquad\qquad \text{violet}$$

$$[Pc_2LuH]^- \text{(s)} + e^- \rightarrow [Pc_2LuH]^{2-} \text{(s)}. \quad (7.4)$$

The lutetium bis(phthalocyanine) system is a truly electropolychromic one,[23] but usually only the blue-to-green transition is used in ECDs. Although prototypes have been constructed,[34] no ECD incorporating [Lu(Pc)$_2$] has yet been marketed, owing to experimental difficulties such as film disintegration caused by constant counter-anion ingress/egress on colour switching.[24] For this reason, larger anions are best avoided to minimise the mechanical stresses. A second, related, handicap of metallophthalocyanine electrochromic devices is their relatively long response times. Nicholson and Pizzarello[30] investigated the kinetics of colour reversal and found that small anions like chloride and bromide allow faster colour switching. Sammells and Pujare overcame the problem of slow penetration of anions into solid lattices by using an ECD containing an electrochrome suspension in semi-solid poly(AMPS) – AMPS = 2-acrylamido-2-methyl propane sulfonic acid) electrolyte.[34] While the response times are still somewhat long, the open-circuit life times ('memory' times) of all colours were found to be very good.[30] Films in chloride, bromide, iodide and sulfate-containing solutions were found to be especially stable in this respect.

7.3.3 Other metal phthalocyanines

Moskalev *et al.* prepared the phthalocyanine complexes of neodymium, americium, europium, thorium and gallium (the latter as the half acetate).[35] Collins and Schiffrin[24] have reported the electrochromic behaviour of the phthalocyanine complexes CoPc, SnCl$_2$Pc, SnPc$_2$, MoPc, CuPc and the metal-free H$_2$Pc. No electrochromism was observed for either the metal-free or for the copper phthalocyanines in the potential ranges employed; all of the other complexes showed limited electrochromism. Both SnCl$_2$Pc and SnPc$_2$

7.3 Metallophthalocyanines and porphyrins

could be readily reduced, but showed no anodic electrochromism. Other molecular phthalocyanine electrochromes studied include complexes of aluminium,[36] copper,[37] chromium,[36,38] erbium,[39] europium,[40] iron,[41] magnesium,[42] manganese,[38,42] titanium,[43] uranium,[44] vanadium,[43] ytterbium,[45,46] zinc[47] and zirconium.[36,40,48] Mixed phthalocyanine systems have also been prepared by reacting mixed-metal precursors comprising the lanthanide metals dysprosium, holmium, erbium, thulium, ytterbium, lutetium, yttrium and small amounts of others;[49] the response times for such mixtures are reportedly superior to those for single-component films. Walton et al. have compared the electrochemistry of lutetium and ytterbium bis(phthalocyanines), finding them to be essentially identical.[50] Both chromium and manganese mono-phthalocyanine complexes undergo metal-centred oxidation and reduction processes.[38] In contrast, the redox reactions of LuPc$_2$ occur on the ligand; electron transfer to the central lutetium causes molecular dissociation.[51] Lever and co-workers have studied cobalt phthalocyanine systems in which two or four Co(Pc) units are connected via chemical links.[52,53,54,55] This group has also studied tetrasulfonated cobalt and iron phthalocyanines.[56] Finally, polymeric ytterbium bis(phthalocyanine) has been investigated[57,58,59] using a plasma to effect the polymerisation.

7.3.4 Electrochemical routes to metallophthalocyanine electrochromic films

For complexes with pendant aniline and hydroxy-substituted ligands, oxidative electropolymerisation is an alternative route to metallophthalocyanine electrochromic films. Although polymer films prepared from [Lu(L^{18})$_2$] monomer show loss of electroactivity on being cycled to positive potentials, in dimethyl sulfoxide (DMSO) the electrochemical response at negative potentials is stable, with the observation of two broad quasi-reversible one-electron redox couples.[60] Spectroelectrochemical measurements revealed switching times of <2 s for the observed green–grey–blue colour transitions in this region. Oxidative electropolymerisation using pendant aniline substituents has also been applied to monophthalocyaninato transition-metal complexes;[61] the redox reactions and colour changes of two of the examples studied are given in Eqs. (7.5)–(7.8).

$$\text{poly}[\text{Co}^{II}(\text{L}^{18})] + ne^- \rightarrow \text{poly}[\text{Co}^{I}(\text{L}^{18})]^-; \qquad (7.5)$$
blue-green   yellow-brown

$$\text{poly}[\text{Co}^{I}(\text{L}^{18})]^- + ne^- \rightarrow \text{poly}[\text{Co}^{I}(\text{L}^{18})]^{2-}; \qquad (7.6)$$
yellow-brown   red-brown(thick films), deep pink(thin films)

$$\text{poly}[\text{Ni}^{II}(L^{18})] + ne^- \rightarrow \text{poly}[\text{Ni}^{II}(L^{18})]^-; \qquad (7.7)$$
green $\qquad\qquad\qquad\qquad$ blue

$$\text{poly}[\text{Ni}^{II}(L^{18})]^- + ne^- \rightarrow \text{poly}[\text{Ni}^{II}(L^{18})]^{2-}. \qquad (7.8)$$
blue $\qquad\qquad\qquad\qquad$ purple

The first reduction in the cobalt-based polymer is metal-centred, resulting in the appearance of a new MLCT transition, the second reduction being ligand-centred. By contrast, for the nickel-based polymer both redox processes are ligand-based.

L^{18} R=NH$_2$
L^{19} R=O–(2-C$_6$H$_4$OH)
L^{20} R=CO$_2$–CH$_2$CH$_2$CMe$_3$

NB These complexes exist as a mixture of isomers with the substituents attached at either of the positions labelled • on the benzyl rings.

Electrochromic polymer films have been prepared by oxidative electropolymerisation of the monomer [Co(L^{19})].[62] The technique involved voltammetric cycling from -0.2 to $+1.2$ V vs. SCE at 100 mV s^{-1} in dry acetonitrile, resulting in the formation of a fine green polymer. Cyclic voltammograms during polymer growth showed the irreversible phenol oxidation peak at $+0.58$ V and a reversible phthalocyanine-ring oxidation peak at $+0.70$ V. Polymer-modified electrodes gave two distinct redox processes with half-wave potentials at -0.35 [from Co$^{II} \rightarrow$ Co^{I}] and -0.87 V (from ring reduction). The coloration switched from transparent light green [Co^{II} state] to yellowish green [Co^{I} state] to dark yellow (reduced ring).

7.3.5 Langmuir–Blodgett metallophthalocyanine electrochromic films

The electrochemical properties of a variety of metallophthalocyanines have been studied as multilayer Langmuir–Blodgett (LB) films. For example, LB films of alkyloxy-substituted [Lu(Pc)$_2$] exhibited a one-electron reversible reduction and a one-electron reversible oxidation corresponding to a transition from green to orange and blue forms respectively, with the electron transport through the multilayers being at least in part diffusion controlled.[63]

An explanation of the relatively facile redox reaction in such multilayers is that the Pc ring is large compared with the alkyl tail, and there is enough space and channels present in the LB films to allow the necessary charge-compensating ion transport. More recently, the structure, electrical conductivity and electrochromism in thin films of substituted and unsubstituted lanthanide bis-phthalocyanine derivatives have been investigated with particular reference to the differences between unsubstituted and butoxy-substituted [Lu(Pc)$_2$] materials.[64] Scanning tunnelling microscopy (STM) images on graphite reveal the differences in the two structures, giving molecular dimensions of 1.5×1.0 nm and 2.8×1.1 nm respectively. The in-plane dc conductivity was studied as a function of film thickness and temperature, with unsubstituted [Lu(Pc)$_2$] being approximately 10^6 times more conductive than the substituted material. The green–red oxidative step is seen for both cases but the green–blue reductive step is absent in the butoxy-substituted material. High-quality LB films of [M(L^{20})] (M = Cu, Ni) have also been reported.[65] Ellipsometric and polarised optical absorption measurements suggest that the Pc rings are oriented with their large faces perpendicular to the immersion direction and to the substrate plane.

The LB technique may be used for the fabrication of ECDs: an LB thin-film display based on bis(phthalocyaninato)praseodymium(III) has been reported.[66] The electrochromic electrode was fabricated by deposition of multilayers (10–20 layers, ≈ 100–200 Å) of the complex onto ITO-coated glass (7×4 cm^2) slides. The display exhibited blue–green–yellow–red electropolychromism over a potential range of -2 to $+2$ V. After 10^5 cycles no significant changes are observed in the spectra of these colour states, again approaching exploitability. The high stability of the device was ascribed to the preparation, by the LB technique, of well-ordered monolayers that allow better diffusion of the counter ions into the film, which improves reversibility. Unless these structures provide ion channels, ordered structures might be considered to favour electronic rather than ionic motion, as the latter could benefit more from defects arising from disorder.

7.3.6 Species related to metallophthalocyanines

Naphthalocyanine (nc) species are structurally similar to the simpler phthalocyanines described above and have two isomers, denoted here (2,3-nc) and (1,2-nc).

Naphthalocyanines show an intense optical absorption at long wavelengths ($700 < \lambda < 900$ nm) owing to electronic processes within the extended conjugated system of the ligand.[67,68] Thin-film [Co(2,3-nc)$_2$] is green and is readily oxidised to form a violet-coloured species. Thin-film [Zn(2,3-nc)$_2$] is also green

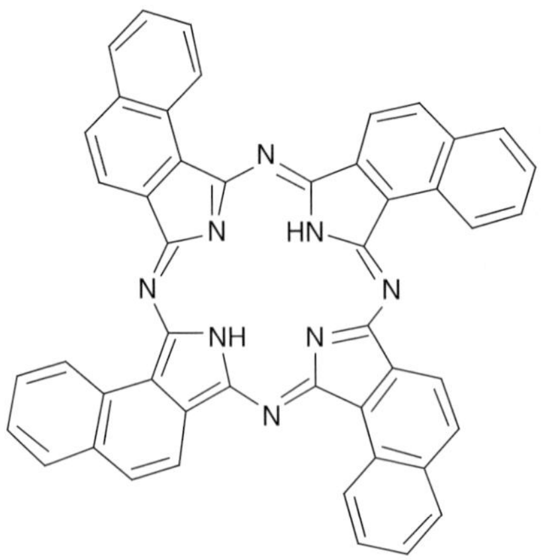

1,2-Naphthalocyanine

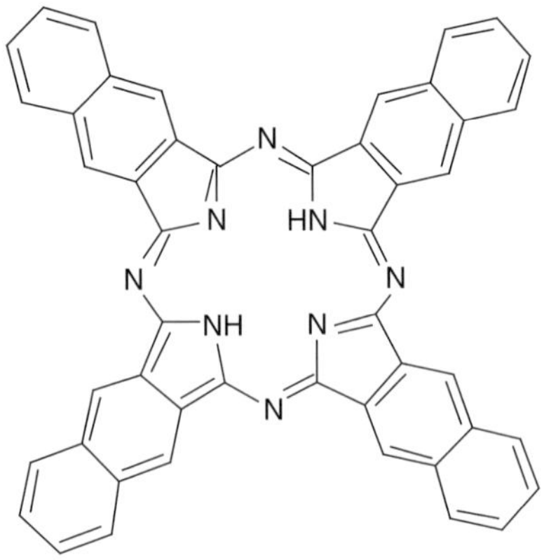

2,3-Naphthalocyanine

when neutral. A 'triple-decker' naphthalocyanine compound [(1,2-nc)Lu(1,2-nc)Lu(1,2-nc)] has been reported.[69] Electrochromism in the pyridinoporphyrazine system and its cobalt complex has also received some attention.[70] Here, the ligand is similar to a phthalocyanine but with quaternised pyridyl residues replacing all four fused benzo groups.

It is not only homoleptic (i.e. all ligands similar) phthalocyanine complexes that can form sandwich structures; recently a substantial number of heteroleptic sandwich-type metal complexes, with mixed phthalocyaninato and/or porphyrinato ligands, have been synthesised and are likely to show interesting electrochromic properties.[71] Although considerable progress has been made in this field, there is clearly much room for further investigation. By attaching functional groups or special (donor or acceptor) moieties to these compounds, it may be possible to tune their electronic properties without altering the ring-to-ring separation. The properties associated with these units may also be imparted to the parent sandwich compounds. The electrochromic properties of some silicon–phthalocyanine thin films, in which a redox active ferrocene-carboxylato unit is appended to the electrochromic centre, have been studied.[72]

7.3.7 Electrochromic properties of porphyrins

Early results suggest that an investigation into porphyrin electrochromism is warranted, although there has been little systematic study to date. Thus, the spectra of the chemical reduction products of Zn(TPP) have been reported,[73,74] with colours changing between a pink (parent complex), green (mono-negative ion), and amber (di-negative ion).[73] Felton and Linschitz[75] reported that the electrochemically produced monoanion spectrum is similar to that produced chemically. Fajer *et al.*[76] showed that Zn(TPP) changes colour to green upon one-electron oxidation by controlled potential electrolysis. Felton *et al.*[77] reported that the electrolysis of Mg(OEP) yielded a blue–green solution. The recently reported[78] green–pink colour change of a porphyrin monomer appears to be a pH-change-induced transformation of the J-aggregate (ordered molecular arrangement, excited state spread over N molecules in one dimension) to the monomer, and is therefore electrochromic only indirectly, from the redox viewpoint.

Recently it has been found that oxidative electropolymerisation of substituted porphyrins could be useful towards the development of electrochromic porphyrin devices.[79]

7.4 Near-infrared region electrochromic systems

7.4.1 Significance of the near-infrared region

The metal complexes described so far in this chapter have been of interest for their electrochromism in the visible region of the spectrum, a property which is of obvious interest for use in display devices and windows. Electrochromism in the near-infrared (NIR) region of the spectrum (*ca.* 800–2000 nm) is an area which has also attracted much recent interest[3] because of the considerable technological importance of this region of the spectrum. Near-infrared radiation finds use in applications as diverse as optical data storage,[1] in medicine, where photodynamic therapy exploits the relative transparency of living tissue to NIR radiation around 800 nm,[80] and in telecommunications, where fibre-optic signal transmission through silica fibres exploits the 'windows of transparency' of silica in the 1300–1550 nm region. Near-infrared radiation is also felt as radiant heat, so NIR-absorbing or reflecting materials could have use in smart windows that allow control of the environment inside buildings; and the fact that much of the solar emission spectrum is in the NIR region means that effective light-harvesting compounds for use in solar cells need to capture NIR as well as visible light.[81]

Many molecules with strong NIR absorptions have been investigated, often with a view to examining their performance as dyes in optical data-storage media.[82,83,84] The majority of these are highly conjugated organic molecules that are not redox active. A minority however are based on transition-metal complexes and it is generally these which have the redox activity necessary for electrochromic behaviour and which are discussed in the following sections. One such set of complexes has already been discussed in this chapter: spectral modulation in the NIR region has been reported for a variety of [Ru(bipy)$_3$]$^{2+}$ derivatives, whose reduced forms contain ligand radical anions that show intense, low-energy electronic transitions. Since these are also electrochromic in the visible region, they were discussed earlier in Section 7.2.1.[7] Near-infrared electrochromic materials based on doped metal oxides[85] (see Chapter 6) and conducting polymeric films[86] (see Chapter 10) are also extensively studied.

7.4.2 Planar dithiolene complexes of Ni, Pd and Pt

One of the earliest series of metal complexes which showed strong, redox-dependent NIR absorptions is the well-studied set of square-planar bis-dithiolene complexes of Ni, Pd and Pt (see below). Extensive delocalisation between metal and ligand orbitals in these 'non-innocent' systems means that assignment of oxidation states is problematic, but it does result in intense electronic transitions. These complexes have two reversible redox processes connecting the neutral, monoanionic and dianionic species.

Generic structure of planar bis-dithiolene complexes: M = Ni, Pd, Pt; n = 0, 1, 2

Complexes of dialkyl-substituted imidazolidine-2,4,5-trithiones (M = Ni, Pd) (refs. 90, 91, 92, 93, 94, 95, 96)

The structures and redox properties of these complexes have been extensively reviewed;[87,88] of interest here is the presence of an intense NIR transition in the neutral and monoanionic forms, but not the dianionic forms, i.e. the complexes are electropolychromic. The positions of the NIR absorptions are highly sensitive to the substituents on the dithiolene ligands. A large number of substituted dithiolene ligands have been prepared and used to prepare complexes of Ni, Pd and Pt which show comparable electrochromic properties with absorption maxima at wavelengths up to *ca.* 1400 nm and extinction coefficients up to *ca.* 40 000 dm^3 mol^{-1} cm^{-1} (see refs. 87 and 88 for an extensive listing).

The main application of the strong NIR absorbance of these complexes, pioneered by Müller-Westerhoff and co-workers,[88,89] is for use in the neutral state as dyes to induce Q-switching of NIR lasers such as the Nd-YAG

(1064 nm), iodine (1310 nm) and erbium (1540 nm) lasers. This relies on a combination of very high absorbance at the laser wavelengths, an appropriate excited-state lifetime following excitation, and good long-term thermal and photochemical stability. The use of a range of metal dithiolene complexes in this respect has been reviewed.[88,89] The strong NIR absorptions of these complexes have continued to attract attention since these reviews appeared. A new series of neutral, planar dithiolenes of Ni and Pd has been prepared based on the ligands [R$_2$timdt]$^-$ which contain the dialkyl-substituted imidazolidine-2,4,5-trithione core (see above).[90,91,92,93,94,95,96] In these ligands the peripheral ring system ensures that the electron-donating N substituents are coplanar with the dithiolene unit, maximising the electronic effect. This shifts the NIR absorptions of the [M(R$_2$timdt)$_2$] complexes to lower energy than found in the 'parent' dithiolene complexes. The result is that the NIR absorption maximum occurs at around 1000 nm and has a remarkably high extinction coefficient (up to 80 000 dm^3 mol^{-1} cm^{-1}). The high thermal and photochemical stabilities of these complexes make them excellent candidates for Q-switching of the 1064 nm Nd-YAG laser. In addition, one-electron reduction to the monoanionic species [M(R$_2$timdt)$_2$]$^-$ results in a shift of the NIR absorption maximum to *ca.* 1400 nm, indicating possible exploitation of their electrochromism.[96]

7.4.3 Mixed-valence dinuclear complexes of ruthenium

Another well-known class of metal complexes showing NIR electrochromism is the extensive series of dinuclear mixed-valence complexes based principally on ruthenium–ammine or ruthenium–polypyridine components, in which a strong electronic coupling between the metal centres makes a stable Ru^{II}–Ru^{III} mixed-valence state possible. Such complexes generally show a Ru^{II}→Ru^{III} IVCT transition which is absent in both the Ru^{II}–Ru^{II} and Ru^{III}–Ru^{III} forms. These complexes have primarily been of interest because the characteristics of the IVCT transition provide quantitative information on the magnitude of the electronic coupling between the metal centres, and is accordingly an excellent diagnostic tool. Nevertheless, the position and intensity of the IVCT transition in some cases mean that complexes of this sort could be exploited for their optical properties. Table 7.2 shows a small, representative selection of recent examples which show electrochromic behaviour (in terms of the intensity of the IVCT transitions) typical of this class of complex.[97,98,99,100] The main purpose of this selection is to draw the reader's attention to the fact that these complexes which, as a class, are so familiar, in a different context could be equally valuable for their electrochromic properties. Of course the field

Table 7.2. *Examples of mixed-valence dinuclear complexes showing NIR electrochromism.*

Complex	λ/nm (ε/dm^3 mol^{-1} cm^{-1})	Ref.
[3+] (bipy)$_2$Ru–O,N / Ru(bipy)$_2$ bridged biphenyl-bis(pyridyl-phenoxide)	2000 14 000	97
[3+] (bipy)$_2$Ru / Ru(bipy)$_2$ bis(benzoylamide) bridge	1600 11 700	98
[−] (H$_3$N)$_5$Ru^{II}–N,N'-benzotriazolate–Fe^{III}(CN)$_5$	1210 3 900	99
[3+] (bpy)(terpy)Ru–NC–C$_6$H$_4$–N=N–C$_6$H$_4$–CN–Ru(terpy)(bpy)	1920 10 000	100

is not limited to ruthenium complexes, although these have been the most extensively studied because of their synthetic convenience and ideal electrochemical properties; analogous complexes of other metals have also been prepared and could be equally effective NIR electrochromic dyes.

Very recently, a trinuclear Ru^{II} complex has been reported which shows a typical IVCT transition at 1550 nm in the mixed-valence Ru^{II}–Ru^{III} form. The complex has pendant hydroxyl groups which react with a tri-isocyanate to give a crosslinked polymer which was deposited on an ITO substrate. Good electrochromic switching of 1550 nm radiation was maintained, with fast switching times (of the order of 1 second), over several thousand redox cycles.[101]

7.4.4 Tris(pyrazolyl)borato-molybdenum complexes

In the last few years McCleverty, Ward and co-workers have reported the NIR electrochromic behaviour of a series of mononuclear and dinuclear complexes containing the oxo-Mo^V core unit [Mo(Tp*)(O)Cl(OAr)], where 'Ar' denotes a phenyl or naphthyl ring system and [Tp* = hydrotris(3,5-dimethylpyrazolyl) borate].[102,103,104,105,106,107] Mononuclear complexes of this type undergo reversible Mo^{IV}–Mo^V and Mo^V–Mo^{VI} redox processes with all three oxidation states accessible at modest potentials. Whilst reduction to the Mo^{IV} state results in unremarkable changes in the electronic spectrum, oxidation to Mo^{VI} results in the appearance of a low-energy phenolate- (or naphtholate)-to-Mo^{VI} LMCT process.[102,103]

In mononuclear complexes these transitions are at the low-energy end of the visible region and of moderate intensity: for [Mo(Tp*)(O)Cl(OPh)] for example the LMCT transition is at 681 nm with $\varepsilon = 13\,000$ dm^3 mol^{-1} cm^{-1}.[103] However in many dinuclear complexes of the type [{Mo(Tp*)(O)Cl}$_2$(μ-OC$_6$H$_4$EC$_6$H$_4$O)], in which two oxo-Mo(V) fragments are connected by a bis-phenolate bridging ligand in which a conjugated spacer 'E' separates the two phenyl rings, the NIR electrochromism is much stronger. In these complexes an electronic interaction between the two metals results in a separation of the two Mo^V–Mo^{VI} couples, such that the complexes can be oxidised from the Mo^V–Mo^V state to Mo^V–Mo^{VI} and then Mo^{VI}–Mo^{VI} in two distinct steps. The important point here is that in the oxidised forms, containing one or two Mo^{VI} centres, the LMCT transitions are at lower energy and of much higher intensity than in the mononuclear complexes (Figure 7.1 gives a representative example).[103,104,105] Depending on the nature of the group E in the bridging ligand, the absorption maxima can span the range 800–1500 nm, with extinction coefficients of up to 50 000 dm^3 mol^{-1} cm^{-1} (see Table 7.3)[103,104,105]

A prototypical device to illustrate the possible use of these complexes for modulation of NIR radiation has been described.[106] A thin-film cell was prepared containing a solution of an oxo-Mo^V dinuclear complex and base electrolyte between transparent, conducting-glass slides. The complex used has the spacer E = bithienyl between the two phenolate termini (sixth entry in Table 7.3); this complex develops an LMCT transition (centred at 1360 nm, with $\varepsilon = 30\,000$ dm^3 mol^{-1} cm^{-1}) on one-electron oxidation to the Mo^V–Mo^{VI} state which is completely absent in the Mo^V–Mo^V state. Application of an alternating potential, stepping between +1.5 V and 0 V for a few seconds each, resulted in fast switching on/off of the NIR absorbance reversibly over several thousand cycles. A larger cell was used to show how a steady increase in the potential applied to the solution, which resulted in a larger proportion of the

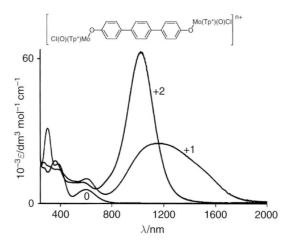

Figure 7.1 Electrochromic behaviour of [{Mo(Tp*)(O)Cl}$_2$(μ-OC$_6$H$_4$ C$_6$H$_4$ C$_6$H$_4$O)]$^{n+}$ in the oxidation states Mo^V–Mo^V ($n=0$), Mo^V–Mo^{VI} ($n=1$), Mo^{VI}–Mo^{VI} ($n=2$). Spectra were measured at 243 K in CH$_2$Cl$_2$. (Figure reproduced from Harden, N. C., Humphrey, E. R., Jeffrey, J. C. et al. 'Dinuclear oxomolybdenum(V) complexes which show strong electrochemical interactions across bis-phenolate bridging ligands: a combined spectro-electrochemical and computational study.' J. Chem. Soc., Dalton Trans. 1999, 2417–26, with permission of The Royal Society of Chemistry.)

material being oxidised, allowed the intensity of a 1300 nm laser to be attenuated reversibly and controllably over a dynamic range of 50 dB (a factor of ca. 10^5): the cell accordingly acts as a NIR variable optical attenuator.[106] The disadvantage of this prototype is that, being solution-based, switching is relatively slow compared to thin films or solid-state devices, but the optical properties of these complexes show great promise for further development.

Some nitrosyl–Mo^I complexes of the form [Mo(Tp*)(NO)Cl(py-R)] (where py-R is a substituted pyridine) also undergo moderate NIR electrochromism on reversible reduction to the Mo^0 state. In these complexes reduction of the metal centre results in appearance of a Mo$^0 \rightarrow$ py(π*) MLCT transition at the red end of the spectrum (for R = 4-CH(^{n}Bu)$_2$, $\lambda_{max} = 830$ nm with $\varepsilon = 12\,000$ dm^3 mol^{-1} cm^{-1}). However, when the pyridyl ligand contains an electron-withdrawing substituent *meta* to the N atom (R = 3-acetyl or 3-benzoyl) an additional MLCT transition at much longer wavelength develops ($\lambda_{max} = 1274$ and 1514 nm, respectively, with ε ca. 2400 dm^3 mol^{-1} cm^{-1} in each case).[107]

7.4.5 Ruthenium and osmium dioxolene complexes

Lever and co-workers described in 1986 how the mononuclear complex [Ru(bipy)$_2$(CAT)], which has no NIR absorptions, undergoes two reversible

7.4 Near-infrared region electrochromic systems

Table 7.3. *Principal low-energy absorption maxima of dinuclear complexes [{Mo(Tp*)(O)Cl}$_2$ (μL)]$^{n+}$ in their oxidised forms (n = 1, 2).*

Bridging ligand L	λ_{max}/nm (10^{-3} ε/dm^3 mol^{-1}cm^{-1})	
	Mo(V)–Mo(VI)	Mo(VI)–Mo(VI)
biphenyl-4,4'-diolate	1096 (50)	1017 (48)
methyl-substituted biphenyl-diolate	1245 (19)	832 (32)
terphenyl-4,4''-diolate	1131 (25)	1016 (62)
quaterphenyl-diolate	1047 (24)	1033 (50)
phenyl-thiophene-phenyl diolate	1197 (35)	684 (54)
phenyl-bithiophene-phenyl diolate	1360 (30)	(Not stable)
bis(phenol)sulfide	900 (10)	900 (20)
distyryl diolate	1210 (41)	(Not stable)
azobenzene-4,4'-diolate	1268 (35)	409 (38)
distyrylbenzene diolate	1554 (23)	978 (37)

oxidations which are ligand-centred CAT–SQ and SQ–Q couples (where CAT, SQ and Q are catecholate, 1,2-benzosemiquinone monoanion, and 1,2-benzoquinone, respectively; see Scheme 7.2).[108] In the two oxidised forms the presence of a 'hole' in the dioxolene ligand results in the appearance of Ru^{II}→ SQ and Ru^{II}→Q MLCT transitions, the former at 890 nm and the latter at 640 nm with intensities of *ca.* 10^4 dm^3 mol^{-1} cm^{-1}. The CAT–SQ and SQ–Q couples accordingly result in modest NIR electrochromic behaviour (see structures L^{21}–L^{23}).

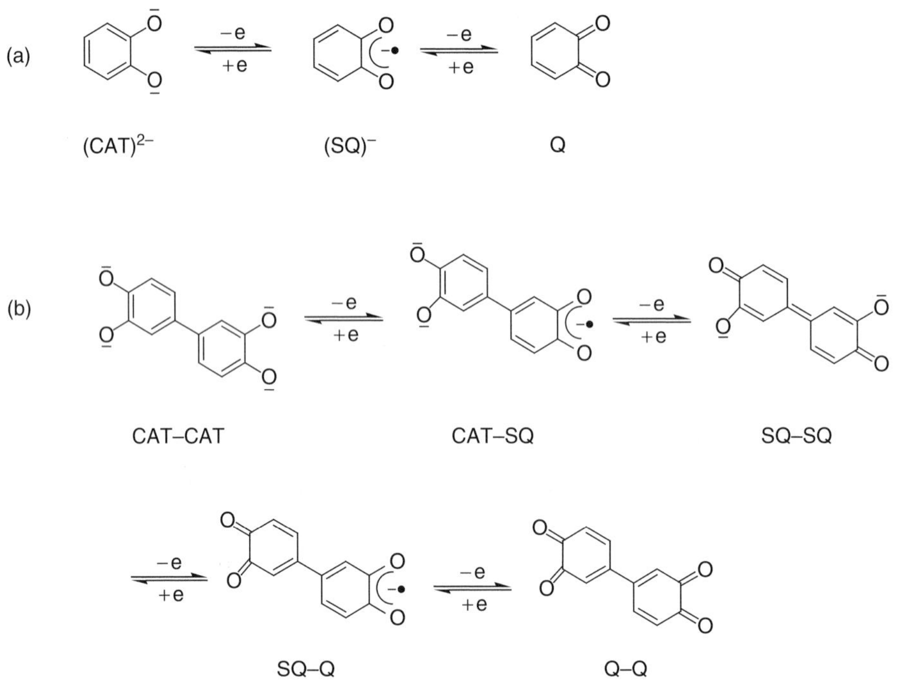

Scheme 7.2 Ligand-based redox activity of (a) the CAT–SQ–Q series; (b) [L^{21}]$^{n-}$ ($n = 4$–0).

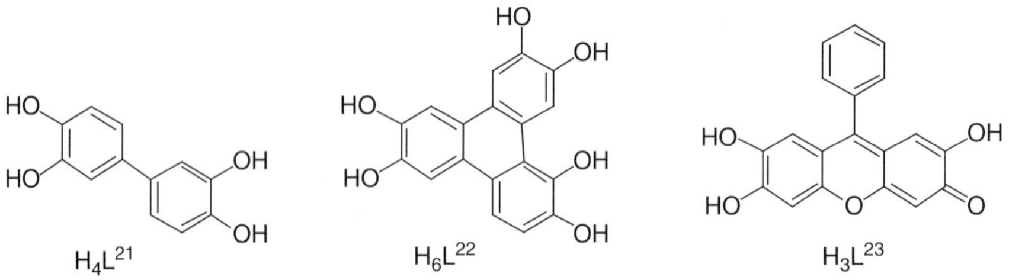

As with the oxo-Mo^V complexes mentioned in the previous section, the NIR transitions become far more impressive when two or more of these chromophores are linked by a conjugated bridging ligand, as in [{Ru(bipy)$_2$}$_2$(μ–L^{21})]$^{n+}$ ($n = 0$–4), which exhibits a five-membered redox chain, with reversible conversions between the fully reduced (bis-catecholate) and fully oxidised (bis-quinone) states all centred on the bridging ligand (Scheme 7.2). In the state $n = 2$, the NIR absorption is at 1080 nm with $\varepsilon = 37\,000$ dm^3 mol^{-1} cm^{-1}; this disappears in the fully reduced form and moves into the visible region in the fully oxidised form.[109] Likewise, the trinuclear complex [{Ru(bipy)$_2$}$_3$(μ–L^{22})]$^{n+}$ ($n = 3$–6) exists in four stable redox

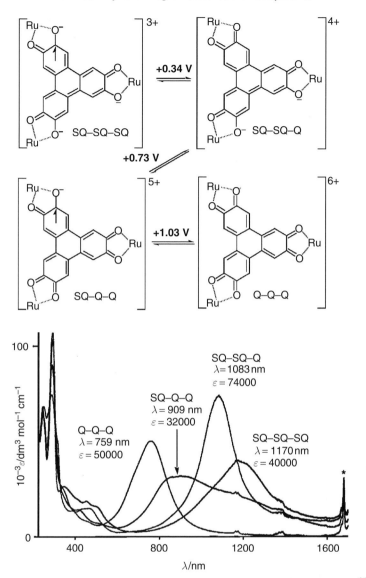

Figure 7.2 Ligand-centred redox interconversions of $[\{Ru(bipy)_2\}_3(\mu\text{-}L^{22})]^{n+}$ ($n = 3$–6) (potentials vs. SCE), and the resulting electrochromic behaviour. Spectra were measured at 243 K in MeCN. (Figure reproduced from Barthram, A. M., Cleary, R. L., Kowallick, R. and Ward, M. D. 'A new redox-tunable near-IR dye based on a trinuclear ruthenium(II) complex of hexahydroxy-triphenylene.' Chem. Commun. 1998, 2695–6, with permission of The Royal Society of Chemistry.)

states based on redox interconversions of the bridging ligand (from SQ–SQ–SQ to Q–Q–Q; Figure 7.2).[110] Thus the complexes are electropolychromic, with a large number of stable oxidation states accessible in which the intense NIR MLCT transitions involving the oxidised forms of the bridging ligand are redox-dependent. In this (typical) example, the NIR transitions vary in wavelength between 759 and 1170 nm over these four oxidation states, with intensities of up to $70\,000\,\text{dm}^3\,\text{mol}^{-1}\,\text{cm}^{-1}$. Other polydioxolene bridging ligands such as $[L^{23}]^{3-}$ have been investigated and their $\{Ru(bipy)_2\}^{2+}$ complexes show comparable electropolychromic behaviour in the NIR region.[111,112] The analogous complexes with osmium have also been characterised and, despite the differences in formal oxidation state assignment of the components (e.g. Os^{III}–catecholate instead of Ru^{II}–semiquinone), also show similar NIR electrochromic behaviour over several oxidation states.[113] Incorporation of these complexes into films or conducting solids, for faster switching, has yet to be described.

Recently, a mononuclear $[Ru(bipy)_2(cat)]$ derivative bearing carboxylate substituents that anchor it to a nanocrystalline Sb-doped tin oxide surface has been reported.[114] Redox cycling of the catecholate–semiquinone couple results in fast electrochromic switching (of the order of one second) of the film at 940 nm as the $Ru^{II} \rightarrow$ SQ MLCT transition appears and disappears.[114]

References

1. Mortimer, R. J. and Rowley, N. M. Metal complexes as dyes for optical data storage and electrochromic materials. In *Comprehensive Coordination Chemistry II: From Biology to Nanotechnology*, McCleverty, J. A. and Meyer, T. J. (eds.), Oxford, Elsevier, 2003, vol. 9, pp. 581–619.
2. Zhang, S. S., Qui, X. P., Chou, W. H., Liu, Q. G., Lang, L. L. and Xing, B. Q. Ferroin-based solid-state electrochromic display. *Solid State Ionics*, **52**, 1992, 287–9.
3. Ward, M. D. and McCleverty, J. A. Non-innocent behaviour in mononuclear and polynuclear complexes: consequences for redox and electronic spectroscopic properties. *J. Chem. Soc., Dalton Trans.*, 2002, 275–88.
4. Juris, A., Balzani, V., Barigelletti, F., Campagna, S., Belser, P. and von Zelewsky, A. Ru(II)-polypyridine complexes – photophysics, photochemistry, electrochemistry and chemi-luminescence. *Coord. Chem. Rev.*, **84**, 1988, 85–277.
5. Pichot, F., Beck, J. H. and Elliott, C. M. A series of multicolour electrochromic ruthenium(II) trisbipyridine complexes: synthesis and electrochemistry. *J. Phys. Chem. A*, **103**, 1999, 6263–7.
6. Elliott, C. M. and Redepenning, J. G. Stability and response studies of multicolour electrochromic polymer modified electrodes prepared from tris(5,5'-dicarboxyester-2,2'-bipyridine)ruthenium(II). *J. Electroanal. Chem.*, **197**, 1986, 219–32.
7. Elliott, C. M. Electrochemistry and near infrared spectroscopy of tris(4,4'-dicarboxyethyl-2,2'-bipyridine)ruthenium(II). *J. Chem. Soc., Chem. Commun.*, 1980, 261–2.

8. Elliott, C. M. and Hershenhart, E. J. Electrochemical and spectral investigations of ring-substituted bipyridine complexes of ruthenium. *J. Am. Chem. Soc.*, **104**, 1982, 7519–26.
9. Mortimer, R. J. Dynamic processes in polymer modified electrodes. In *Research in Chemical Kinetics*, Compton, R. G. and Hancock, G. (eds.), vol. 2, Amsterdam, Elsevier, 1994, pp. 261–311.
10. Ellis, C. D., Margerum, L. D., Murray, R. W. and Meyer, T. J. Oxidative electropolymerization of polypyridyl complexes of ruthenium. *Inorg. Chem.*, **22**, 1983, 1283–91.
11. Horwitz, C. P. and Zuo, Q. Oxidative electropolymerization of iron and ruthenium complexes containing aniline-substituted 2,2'-bipyridine ligands. *Inorg. Chem.*, **31**, 1992, 1607–13.
12. Hanabusa, K., Nakamura, A., Koyama, T. and Shirai, H. Electropolymerization and characterization of terpyridinyl iron(II) and ruthenium(II) complexes. *Polym. Int.*, **35**, 1994, 231–8.
13. Zhang, H.-T., Subramanian, P., Fussa-Rydal, O., Bebel, J. C. and Hupp, J. T. Electrochromic devices based on thin metallopolymeric films. *Sol. Energy Mater. Sol. Cells*, **25**, 1992, 315–25.
14. Beer, P. D., Kocian, O., Mortimer, R. J., Ridgway, C. and Stradiotto, N. R. Electrochemical polymerisation studies of aza-1 5-crown-5 vinyl-2,2'-bipyridine ruthenium(II) complexes. *J. Electroanal. Chem.*, **408**, 1996, 61–6.
15. Beer, P. D., Kocian, O. and Mortimer, R. J. Novel mono- and di-ferrocenyl bipyridyl ligands: syntheses, electrochemistry and electropolymerisation studies of their ruthenium(II) complexes. *J. Chem. Soc., Dalton Trans.* 1990, 3283–8.
16. Beer, P. D., Kocian, O., Mortimer, R. J. and Ridgway, C. Cyclic voltammetry of benzo-1 5-crown-5 ether vinyl-bipyridyl ligands, their ruthenium(II) complexes and bismethoxyphenyl vinyl-bipyridyl ruthenium(II) complexes. Electrochemical polymerisation studies and supporting electrolyte effects. *J. Chem. Soc., Faraday Trans.*, **89**, 1993, 333–8.
17. Beer, P. D., Kocian, O., Mortimer, R. J. and Ridgway, C. New alkynyl- and vinyl-linked benzo- and aza-crown ether-bipyridyl ruthenium(II) complexes which spectrochemically recognised group IA and IIA metal cations. *J. Chem. Soc., Dalton Trans.*, 1993, 2629–38.
18. Leasure, R. M., Ou, W., Moss, J. A., Linton, R. W. and Meyer, T. J. Spatial electrochromism in metallopolymeric films of ruthenium polypyridyl complexes, *Chem. Mater.*, **8**, 1996, 264–73.
19. Mashiko, T. and Dolphin, D. Porphyrins, hydroporphyrins, azaporphyrins, phthalocyanines, corroles, corrins and related macrocycles. In *Comprehensive Coordination Chemistry*, Wilkinson, G., Gillard, R. D. and McCleverty, J. A. (eds.), Oxford, Pergamon, 1987, vol. 2, ch. 21.1.
20. Leznoff, C. C. and Lever, A. B. P. (eds.) *Phthalocyanines: Properties and Applications*, New York, Wiley, vol. 1 (1989); vol. 2 (1993); vol. 3 (1993); vol. 4 (1996).
21. Gregory, P. Metal complexes as speciality dyes and pigments. In *Comprehensive Coordination Chemistry II: From Biology to Nanotechnology*, McCleverty, J. A. and Meyer, T. J. (eds.), Oxford, Elsevier, 2003, vol. 9, pp. 549–79.
22. Passard, M., Blanc, J. P. and Maleysson, C. Gaseous oxidation and compensating reduction of lutetium bis-phthalocyanine and lutetium phthalo-naphthalocyanine films. *Thin Solid Films*, **271**, 1995, 8–14.
23. P. N. Moskalev and I. S. Kirin. Effects of electrode potential on the absorption spectrum of a rare-rarth diphthalocyanine layer. *Opt. Spectros.*, **29**, 1970, 220.

24. Collins, G. C. S. and Schiffrin, D. J. The electrochromic properties of lutetium and other phthalocyanines. *J. Electroanal. Chem.*, **139**, 1982, 335–69.
25. Collins, G. C. S. and Schiffrin, D. J. The properties of electrochromic film electrodes of lanthanide diphthalocyanines in ethylene-glycol. *J. Electrochem. Soc.*, **132**, 1985, 1835–42.
26. Nicholson, M. M. and Pizzarello, F. A. Charge transport in oxidation product of lutetium diphthalocyanine. *J. Electrochem. Soc.*, **126**, 1979, 1490–5.
27. Nicholson, M. M. and Pizzarello, F. A. Galvanic transients in lutetium diphthalocyanine films. *J. Electrochem. Soc.*, **127**, 1980, 821–7.
28. Nicholson, M. M. and Pizzarello, F. A. Effects of the gaseous environment on propagation of anodic reaction boundaries in lutetium diphthalocyanine films. *J. Electrochem. Soc.*, **127**, 1980, 2617–20.
29. Nicholson, M. M. and Pizzarello, F. A. Cathodic electrochromism of lutetium diphthalocyanine films. *J. Electrochem. Soc.*, **128**, 1981, 1740–3.
30. Pizzarello, F. A. and Nicholson, M. M. Kinetics of colour reversal in lutetium diphthalocyanine oxidation-products formed with different anions. *J. Electrochem. Soc.*, **128**, 1981, 1288–90.
31. Nicholson, M. M. Lanthanide diphthalocyanines – electrochemistry and display applications. *Ind. Eng. Chem., Prod. Res. Develop.*, **21**, 1982, 261–6.
32. Chang, A. T. and Marchon, J. C. Preparation and characterization of oxidized and reduced forms of lutetium diphthalocyanine. *Inorg. Chim. Acta*, **53**, 1981, L241–3.
33. Moskalev, P. N. and Shapkin, G. N. Electrochemical properties of the diphthalocyanines of lanthanides. *Sov. Electrochem.*, **14**, 1978, 486–8.
34. Sammells, A. F. and Pujare, N. U. Solid-state electrochromic cell using lutecium diphthalocyanine. *J. Electrochem. Soc.*, **133**, 1986, 1065–6.
35. Moskalev, P. N., Shapkin, G. N. and Darovskikh, A. N. Preparation and properties of electrochemically oxidised rare-earth element and americium diphthalocyanine. *Russ. J. Inorg. Chem.*, **24**, 1979, 188–92.
36. Green, J. M. and Faulkner, L. R. Reversible oxidation and re-reduction of entire thin-films of transition-metal phthalocyanines. *J. Am. Chem. Soc.*, **105**, 1983, 2950–5.
37. Kohno, Y., Masui, M., Ono, K., Wada, T. and Takeuchi, M. Electrochromic behavior of amorphous copper phthalocyanine thin-films. *Jpn. J. Appl. Phys.*, **31**, 1992, L252–3.
38. Silver, J., Lukes, P., Hey, P. and Ahmet, M. T. Electrochromism in the transition-metal phthalocyanines. 2. Structural-changes in the properties of Cr(Pc) and [Mn(Pc)] films. *J. Mater. Chem.*, **2**, 1992, 841–7.
39. Starke, M., Androsche, I. and Hamann, C. A solid-state electrochromic cell using erbium-diphthalocyanine. *Phys. Status Solidi A*, **120**, 1990, K95–9.
40. Silver, J., Billingham, J. and Barber, D. J. Thin films of zirconium and rare-earth element bis-phthalocyanines: changes in structure caused by gas adsorption/reaction. In Shi, C., Li, H. and Scott, A. (eds.), *The First Pacific Rim International Conference on Advanced Materials and Processing*. Warrendale, PA, The Minerals, Metals and Materials Society, 1992, 521–5.
41. Silver, J., Lukes, P., Houlton, A., Howe, S., Hey, P. and Ahmet, M. T. Electrochromism in the transition-metal phthalocyanines, 3: molecular-organization, reorganization and assembly under the influence of an applied electric-field – response of Fe(Pc) and [Fe(Pc)Cl]. *J. Mater. Chem.*, **2**, 1992, 849–55.
42. Kahl, J. L., Faulkner, L. R., Dwarakanath, K. and Tackikawa, H. Reversible oxidation and re-reduction of magnesium phthalocyanine

electrodes – electrochemical-behavior and in situ Raman spectroscopy. *J. Am. Chem. Soc.*, **108**, 1986, 5438–40.
43. Silver, J., Lukes, P., Hey, P. and Ahmet, M. T. Electrochromism in titanyl and vanadyl phthalocyanine thin-films. *J. Mater. Chem.*, **1**, 1991, 881–8.
44. Corbeau, P., Riou, M. T., Clarisse, C., Bardin, M. and Plichon, V. Spectroelectrochemical properties of uranium diphthalocyanine. *J. Electroanal. Chem.*, **274**, 1989, 107–15.
45. Petty, M., Lovett, D. R., Miller, J. and Silver, J. Electrochemical salt formation in bis(phthalocyaninato)ytterbium(III)-stearic acid Langmuir Blodgett films. *J. Mater. Chem.*, **1**, 1991, 971–6.
46. Lukas, B., Lovett, D. R. and Silver, J. Electrochromism in mixed Langmuir Blodgett films containing rare-earth bisphthalocyanines. *Thin Solid Films*, **210–11**, 1992, 213–15.
47. Muto, J. and Kusayanagi, K. Electrochromic properties of zinc phthalocyanine with solid electrolyte. *Phys. Status Solidi A*, **126**, 1991, K129–32.
48. Silver, J., Lukes, P., Howe, S. D. and Howlin, B. Synthesis, structure, and spectroscopic and electrochromic properties of bis(phthalocyaninato)-zirconium(IV). *J. Mater. Chem.*, **1**, 1991, 29–35.
49. Frampton, C. S., O'Connor, J. M., Peterson, J. and Silver, J. Enhanced colours and properties in the electrochromic behavior of mixed rare-earth-element bisphthalocyanines. *Displays*, **9**, 1988, 174–8.
50. Walton, D., Ely, B. and Elliott, G. Investigations into the electrochromism of lutetium and ytterbium diphthalocyanines. *J. Electrochem. Soc.*, **128**, 1981, 2479–84.
51. Irvine, J. T. S., Eggins, B. R. and Grimshaw, J. The cyclic voltammetry of some sulfonated transition-metal phthalocyanines in dimethylsulfoxide and in water. *J. Electroanal. Chem.*, **271**, 1989, 161–72.
52. Leznoff, C. C., Lam, H., Marcuccio, S. M., Newin, W. A., Janda, P., Kobayashi, N. and Lever, A. B. P. A planar binuclear phthalocyanine and its dicobalt derivatives. *J. Chem. Soc., Chem. Commun.*, 1987, 699–701.
53. Nevin, W. A., Hempstead, M. R., Liu, W., Leznoff, C. C. and Lever, A. B. P. Electrochemistry and spectroelectrochemistry of mononuclear and binuclear cobalt phthalocyanines. *Inorg. Chem.*, **26**, 1987, 570–7.
54. Nevin, W. A., Liu, W., Greenberg, S., Hempstead, M. R., Marcuccio, S. M., Melnik, M., Leznoff, C. C. and Lever, A. B. P. Synthesis, aggregation, electrocatalytic activity, and redox properties of a tetranuclear cobalt phthalocyanine. *Inorg. Chem.*, **26**, 1987, 891–9.
55. Nevin, W. A., Liu, W. and Lever, A. B. P. Dimerization of mononuclear and binuclear cobalt phthalocyanines. *Can. J. Chem.*, **65**, 1987, 855–8.
56. Nevin, W. A., Liu, W., Melnik, M. and Lever, A. B. P. Spectroelectrochemistry of cobalt and iron tetrasulfonated phthalocyanines. *J. Electroanal. Chem.*, **213**, 1986, 217–34.
57. Yamana, M., Kanda, K., Kashiwazaki, N., Yamamoto, M., Nakano, T. and Walton, C. Preparation of plasma-polymerized $YbPc_2$ films and their electrochromic properties. *Jpn. J. Appl. Phys.*, **28**, 1989, L1592–4.
58. Kashiwazaki, N. New complementary electrochromic display utilizing polymeric $YbPc_2$ and Prussian blue films. *Sol. Energy Mater. Sol. Cells*, **25**, 1992, 349–59.
59. Kashiwazaki, N. Iodized polymeric Yb-diphthalocyanine films prepared by plasma polymerization method. *Jpn. J. Appl. Phys.*, **31**, 1992, 1892–6.

60. Moore, D. J. and Guarr, T. F. Electrochromic properties of electrodeposited lutetium diphthalocyanine thin-films. *J. Electroanal. Chem.*, **314**, 1991, 313–21.
61. Li, H. F. and Guarr, T. F. Reversible electrochromism in polymeric metal phthalocyanine thin-films. *J. Electroanal. Chem.*, **297**, 1991, 169–83.
62. Kimura, M., Horai, T., Hanabusa, K. and Shirai, H. Electrochromic polymer derived from oxidized tetrakis(2-hydroxyphenoxy) phthalocyaninatocobalt(II) complex. *Chem. Lett.*, **7**, 1997, 653–4.
63. Besbes, S., Plichon, V., Simon, J. and Vaxiviere, J. Electrochromism of octaalkoxymethyl-substituted lutetium diphthalocyanine. *J. Electroanal. Chem.*, **237**, 1987, 61–8.
64. Jones, R., Krier, A. and Davidson, K. Structure, electrical conductivity and electrochromism in thin films of substituted and unsubstituted lanthanide bisphthalocyanines. *Thin Solid Films*, **298**, 1997, 228–36.
65. Granito, C., Goldenberg, L. M., Bryce, M. R., Monkman, A. P., Troisi, L., Pasimeni, L. and Petty, M. C. Optical and electrochemical properties of metallophthalocyanine derivative Langmuir–Blodgett films. *Langmuir*, **12**, 1996, 472–6.
66. Rodríguez-Méndez, M. L., Souto, J., de Saja, J. A. and Aroca, R. Electrochromic display based on Langmuir Blodgett films of praseodymium bisphthalocyanine. *J. Mater. Chem.*, **5**, 1995, 639–42.
67. Schlettwein, D., Kaneko, M., Yamada, A., Wöhrle, D. and Jaeger, N. I. Light-induced dioxygen reduction at thin-film electrodes of various porphyrins. *J. Phys. Chem.*, **95**, 1991, 1748–55.
68. Yanagi, H. and Toriida, M. Electrochromic oxidation and reduction of cobalt and zinc naphthalocyanine thin-films. *J. Electrochem. Soc.*, **141**, 1994, 64–70.
69. Guyon, F., Pondaven, A. and L'Her, M. Synthesis and characterization of a novel lutetium(III) triple-decker sandwich compound – a tris(1,2-naphthalocyaninato) complex. *J. Chem. Soc., Chem. Commun.*, 1994, 1125–6.
70. Yamada, Y., Kashiwazaki, N., Yamamoto, M. and Nakano, T. Electrochromic effects on polymeric co-pyridinoporphyrazine films prepared by electrochemical polymerisation. *Displays*, **9**, 1988, 190–8.
71. Ng, D. K. P. and Jiang, J. Sandwich-type heteroleptic phthalocyaninato and porphyrinato metal complexes. *Chem. Soc. Rev.*, **26**, 1997, 433–42.
72. Silver, J., Sosa-Sanchez, J. L. and Frampton, C. S. Structure, electrochemistry, and properties of bis(ferrocenecarboxylato)(phthalocyaninato)silicon(IV) and its implications for $(Si(Pc)O)_n$ polymer chemistry. *Inorg. Chem.*, **37**, 1988, 411–17.
73. Dodd, J. W. and Hush, N. S. The negative ions of some porphin and phthalocyanine derivatives, and their electronic spectra. *J. Chem. Soc,.* 1964, 4607–12.
74. Closs, G. L. and Closs, L. E. Negative ions of porphin metal complexes. *J. Am. Chem. Soc.*, **85**, 1963, 818–19.
75. Felton, R. H. and Linschitz, H. Polarographic reduction of porphyrins and electron spin resonance of porphyrin anions. *J. Am. Chem. Soc.*, **88**, 1966, 1112–16.
76. Fajer, J., Borg, D. C., Forman, A., Dolphin, D. and Felton, R. H. π-Cation radicals and dications of metalloporphyrins. *J. Am. Chem. Soc.*, **92**, 1970, 3451–9.
77. Felton, R. H., Dolphin, D., Borg, D. C. and Fajer, J. Cations and cation radicals of porphyrins and ethyl chlorophyllide. *J. Am. Chem. Soc.*, **91**, 1969, 196–8.

78. Aziz, A., Narasimhan, K. L., Periasamy, N. and Maiti, N. C. Electrical and optical properties of porphyrin monomer and its J-aggregate. *Philos. Mag. B*, **79**, 1999, 993–1004.
79. Mortimer, R. J., Rowley, N. M. and Vickers, S. J. Unpublished work.
80. Bonnett, R. Metal complexes for photodynamic therapy. In *Comprehensive Coordination Chemistry II: From Biology to Nanotechnology*, McCleverty, J. A. and Meyer, T. J. (eds.), Oxford, Elsevier, 2003, vol. 9, pp. 945–1003.
81. Nazeeruddin, M. K. and Grätzel, M. Conversion and storage of solar energy using dye-sensitized nanocrystalline TiO_2 cells. In *Comprehensive Coordination Chemistry II: From Biology to Nanotechnology*, McCleverty, J. A. and Meyer, T. J. (eds.), Oxford, Elsevier, 2003, vol. 9, pp. 719–58.
82. Fabian, J., Nakazumi, H. and Matsuoka, M. Near-infrared absorbing dyes. *Chem. Rev.*, **92**, 1992, 1197–226.
83. Emmelius, M., Pawlowski, G. and Vollmann, H. W. Materials for optical data storage. *Angew. Chem. Int. Ed. Engl.*, **28**, 1989, 1445–71.
84. Fabian, J. and Zahradnik, R. The search for highly-coloured organic compounds. *Angew. Chem. Int. Ed. Engl.*, **28**, 1989, 677–94.
85. Franke, E. B., Trimble, C. L., Hale, J. S., Schubert, M. and Woollam, J. A. Infrared switching electrochromic devices based on tungsten oxide. *J. Appl. Phys.*, **88**, 2000, 5777–84.
86. Schwendeman, I., Hwang, J., Welsh, D. M., Tanner, D. B. and Reynolds, J. R. Combined visible and infrared electrochromism using dual polymer devices. *Adv. Mater.*, **13**, 2001, 634–7.
87. McCleverty, J. A. Metal 1,2-dithiolene and related complexes. *Prog. Inorg. Chem.*, **10**, 1968, 49–221.
88. Mueller-Westerhoff, U. T. and Vance, B. Dithiolenes and related species. In *Comprehensive Coordination Chemistry*, Wilkinson, G., Gillard, R. D. and McCleverty, J. A. (eds.), Oxford, Pergamon, 1987; vol. 2, pp. 595–631.
89. Mueller-Westerhoff, U. T., Vance, B. and Yoon, D. I. The synthesis of dithiolene dyes with strong near-IR absorption. *Tetrahedron*, **47**, 1991, 909–32.
90. Bigoli, F., Deplano, P., Devillanova, F. A., Lippolis, V., Lukes, P. J., Mercuri, M. L., Pellinghelli, M. A. and Trogu, E. F. New neutral nickel dithiolene complexes derived from 1,3-dialkylimidazolidine-2,4,5-trithione, showing remarkable near-IR absorption. *J. Chem. Soc., Chem. Commun.*, 1995, 371–2.
91. Bigoli, F., Deplano, P., Mercuri, M. L., Pellinghelli, M. A., Pintus, G., Trogu, E. F., Zonneda, G., Wong, H. H. and Williams, J. M. Novel oxidation and reduction products of the neutral nickel-dithiolene [Ni(^{i}Pr$_2$ timdt)$_2$] (^{i}Pr$_2$timdt is the monoanion of 1,3-diisopropylimidazolidine-2,4,5-trithione). *Inorg. Chim. Acta*, **273**, 1998, 175–83.
92. Bigoli, F., Deplano, P., Devillanova, F. A., Ferraro, J. R., Lippolis, V., Lukes, P. J., Mercuri, M. L., Pellinghelli, M. A., Trogu, E. F. and Williams, J. M. Syntheses, X-ray crystal structures, and spectroscopic properties of new nickel dithiolenes and related compounds. *Inorg. Chem.*, **36**, 1997, 1218–26.
93. Arca, M., Demartin, F., Devillanova, F. A., Garau, A., Isaia, F., Lelj, F., Lippolis, V., Pedraglio. S. and Verani, G. Synthesis, X-ray crystal structure and spectroscopic characterisation of the new dithiolene [Pd(Et$_2$timdt)$_2$] and of its adduct with molecular diiodine [Pd(Et$_2$timdt) $_2$]·I$_2$·CHCl$_3$ (Et$_2$ timdt = monoanion of 1,3-diethylimidazolidine-2,4,5-trithione. *J. Chem. Soc., Dalton Trans.*, 1998, 3731–6.
94. Aragoni, M. C., Arca, M., Demartin, F., Devillanova, F. A., Geran, A., Isaia, F., Lelj, F., Lippolis, V. and Verani, G. New [M(R,R'timdt)$_2$] metal-dithiolenes and

related compounds (M = Ni, Pd, Pt; R,R'timdt = monoanion of disubstituted imidazolidine-2,4,5-trithiones): an experimental and theoretical investigation. *J. Am. Chem. Soc.*, **121**, 1999, 7098–107.

95. Bigoli, F., Cassoux, P., Deplano, P., Mercuri, M. L., Pellinghelli, M. A., Pintus, G., Serpe, A. and Trogu, E. F. Synthesis, structure and properties of new unsymmetrical nickel dithiolene complexes useful as near-infrared dyes. *J. Chem. Soc., Dalton Trans.*, 2000, 4639–44.

96. Deplano, P., Mercuri, M. L., Pintus, G. and Trogu, E. F. New symmetrical and unsymmetrical nickel-dithiolene complexes useful as near-IR dyes and precursors of sulfur-rich donors. *Comments Inorg. Chem.*, **22**, 2001, 353–74.

97. Laye, R. H., Couchman, S. M. and Ward, M. D. Comparison of metal–metal electronic interactions in an isomeric pair of dinuclear ruthenium complexes with different bridging pathways: effective hole-transfer through a bis-phenolate bridge. *Inorg. Chem.*, **40**, 2001, 4089–92.

98. Kasack, V., Kaim, W., Binder, H., Jordanov, J. and Roth, E. When is an odd-electron dinuclear complex a mixed-valent species? – tuning of ligand-to-metal spin shifts in diruthenium(III, II) complexes of noninnocent bridging ligands O=C(R)N–NC(R)=O. *Inorg. Chem.*, **34**, 1995, 1924–33.

99. Rocha, R. C. and Toma, H. E. Intervalence transfer in a new benzotriazolate bridged ruthenium–iron complex. *Can. J. Chem.*, **79**, 2001, 145–56.

100. Mosher, P. J., Yap, G. P. A. and Crutchley, R. J. A donor-acceptor bridging ligand in a class III mixed-valence complex. *Inorg. Chem.*, **40**, 2001, 1189–95.

101. Qi, Y., Desjardins, P. and Wang, Z. Y. Novel near-infrared active dinuclear ruthenium complex materials: effects of substituents on optical attenuation. *J. Opt. A: Pure Appl. Opt.*, **4**, 2002, S273–7.

102. Lee, S.-M., Marcaccio, M., McCleverty, J. A. and Ward, M. D. Dinuclear complexes containing ferrocenyl and oxomolybdenum(V) groups linked by conjugated bridges: a new class of electrochromic near-infrared dye. *Chem. Mater.*, **10**, 1998, 3272–4.

103. Harden, N. C., Humphrey, E. R., Jeffrey, J. C., Lee, S.-M., Marcaccio, M., McCleverty, J. A., Rees, L. H. and Ward, M. D. Dinuclear oxomolybdenum(V) complexes which show strong electrochemical interactions across bis-phenolate bridging ligands: a combined spectroelectrochemical and computational study. *J. Chem. Soc., Dalton Trans.*, 1999, 2417–26.

104. Bayly, S. R., Humphrey, E. R., de Chair, H., Paredes, C. G., Bell, Z. R., Jeffrey, J. C., McCleverty, J. A., Ward, M. D., Totti, F., Gatteschi, D., Courric, S., Steele, B. R. and Screttas, C. G. Electronic and magnetic metal–metal interactions in dinuclear oxomolybdenum(V) complexes across bis-phenolate bridging ligands with different spacers between the phenolate termini: ligand-centred *vs.* metal-centred redox activity. *J. Chem. Soc., Dalton Trans.*, 2001, 1401–14.

105. McDonagh, A. M., Ward, M. D. and McCleverty, J. A. Redox and UV/VIS/NIR spectroscopic properties of tris(pyrazolyl)borato-oxo-molybdenum(V) complexes with naphtholate and related co-ligands. *New J. Chem.*, **25**, 2001, 1236–43.

106. McDonagh, A. M., Bayly, S. R., Riley, D. J., Ward, M. D., McCleverty, J. A., Cowin, M. A., Morgan, C. N., Verrazza, R., Penty, R. V. and White, I. H. A variable optical attenuator operating in the near-infrared region based on an electrochromic molybdenum complex. *Chem. Mater.*, **12**, 2000, 2523–4.

107. Kowallick, R., Jones, A. N., Reeves, Z. R., Jeffrey, J. C., McCleverty, J. A. and Ward, M. D. Spectroelectrochemical studies and molecular orbital calculations on mononuclear complexes [Mo(Tp^{Me,Me})(NO)Cl(py)] (where py is a substituted

pyridine derivative): electrochromism in the near-infrared region of the electronic spectrum. *New J. Chem.*, **23**, 1999, 915–21.
108. Haga, M., Dodsworth, E. S. and Lever, A. B. P. Catechol–quinone redox series involving bis(bipyridine)ruthenium(II) and tetrakis(pyridine)ruthenium(II). *Inorg. Chem.*, **25**, 1986, 447–53.
109. Joulié, L. F., Schatz, E., Ward, M. D., Weber, F. and Yellowlees, L. J. Electrochemical control of bridging ligand conformation in a binuclear complex – a possible basis for a molecular switch. *J. Chem. Soc., Dalton Trans.*, 1994, 799–804.
110. Barthram, A. M., Cleary, R. L., Kowallick, R. and Ward, M. D. A new redox-tunable near-IR dye based on a trinuclear ruthenium(II) complex of hexahydroxytriphenylene. *Chem. Commun.*, 1998, 2695–6.
111. Barthram, A. M. and Ward, M. D. Synthesis, electrochemistry, UV/VIS/NIR spectroelectrochemistry and ZINDO calculations of a dinuclear ruthenium complex of the tetraoxolene bridging ligand 9-phenyl-2,3,7-trihydroxy-6-fluorone. *New J. Chem.*, **24**, 2000, 501–4.
112. Barthram, A. M., Cleary, R. L., Jeffery, J. C., Couchman, S. M. and Ward, M. D. Effects of ligand topology on the properties of dinuclear ruthenium complexes of bis-semiquinone bridging ligands. *Inorg. Chim. Acta*, **267**, 1998, 1–5.
113. Barthram, A. M., Reeves, Z. R., Jeffrey, J. C. and Ward, M. D. Polynuclear osmium–dioxolene complexes: comparison of electrochemical and spectroelectrochemical properties with those of their ruthenium analogues. *J. Chem. Soc., Dalton Trans.*, 2000, 3162–9.
114. García-Cañadas, J., Meacham, A. P., Peter, L. M. and Ward, M. D. Electrochromic switching in the visible and near IR with a Ru–dioxolene complex adsorbed on to a nanocrystalline SnO_2 electrode. *Electrochem. Commun.*, **5**, 2003, 416–20.

8

Electrochromism by intervalence charge-transfer coloration: metal hexacyanometallates

8.1 Prussian blue systems: history and bulk properties

Prussian blue – PB; ferric ferrocyanide, or iron(III) hexacyanoferrate(II) – first made by Diesbach in Berlin in 1704,[1] is extensively used as a pigment in the formulation of paints, lacquers and printing inks.[2,3] Since the first report[4] in 1978 of the electrochemistry of PB films, numerous studies concerning the electrochemistry of PB and related analogues have been made,[5,6,7] with, in addition to electrochromism, proposed applications in electroanalysis and electrocatalysis.[8,9,10,11] Fundamental studies[12,13,14] on basic PB properties (electronic structure, spectra and conductimetry) underlie the elaborations that follow.

Prussian blue is the prototype of numerous polynuclear transition-metal hexacyanometallates, which form an important class of insoluble mixed-valence compounds.[15,16,17] They have the general formula $M'_k[M''(CN)_6]_l$ (k, l integral) where M' and M'' are transition metals with different formal oxidation numbers. These materials can contain ions of other metals and varying amounts of water. In PB the two transition metals in the formula are the two common oxidation states of iron, Fe^{III} and Fe^{II}. Prussian blue is readily prepared by mixing aqueous solutions of a hexacyanoferrate(III) salt with iron(II), the preferred industrial-production route (rather than iron(III) with a hexacyanoferrate(II) salt). In the PB chromophore, the distribution of oxidation states is Fe^{III}–Fe^{II} respectively; i.e. it contains Fe^{3+} and $[Fe^{II}(CN)_6]^{4-}$, as established by the CN stretching frequency in the IR spectrum and confirmed by Mössbauer spectroscopy.[18] The chromophore alone thus has a negative charge, therefore in the solid a counter cation is to be incorporated. The Fe^{III} is usually high spin with H_2O coordinated, whereas the Fe^{II} is low spin. While the precise composition of any PB solid is extraordinarily preparation-sensitive, the major classification of extreme cases delineates 'insoluble' PB (abbreviated to i-PB) which is $Fe^{3+}[Fe^{3+}\{Fe^{II}(CN)_6\}^{4-}]_3$, and 'soluble' PB ($s$-PB), in full $K^+Fe^{3+}[Fe^{II}(CN)_6]^{4-}$, i.e. dependent on the counter cation. All forms of PB are in fact highly insoluble in water ($K_{sp} \sim 10^{-40}$),[19] the 'solubility' attributed to the latter form being an illusion caused by its easy dispersion as colloidal particles, forming a blue sol in water that looks like a true solution.

The $Fe^{3+}[Fe^{II}(CN)_6]^{4-}$ chromophore falls into Group II of the Robin–Day mixed-valence classification, the blue IVCT band on analysis of the intensity indicating ~1% delocalisation of the transferable electron in the ground state (i.e., before any optical CT).[20] X-Ray powder diffraction patterns for s-PB indicate a face-centred cubic lattice, with the high-spin Fe^{III} and low-spin Fe^{II} ions coordinated octahedrally by the N or C of the cyanide ligands, with K^+ ions occupying interstitial sites.[21] In i-PB, Mössbauer spectroscopy confirms the interstitial ions to be the Fe^{3+} counter cation.[18] Single-crystal X-ray diffraction patterns of i-PB indicate however a primitive cubic lattice, where one quarter of the Fe^{II} sites are vacant.[22] This proposed structure contains no interstitial ions, with one quarter of the Fe^{III} centres being coordinated by six N-bound cyanide ligands, the remainder by four N-bound cyanides, and every Fe^{II} centre surrounded by six C-bound cyanides ligands. The Fe^{II} vacancies are randomly distributed, and occupied by water molecules, which complete the octahedral coordination about Fe^{III}. The widespread assumption of Ludi et al.'s model[22] for i-PB is highly questionable[23] in view of the substantial differences between the (very slowly grown) single crystals[22] and the more usual polycrystalline forms arising from relatively rapid growth, as in the electrodeposition for electrochromic use. Other (bivalent) counter cations also appear to be interstitial.[24]

8.2 Preparation of Prussian blue thin films

Prussian blue thin films are generally prepared by the original method based on electrochemical deposition,[4] although electroless deposition,[25] sacrificial-anode (SA) methods,[26,27] the extensive redox cycling of hexacyanoferrate(II)-containing solutions,[28] the embedding of micrometre-sized crystals directly into electrode surfaces using powder abrasion,[29] and a method using catalytic silver paint[30,31] have all been described. Thus PB films can be electrochemically deposited onto a variety of inert electrode substrates by electroreduction of solutions containing iron(III) and hexacyanoferrate(III) ions as the adduct $Fe^{3+}[Fe^{III}(CN)_6]^{3-}$, Eq. (8.1). Prussian blue electrodeposition has been studied by numerous techniques. Voltammetry[32,33,34] and galvanostatic studies[35] have indicated that reduction of iron(III) hexacyanoferrate(III) is the principal electron-transfer process in PB electrodeposition. This brown–yellow soluble complex dominates in solutions containing iron(III) and hexacyanoferrate(III) ions as a result of the equilibrium in Eq. (8.1):

$$Fe^{3+} + \left[Fe^{III}(CN)_6\right]^{3-} = \left[Fe^{III}Fe^{III}(CN)_6\right]^{0}. \qquad (8.1)$$

Chronoabsorptiometric studies[36] for galvanostatic PB electrodeposition onto ITO electrodes have shown that the absorbance due to the IVCT band of the growing PB film is proportional to the charge passed. Electrochemical quartz-crystal microbalance (EQCM) measurements for potentiostatic PB electrodeposition onto gold have revealed that the mass gain per unit area is proportional to the charge passed.[37] Ellipsometric measurements for potentiostatic PB electrodeposition onto platinum indicated that the level of hydration was around 34 H_2O per PB unit cell.[38] Hydration is in fact variable and, for bulk PB taken out of solution, depends on ambient humidity.[39]

Changes in the ellipsometric parameters during PB electrodeposition revealed initial growth of a single homogeneous film for the first 80 seconds, followed by growth of a second, outer, more porous film on top of the relatively compact inner film.[38] Chronoamperometric measurements (over a scale of several seconds) supported by scanning electron microscopy (SEM) for the electrodeposition of PB onto ITO and platinum by electroreduction from solutions of iron(III) hexacyanoferrate(III) have been performed.[40] In earlier preparations in the 'zeroth' step the deposition electrode was first made *positive* during addition of solutions in order to preclude spontaneous or uncontrolled deposits of PB, but this was later shown to cause initial deposition of the solid $Fe^{III}\,Fe^{III}$ complex, which, when the electrode was made cathodic, persisted briefly before being incorporated into the growing PB.[41] A solubility of the $Fe^{III}\,Fe^{III}$ complex was estimated[41] as ca. 10^{-3} mol dm^{-3}.

Variation of electrode potential, supporting electrolyte and concentrations of electroactive species have established a subsequent three-stage electrodeposition mechanism. In the early growth phase[40] the surface becomes uniformly covered as small PB nuclei form and grow on electrode substrate sites. In the second growth phase there is an increase in rate towards maximal roughness, as the electroactive area increases by formation and three-dimensional growth of PB nuclei attached to the PB interface formed in the initial stage. In the final growth phase, diffusion of locally depleted electroactive species to the now three-dimensional PB interface plays an increasingly dominant role and limits electron transfer, resulting in a fall in growth rate. (If through-film electron transfer to the film–electrolyte interface wanes with growth, the seeping in of reactant solution between the PB film and electrode substrate for later growth phases is not precluded.)

More recently, a new method of assembling multilayers of PB on surfaces has been described.[42,43] In contrast to the familiar process of self-assembly, which is spontaneous and can lead to single monolayers, 'directed assembly' is driven by the experimenter and leads to extended multilayers. In a proof-of-concept experiment, the generation of multilayers of Prussian blue (and the mixed Fe^{III}–Ru^{II} analogue 'Ruthenium purple') on gold surfaces, by exposing them alternately to positively charged iron(III) cations and $[Fe^{II}(CN)_6]^{4-}$ or $[Ru^{II}(CN)_6]^{4-}$ anions, has been demonstrated.[42] Tieke and co-workers[43,44,45] have investigated the optical, electrochemical, structural and morphological properties of such multilayer systems, and have also demonstrated their application as ion-sieving membranes. They take care to note that 'because metal hexacyanoferrate salts are known to organise in a cubic crystal lattice structure, a normal layering of metal cations and hexacyanoferrate anions is highly unlikely'. They avoid the term 'layer-by-layer' deposition and instead use 'multiple sequential deposition'.

8.3 Electrochemistry, *in situ* spectroscopy and characterisation of Prussian blue thin films

Electrodeposited PB films may be partially oxidised[32,33,34] to Prussian green (PG), a species historically also known as Berlin green and assigned the fractional composition shown, Eq. (8.2):

$$\left[Fe^{III}Fe^{II}(CN)_6\right]^- \rightarrow Fe^{III}\left[\{Fe^{III}(CN)_6\}_{2/3}\{Fe^{II}(CN)_6\}_{1/3}\right]^{1/3-} + 2/3\,e^-. \quad (8.2)$$
$$\text{PB} \qquad\qquad\qquad\qquad \text{PG}$$

The fractions $2/3$ and $1/3$ are illustrative rather than precise. Thus, although in bulk form PG is believed to have a fixed composition with the anion composition shown above, it has been inferred (but with reservations, below) that there is a continuous composition range in thin films from PB, *via* the partially oxidised PG form, to the fully oxidised all-Fe^{III} form Prussian brown (PX).[34] Prussian brown appears brown as a bulk solid, brown–yellow in solution, and golden yellow as a particularly pure form that is prepared on electro-oxidation of thin-film PB – Eq. (8.3):[33,34]

$$\left[Fe^{III}Fe^{II}(CN)_6\right]^- \rightarrow \left[Fe^{III}Fe^{III}(CN)_6\right]^0 + e^-. \quad (8.3)$$
$$\text{PB} \qquad\qquad\qquad \text{PX}$$

Redox in the other direction, that is, reduction of PB, yields Prussian white (PW), also known as Everitt's salt, which appears colourless as a thin film – Eq. (8.4).

$$\left[Fe^{III}Fe^{II}(CN)_6\right]^- + e^- \rightarrow \left[Fe^{II}Fe^{II}(CN)_6\right]^{2-}. \tag{8.4}$$

$$\text{PB} \hspace{4cm} \text{PW}$$

Figure 8.1 shows a cyclic voltammogram of the PB–PW transition.

For all redox reactions above there is concomitant counter-ion movement into or out of the films to maintain overall electroneutrality. The electron transfer occurs at the electrode-substrate–film interface, while counter-ion egress or ingress occurs at the film–electrolyte interface; it is not established which through-film transport, that of electron or ion, determines the rate of coloration.

Whilst s-PB, i-PB, PG and PW are all insoluble in water, PX is slightly soluble in its pure (golden-yellow) form (indeed the electrodeposition technique depends on the solubility of the $[Fe^{III}Fe^{III}(CN)_6]^0$ complex). This implies a positive potential limit of about +0.9 V for a high write-erase efficiency in

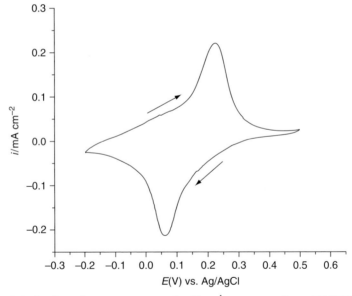

Figure 8.1 Cyclic voltammogram at 5 mV s^{-1} scan rate for a PB|ITO|glass electrode in aqueous KCl supporting electrolyte (0.2 mol dm^{-3}), showing the voltammetric wave for the PB–PW redox switch. The initial potential was +0.50 V vs. Ag|AgCl. The arrows indicate the direction of potential scan. (Figure reproduced from Mortimer, R. J. and Reynolds, J. R. 'In situ colorimetric and composite coloration efficiency measurements for electrochromic Prussian blue'. J. Mater. Chem., 15, 2005, 2226–33, with permission from The Royal Society of Chemistry.)

contact with water. Although practical electrochromic devices based on PB have primarily exploited the PB–PW transition, this does not rule out the prospect of four-colour PB electropolychromic ECDs, as other solvent systems might not dissolve PX. The spectra of the yellow, green, blue and clear ('white') forms of PB and its redox variants are shown in Figure 8.2, together with spectra of possibly two intermediate states between the blue and the yellow forms.

The yellow absorption band corresponds with that of $[Fe^{III}Fe^{III}(CN)_6]^0$ in solution, both maxima being at 425 nm and coinciding with the (weaker) $[Fe^{III}(CN)_6]^{3-}$ absorption maximum. On increase from +0.50 V to more oxidising potentials, the original 690 nm PB peak continuously shifts to longer wavelengths with diminishing absorption, while the peak at 425 nm steadily increases, owing to the increasing $[Fe^{III}Fe^{III}(CN)_6]^0$ absorption. The reduction of PB to PW is by contrast abrupt, with transformation to all PW or all PB without pause, depending on the applied potential. One broad voltammetric peak usually seen for PB → PX, in contrast with the sharply peaked PB → PW transition, apparently indicates a range of compositions to be involved. The contrast (broad vs. sharp) behaviour, supported by ellipsometric measurements,[38] could imply continuous mixed-valence compositions over the blue-

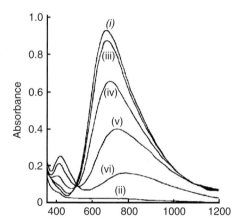

Figure 8.2 Spectra of iron hexacyanoferrate films on ITO-coated glass at various potentials [(i) + 0.50 (PB, blue), (ii) − 0.20 (PW, transparent), (iii) + 0.80 (PG, green), (iv) + 0.85 (PG, green), (v) + 0.90 (PG, green) and (vi) + 1.20 V (PX, yellow) (potentials vs. SCE)] with KCl 0.2 mol dm^{-3} + HCl 0.01 mol dm^{-3} as supporting electrolyte. Wavelengths (abscissa) are in nm. (Figure reproduced from Mortimer, R. J. and Rosseinsky, D. R. 'Iron hexacyanoferrate films: spectroelectrochemical distinction and electrodeposition sequence of 'soluble' (K$^+$-containing) and 'insoluble' (K$^+$-free) Prussian blue and composition changes in polyelectrochromic switching'. *J. Chem. Soc., Dalton Trans.*, 1984, 2059–61, by permission of the Royal Society of Chemistry.)

to-yellow range in contrast with the (presumably immiscible) PB and PW, which clearly transform, one into the other, without intermediacy of composition. However, two-peak PB → PX voltammetry pointing to a specific intermediate composition has also been seen, first attributed to the absence of traces of Cl⁻ from those samples,[41] but later also observed in slow voltammetry on PB from KCl-containing preparations.[46] Thus the intermediate green colour observed in PB–PX voltammetry could be a true compound PG rather than either a continuously changing mixed-valence phenomenon, a PB + PX series of solid solutions, or varying PB + PX physical mixtures of microcrystals.

The identity as s-PB or i-PB of the initially electrodeposited PB has been debated in the literature.[34,47,48,49,50,51] Based on changes that take place in the IVCT band on redox cycling, it has been postulated that i-PB is first formed, followed by a transformation to s-PB on potential cycling.[34] Further evidence for this is provided by the difference in the voltammetric response for the PB–PW transition between the first cycle and all succeeding cycles, suggesting structural reorganisation of the film during the first cycle.[33] On soaking s-PB films in saturated FeCl₃ solutions partial reversion of the absorbance maximum and broadening of the spectrum, approaching the values observed for i-PB, is found.[34] Itaya and Uchida,[47] however, claimed that the film is always i-PB. Their argument is based on the ratio of charge passed on oxidation to PX to that passed on reduction to PW, which was 0.708 rather than 1.00; that is, Eqs. (8.5) and (8.6) are applicable:

$$Fe^{3+}\left[Fe^{III}Fe^{II}(CN)_6\right]_3 + 4e^- + 4K^+ \rightarrow K_4Fe^{2+}\left[Fe^{II}Fe^{II}(CN)_6\right]_3; \quad (8.5)$$

i-PB PW

$$Fe^{3+}\left[Fe^{III}Fe^{II}(CN)_6\right]_3 - 3e^- + 3X^- \rightarrow Fe^{3+}\left[Fe^{III}Fe^{III}(CN)_6\right]_3 X_3. \quad (8.6)$$

i-PB PX

In refutation, Emrich et al.[48] using X-ray photoelectron spectroscopy (XPS) data, and Lundgren and Murray[49] using cyclic voltammetry (CV), energy dispersive analysis of X-rays (EDAX), XPS, elemental analytical and spectroelectrochemical measurements, both confirmed i-PB as the initially-deposited form with a 'gradual' transformation to s-PB on potential cycling. Later work[41] established a major (approximately one-third) conversion in the first cycle, but thereafter a much slower introduction of K⁺. Other support for the i-PB to s-PB transformation comes from an ellipsometric study by Beckstead et al.[50] who found that the PB film, after the first and subsequent cycles for the PB–PW transition, developed optical properties that differed from the original PB film. Results from *in situ* Fourier-transform infrared spectroscopy also demonstrated an *i*-PB to *s*-PB transformation on repeated reductive cycling.[51] The EQCM mass-change measurements on voltammetrically scanned PB films reinforce the theory of lattice reorganisation during the initial film reduction.[37]

Only about one third of the three K^+ ions, expected to replace the counter-cationic Fe^{3+}, are found to be incorporated in the first substitutive voltammetric cycle. It has been suggested that this follows from reduction in PB → PW of the counter-cationic Fe^{3+} to Fe^{2+} which is retained on re-oxidation of the PW to the PB, so requiring $Fe^{2+} K^+$ as counter cations; the now somewhat dispersed counter-cation population does not subsequently drive K^+ incorporation as strongly as happens with solely Fe^{3+} as (charge-concentrated) counter cation.[41] Lattice-energy calculations support most of this argument.[52]

In detail, a further EQCM study[53] shows mass changes, following one PB → PW → PB cycle of KCl-prepared PB in different M^+Cl^- solutions, in the sequence of counter cations $Na^+ > K^+ < Rb^+ < Cs^+$. This sequence correlates with the wavelengths of the maximum in each case of the PB absorption in the region of 700 nm. Together with related observations on PB samples that contained sundry M^{2+} counter cations, varied M^+ or M^{2+} lattice interactions with the chromophore were concluded to affect the optical absorptions commensurately.[53,54]

Whilst PB film stability is frequently discussed in the preceding papers, Stilwell *et al.*[55] have studied in detail the factors that influence the cycle stability of PB films. They found that electrolyte pH was the overwhelming factor in film stability; cycle numbers in excess of 100 000 were easily achieved in solutions of pH 2–3, though other conclusions regarding stabilisation by pH have since been reached.[41] Concurrently with this increase in stability at lower pH was a considerable increase in switching rate. Furthermore, films grown from chloride-containing solutions were said to be slightly more stable, in terms of cycle life, compared to those grown from chloride-free solutions,[55] but again with contrary conclusions.[41]

8.4 Prussian blue electrochromic devices

Early PB-based ECDs employed PB as the sole electrochromic material. Examples include a seven-segment display using PB-modified SnO_2 working and counter electrodes at 1 mm separation,[56] and an ITO | PB–Nafion® | ITO solid-state device.[57,58] For the solid-state system, device fabrication involved chemical (rather than electrochemical) formation of the PB, by immersion of a membrane of the solid polymer electrolyte Nafion® (a sulfonated poly(tetrafluoroethane)polymer) in aqueous solutions of $FeCl_2$, then $K_3Fe(CN)_6$. The resulting PB-containing Nafion® composite film was sandwiched between the two ITO plates. The construction and optical behaviour of an ECD utilising a single film of PB, without addition of a conventional electrolyte, has also been described.[59] In this design, a film of PB is sandwiched between two optically transparent electrodes (OTEs). Upon application of an appropriate potential across the film, oxidation occurs near the positive electrode and reduction near the negative electrode to yield PX and PW respectively. The conversion of the outer portions of the film results in a net half-bleaching of the device. The functioning of the device relies on the fact that PB can be bleached both anodically – to the yellow state, Eq. (8.3) – and cathodically – to a transparent state, Eq. (8.4) – and that it is a mixed conductor through which potassium cations can move to provide charge compensation required for the electrochromic redox reactions. However, at the conjunction of the (II)(II) state with the (III)(III) state, their comproportionation reaction results in half the material remaining in the device centre as the (III)(II) form, PB.

Since PB and WO_3 (see Chapter 6) are respectively anodically and cathodically colouring electrochromic materials, they can be used together in a single device[60,61,62,63,64] so that their electrochromic reactions are complementary, Eqs. (8.7) and (8.8):

$$[Fe^{III}Fe^{II}(CN)_6]^- + e^- \rightarrow [Fe^{II}Fe^{II}(CN)_6]^{2-}; \qquad (8.7)$$

blue   transparent

$$WO_3 + x(M^+ + e^-) \rightarrow M_xW^{VI}_{(1-x)}W^V_xO_3. \qquad (8.8)$$

transparent   blue

In an example of the construction of such a device, thin films of these materials are deposited on OTEs that are separated by a layer of a transparent ionic conductor such as KCF_3SO_3 in poly(ethylene oxide).[64] The films can be coloured simultaneously (giving deep blue) when a sufficient voltage is applied between them such that the WO_3 electrode is the cathode and the PB electrode the anode. On appropriate switching, the coloured films can be bleached to transparency when the polarity is reversed, returning the ECD to a transparent state.

Numerous workers[65,66,67,68,69,70,71] have combined PB with the conducting polymer poly(aniline) in complementary ECDs that exhibit deep blue-to-light green electrochromism. Electrochromic compatibility is obtained by combining the coloured oxidised state of the polymer (see Chapter 10) with the blue PB, and the bleached reduced state of the polymer with PG, Eq. (8.9):

$$\text{Oxidised poly(aniline)} + \text{PB} \rightarrow \text{Emeraldine poly(aniline)} + \text{PG}. \qquad (8.9)$$

$$\text{coloured} \qquad\qquad\qquad \text{bleached}$$

Jelle and Hagen[68,69,71,72] have developed an electrochromic window for solar modulation using PB, poly(aniline) and WO_3. They took advantage of the symbiotic relationship between poly(aniline) and PB, and incorporated PB together with poly(aniline), and WO_3, in a complete solid-state electrochromic window. The total device comprised Glass | ITO | poly(aniline) | PB | poly(AMPS) | WO_3 | ITO| Glass. Compared with their earlier results with a poly(aniline)–WO_3 window, Jelle and Hagen were able to block off much more of the light by inclusion of PB within the poly(aniline) matrix, while still regaining about the same transparency during the bleaching of the window.

As noted in Chapter 10, a new complementary ECD has recently been described,[73] based on the assembly of the cathodically colouring conducting polymer, poly[3,4-(ethylenedioxy)thiophene] – PEDOT – on ITO glass and PB on ITO glass substrates with a poly(methyl methacrylate) – PMMA-based gel polymer electrolyte. The colour states of the PEDOT (blue-to-colourless) and PB (colourless-to-blue) films fulfil the requirement of complementarity. The ECD exhibited deep blue–violet at −2.1 V and light blue at 0.6 V.

Kashiwazaki[74] has fabricated a complementary ECD using plasma-polymerised ytterbium bis(phthalocyanine) (pp–Yb(Pc)$_2$) and PB films on ITO with an aqueous solution of KCl (4 mol dm^{-3}) as electrolyte. Blue-to-green electrochromicity was achieved in a two-electrode cell by complementing the green-to-blue colour transition (on reduction) of the pp–Yb(Pc)$_2$ film with the blue (PB)-to-colourless (PW) transition (oxidation) of the PB. A three-colour display (blue, green and red) was fabricated in a three-electrode cell in which a third electrode (ITO) was electrically connected to the PB electrode. A reduction reaction at the third electrode, as an additional counter electrode, provides adequate oxidation of the pp–Yb(Pc)$_2$ electrode, resulting in the red colouration of the pp–Yb(Pc)$_2$ film.

8.5 Prussian blue analogues

Prussian blue analogues, comprising other polynuclear transition-metal hexacyanometallates,[12,13,14] which have been prepared and investigated as thin films, are surveyed in this section. The majority are expected to be electrochromic, although this property has only been studied in any depth in a few cases. The field therefore appears to be open for further investigation and exploitation, although it is to be noted that from the qualitative description of colour states, contrast ratios are likely to be low.

8.5.1 Ruthenium purple

Bulk ruthenium purple – RP; ferric ruthenocyanide, iron(III) hexacyanoruthenate(II) – is synthesised via precipitation from solutions of the appropriate iron and hexacyanoruthenate salts. The visible absorption spectrum of a colloidal suspension of bulk synthesised RP with potassium as counter cation confirms the Fe^{3+} $[Ru^{II}(CN)_6]^{4-}$ combination as the chromophore.[12] The X-ray powder pattern with iron(III) as counter cation gives a lattice constant of 10.42 Å as compared to 10.19 Å for the PB analogue.[12] However, although no single-crystal studies have been made, RP could have a disordered structure similar to that reported for single-crystal PB.[13] The potassium and ammonium salts give cubic powder patterns similar to their PB analogues.[14]

Ruthenium purple films have been prepared by electroreduction of the soluble iron(III) hexacyanoruthenate(III) complex potentiostatically, galvanostatically or by using a copper wire as sacrificial anode.[75,76] The visible absorption spectrum of RP prepared in the presence of excess of potassium ion showed a broad CT band, as for bulk synthesised RP, with a maximum at approximately 550 nm.[75] Ruthenium purple films can be reversibly reduced to the colourless iron(II) hexacyanoruthenate(II) form, but no partial electrooxidation to the Prussian green analogue is observed. The large background oxidation current observed in chloride-containing electrolyte suggests electrocatalytic activity of RP for either oxygen or chlorine evolution.[76]

8.5.2 Vanadium hexacyanoferrate

Vanadium hexacyanoferrate (VHCF) films have been prepared on Pt or fluorine-doped tin oxide (FTO) electrodes by potential cycling from a solution containing Na_3VO_4 and $K_3Fe(CN)_6$ in H_2SO_4 (3.6 mol dm^{-3}).[77,78] Carpenter et al.,[77] by correlation with CVs for solutions containing only one of the individual electroactive ions, have proposed that electrodeposition involves the reduction of the dioxovanadium ion VO_2^+ (the stable form of vanadium(V) in these acidic conditions), followed by precipitation with hexacyanoferrate(III) ion. While the reduction of the hexacyanoferrate(III) ion in solution probably also occurs when the electrode is swept to more negative potentials, this reduction does not appear to be critical to film formation, since VHCF films can be successfully deposited by potential cycling over a range positive of that required for hexacyanoferrate(III) reduction.

No evidence was obtained for the formation of a vanadium(V)–hexacyanoferrate(III) type complex analogous to iron(III) hexacyano-ferrate(III), the visible absorption spectrum of the mixed solution being a simple summation of spectra of the single-component solutions. While VHCF films are visually electrochromic, switching from green in the oxidised state to yellow in the reduced state, Carpenter et al. show that most of the electrochromic modulation occurs in the ultraviolet (UV) region.[77] From electrochemical data and XPS they conclude that the electrochromism involves only the iron centres in the film. The vanadium ions, found to be present predominantly in the +IV oxidation state, are not redox active under these conditions.

8.5.3 Nickel hexacyanoferrate

Nickel hexacyanoferrate (NiHCF) films can be prepared by electrochemical oxidation of nickel electrodes in the presence of hexacyanoferrate(III) ions,[79] or by voltammetric cycling of inert substrate electrodes in solutions containing nickel(II) and hexacyanoferrate(III) ions.[80] The NiHCF films do not show low-energy IVCT bands, but when deposited on ITO they are observed to switch reversibly from yellow to colourless on electroreduction.[81]

A more dramatic colour change can be observed by substitution of two iron-bound cyanides by a suitable bidentate ligand.[82] Thus, 2,2′-bipyridine can be indirectly attached to nickel metal via a cyano–iron complex to form a derivatised electrode. When 2,2′-bipyridine is employed as the chelating agent, the complex $[Fe^{II}(CN)_4(bipy)]^{2-}$ is formed which takes on an intense red colour associated with a MLCT absorption band centred at 480 nm. This optical transition is sensitive to both the iron oxidation state, only arising in the Fe^{II} form of the complex, and to the environment of the cyanide-nitrogen lone pair. Reaction of the complex with Ni^{2+} either under bulk conditions or at a nickel electrode surface generates a bright red material. By analogy with the parent iron complex this red colour is associated with the $(d\pi)Fe^{II} \rightarrow (\pi^*)$bipy CT transition. For bulk samples, chemical oxidation to the Fe^{III} state yields a light-orange material, while modified electrodes can be reversibly cycled between the intensely red and transparent forms, a process which correlates well with the observed CV response.[82] In principle, orange–transparent and green–transparent electrochromism could be available, using the complexes [Ru^{II}(CN)$_4$(bipy)]$^{2-}$ and [Os^{II}(CN)$_4$(bipy)]$^{2-}$ respectively.

8.5.4 Copper hexacyanoferrate

Copper hexacyanoferrate (CuHCF) films can be prepared voltammetrically by electroplating a thin film of copper on glassy carbon (GC) or ITO electrodes in the presence of hexacyanoferrate(II) ions.[83,84,85,86] Films are deposited by first cycling between +0.40 and +0.05 V in a solution of cupric nitrate in aqueous KClO$_4$. Copper is then deposited on the electrode by stepping the potential from +0.03 to −0.50 V, and subsequently removed (stripped) by linearly scanning the potential from −0.50 to +0.50 V. The deposition and removal sequence was repeated until a reproducible CV was obtained during the stripping procedure. The CuHCF film was then formed by stepping the electrode potential in the presence of cupric ion from +0.03 to −0.50 V followed by injection of an aliquot of K$_4$Fe(CN)$_6$ solution (a red–brown hexacyanoferrate(II) sol formed immediately) into the cell. The CuHCF film formation mechanism has not been elucidated but the co-deposition of copper is important in the formation of stable films. Films formed by galvanostatic or potentiostatic methods from solutions of cupric ion and hexacyanoferrate(III) ion showed noticeable deterioration within a few CV scans. The co-deposition procedure provides a fresh copper surface for film adhesion and the resulting films are able to withstand ∼1000 voltammetric cycles. Such scanning of a CuHCF film in K$_2$SO$_4$ (0.5 mol dm^{-3}) gave a well-defined reversible couple at +0.69 V, characteristic of an adsorbed species. Copper hexacyanoferrate films exhibit red-brown to yellow electrochromicity.[86] For the reduced film, a broad visible absorption band associated with the iron-to-copper CT in cupric hexacyanoferrate(II) was observed (λ_{max} = 490 nm, $\varepsilon = 2 \times 10^3$ dm^3 mol^{-1} cm^{-1}). This band was absent in the spectrum of the oxidised film, the yellow colour arising from the CN$^-\rightarrow$ Fe^{III} CT band at 420 nm for the hexacyanoferrate(III) species (arrow denoting electron transfer).

8.5.5 Palladium hexacyanoferrate

The preparation of electrochromic palladium hexacyanoferrate (PdHCF) films by simple immersion of the electrode substrate for at least one hour, or potential cycling of conducting substrates (Ir, Pd, Au, Pt, GC), in a mixed solution of PdCl$_2$ and K$_3$Fe(CN)$_6$ has been reported.[87] The resulting modified electrodes gave broad CV responses, assigned to Fe^{III}(CN)$_6$–Fe^{II}(CN)$_6$, the Pd^{II} sites being electro-inactive. Films were orange at >1.0 V and yellow-green at <0.2V. More recently, potentiodynamically grown PdHCF films have been studied using cyclic voltammetry, *in situ* infrared and UV-visible spectroelectrochemistry.[88] UV-visible reflectance spectra of films on platinum demonstrated the reversible progressive conversion of PdHCF between its reduced (light yellow) and oxidised (yellow green) states.

8.5.6 Indium hexacyanoferrate and gallium hexacyanoferrate

Indium hexacyanoferrate films[89,90,91,92] have been grown by potential cycling in a mixed solution containing $InCl_3$ and $K_3Fe(CN)_6$. The electrodeposition occurs during the negative scans as sparingly soluble deposits of In^{3+} with $[Fe(CN)_6]^{4-}$ were formed.[89] The resulting films are electrochromic, being white when reduced and yellow when oxidised.[92]

Solid films of gallium hexacyanoferrate have been prepared by direct modification of a gallium electrode surface in an aqueous solution of 5 mmol dm^{-3} potassium hexacyanoferrate(III) in KCl (0.1 mol dm^{-3}).[93] This one-step electroless deposition proceeds via a chemical oxidation reaction of the metallic gallium to Ga^{3+} in the aqueous solution, followed by reaction with the hexacyanoferrate(III) ions. To date, the electrochromic properties of the films have not been investigated.

8.5.7 Miscellaneous Prussian blue analogues

Prussian blue analogues investigated include thin films of cadmium hexacyanoferrate[94] (reversibly white to colourless on reduction[81]), chromium hexacyanoferrate[95] (reversibly blue to pale blue-grey on reduction[81]), cobalt hexacyanoferrate[96] (reversibly green-brown to dark green on reduction[81]), manganese hexacyanoferrate[97] (reversibly pale yellow to colourless on reduction[81]), molybdenum hexacyanoferrate[98] (pink to red on reduction[81]), osmium hexacyanoferrate,[99] osmium(IV) hexacyanoruthenate,[100] platinum hexacyanoferrate[101] (pale blue to colourless on reduction[81]), rhenium hexacyanoferrate[81] (pale yellow to colourless on reduction[81]), rhodium hexacyanoferrate[81] (pale yellow to colourless on reduction[81]), ruthenium oxide–hexacyanoruthenate,[102] mixed films of ruthenium oxide–hexacyanoferrate and ruthenium hexacyanoferrate,[103] silver hexacyanoferrate,[5] silver–'crosslinked' nickel hexacyanoferrate[104] (reversibly yellow to white on reduction[81]), titanium hexacyanoferrate[105] (reversibly brown to pale yellow on reduction[81]), zinc hexacyanoferrate[106] and zirconium hexacyanoferrate.[107]

Mixed-ligand Prussian blue analogues reported as redox-active thin films include copper heptacyanonitrosylferrate,[108] iron(III) carbonylpentacyanoferrate,[5] and iron(III) pentacyanonitroferrate.[5]

Of the lanthanoids and actinoids, lanthanum hexacyanoferrate,[109] samarium hexacyanoferrate[110] and uranium hexacyanoferrate,[111] as thin redox-active films have been studied.

8.5.8 Mixed-metal hexacyanoferrates

Glassy carbon electrodes have been modified with films of *mixed* metal hexacyanoferrates.[97] Cyclic voltammograms of PB–nickel hexacyanoferrate and PB–manganese hexacyanoferrate films show electroactivity of both metal hexacyanoferrate components in each mixture. It is suggested that the mixed-metal hexacyanoferrates have a structure in which some of the outer sphere iron centres in the PB lattice are replaced by Ni^{2+} or Mn^{2+}, rather than being a co-deposited mixture of PB and nickel or manganese hexacyanoferrate.[97] Although film colours are not reported, it seems likely that variation of metal hexacyanoferrate and compositions of electrodeposition solution could allow colour choice in the anticipated electropolychromic systems. The approach seems general, with PB–metal hexacyanoferrate (metal = Co, Cu, In, Cr, Ru) modified electrodes also being successfully prepared. Thin films of mixed nickel–palladium hexacyanoferrates have been prepared and characterised, and spectral measurements show them to be electrochromic, although colours have not been reported.[112]

References

1. Diesbach (1704), cited in *Gmelin, Handbuch der Anorganischen Chemie*, Frankfurt am Main, Deutsche Chemische Gesellschaft, 1930, vol. 59, Eisen B, p. 671.
2. Fukuda, K. In *Pigment Handbook*, 2nd edn, Lewis, P. A. (ed.), New York, Wiley Interscience, 1988, vol. 1, pp. 357–65.
3. *Colour Index*, 3rd edn, Bradford, Society of Dyers and Colourists, 1971, vol. 4, p. 4673.
4. Neff, V. D. Electrochemical oxidation and reduction of thin films of Prussian blue. *J. Electrochem. Soc.*, **125**, 1978, 886–7.
5. Itaya, K., Uchida, I. and Neff, V. D. Electrochemistry of polynuclear transition-metal cyanides – Prussian blue and its analogs. *Acc. Chem. Res.*, **19**, 1986, 162–168.
6. Monk, P. M. S., Mortimer, R. J. and Rosseinsky, D. R. *Electrochromism: Fundamentals and Applications*, Weinheim, VCH, 1995, ch. 6.
7. de Tacconi, N. R., Rajeshwar, K. and Lezna, R. O. Metal hexacyanoferrates: electrosynthesis, in situ characterization, and applications. *Chem. Mater.*, **15**, 2003, 3046–62.

8. Cox, J. A., Jaworski, R. K. and Kulesza, P. J. Electroanalysis with electrodes modified by inorganic films. *Electroanalysis*, **3**, 1991, 869–77.
9. Karyakin, A. A. Prussian blue and its analogues: electrochemistry and analytical applications. *Electroanalysis*, **13**, 2001, 813–19.
10. Koncki, R. Chemical sensors and biosensors based on Prussian blues. *Crit. Rev. Anal. Chem.*, **32**, 2002, 79–96.
11. Ricci, F. and Palleschi, G. Sensor and biosensor preparation, optimisation and applications of Prussian blue modified electrode. *Biosens. Bioelectron.* **21**, 2005, 389–407.
12. Robin, M. B. The colour and electronic configuration of Prussian blue. *Inorg. Chem.*, **1**, 1962, 337–42.
13. Inoue, H. and Yanagisawa, S. Bonding nature and semiconductivity of Prussian blue and related compounds. *J. Inorg. Nucl. Chem.*, **36**, 1974, 1409–11.
14. Wilde, R. E., Ghosh, S. N. and Marshall, B. J. The Prussian blues. *Inorg. Chem.*, **9**, 1970, 2512–16.
15. Chadwick, B. M. and Sharpe, A. G. Transition metal cyanides and their complexes. *Adv. Inorg. Chem. Radiochem*, **8**, 1966, 83–176.
16. Sharpe, A. G. *The Chemistry of Cyano Complexes of the Transition Metals*, New York, Academic Press, 1976.
17. Dunbar, K. R. and Heintz, R. A. Chemistry of transition metal cyanide compounds: modern perspectives. *Prog. Inorg. Chem.*, **45**, 1997, 283–391.
18. Bonnette, A. K., jr., and Allen, J. F. Isotopic labelling for Mössbauer studies, an application to the iron cyanides. *Inorg. Chem.*, **10**, 1971, 1613–16.
19. Sillen, L. G. and Martell, A. E. *Stability Constants – Supplement No. 1*. London, The Chemical Society, 1971, Special Publication No. 25.
20. Mayoh, B. and Day, P. Charge transfer in mixed valence solids. Part VII. Perturbation calculations of valence delocalisation in iron(II, III) cyanides and silicates. *J. Chem. Soc., Dalton Trans.*, 1974, 846–52.
21. Keggin, J. F. and Miles, F. D. Structures and formulae of the Prussian blues and related compounds. *Nature (London)*, **137**, 1936, 577–8.
22. Buser, H. J., Schwarzenbach, D., Petter, W. and Ludi, A. The crystal structure of Prussian blue: $Fe_4[Fe(CN)_6]_3 \cdot xH_2O$. *Inorg. Chem.*, **16**, 1977, 2704–10.
23. Widmann, A., Kahlert, H., Petrovic-Prelevic, H., Wulff, H., Yakshmi, J. V., Bagkar, N. and Scholz, F. Structure, insertion electrochemistry, and magnetic properties of a new type of substitutional solid solutions of copper, nickel, and iron hexacyanoferrates/hexacyanocobaltates. *Inorg. Chem.*, **41**, 2002, 5706–15.
24. Rosseinsky, D. R., Lim, H., Jiang, H. and Chai, J. W. Optical charge-transfer in iron(III)hexacyanoferrate(II): electro-intercalated cations induce lattice-energy-dependent ground-state energies. *Inorg. Chem.*, **42**, 2003, 6015–23.
25. Yano, Y., Kinugasa, N., Yoshida, H. K., Fujino, K. and Kawahara, H. Electrochemical properties of amorphous Prussian blue films chemically deposited from aqueous solutions. *Proc. Electrochem. Soc.*, **90–2**, 1990, 125–36.
26. Ellis, D., Eckhoff, M. and Neff, V. D. Electrochromism in the mixed-valence hexacyanides. 1. Voltammetric and spectral studies of the oxidation and reduction of thin-films of Prussian blue. *J. Phys. Chem.*, **85**, 1981, 1225–31.
27. Ho, K.-C. On the deposition of Prussian blue by the sacrificial anode method. *Proc. Electrochem. Soc.*, **94–2**, 1994, 170–84.
28. Gomathi, H. and Rao, G. P. Simple electrochemical immobilization of the ferro ferricyanide redox couple on carbon electrodes. *J. Appl. Electrochem.*, **20**, 1990, 454–6.

29. Scholz, F. and Meyer, B. Voltammetry of solid microparticles immobilized on electrode surfaces. In *Electroanalytical Chemistry: A Series of Advances*, Bard, A. J. and Rubinstein, I. (eds.), New York, Marcel Dekker, 1998, vol. 20, pp. 1–86.
30. Kellawi, H. and Rosseinsky, D. R. Electrochemical bichromic behaviour of ferric ferrocyanide (Prussian blue) in thin film redox processes. *J. Electroanal. Chem.*, **131**, 1982, 373–6.
31. Barton, R. T., Kellawi, H., Marken, F., Mortimer, R. J. and Rosseinsky, D. R. Unpublished results.
32. Goncalves, R. M. C., Kellawi, H. and Rosseinsky, D. R. Electron-transfer processes and electrodeposition involving the iron hexacyanoferrates studied voltammetrically. *J. Chem. Soc., Dalton Trans.*, 1983, 991–4.
33. Mortimer, R. J. and Rosseinsky, D. R. Electrochemical polychromicity in iron hexacyanoferrate films, and a new film form of ferric ferricyanide. *J. Electroanal. Chem.*, **151**, 1983, 133–47.
34. Mortimer, R. J. and Rosseinsky, D. R. Iron hexacyanoferrate films: spectroelectrochemical distinction and electrodeposition sequence of 'soluble' (K^+-containing) and 'insoluble' (K^+-free) Prussian blue and composition changes in polyelectrochromic switching. *J. Chem. Soc., Dalton Trans.*, 1984, 2059–61.
35. Itaya, K., Ataka, T. and Toshima, S. Spectroelectrochemistry and electrochemical preparation method of Prussian blue modified electrodes. *J. Am. Chem. Soc.*, **104**, 1982, 4767–72.
36. Cheng, G. J. and Dong, S. J. Chronoabsorptometric study of Prussian blue modified film electrode. *Electrochim. Acta*, **32**, 1987, 1561–5.
37. Feldman, B. J. and Melroy, O. R. Ion flux during electrochemical charging of Prussian blue films. *J. Electroanal. Chem.*, **234**, 1987, 213–27.
38. Hamnett, A., Higgins, S., Mortimer, R. J. and Rosseinsky, D. R. A study of the electrodeposition and subsequent potential cycling of Prussian blue films using ellipsometry. *J. Electroanal. Chem.*, **255**, 1988, 315–24.
39. Rosseinsky, D. R. and Tonge, J. S. Electron transfer rates by dielectric relaxometry and the DC conductivities of solid homonuclear and heteronuclear mixed valence metal cyanometallates, and of the methylene-blue/iron-dithiolate adduct. *J. Chem. Soc., Faraday Trans.*, 1, **83**, 1987, 245–55.
40. Mortimer, R. J., Rosseinsky, D. R. and Glidle, A. Polyelectrochromic Prussian blue: a chronoamperometric study of the electrodeposition. *Sol. Energy Mater. Sol. Cells*, **25**, 1992, 211–23.
41. Rosseinsky, D. R. and Glidle, A. EDX, spectroscopy, and composition studies of electrochromic iron(III) hexacyanoferrate(II) deposition. *J. Electrochem. Soc.*, **150**, 2003, C641–5.
42. Millward, R. C., Madden, C. E., Sutherland, I., Mortimer R. J., Fletcher, S. and Marken, F. Directed assembly of multi-layers: the case of Prussian blue. *Chem. Commun*, 2001, 1994–5.
43. Pyrasch, M. and Tieke, B. Electro- and photoresponsive films of Prussian blue prepared upon multiple sequential adsorption. *Langmuir*, **17**, 2001, 7706–9.
44. Pyrasch, M., Toutianoush, A., Jin, W., Schnepf, J. and Tieke, B. Self-assembled films of Prussian blue and analogues: optical and electrochemical properties and application as ion-sieving membranes. *Chem. Mater.*, **15**, 2003, 245–54.
45. Jin, W., Toutianoush, A., Pyrasch, M., Schnepf, J., Gottschalk, H., Rammensee, W. and Tieke, B. *J. Phys. Chem. B*, **107**, 2003, 12062–70.
46. Mortimer, R. J. Unpublished observations.

47. Itaya, K. and Uchida, I. Nature of intervalence charge-transfer bands in Prussian blues. *Inorg. Chem.*, **25**, 1986, 389–92.
48. Emrich, R. J., Traynor, L., Gambogi, W. and Buhks, E. Surface-analysis of electrochromic displays of iron hexacyanoferrate films by X-ray photoelectron spectroscopy. *J. Vac. Sci. Technol. A*, **5**, 1987, 1307–10.
49. Lundgren, C. A. and Murray, R. W. Observations on the composition of Prussian blue films and their electrochemistry. *Inorg. Chem.*, **27**, 1988, 933–9.
50. Beckstead, D. J., De Smet, D. J. and Ord, J. L. An ellipsometric investigation of the formation and conversion of Prussian blue films. *J. Electrochem. Soc.*, **136**, 1989, 1927–32.
51. Christensen, P. A., Hamnett, A. and Higgins, S. J. A study of electrochemically grown Prussian blue films using Fourier-transform infrared spectroscopy. *J. Chem. Soc., Dalton Trans.*, 1990, 2233–8.
52. Rosseinsky, D. R., Glasser, L. and Jenkins, H. D. B. Thermodynamic clarification of the curious ferric/potassium ion exchange accompanying the electrochromic redox reactions of Prussian Blue, iron(III) hexacyanoferrate(II). *J. Am. Chem. Soc.*, **126**, 2004, 10473–7.
53. Lee, O. H., Yang, H. and Kwak, J. Ion and water transports in Prussian blue films investigated with electrochemical quartz crystal microbalance. *Electrochem. Commun.*, **3**, 2001, 274–80.
54. Rosseinsky, D. R., Lim, H., Zhang, X., Jiang, H. and Chai, J. W. Charge-transfer band shifts in iron(III)hexacyanoferrate(II) by electro-intercalated cations *via* groundstate-energy/lattice-energy link. *Chem. Commun.*, 2002, 2988–9.
55. Stilwell, D. E., Park, K. W. and Miles, M. H. Electrochemical studies of the factors influencing the cycle stability of Prussian blue films. *J. Appl. Electrochem.*, **22**, 1992, 325–31.
56. Itaya, K., Shibayama, K., Akahoshi, H. and Toshima, S. Prussian-blue-modified electrodes – an application for a stable electrochromic display device. *J. Appl. Phys.*, **53**, 1982, 804–5.
57. Honda, K., Ochiai, J. and Hayashi, H. Polymerization of transition-metal complexes in solid polymer electrolytes. *J. Chem. Soc., Chem. Commun.*, 1986, 168–70.
58. Honda, K. and Kuwano, A. Solid-state electrochromic device using polynuclear metal complex-containing solid polymer electrolyte. *J. Electrochem. Soc.*, **133**, 1986, 853–4.
59. Carpenter, M. K. and Conell, R. S. A single-film electrochromic device. *J. Electrochem. Soc.*, **137**, 1990, 2464–7.
60. Honda, K., Fujita, M., Ishida, H., Yamamoto, R. and Ohgaki, K. Solid-state electrochromic devices composed of Prussian blue, WO_3, and poly(ethylene oxide)–polysiloxane hybrid-type ionic conducting membrane. *J. Electrochem. Soc.*, **135**, 1988, 3151–4.
61. Habib, M. A., Maheswari, S. P. and Carpenter, M. K. A tungsten-trioxide Prussian blue complementary electrochromic cell with a polymer electrolyte. *J. Appl. Electrochem.*, **21**, 1991, 203–7.
62. Habib, M. A. and Maheswari, S. P. Effect of temperature on a complementary WO_3–Prussian blue electrochromic system. *J. Electrochem. Soc.*, **139**, 1992, 2155–7.
63. Béraud, J.-G. and Deroo, D. Some novel prospective polymer electrolytes containing potassium-ion for electrochromic devices, with preliminary tests on Prussian blue/K_xWO_3 electrochromic windows. *Sol. Energy Mater. Sol. Cells*, **31**, 1993, 263–75.

64. Ho, K. C. Cycling and at-rest stabilities of a complementary electrochromic device based on tungsten oxide and Prussian blue thin films. *Electrochim. Acta*, **44**, 1999, 3227–35.
65. Duek, E. A. R., De Paoli, M.-A. and Mastragostino, M. An electrochromic device based on polyaniline and Prussian blue. *Adv. Mater.*, **4**, 1992, 287–91.
66. Duek, E. A. R., De Paoli, M.-A. and Mastragostino, M. A. A solid-state electrochromic device based on polyaniline, Prussian blue and an elastomeric electrolyte. *Adv. Mater.*, **5**, 1993, 650–2.
67. Morita, M. Electrochromic behavior and stability of polyaniline composite films combined with Prussian blue. *J. Appl. Polym. Sci.*, **52**, 1994, 711–19.
68. Jelle, B. P., Hagen, G. and Nødland, S. Transmission spectra of an electrochromic window consisting of polyaniline, Prussian blue and tungsten oxide. *Electrochim. Acta*, **38**, 1993, 1497–500.
69. Jelle, B. P. and Hagen, G. J. Transmission spectra of an electrochromic window based on polyaniline, Prussian blue and tungsten oxide. *J. Electrochem. Soc.*, **140**, 1993, 3560–4.
70. Leventis, N. and Chung, Y. C. Polyaniline–Prussian blue novel composite-material for electrochromic applications. *J. Electrochem. Soc.*, **137**, 1990, 3321–2.
71. Jelle, B. P. and Hagen, G. Correlation between light absorption and electric charge in solid state electrochromic windows. *J. Appl. Chem.*, **29**, 1999, 1103–10.
72. Jelle, B. P. and Hagen, G. Performance of an electrochromic window based on polyaniline, Prussian blue and tungsten oxide. *Sol. Energy Mater. Sol. Cells*, **58**, 1999, 277–86.
73. Tung, T.-S. and Ho, K.-C. Cycling and at-rest stabilities of a complementary electrochromic device containing poly(3,4-ethylenedioxythiophene) and Prussian blue. *Sol. Energy Mater. Sol. Cells*, **90**, 2006, 521–37.
74. Kashiwazaki, N. New complementary electrochromic display utilizing polymeric YbPc$_2$ and Prussian blue films. *Sol. Energy Mater. Sol. Cells*, **25**, 1992, 349–59.
75. Rajan, K. P. and Neff, V. D. Electrochromism in the mixed-valence hexacyanides. 2. Kinetics of the reduction of ruthenium purple and Prussian blue. *J. Phys. Chem.*, **86**, 1982, 4361–8.
76. Itaya, K., Ataka, T. and Toshima, S. Electrochemical preparation of a Prussian blue analog – iron–ruthenium cyanide. *J. Am. Chem. Soc.*, **104**, 1982, 3751–2.
77. Carpenter, M. K., Conell, R. S. and Simko, S. J. Electrochemistry and electrochromism of vanadium hexacyanoferrate. *Inorg. Chem.*, **29**, 1990, 845–50.
78. Dong, S. J. and Li, F. B. Researches on chemically modified electrodes.16. Electron-diffusion coefficient in vanadium hexacyanoferrate film. *J. Electroanal. Chem.*, **217**, 1987, 49–63.
79. Bocarsly, A. B. and Sinha, S. Chemically derivatized nickel surfaces – synthesis of a new class of stable electrode interfaces. *J. Electroanal. Chem.*, **137**, 1982, 157–62.
80. Joseph, J., Gomathi, H. and Rao, G. P. Electrochemical characteristics of thin-films of nickel hexacyanoferrate formed on carbon substrates. *Electrochim. Acta*, **36**, 1991, 1537–41.
81. Dillingham, J. L. Investigation of bipyridilium and Prussian blue systems for their potential application in electrochromic devices. Ph.D. Thesis, Loughborough University, 1999, ch. 5 (A survey of the transition metal hexacyanoferrates).

82. Sinha, S., Humphrey, B. D., Fu, E. and Bocarsly, A. B. The coordination chemistry of chemically derivatized nickel surfaces – generation of an electrochromic interface. *J. Electroanal. Chem.*, **162**, 1984, 351–7.
83. Siperko, L. M. and Kuwana, T. Electrochemical and spectroscopic studies of metal hexacyanometalate films. 1. Cupric hexacyanoferrate. *J. Electrochem. Soc.*, **130**, 1983, 396–402.
84. Siperko, L. M. and Kuwana, T. Electrochemical and spectroscopic studies of metal hexacyanoferrate films. 2. Cupric hexacyanoferrate and Prussian blue layered films. *J. Electrochem. Soc.*, **133**, 1986, 2439–40.
85. Siperko, L. M. and Kuwana, T. Electrochemical and spectroscopic studies of metal hexacyanometalate films. 3. Equilibrium and kinetic studies of cupric hexacyanoferrate. *Electrochim. Acta*, **32**, 1987, 765–71.
86. Siperko, L. M. and Kuwana, T. Studies of layered thin-films of Prussian-blue-type compounds. *J. Vac. Sci. Technol. A*, **5**, 1987, 1303–6.
87. Jiang, M. and Zhao, Z. F. A novel stable electrochromic thin-film – a Prussian blue analog based on palladium hexacyanoferrate. *J. Electroanal. Chem.*, **292**, 1990, 281–7.
88. Lezna, R. O., Romagnoli, R., de Tacconi, N. R. and Rajeshwar, K. Spectroelectrochemistry of palladium hexacyanoferrate films on platinum substrates. *J. Electroanal. Chem.*, **544**, 2003, 101–6.
89. Kulesza, P. J. and Faszynska, M. Indium(III) hexacyanoferrate as a novel polynuclear mixed-valent inorganic material for preparation of thin zeolitic films on conducting substrates. *J. Electroanal. Chem.*, **252**, 1988, 461–6.
90. Kulesza, P. J. and Faszynska, M. Indium(III)–hexacyanoferrate(III, II) as an inorganic material analogous to redox polymers for modification of electrode surfaces. *Electrochim. Acta*, **34**, 1989, 1749–53.
91. Dong, S. J. and Jin, Z. Electrochemistry of indium hexacyanoferrate film modified electrodes. *Electrochim. Acta*, **34**, 1989, 963–8.
92. Jin, Z. and Dong, S. J. Spectroelectrochemical studies of indium hexacyanoferrate film modified electrodes. *Electrochim Acta*, **35**, 1990, 1057–60.
93. Eftekhari, A. Electrochemical behavior of gallium hexacyanoferrate film directly modified electrode in a cool environment. *J. Electrochem. Soc.*, **151**, 2004, E297–301.
94. Luangdilok, C. H., Arent, D. J., Bocarsly, A. B. and Wood, R. Investigation of the structure reactivity relationship in the $Pt/M_xCdFe(CN)_6$ modified electrode system. *Langmuir*, **8**, 1992, 650–7.
95. Jiang, M., Zhou, X. and Zhao, Z. A new zeolitic thin-film based on chromium hexacyanoferrate on conducting substrates. *J. Electroanal. Chem.*, **287**, 1990, 389–94.
96. Joseph, J., Gomathi, H. and Prabhakar Rao, G. Electrodes modified with cobalt hexacyanoferrate. *J. Electroanal. Chem.*, **304**, 1991, 263–9.
97. Bharathi, S., Joseph, J., Jeyakumar, D. and Prabhakara Rao, G. Modified electrodes with mixed metal hexacyanoferrates. *J. Electroanal. Chem.*, **319**, 1991, 341–5.
98. Dong, S. and Jin, Z. Molybdenum hexacyanoferrate film modified electrodes. *J. Electroanal. Chem.*, **256**, 1988, 193–8.
99. Chen, S.-M. and Liao, C.-J. Preparation and characterization of osmium hexacyanoferrate films and their electrocatalytic properties. *Electrochim. Acta*, 2004, **50**, 115–25.
100. Cox, J. A. and Das, B. K. Characteristics of a glassy-carbon electrode modified in a mixture of osmium-tetroxide and hexacyanoruthenate. *J. Electroanal. Chem.*, **233**, 1987, 87–98.

101. Liu, S. Q., Li, H. L., Jiang, M. and Li, P. B. Platinum hexacyanoferrate: a novel Prussian blue analogue with stable electroactive properties. *J. Electroanal. Chem.*, **426**, 1997, 27–30.
102. Kulesza, P. J. A polynuclear mixed-valent ruthenium oxide cyanoruthenate composite that yields thin coatings on a glassy-carbon electrode with high catalytic activity toward methanol oxidation. *J. Electroanal. Chem.*, **220**, 1987, 295–309.
103. Chen, S.-M., Lu, M.-F. and Lin, K.-C. Preparation and characterization of ruthenium oxide/hexacyanoferrate and ruthenium hexacyanoferrate mixed films and their electrocatalytic properties. *J. Electroanal. Chem.*, **579**, 2005, 163–74.
104. Kulesza, P. J., Jedral, T. and Galus, Z. A new development in polynuclear inorganic films – silver(I) crosslinked nickel(II) hexacyanoferrate(III, II) microstructures. *Electrochim. Acta*, **34**, 1989, 851–3.
105. Jiang, M., Zhou, X. Y. and Zhao, Z. F. Preparation and characterization of mixed-valent titanium hexacyanoferrate film modified glassy-carbon electrode. *J. Electroanal. Chem.*, **292**, 1990, 289–96.
106. Joseph, J., Gomathi, H. and Rao, G. P. Modification of carbon electrodes with zinc hexacyanoferrate. *J. Electroanal. Chem.*, **431**, 1997, 231–5.
107. Liu, S.-Q., Chen, Y. and Chen, H.-Y. Studies of spectroscopy and cyclic voltammetry on a zirconium hexacyanoferrate modified electrode. *J. Electroanal. Chem.*, **502**, 2001, 197–203.
108. Gao, Z., Zhang, Y., Tian, M. and Zhao, Z. Electrochemical study of copper heptacyanonitrosylferrate film modified electrodes: preparation, properties and applications. *J. Electroanal. Chem.*, **358**, 1993, 161–76.
109. Liu, S.-Q. and Chen, H.-Y. Spectroscopic and voltammetric studies on a lanthanum hexacyanoferrate modified electrode. *J. Electroanal. Chem.*, **528**, 2002, 190–5.
110. Wu, P., Lu, S. and Cai, C. Electrochemical preparation and characterization of a samarium hexcyanoferrate modified electrode. *J. Electroanal. Chem.*, **569**, 2004, 143–50.
111. Jiang, M., Wang, M and Zhou, X. Facile attachment of uranium hexacyanoferrate to carbon electrode by reductive electrodeposition. *Chem. Lett.*, 1992, 1709–12.
112. Kulesza, P. J., Malik, M. A., Schmidt, R., Smolinska, A., Miecznikowski, K., Zamponi, S., Czerwinski, A., Berrettoni, M. and Marassi, R. Electrochemical preparation and characterization of electrodes modified with mixed hexacyanoferrates of nickel and palladium. *J. Electroanal. Chem.*, **487**, 2000, 57–65.

9

Miscellaneous inorganic electrochromes

9.1 Fullerene-based electrochromes

The electrochromism of thin films of Buckminsterfullerene C_{60} was first demonstrated in 1993 by Rauh and co-workers.[1] The electro-coloration occurs during reduction to form lithium fulleride, Li_xC_{60}:

$$C_{60} + x(Li^+ + e^-) \rightarrow Li_xC_{60}. \quad (9.1)$$

           light brown   dark brown

The reduced form develops a band maximum in the near infrared, in the range 1060–1080 nm. A band also forms in the UV. Figure 9.1 shows the spectrum of C_{60} as a function of applied potential. Electrochemically formed Li_xC_{60} is identical with the fulleride salt formed by exposing C_{60} to alkali-metal vapour.

Konesky[2] has shown that Ag^+, Cr^{3+}, Cu^{2+}, Mg^{2+} and Ba^{2+} ions, in addition to Li^+, can be electro-intercalated into such fulleride films during coloration as counter ions from solvents γ-butyrolactone or water. As fullerene and fulleride films are partially soluble in the polar organic electrolytes used, the cycle life is depleted by prolonged exposure to such electrolytes.[3] The solubility increases with higher insertion coefficient, x.[4] Furthermore, the higher-x outer layers of the film can peel away from the electrode.[4]

The electrochromism is reversible with electrochemically intercalated alkali-metal or alkaline-earth ions, although the extent of reversibility depends on the insertion coefficient x: reversibility is lost if x is too high,[3] as found with tungsten oxide (*cf.* p. 114).

Steep concentration gradients form in the fulleride films during electrochromic operation. Analysis is complicated since ionic mobilities are a function of insertion coefficient.[4,5] Applying pulsed potentials improves both the durability of the film and the extent of electro-reversibility, presumably by allowing such gradients to dissipate during the 'off' period between pulses.[3]

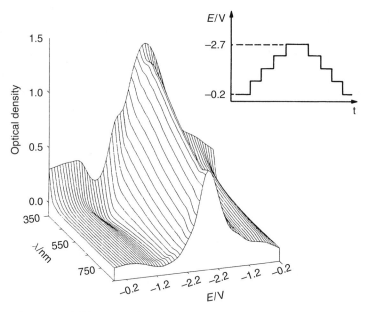

Figure 9.1 UV-visible spectrum of immobilised fullerene on an electrode surface as a function of applied potential E: the C_{60} was on a SnO_2-coated glass electrode immersed in PC containing $LiClO_4$ (1 mol dm^{-3}). (Figure reproduced from de Torresi, S. I. C., Torresi, R. M., Ciampi, G. and Luengo, C. A. 'Electrochromic phenomena in fullerene thin films'. *J. Electroanal. Chem.*, **377**, 1994, 283–5, by permission of Elsevier Science.)

de Torresi *et al.*[6] suggest the coloured form of the electrochrome, Li_xC_{60}, is not stable: electrochromic stability is degraded by residual oxygen in both the electrolyte system and the fullerene film. This rapid reaction yields C_{60}, Li^+ and oxide ion.[5] Reaction with water is also rapid. Additionally Li_xC_{60} catalyses the electro-decomposition of solvent, which may explain why the coloration efficiency is 20 cm^2 C^{-1} during coloration but 35 cm^2 C^{-1} during bleaching.

Goldenberg[7] has also prepared thin electrochromic films of fullerene via Langmuir–Blodgett techniques.

9.2 Other carbon-based electrochromes

Pfluger *et al.*[8] have reported an ECD with graphite as a solid-solution intercalation electrode. Many alkali-metal cations may be inserted into graphite sheets from aprotic solutions, lithium apparently giving the best speed and electro reversibility. This ECD is electropolychromic switching from brassy black → deep blue → light green → golden yellow within the potential range 3–5 V. When the potential was reversed, the ECD reverted back to the brassy black colour, with τ of about 0.2 s. Kuwabara and Noda[9] and White and co-workers[10] have also used graphite as counter-electrode layer in an ECD.

Diamond,[11] electrodeposited by the oxidation of lithium acetylide, is yellow, but becomes brown following reductive ion insertion, showing a new band in the UV.

Other forms of carbon have been used as counter electrodes: screen-printed carbon black,[12] 'carbon'[9,13,14,15] and 'carbon-based' electrodes.[16,17] No colour change is mentioned regarding these materials.

9.3 Reversible electrodeposition of metals

Comparatively few *inorganic* type-II electrochromes have been reported. Of these few, the only viable systems are those in which finely divided metal is electrodeposited onto an OTE, as reviewed by Ziegler (in 1999).[18]

In all these systems, reduction of a dissolved metal cation results in the deposition of finely divided metal, so the 'electrochromism' results not from photon absorption but rather from the film becoming opaque or even optically reflective (by specular reflection). The three systems studied for electrochromism are listed below.

Bismuth

In recent work on the electrodeposition of metallic bismuth from aqueous solution,[19,20,21,22,23] the deposition/coloration reaction is cited[22] as Eq. (9.2):

$$2\,Bi^{3+}(soln) + 9\,Br^-(soln) \rightarrow 2\,Bi^0(s) + 3\,Br_3^-(soln). \qquad (9.2)$$

      colourless   opaque

The deposition of particulate bismuth, rather than a continuous metal film, is achieved by underpotential deposition, the solution containing traces of copper to act as an electron mediator. The reaction sequence has not yet been detailed.

Gelling an aqueous–organic electrolyte makes the image less patchy.[24] The pH of the deposition solution must be relatively low in order to maintain high solubility of the bismuth cation precursor, but not so low as to cause deterioration of the ITO layer of the transparent electrode (the OTE).

Despite experimental problems, however, electrodeposited particulate bismuth exhibiting opacity has shown[18] a cycle-life of 5×10^7. Thus the 'spectrum' of such bismuth on an OTE is invariant with wavelength, lacking absorption peaks, appearing as an almost horizontal line that increases in height with thickness of electrodeposited bismuth; see Figure 9.2. Accordingly,

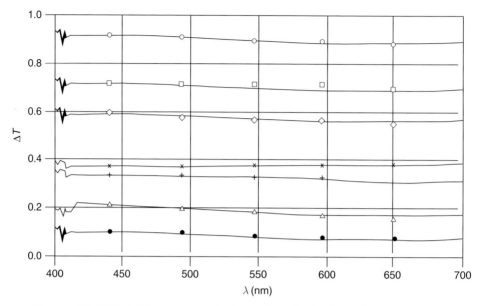

Figure 9.2 UV-visible spectra of electrodeposited bismuth on ITO. The bismuth was deposited reductively from a solution initially comprising aqueous Bi^{3+} (0.02 mol dm^{-3}). This is not a true 'spectrum' because the bismuth is reflective, rather than optically absorbing. (Figure reproduced from Ziegler, J. P. and Howard, B. M. 'Spectroelectrochemistry of reversible electrodeposition electrochromic materials'. *Proc. Electrochem. Soc.*, **94(2)**, 1994, 158–69, by permission of The Electrochemical Society, Inc.)

the 'coloration efficiency' for such systems is also little dependent on λ, varying only between 73 cm^2 C^{-1} at 550 nm and 77 cm^2 C^{-1} at 700 nm, with a fairly high contrast ratio of 25:1,[18] reflecting as much as 60% of all incident visible light.[20]

A bismuth-based ECD has been marketed commercially by the Polyvision Corporation.[20,23]

Lead

Metallic lead may be electrodeposited[25,26] onto ITO from aqueous solutions of Pb(NO$_3$)$_2$; see Eq. (9.3):

$$Pb^{2+}(aq) + 2e^- \rightarrow Pb^0(s). \quad (9.3)$$

colourless   opaque

Similarly to bismuth, traces of copper are added to the colourless precursor solution as a mediator.[26] However, the Cu^{2+} is not merely a mediator, it also affects the morphology of the deposit, effecting increased transmittance changes by up to 60%. Copper(II) chloride in the electrolyte also leads to a more homogeneous deposit on the ITO surface. The use of bromide ion to mediate the underpotential deposition of Pb has also been investigated.[27]

Silver

Thin films of silver have also been prepared by electrodeposition from Ag^+ ion onto OTEs,[28] Eq. (9.4).

$$Ag^+(aq) + e^- \rightarrow Ag^0(s). \tag{9.4}$$

A thin film of non-particulate, continuous metallic plate is formed. ('Electrochromism' was not referred to in this 1962 work, done prior to Deb's use of the term in 1969.)

9.4 Reflecting metal hydrides

An impressive example of electrochromes showing specular reflectance are the lanthanide hydride devices, sometimes called 'switchable mirrors'.[29] The reflective properties are those of the electrochrome, not any underlying substrate. Thin-film LaH_2 exhibits specular reflection of this sort, but chemical oxidation to form LaH_3 results in a loss of the metallicity and hence the reflectivity. Chemical reaction therefore causes switching between reflective and non-reflective states, Eq. (9.5):

$$LaH_2(s) + H^-(soln.) \rightarrow LaH_3(s) + e^-. \tag{9.5}$$

      reflective   non-reflective

The cause of the change in reflectivity is a metal–insulator transition. Although dramatic changes in optical and electrical properties accompany such transitions, their interpretation is complicated by attendant changes in crystallographic structure; such changes are expected as such electronic transitions require changes in nuclear spin. For these reasons, Eq. (9.5) is not a mechanistically comprehensive representation of the redox reaction.

Yttrium, lanthanum and the trivalent rare-earth elements all form hydrides that exhibit such transitions. The transition time scale is about a few seconds. The transition from a metallic state (YH_2 or LaH_2) to a semiconducting state (YH_3 or LaH_3) occurs during the continuous absorption of hydrogen, accompanied by profound changes in their optical properties.

The extreme reactivity and fragility of these materials preclude their ready utilisation. To overcome these problems, thin films of hydride are coated with a thin layer of palladium, through which hydrogen can diffuse, presumably forming atomic hydrogen. While the palladium layer also catalyses the adsorption and desorption of hydrogen,[30,31] it also limits the maximum visible transmittance of the hydride layer to about 35–40%.[32]

Alloys of lanthanum also show this reflective transition. For example, magnesium–lanthanide alloys can pass through three different optical states: a colour-neutral, transparent state at high pressures of hydrogen; a dark, non-transparent state at intermediate pressures of hydrogen; and a highly reflective metallic state at low pressures of hydrogen. The optical properties of alloys are also preferred because their colours contrast with the red–yellow colour of the transparent lanthanide states,[33] thereby lending them a 'neutral hue'.[29,34] Furthermore, the La–Mg alloy has virtually no transmittance at high pressures of hydrogen. von Rottkay suggests the change in reflectivity is about 50% for Mg–La hydride.[32] The coloration efficiency η of thin-film $Sm_{0.3}Mg_{0.7}H_x$ is slightly lower than for H_xWO_3.[35]

The use of hydrogen gas effects a very rapid optical transition, but elemental H_2 is neither safe nor an attractive proposal for a viable device. Notten et al.[36] have more recently shown how the same effect can be observed with the lanthanum film immersed in aqueous KOH (1 mol dm^{-3}), depicted in Eq. (9.6) for lanthanum hydride via an electrochemical reaction:

$$LaH_x(s) + yOH^-(aq) \rightarrow LaH_{(x-y)}(s) + yH_2O + ye^-. \quad (9.6)$$

In this way, more typical ECDs can be fabricated in which a clear, solid electrolyte layer allows the transport of hydrogen.[29] The main technological drawbacks at present are the formation of an oxide layer between the lanthanum and the palladium top-coat (cf. the operation of palladium oxides electrochromes on p. 178) and slower colouration kinetics than with H_2 gas.[29,37] Alternatively, van der Sluis et al.[38] show that thin films of lanthanide hydride can be switched from absorbing to transparent with aqueous $NaBH_4$ solution. The reverse reaction can be accomplished with an aqueous H_2O_2 solution. The optical properties of these films are similar to those of films switched electrochemically or exposed to hydrogen gas.

No yttrium-based reflective devices are ready for marketing, but rapid technological advances are likely. Janner et al.[37] have examined the durability of lanthanide hydride films immersed in aqueous KOH solution. Typically, the macroscopic effects of degeneration upon cycling of the switchable mirror include slower rates of coloration and bleaching, irreversible oxidation of the metal hydride films, and delamination as the films peel from their substrates. Of the various attempts to improve the cycle lifetime, the best results were obtained with switchable mirrors pre-loaded with hydrogen during deposition.

9.5 Other miscellaneous inorganic electrochromes

Electrochromism has also been reported for the other miscellaneous inorganic materials such as nickel-doped strontium titanate, $SrTiO_3$;[39] indium nitride,[40] ruthenium dithiolene,[41] phosphotungstic acid,[42,43,44] organic ruthenium complexes,[45] and ferrocene–naphthalimides dyads.[46]

References

1. Klein, J. D., Yen, A., Rauh, R. D. and Causon, S. L. Near-infrared electrochromism in Li_xC_{60} films. *Appl. Phys. Lett.*, **63**, 1993, 599–601.
2. Konesky, G. A. Reversible electrochromic effect in fullerene thin films utilizing alkali and transition metals. *Mater. Res. Soc. Symp. Proc.*, **417**, 1996, 407–13.
3. Konesky, G. A. Pulse-width modulation effects on fullerene electrochromism. *Proc. SPIE*, **3788**, 1999, 14–21.
4. Konesky, G. Fullerene electrochromism under high pulsed fields. Proceedings of the Annual Technical Conference: Society of Vacuum Coaters, Boston, MA, 18–23 April 1998, pp. 144–6.
5. Konesky, G. Stability and reversibility of the electrochromic effect in fullerene thin films. *Proc. SPIE*, **3142**, 1997, 205–15.
6. de Torresi, S. I. C., Torresi, R. M., Ciampi, G. and Luengo, C. A. Electrochromic phenomena in fullerene thin films. *J. Electroanal. Chem.*, **377**, 1994, 283–5.
7. Goldenberg, L. M. Electrochemical properties of Langmuir–Blodgett films. *J. Electroanal. Chem.*, **379**, 1994, 3–19.
8. Pfluger, P., Künzi, H. U. and Güntherodt, H. J. Discovery of a new reversible electrochromic effect. *Appl. Phys. Lett*, **35**, 1979, 771–2.
9. Kuwabara, K. and Noda, Y. Potential wave-form measurements of an electrochromic device, $WO_3/Sb_2O_5/C$, at coloration–bleaching processes using a new quasi-reference electrode. *Solid State Ionics*, **61**, 1993, 303–8.
10. Yu, P., Popov, B. N., Ritter, J. A. and White, R. E. Determination of the lithium ion diffusion coefficient in graphite. *J. Electrochem. Soc.*, **146**, 1999, 8–14.
11. Kulak, A. I., Kokorin, A. I., Meissner, D., Ralcherko, V. G., Vlasou, I. I., Kondratyuk, A. V. and Kulak, T. I. Electrodeposition of nanostructured diamond-like films by oxidation of lithium acetylide. *Electrochem. Commun.*, **5**, 2003, 301–5.
12. Wang, J., Tian, B. M., Nascomento, V. B. and Angnes, L. Performance of screen-printed carbon electrodes fabricated from different carbon inks. *Electrochim. Acta*, **43**, 1998, 3459–65.
13. Edwards, M. O. M., Andersson, M., Gruszecki, T., Pettersson, H., Thunman, R., Thuraisingham, G., Vestling, L. and Hagfeldt, A. Charge–discharge kinetics of electric-paint displays. *J. Electroanal. Chem.*, **565**, 2004, 175–84.
14. Edwards, M. O. M., Boschloo, G., Gruszecki, T., Pettersson, H., Sohlberg, R. and Hagfeldt, A. 'Electric-paint displays' with carbon counter electrodes. *Electrochim. Acta*, **46**, 2001, 2187–93.
15. Edwards, M. O. M., Gruszecki, T., Pettersson, H., Thuraisingham, G. and Hagfeldt, A. A semi-empirical model for the charging and discharging of electric-paint displays. *Electrochem. Commun.*, **4**, 2002, 963–7.

16. Asano, T., Kubo, T. and Nishikitani, Y. Durability of electrochromic windows fabricated with carbon-based counterelectrode. *Proc. SPIE*, **3788**, 1999, 84–92.
17. Nishikitani, Y., Asano, T., Uchida, S. and Kubo, T. Thermal and optical behavior of electrochromic windows fabricated with carbon-based counterelectrode. *Electrochim. Acta*, **44**, 1999, 3211–17.
18. Ziegler, J. P. Status of reversible electrodeposition electrochromic devices. *Sol. Energy Mater. Sol. Cells*, **56**, 1999, 477–93.
19. Ziegler, J. P. and Howard, B. M. Applications of reversible electrodeposition electrochromic devices. *Sol. Energy Mater. Sol. Cells*, **39**, 1995, 317–31.
20. Howard, B. M. and Ziegler, J. P. Optical properties of reversible electrodeposition electrochromic materials. *Sol. Energy Mater. Sol. Cells*, **39**, 1995, 309–16.
21. de Torresi, S. I. C. and Carlos, I. A. Optical characterization of bismuth reversible electrodeposition. *J. Electroanal. Chem.*, **414**, 1996, 11–16.
22. Ziegler, J. P. and Howard, B. M. Spectroelectrochemistry of reversible electrodeposition electrochromic materials. *Proc. Electrochem. Soc.*, **94–2**, 1994, 158–69.
23. Richards, T. C. and Brzezinski, M. R. Oxidation mechanism for reversibly electrodeposited bismuth in electrochromic devices. 121st Electrochemical Society Meeting, Montreal, Canada, 6 May 1997, abstract 945.
24. de Oliveira, S. C., de Morais, L. C., da Silva Curvelo, A. A. and Torresi, R. M. Improvement of thermal stability of an organic–aqueous gel electrolyte for bismuth electrodeposition devices. *Sol. Energy Mater. Sol. Cells*, **85**, 2005, 489–97.
25. Mascaro, L. H., Kaibara, E. K. and Bulhões, L. A. An electrochromic system based on redox reactions. *Proc. Electrochem. Soc.*, **96–24**, 1996, 96–105.
26. Mascaro, L. H., Kaibara, E. K. and Bulhões, L. A. An electrochromic system based on the reversible electrodeposition of lead. *J. Electrochem. Soc.*, **144**, 1997, L273–4.
27. Marković, N. M., Grgur, B. N., Lucas, C. A., and Ross, jr, P. N. Underpotential deposition of lead on Pt(111) in the presence of bromide: $RRD_{Pt(111)}$ E and X-ray scattering studies. *J. Electroanal. Chem.*, **448**, 1998, 183–8.
28. Mantell, J. and Zaromb, S. Inert electrode behaviour of tin oxide-coated glass on repeated plating–deplating cycling in concentrated NaI–AgI solutions. *J. Electrochem. Soc.*, **109**, 1962, 992–3.
29. van der Sluis, P. and Mercier, V. M. M. Solid state Gd–Mg electrochromic devices with $ZrO_2 H_x$ electrolyte. *Electrochim. Acta*, **46**, 2001, 2167–71.
30. Huiberts, J. N., Griessen, R., Rector, J. H., Wijngaarden. R. J., Decker, J. P., de Groot, D. G. and Koeman, N. J. Yttrium and lanthanum hydride films with switchable optical properties. *Nature (London)*, **380**, 1996, 231–4.
31. Huiberts, J. N., Rector, J. H., Wijngaarden, R. J., Jetten, S., de Groot, D. G., Dan, B., Koeman, N. J., Griessen, R., Hjörvarsson, B., Olafsson, S. and Cho, Y. S. Synthesis of yttrium trihydride films for *ex-situ* measurements. *J. Alloys Compd.*, **239**, 1996, 158–71.
32. von Rottkay, K., Rubin, M., Michalak, F., Armitage, R., Richardson, T., Slack, J. and Duine, P. A. Effect of hydrogen insertion on the optical properties of Pd-coated magnesium lanthanides. *Electrochim. Acta*, **44**, 1999, 3093–100.
33. Kooij, E. S., van Gogh, A. T. M. and Griessen, R. *In situ* resistivity measurements and optical transmission and reflection spectroscopy of electrochemically loaded switchable YH_x films. *J. Electrochem. Soc.*, **146**, 1999, 2990–4.
34. van der Sluis, P., Ouwerkerk, M. and Duine, P. A. Optical switches based on magnesium lanthanide alloy hydrides. *Appl. Phys. Lett.*, **70**, 1997, 3356–8.

35. Ouwerkerk, M. Electrochemically induced optical switching of $Sm_{0.3}Mg_{0.7}H_x$ thin layers. *Solid State Ionics*, **113–15**, 1998, 431–7.
36. Notten, P. L. H., Kremers, M. and Griessen, T. R. Optical switching of Y-hydride thin film electrodes: a remarkable electrochromic phenomenon. *J. Electrochem. Soc.*, **143**, 1996, 3348–53.
37. Janner, A.-M., van der Sluis, P. and Mercier, V. Cycling durability of switchable mirrors. *Electrochim. Acta*, **46**, 2001, 2173–8.
38. van der Sluis, P. Chemochromic optical switches based on metal hydrides. *Electrochim. Acta*, **44**, 1999, 3063–6.
39. Mohapatra, S. K. and Wagner, S. Electrochromism in nickel-doped strontium titanate. *J. Appl. Phys.*, **50**, 1979, 5001–6.
40. Ohkubo, M., Nonomura, S., Watanabe, H., Gotoh, T., Yamamoto, K. and Nitta, S. Optical properties of amorphous indium nitride films and their electrochromic and photodarkening effects. *Appl. Surf. Sci.*, **1130–14**, 1997, 476–9.
41. García-Canãdas, J., Meacham, A. P., Peter, L. M. and Ward, M. D. Electrochromic switching in the visible and near IR with a Ru–dioxolene complex adsorbed on a nanocrystalline SnO_2 electrode. *Electrochem. Commun.*, **5**, 2003, 416–20.
42. Tell, B. Electrochromism in solid phosphotungstic acid. *J. Electrochem. Soc.*, **127**, 1980, 2451–4.
43. Medina, A., Solis, J. L., Rodriguez, J. and Estrada, W. Synthesis and characterization of rough electrochromic phosphotungstic acid films obtained by spray-gel process. *Sol. Energy Mater. Sol. Cells*, **80**, 2003, 473–81.
44. Tell, B. and Wudl, F. Electrochromic effects in solid phosphotungstic acid and phosphomolybdic acid. *J. Appl. Phys.*, **50**, 1979, 5944–6.
45. Qi, Y. H., Desjardins, P., Meng, X. S. and Wang, Z. Y. Electrochromic ruthenium complex materials for optical attenuation. *Opt. Mater.*, **21**, 2003, 255–63.
46. Gan, J., Tian, H., Wang, Z., Chen, K., Hill, J., Lane, P. A., Rahn, M. D., Fox, A. M. and Bradley, D. D. C. Synthesis and luminescence properties of novel ferrocene–naphthalimides dyads. *J. Organometallic Chem.*, **645**, 2002, 168–75.

10

Conjugated conducting polymers

10.1 Introduction to conjugated conducting polymers

10.1.1 Historical background and applications

The history of conjugated conducting polymers or 'synthetic metals' can be traced back to 1862, when Letheby, a professor of chemistry in the College of London Hospital, reported the electrochemical synthesis of a 'thick layer of dirty bluish-green pigment' (presumably a form of 'aniline black' or poly(aniline)) by oxidation of aniline in sulfuric acid at a platinum electrode.[1] However, widespread interest in these fascinating materials did not take place until after 1977, following the discovery[2,3,4] of the metallic properties of poly(acetylene), which led to the award of the 2000 Nobel Prize in Chemistry to Shirakawa, Heeger and MacDiarmid.[5,6] Since 1977, electroactive conducting polymers have been intensively investigated for their conducting, semiconducting and electrochemical properties. Numerous electronic applications have been proposed and some realised, including electrochromic devices (ECDs), electroluminescent organic light-emitting diodes (OLEDs),[7,8] photovoltaic elements for solar-energy conversion,[9] sensors[10] and thin-film field-effect transistors.[11]

10.1.2 Types of electroactive conducting polymers

Poly(acetylene), $(CH)_x$, is the simplest form of conjugated conducting polymer, with a conjugated π system extending over the polymer chain. Its electrical conductivity exhibits a twelve order of magnitude increase when doped with iodine.[2] However, due to its intractability and air sensitivity, poly(acetylene) has seen few applications and most research on conjugated conductive polymers has been carried out with materials derived from aromatic and heterocyclic aromatic structures. Thus, chemical or electrochemical oxidation of numerous resonance-stabilised aromatic molecules, such as pyrrole, thiophene, 3,4-(ethylenedioxy)thiophene (EDOT), aniline, furan, carbazole, azulene, indole (see structures below), and others, produces electroactive conducting polymers.[12,13,14,15,16,17,18,19]

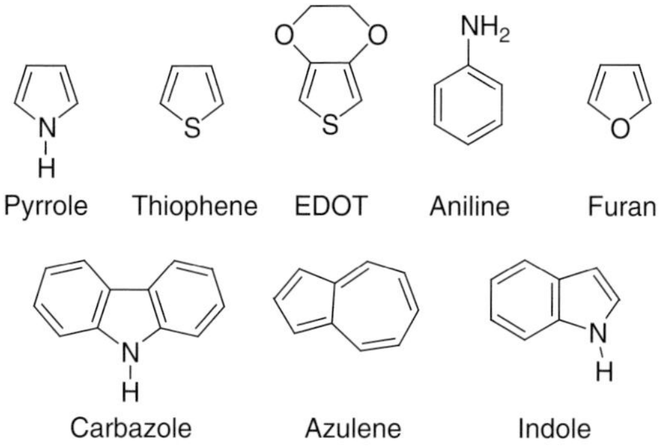

Pyrrole Thiophene EDOT Aniline Furan

Carbazole Azulene Indole

Of the resulting polymers, the poly(thiophene)s, poly(pyrrole)s and poly(aniline)s have received the most attention in regard to their electrochromic properties, and will be discussed in this chapter.

Note that 'electroactive' denotes the capability of interfacial electron transfer in one or other direction (oxidation and/or reduction, i.e. a redox capability that allows of colour change). On the other hand, the enhanced conductivity of a charged state (oxidised or reduced) relative to an uncharged state is an accompaniment that is useful in assisting towards rapid redox change, hence rapid colour change. However, the relation between redox properties and conductivity is not necessarily straightforward and varies from polymer to polymer.

10.1.3 Mechanism of oxidative polymerisation of resonance-stabilised aromatic molecules

Polymerisation begins with the formation of an oxidatively generated monomer radical cation. The succeeding mechanism is believed to involve either coupling between radical cations, or reaction of a radical cation with a neutral monomer. As an example, the electropolymerisation mechanism for the five-membered heterocycle, pyrrole, showing radical cation–radical cation coupling is given in Scheme 10.1.

After the loss of two protons and re-aromatisation, the pyrrole dimer forms from the corresponding dihydro dimer dication. The dimer (and succeeding oligomers) are more easily oxidised than the monomer and the resulting dimer radical cation undergoes further coupling reactions, proton loss and

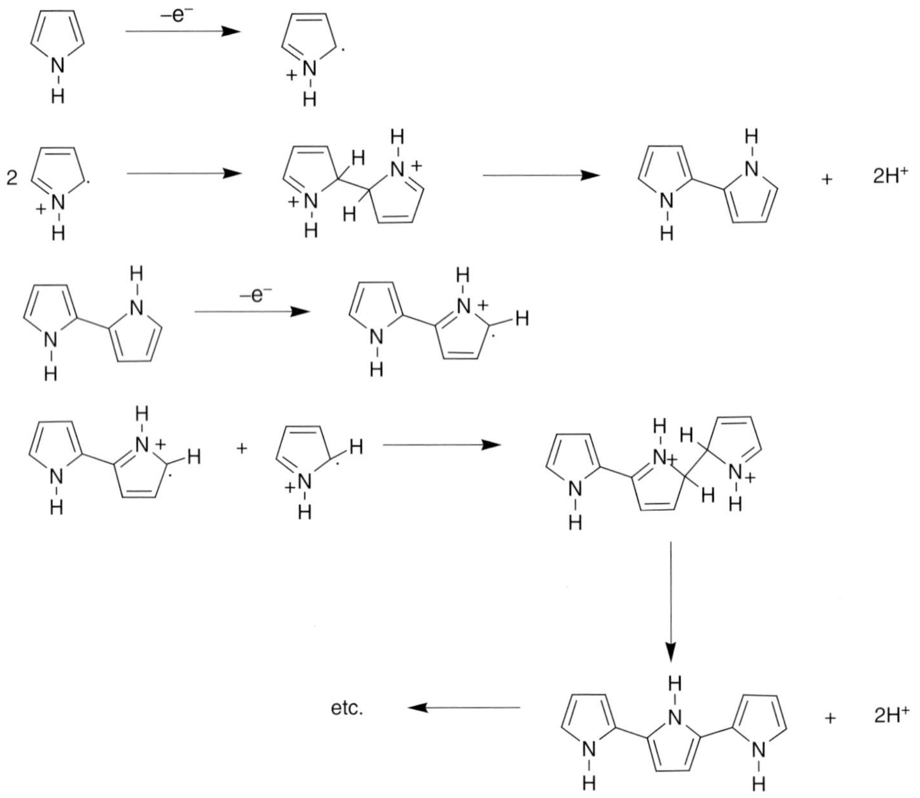

Scheme 10.1 Proposed mechanism of the electropolymerisation of pyrrole. The case of radical cation–radical cation coupling is shown.

re-aromatisation. Electropolymerisation proceeds through successive electrochemical and chemical steps according to a general E(CE)$_n$ scheme,[20] until the oligomers become insoluble in the electrolyte solution and precipitate (like a salt) as the electroactive conducting polymer. Films of high-quality oxidised polymer can be formed directly onto electrode surfaces.[16]

10.1.4 Conductivity and optical properties

Electronic conductivity in electroactive polymers results from the extended conjugation within the polymer, longer chains promoting high conductivity. The average number of linked monomer units within a conducting polymer is often termed the 'conjugation length'. X-Ray diffraction of pyrrole oligomers suggests the poly(pyrrole) rings to be coplanar[21] but substitution at nitrogen and the β-carbon introduces a significant twist in the polymer backbone, imposing a non-zero dihedral angle ϕ. Note that $\phi \neq 0$ if $R^1 \neq H$ and $R^2 \neq H$.

10.1 Introduction

In the conducting oxidised state with positive charge carriers, electroactive conducting polymers are charge-balanced (doped) with counter anions ('p-doping') and have delocalised π-electron band structures,[16] with typical conductivity values in the range 10^1–10^5 S cm^{-1}. Figure 10.1 shows illustrative conductivity ranges for poly(acetylene), poly(thiophene) and poly(pyrrole). Values of σ are compared with those for common metals, semiconductors and insulators. Reduction of such p-doped conducting polymers, with concurrent

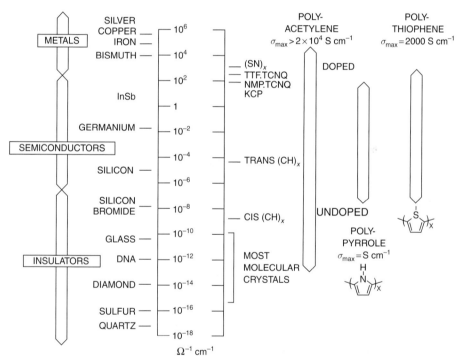

Figure 10.1 The conductivity range available with electroactive conducting polymers spans those common for metals through to insulators. (Figure reproduced from Thomas, C. A. 'Donor–Acceptor methods for band gap reduction in conjugated polymers: the role of electron rich donor heterocycles'. Ph.D. Thesis, Department of Chemistry, University of Florida, 2002, p. 17, by permission of the author, who adapted it from the *Handbook of Conducting Polymers*.[18])

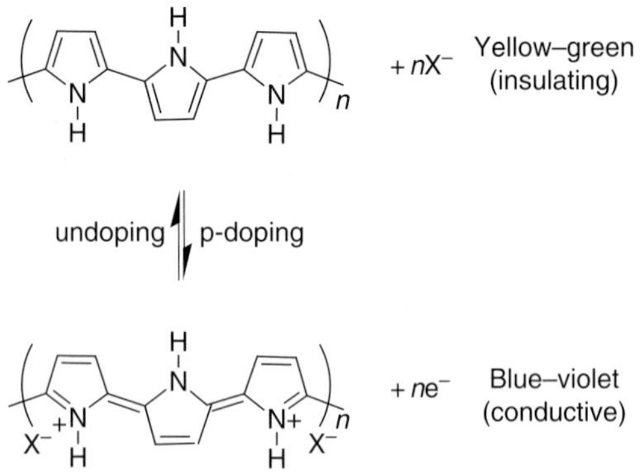

Scheme 10.2 Electrochromism in poly(pyrrole) thin films. The yellow-green (undoped) form undergoes reversible oxidation to the blue-violet (conductive) form, with insertion of charge-compensating anions.

counter-anion egress to, or cation ingress from, the electrolyte, removes the electronic conjugation, that results in the undoped (that is to say, electrically neutral) insulating form. The magnitude of the conductivity change depends on the extent of doping, which, when under electrochemical control, can be adjusted by the applied potential.

The energy gap E_g, the electronic bandgap between the highest-occupied π-electron band (the valence band) and the lowest-unoccupied band (the conduction band), determines the intrinsic optical properties of these materials. This is illustrated in Scheme 10.2, which gives the electrochromic colour states in thin films of poly(pyrrole): the non-conjugation of the oxidised form, that allows visibly evident photo-excitation, provides the coloured structure, as explained in detail towards the end of this section. In the reduced form, such neutral polymers are typically semiconductors and exhibit an aromatic form with alternating double and single bonds in the polymer backbone. On oxidative doping, radical cation charge carriers (polarons) are generated, and the polymer assumes a quinoidal bonding state that facilitates charge transfer along the backbone. Further oxidation results in the formation of dication charge carriers (bipolarons).

In some instances, the undoped (electrically neutral) state of electroactive conducting polymers can undergo reductive cathodic doping or n-doping, with accompanying cation insertion to balance the injected charge. This doping has been exploited in the development of a model ECD using poly{cyclopenta[2,1-b;4,3-b']dithiophen-4-(cyanononafluorobutylsulfonyl)methylidene} (PCNFBS), a low-bandgap conducting polymer that is both p- and n-dopable, as both the anode and the cathode material.[22] The polymer PCNFBS is one of a series of fused bithiophene polymers whose E_g values can be controlled by inclusion (initially in the precursor monomers) of electron-withdrawing substituents. Electrochemically polymerised films of the polymer switch from red in the neutral state to purple in both the p- and n-doped states.[22] The spectral changes observed in an electrochemical cell assembled from two polymer-coated transparent electrodes were a combination of those seen in the separate p- and n-doped films.[22] Although this is a fascinating example, the stability of negatively charged polymer states is generally limited, and n-doping is difficult to achieve.

It is to be noted that the 'p-doping' and 'n-doping' nomenclature comes from classical semiconductor theory. The supposed similarity between conducting polymers and doped semiconductors arises from the manner in which the redox changes in the polymer alter its optoelectronic properties. In fact, the suitability of the terms 'doping' and 'dopant' has been criticised[23] when they refer to the movement of counter ions and electronic charge through these polymers, because in its initial sense doping involved minute (classically, below ppm) amounts of dopant. However, 'doping' and similar terms are now so widely used in connection with conjugated conducting polymers that attempts to change the terminology could cause confusion.

As already noted in the case of poly(pyrrole), in fact all thin films of electroactive conducting polymers have electrochromic possibilities, since redox switching involving ingress or egress of counter ions gives rise to new optical absorption bands and allows transport of electronic charge in the polymer matrix. Electroactive conducting polymers are type-III electrochromes since they are permanently solid. Oxidative p-doping shifts the optical absorption band towards the lower energy part of the spectrum. The colour change or contrast between doped and undoped forms of the polymer depends on the magnitude of the bandgap of the undoped polymer. Thin films of conducting polymers with E_g greater than 3 eV,[a] which gives a corresponding spectroscopic value of λ_{max} of ~400 nm, are colourless and transparent in the undoped form, while in the doped form they generally absorb in the visible region. Those with E_g equal to or less than 1.5 eV (~800 nm) are highly absorbing in the undoped form but, after doping, the free carrier absorption is relatively weak in the visible region as it is transferred to the near infrared (NIR) part of the spectrum. Polymers with a bandgap of intermediate magnitude have distinct optical changes throughout the visible region, and can be made to induce many colour changes.

---

[a] $1 eV = 1.602 \times 10^{-19} J$.

10.1.5 Previous reviews of electroactive conducting polymer electrochromes

A vast literature encompasses the electrochromism of electroactive conducting polymers, and many reviews are available, including 'Application of polyheterocycles to electrochromic display devices' by Gazard[24] (in 1986), 'Electrochromic devices' by Mastragostino[25] (in 1993), 'Electrochromism of conducting polymers' by Hyodo[26] (in 1994), Chapter 9 of *Electrochromism: Fundamentals and Applications* by Monk, Mortimer and Rosseinsky[12] (in 1995), 'Organic electrochromic materials' by Mortimer (in 1999),[27] 'Electrochromic polymers' by Mortimer (in 2004),[28] 'Polymeric electrochromics' by Sonmez (in 2005)[29] and 'Electrochromic organic and polymeric materials for display applications' by Mortimer *et al.* (in 2006).[30]

10.2 Poly(thiophene)s as electrochromes

10.2.1 Introduction to poly(thiophene)s

Poly(thiophene)s[16,19,31] are of interest as electrochromes due to their relative ease of chemical and electrochemical synthesis, environmental stability, and processability.[31] A vast number of substituted thiophenes has been synthesised, which has led to the study of numerous novel poly(thiophene)s, with particular emphasis on poly(3-substituted thiophene)s and poly(3,4-disubstituted thiophene)s.[16] Thin polymeric films of the parent poly(thiophene) are blue ($\lambda_{max} = 730$ nm) in the doped (oxidised) state and red ($\lambda_{max} = 470$ nm) in the undoped form. However, due to its lower oxidation potential,[b] the electropolymerisation and switching of β-methylthiophene has been more intensively studied than the unsubstituted parent thiophene. Furthermore, the introduction of a methyl group at the 3-position of the thiophene ring leads to a significant increase of the polymer conjugation length and hence electronic conductivity.[16] This effect has been attributed to the statistical decrease in the number of insulative α–β′ couplings and also to the decrease of the oxidation potential caused by the inductive (electron-donating) effect of the methyl group.[16] Poly(3-methylthiophene) is purple when neutral with an absorption maximum at 530 nm (2.34 eV), and turns pale blue upon oxidation.[32]

---

[b] When oxidation processes predominate in discussion, it is convenient to cite *oxidation* potentials, which are for processes that are the reverse of the conventional half reactions (i.e. reductions) of Chapter 3. In the present chapter, positive values are implied: the greater the value, the more positive (and the more oxidising) is the potential that is applied to the electrode under consideration.

10.2 Poly(thiophene)s as electrochromes

The evolution of the electronic band structure during electrochemical p-doping of electrochromic polymers can be followed by recording *in situ* visible and NIR spectra as a function of applied electrode potential. Figure 10.2 shows the spectroelectrochemical series for an alkylenedioxy-substituted thiophene polymer, poly[3,4-(ethylenedioxy)thiophene] – PEDOT, which exhibits a deep blue colour in its neutral state and a light blue transmissive state upon oxidation.[33] The strong absorption band of the undoped polymer, with a maximum at 621 nm (2.0 eV), is characteristic of a π–π* interband transition. Upon doping, the interband transition decreases, and two new optical transitions (at ~1.25 and ~0.80 eV) appear at lower energy, corresponding to the presence of a polaronic charge carrier (a single charge of spin ½). Further oxidation leads

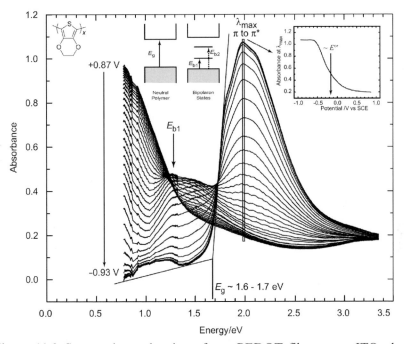

Figure 10.2 Spectroelectrochemistry for a PEDOT film on an ITO–glass substrate. The film had been deposited from EDOT (0.3 mol dm^{-3}) in propylene carbonate solution containing tetrabutylammonium perchlorate (0.1 mol dm^{-3}) and spectra are shown on switching in tetrabutylammonium perchlorate (0.1 mol dm^{-3}) in acetonitrile. The inset shows absorbance vs. potential. The bandgap is determined by extrapolating the onset of the π to π* absorbance to the background absorbance. The E_{b1} transition is allowed and is visible at intermediate doping levels. (Figure reproduced from Thomas, C. A. 'Donor–Acceptor methods for band gap reduction in conjugated polymers: the role of electron rich donor heterocycles'. Ph.D. Thesis, Department of Chemistry, University of Florida, 2002, p. 41, by permission of the author.)

to formation of a bipolaron and the absorption is enhanced at lower energies, i.e. the colour shifts towards the characteristic absorption band of the free carrier of the metallic-like state, which appears when the bipolaron bands finally merge with the valence and conduction bands. In such electroactive conducting polymers, the optical and structural changes are often reversible through repeated doping and de-doping over many thousands of redox cycles.

10.2.2 Poly(thiophene)s derived from substituted thiophenes and oligothiophenes

As already noted above in the comparison of poly(thiophene) and poly(3-methylthiophene), tuning of colour states can be achieved by suitable choice of thiophene monomer. This tuning represents a major advantage of using conducting polymers for electrochromic applications. Subtle modifications to the thiophene monomer can significantly alter spectral properties. A recent example is provided by cast films of chemically polymerised thiophene-3-acetic acid, which reversibly switch from red to black on oxidation.[34]

There has been much interest in polymer films derived from electrochemical oxidation of thiophene-based monomers that comprise more than one thiophene heterocyclic unit. The species containing two thiophene units (joined at the α-carbon, i.e. that next to S) is called bithiophene, while compounds containing three or more thiophene units have the general name of 'oligothiophene'. It has been shown[35] that the wavelength maxima of undoped poly(oligothiophene) films decrease as the length of the oligothiophene monomer increases, Table 10.1. The oxidation potentials included in this table do not vary much with oligothiophene.

The colours available with polymer films prepared from 3-methylthiophene-based oligomers are strongly dependent on the relative positions of methyl groups on the polymer backbone.[32,36] As listed in Table 10.2, these include pale blue, blue and violet in the oxidised form, and purple, yellow, red and orange in the reduced form. The colour variations have been ascribed to changes in the effective conjugation length of the polymer chain.

To investigate the effect of the dihedral angle ϕ between thiophene planes, oligothiophenes containing alkyl groups at the β-carbon have been synthesised.[35] Groups at the β-carbon cause steric hindrance, whereas bridged species (exemplified in Scheme 10.3 below) are linear. The results in Table 10.3 show that those polymers with the smallest dihedral angle ϕ generally have the highest wavelength maxima. Oxidation potentials are generally unaffected by variations in ϕ.

Further study of the effects of steric factors is provided by the electronic properties of poly(thiophene)s with 3,4-dialkyl substituents. In principle,

10.2 Poly(thiophene)s as electrochromes

Table 10.1. *Wavelength maxima and oxidation potentials of polymers derived from oligothiophenes (based on ref. 35).*

Monomer[a]	λ_{max}/nm[b] (undoped)	E_{ox}/V
(thiophene)	519	0.95
(bithiophene)	484	1.00
(terthiophene)	356	1.04
(quaterthiophene)	340	0.93

[a] Note that these structures do not represent the molecular stereochemistry. [b] Wavelength maximum refers to the reduced (undoped) redox state of the polymer.

disubstitution at the β, β' positions should provide the synthetic basis to perfectly stereoregular polymers. However, this approach is severely limited by the steric interactions between substituents, which lead to a decrease in polymer conjugation length. In fact, poly(3,4-dialkylthiophene)s have higher oxidation potentials, higher optical bandgaps, and lower conductivities than poly(3-alkylthiophene)s.[16] Alternation between the 3 and 4 positions relieves steric hindrance in thiophenes, but many are harder to electropolymerise than, say, 3-methylthiophene. The electron-donating effect of alkoxy groups offers an answer here, and alkoxy-substituted poly(thiophene)s are being intensively investigated for their electrochromic properties.[37,38]

10.2.3 Poly(thiophene)s derived from 3,4-(ethylenedioxy)thiophenes

Materials based on PEDOT have a bandgap lower than either poly(thiophene) or alkyl-substituted poly(thiophene)s, owing to the presence of the two electron-donating oxygen atoms adjacent to the thiophene unit. Scheme 10.3 shows the

Table 10.2. *Colours of polymers derived from oligomers based on 3-methylthiophene (based on ref. 15).*

Monomer	λ_{max}/nm (undoped)	Polymer colour (reduced form)	Polymer colour (oxidised form)
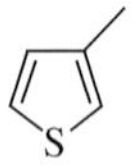	530	Purple	Pale blue
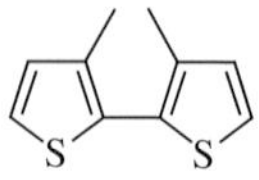	415	Yellow	Violet
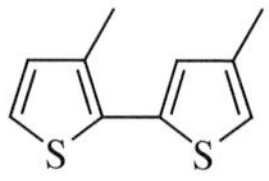	505	Red	Blue
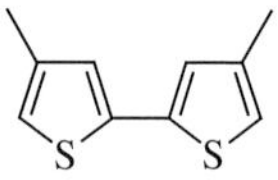	450	Orange	Blue
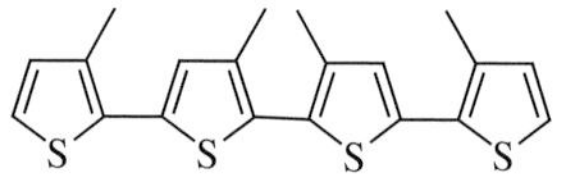	425	Yellow	Blue
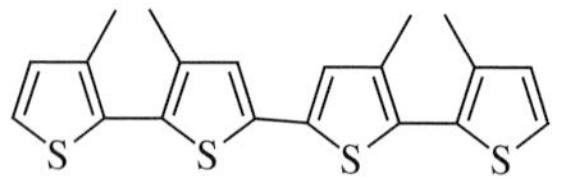	405	Yellow	Violet
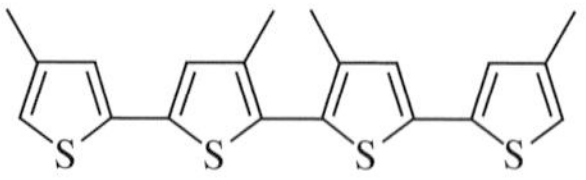	410	Yellow	Blue–violet
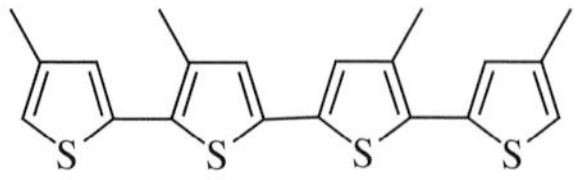	425	Yellow–orange	Blue

structural changes of PEDOT upon reproducible electrochemical oxidation and reduction. The attributes of ethylenedioxy substitution are also pointed out in the figure.

As shown above, the bandgap of PEDOT ($E_g = 1.6–1.7$ eV) itself is 0.5 eV lower than poly(thiophene), which results in an absorbance maximum in the red region of the electromagnetic spectrum. Compared with other substituted poly(thiophene)s, these materials exhibit excellent stability in the doped state, which has a high electronic conductivity. The polymer PEDOT was first

Table 10.3. *Effect of the dihedral angle ϕ: Spectroscopic and electrochemical characteristics of poly(oligothiophene)s (based on ref. 35).*

Monomer	λ_{max}/nm (undoped)	E_{ox}/V
(bithiophene)	484	1.00
(3-methyl bithiophene)	475	0.96
(3,3'-dimethyl bithiophene)	420	0.99
(bridged bithiophene)	413	0.88
(cyclopenta-fused bithiophene)	550	0.90
(terthiophene)	356	1.04
(dimethyl terthiophene)	375	0.94

developed by Bayer AG research laboratories in Germany in an attempt to produce an easily oxidised, soluble and stable conducting polymer.[39,40] Bayer AG now produce the EDOT monomer, 3, 4-(ethylenedioxy)thiophene,[41] on a multi-ton scale and it is available commercially as BAYTRON M. To aid processing, the insolubility of PEDOT can be overcome by the use of a water-soluble polyelectrolyte – poly(styrene sulfonate), PSS – as the counter ion in the doped state, to yield the commercially available product PEDOT:PSS BAYTRON P by Bayer AG and ORGATRON by AGFA Gevaert, which forms a dispersion in water.

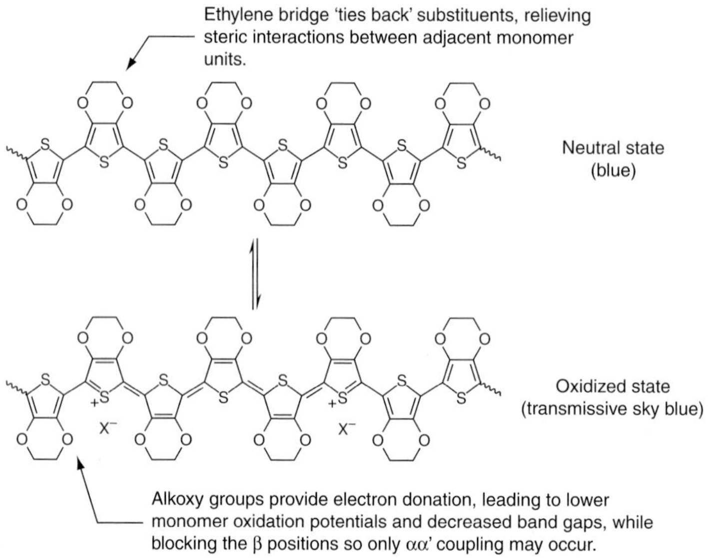

Scheme 10.3 Structural changes of poly[3,4-ethylenedioxythiophene] – PEDOT – upon reproducible electrochemical oxidation and reduction. Attributes of ethylenedioxy substitution are also pointed out. (Figure reproduced from Gaupp, C. L. 'Structure–property relationships of electrochromic 3,4-alkylenedioxyheterocycle-based polymers and co-polymers'. Ph.D. Thesis, Department of Chemistry, University of Florida, 2002, p. 28, by permission of the author.)

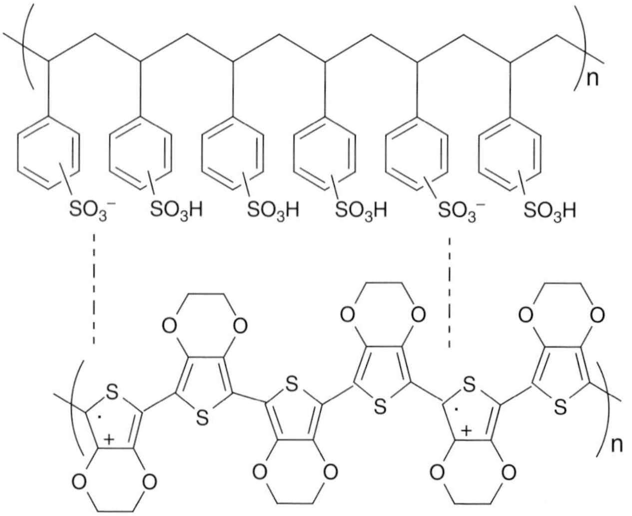

PEDOT:PSS

As PEDOT and its alkyl derivatives are cathodically colouring electrochromic materials, they can be used with anodically colouring conducting polymers

10.2 Poly(thiophene)s as electrochromes

as the other electrode in the construction of dual-polymer ECDs.[42] Changes in the size of the alkylenedioxy ring in general poly[3,4-(alkylenedioxy)thiophene] – PXDOT – materials, and the nature of the substituents on the alkyl bridge, have led to polymers with faster electrochromic switching times,[43,44,45] higher optical contrasts[43,44,45,46] and better processability through increased solubility.[47,48,49,50]

As for thiophene, numerous substituted EDOT monomers have been synthesised, which has led to the study of a range of variable-bandgap PEDOT-based materials.[37,38] The bandgap of such conjugated polymers is controlled by varying the extent of π-overlap along the backbone via steric interactions, and by controlling the electronic character of the π-system with electron-donating or -accepting substituents. The latter is accomplished by using substituents and co-repeat units that adjust the energies of the highest-occupied molecular orbital (HOMO) and lowest-unoccupied molecular orbital (LUMO) of the π-systems.[37,38] An interesting set of materials is the family of EDOT-based polymers which have been prepared with higher energy gaps than the parent PEDOT. From a series of oxidatively polymerisable bis-arylene EDOT monomers (see structures below), polymers with bandgaps in the range 1.4–2.5 eV have been prepared, which exhibit two to three distinct coloured states.[37,38,51,52,53]

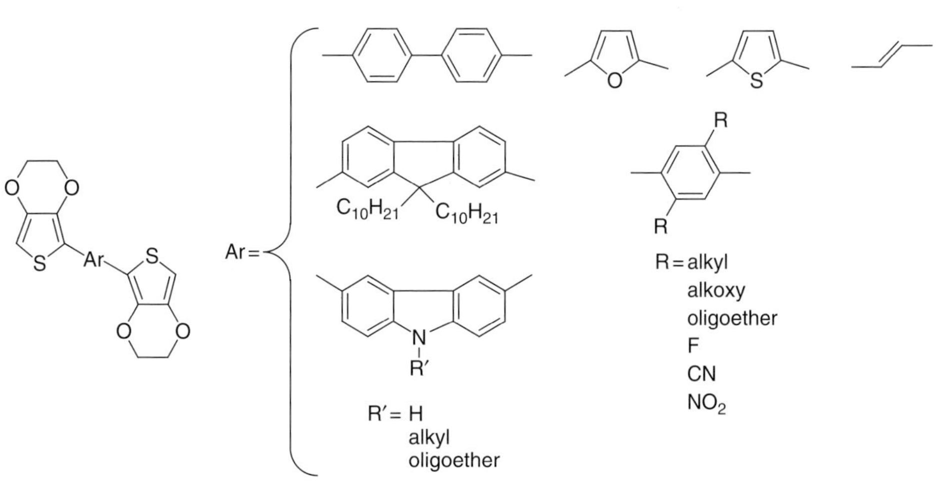

In the neutral polymers, a full 'rainbow' of colours is available, from blue through purple, red, orange, green and yellow as seen in Colour Plate 2. A few examples include bis-arylene EDOT-based polymers, with spacers of vinylene ($E_g = 1.4$ eV) that has a deep-purple neutral state, biphenyl ($E_g = 2.3$ eV) that is orange, p-phenylene ($E_g = 1.8$ eV) that is red, and carbazole ($E_g = 2.5$ eV) that is yellow.[51,52]

Another approach to extend colour choice is electrochemical co-polymerisation from a solution containing two monomers. For example, the ability to adjust the colour of the neutral polymer by electrochemical co-polymerisation has been demonstrated using co-monomer solutions of 2,2′-bis 3,4-ethylenedioxythiophene) – BEDOT – and 3,6-bis[2-(3,4-ethylene-dioxythiophene)]-N-alkylcarbazole – BEDOT-NMeCz.[54] As shown in Colour Plate 3, by varying the ratios of co-monomer concentrations, colours ranging from yellow via red to blue can be evoked in the neutral polymer film.[54] In all co-polymer compositions, the films pass through a green intermediate state to a blue fully oxidised state.[54]

As mentioned previously, some electrochromic conducting polymers also undergo n-type doping. Although n-type doping of most of these polymers results in inherent instability to water and oxygen, the introduction of donor–acceptor units has been shown to increase the stability of this n-type redox state. While incorporation of an electron-rich donor unit allows oxidation for p-doping, the inclusion of an electron-poor acceptor unit allows reduction. This has been shown with EDOT acting as the donor unit and both pyridine (Pyr) and pyrido[3,4-b]pyrazine, i.e. PyrPyr(Ph)$_2$, as the acceptor unit.[55,56] The polymer PBEDOT-Pyr is red in the neutral state. It changes with p-doping to a light-blue colour. Furthermore, it shows a marked blue with n-doping.[55,56] The polymer PBEDOT-PyrPyr(Ph)$_2$ is green when neutral, grey upon p-doping, and magenta upon n-doping.[55,56]

More recently,[57,58,59] a study has been carried out on the development of an electroactive conducting polymer which is green in the neutral state and virtually transparent (very pale brown) in the oxidised state. To achieve this, it was proposed that a polymer backbone be synthesised that contains two well-defined, isolated, conjugated systems which absorb red and blue light. Thus, a 2,3-di(thien-3-yl)-5,7-di(thien-2-yl)thieno[3,4-b]pyrazine (DDTP) monomer that would afford two conjugated chains was designed and synthesised.[57] One chain has electron donor and acceptor groups to decrease the bandgap, which results in absorption of the red light at wavelengths longer than 600 nm; while the other chain absorbs in the blue at wavelengths below 500 nm. Films of poly(DDTP) were synthesised electrochemically on platinum and ITO-coated glass, to obtain the desired green electrochrome in the neutral state. On electrochemical oxidation of the film, the π–π* transitions of both bands are depleted at the expense of an intense absorption band centred in the NIR, which corresponds to low-energy charge carriers. The depletion upon oxidation makes the polymer film more transparent, but, unfortunately, residual absorptions remain in the visible region, giving a transmissive brown colour. The processibility of the poly(DDTP) system has been enhanced by the electrochemical and chemical synthesis of a soluble form of the polymer, using dioctyl-substituted DDTP.[60]

10.2.4 'Star' polymers based on poly(thiophene)s

Star-shaped electroactive conducting polymers, which have a central core with multiple branching points and linear conjugated polymeric arms radiating outward, are now being investigated for electrochromic applications.[61,62,63,64] Examples include star conducting polymers in which the centro-symmetric cores include hyper-branched poly(1,3,5-phenylene) (PP) and poly(triphenylamine) (PTPA), and the radiating arms are regioregular poly(3-hexylthiophene), poly[3,4-(ethylenedioxy)thiophene didodecyloxy-benzene] and poly[dibutyl-3,4-(propylenedioxy)thiophene].[61,62,63,64] These polymers have the advantage that they can be spin coated from a carrier solvent such as tetrahydrofuran (THF), and several can be doped in solution, so that thin films of both doped and undoped forms can be prepared. Despite the branched structure, star polymers self-assemble into thin films with morphological, electrical, and optical properties that reveal a surprisingly high degree of structural order. The polymers, which are smooth and reflecting, all have spectral features that produce a strong band in the visible region for the reduced state and a broad band extending into the NIR for the oxidised state. The colour of the polymers ranges from red via violet to deep blue in the reduced state, and blue to very pale blue in the oxidised state.

10.3 Poly(pyrrole)s and dioxypyrroles as electrochromes

As outlined for poly(thiophene)s, poly(pyrrole)s are also extensively studied for their electrochromic properties, and can easily be chemically or electrochemically synthesised. Again, a wide range of optoelectronic properties are available through alkyl and alkoxy substitution. As noted in Scheme 10.2 above, thin films of the parent poly(pyrrole) are yellow-to-green ($E_g \sim 2.7\,\text{eV}$) in the undoped insulating state and blue-to-violet in the doped conductive state.[65] Poly(pyrrole)s exhibit lower oxidation potentials than their thiophene analogues,[66] and their enhanced compatibility in aqueous electrolytes has led to interest in their use in biological systems.[67]

As for dialkoxy-substituted thiophenes, addition of oxygen at the β-positions lowers the bandgap of the resulting polymer by raising the HOMO level. This fact, combined with the already relatively low oxidation potential for poly(pyrrole), gives the poly(alkylenedioxypyrrole)s the lowest oxidation potential for p-type doping in conducting electrochromic polymers.[68] Poly[3,4-(ethylenedioxy)pyrrole] – PEDOP – exhibits a bright-red colour in its neutral state and a light-blue transmissive state upon oxidation, with a bandgap of 2.05 eV, 0.65 eV lower than that of the parent pyrrole.[69] Furthermore, increasing the ring size of the alkyl bridge has the effect of generating another coloured state at low doping levels.[68] For poly[3,4-(propylenedioxy)pyrrole] – PProDOP – the neutral state is orange, and on intermediate doping passes through brown, and finally to light grey–blue upon full oxidation.[68] Such polychromism is also seen in the substituted PProDOPs and poly[3,4-(butylenedioxy)pyrrole] – PBuDOP.[68]

By effecting substitution at the nitrogen in poly(3,4-alkylenedioxypyrrole)s (i.e. PXDOPs), higher bandgap polymers can be created, which retain their low oxidation potentials.[70] Substitution induces a twist in the polymer backbone, which results in a decrease of the effective π-conjugation, and an increase in the bandgap of the polymer. This bandgap increase results in a blue shift in the π–π* transition absorbance, with the intragap polaron and bipolaron transitions occurring in the visible region.

The nature of the substituent has an effect on the extent to which the π–π* transition is shifted. For N-methyl-PProDOP the bandgap occurs at 3.0 eV, compared to 2.2 eV for PProDOP, and has a purple colour in the neutral state becoming blue when fully oxidised passing through a dark green colour at intermediate extents of oxidation.[70] Both N-[2-(2-ethoxy-ethoxy)ethyl] PProDOP (N-Gly PProDOP) and N-propanesulfonate PProDOP (N-PrS PProDOP) are colourless when fully reduced but coloured upon full oxidation.[70] Both polymers also exhibit multiple coloured states at intermediate extents of oxidation.[70] These two polymers are thus anodically colouring polymers, in that they change from a colourless state to a coloured one upon oxidation, in contrast with cathodically colouring polymers that are coloured in their reduced state and become colourless upon oxidation. These N-substituted polymers have been shown to work effectively in dual-polymer high-contrast absorptive/transmissive ECDs as the anodically colouring material, due to their electrochemical and optical compatibility with various PXDOT polymers.[71]

10.4 Poly(aniline)s as electrochromes

Poly(aniline) films[72] are generally prepared from aqueous solutions of aniline in strong mineral acids.[73] Several redox mechanisms involving protonation–deprotonation and/or anion ingress/egress have been proposed.[74,75,76] Scheme 10.4 gives the composition of the various poly(aniline) redox states.

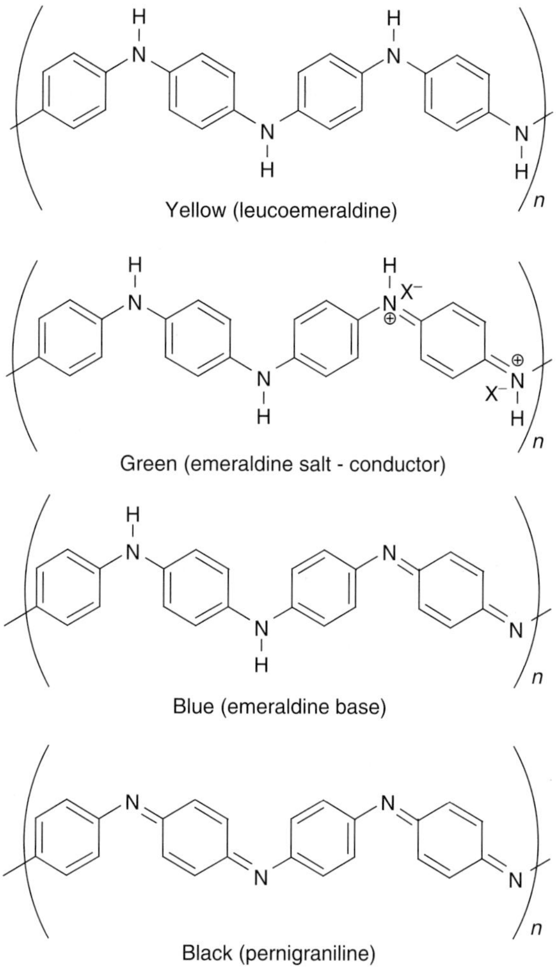

10.4 Poly(aniline)s as electrochromes

Yellow (leucoemeraldine)

Green (emeraldine salt - conductor)

Blue (emeraldine base)

Black (pernigraniline)

Scheme 10.4 Proposed composition of some of the redox states of poly(aniline), from the fully reduced (leucoemeraldine) through to the fully oxidised (pernigraniline) forms; X^- is a charge-balancing anion.

Leucoemeraldine is an insulator since all rings are benzenoid in form and separated by –NH– or (in strong acid solution) –NH$_2^+$– groups, thus preventing conjugation between rings. Emeraldine, as either base or salt, has a ratio of three benzenoid rings to one quinoidal ring, and is electrically conductive. Pernigraniline has equal proportions of quinoidal and benzenoid moieties and shows metallic conductivity. The aniline units within the poly(aniline) backbone are not coplanar, as has been shown by solid-state ^{13}C-NMR spectroscopy.[77] Electrodes bearing such poly(aniline) films are *electropolychromic* and exhibit the following reversible colour changes as the potential is varied: transparent leucoemeraldine to yellow-green emeraldine to dark blue-black pernigraniline, in the potential range −0.2 to +1.0 V vs. SCE.[73] The yellow → green transition is especially durable to repetitive colour switching. Pernigraniline is an intense blue colour, but appears black at very positive potentials if the film is thick. The yellow form of poly(aniline) has an absorbance maximum at 305 nm, but no appreciable absorbance in the visible region. The electrochemistry of poly(aniline) has been shown to involve a two-step oxidation with radical cations as intermediates. At lower applied potentials, the absorbances of poly(aniline) films at 430 and 810 nm are enhanced as the applied potential is made more positive.[78] At higher applied potentials, the absorbance at 430 nm begins to decrease while the wavelength of maximum absorbance shifts from 810 nm to wavelengths of higher energies.[78]

Of the numerous conducting polymers based on substituted anilines that have been hitherto investigated, those with alkyl substituents have drawn much attention. Poly(o-toluidine) and poly(m-toluidine) films have been found to offer enhanced stability of electropolychromic response in comparison with poly(aniline).[79] Absorption maxima and redox potentials shift from values found for poly(aniline) due to the lower conjugation length in poly(toluidine)s. The response times τ for the yellow–green electrochromic transition in the films correlate with the likely differences in the conjugation length implied from the spectroelectrochemical data. The τ values for poly(aniline) are found to be lower than for poly(o-toluidine), which in turn has lower values than poly(m-toluidine). As found for poly(aniline), response times indicate that the reduction process is faster than the oxidation. Electrochemical quartz crystal microbalance (EQCM) studies have demonstrated the complexity of redox switching in poly(o-toluidine) films in aqueous perchloric acid solutions, which occurs in two stages and is accompanied by non-monotonic mass changes that are the result of perchlorate counter ion, proton co-ion, and solvent transfers.[80] The relative extents and rates of each of these transfers depend on electrolyte concentration, experimental time scale, and the switching potential, so that observations in a single electrolyte on a fixed time scale cannot be unambiguously interpreted.

Poly(aniline)-based ECDs include a device that exhibits electrochromism using electropolymerised 1,1′-bis{[p-phenylamino(phenyl)]amido}ferrocene.[81] The monomer consists of a ferrocene group and two flanking polymerisable diphenylamine endgroups linked to the ferrocene by an amide bond. A solid-state aqueous-based ECD was constructed utilising this polymer as the electrochromic material in which the polymer switched from a yellow neutral state to blue upon oxidation.[81]

10.5 Directed assembly of electrochromic electroactive conducting polymers

10.5.1 Layer-by-layer deposition of electrochromes

Following earlier work[82] with poly(viologen) systems, the 'directed-assembly' layer-by-layer deposition of PEDOT:PSS (as the polyanion) with linear poly(ethylene imine) (LPEI) (as the polycation) has been reported.[83] The cathodically colouring PEDOT:PSS/LPEI electrode was then combined with a poly(aniline)–poly(AMPS) anodically colouring layered system to give a blue-green to yellow ECD. More recently, Reynolds et al.[84] have studied the redox and electrochromic properties of films prepared by the 'layer-by-layer' deposition of fully water-soluble, self-doped poly{4-(2,3-dihydrothieno[3,4-b]-[1,4]dioxin-2-yl-methoxy}-1-butanesulfonic acid, sodium salt (PEDOT-S) and poly(allylamine hydrochloride) – PAH – onto unmodified ITO-coated glass. The polymer PEDOT-S is self-doping where oxidation and reduction of the polymer backbone are coupled with cation movement out of, and back into, the polymer film, in its oxidised and reduced forms respectively. Both the film preparation and redox switching of this system are carried out in an aqueous medium. The PEDOT-S/PAH film was found to switch from light blue in the oxidised form to pink-purple in the reduced form.

10.5.2 All-polymer ECDs

The studies outlined in this chapter led to the construction of the *first truly all-polymer ECD*, where the film of ITO has been replaced by PEDOT:PSS as the conducting electrode material, with the glass substrate replaced by plastic.[85] In the construction of this device, electrodes were first prepared by spin coating an aqueous dispersion of PEDOT:PSS (mixed with 5 wt.% N-methylpyrrolidone (NMP) or diethylene glycol (DEG)) onto commercial plastic transparency films for overhead projection. Multiple layers of PEDOT:PSS were achieved by drying the films with hot-air drafts between coatings and subsequent air drying in an oven of the multilayer film. After three coatings, the surface resistivity of the electrodes had decreased to 600 Ω per square (at 300 nm thickness) while remaining highly transmissive throughout the visible region. Following the heat treatment, the PEDOT:PSS multiple-layer film did not return to the non-conducting form over the voltage ranges of the ECD operation.

Two ECDs were reported[85] that employed different complementary pairs of electrochromic polymers. In the first device, poly(3,4-propylenedioxythiophene) – PProDOT-Me$_2$ – and poly{3,6-bis[2-(3,4-ethylenedioxy)thienyl]-N-methylcarbazole} – PBEDOT-N-MeCz – were used respectively as the cathodically and anodically colouring polymers, in a sandwich device, with a polymer-gel electrolyte interposed. In the initial ECD state, PProDOT-Me$_2$ is in its oxidised (sky-blue) form and PBEDOT-N-MeCz is in its neutral (pale-yellow) form, hence the overall colour is an acceptably transmissive green. Application of a voltage (negative bias to PProDOT-Me$_2$) switches the oxidation states of both polymers, causing the device to become blue. In a second all-polymer ECD, two cathodically colouring electrochromic polymers were selected to demonstrate switching between two absorptive colour states (blue and red), with a transmissive intermediate state. The polymer PProDOT-Me$_2$ was again used, together with, as second electrochromic electrode, poly{1,4-bis[2-(3,4-ethylenedioxy)thienyl]-2,5-didodecyloxy-benzene) – PBEDOT-B(OC$_{12}$)$_2$ – showing red to sky-blue electrochromism.

Following this work, an all-plastic ECD has been reported,[86] where PEDOT layers act simultaneously on both electrodes as electrochromes and current collectors, thereby simplifying the construction of electrochromic sandwich devices from seven to five layers. In this research, PEDOT-covered poly(ethylene terephthalate) – PET – foils, commercialised by AGFA under the trademark of ORGACON EL-350, were simply sandwiched together with a poly(ethylene oxide) random co-polymer/lithium triflate polymer electrolyte layer. The contrast ratio for this type of ECD was, however, found to be relatively low, not surprisingly because, as has been noted earlier, both oxidised and reduced forms of a PEDOT are unlikely to be effective electrochromes, but there is clearly scope for improvement. (Several different ORGACON films are available that differ in conductivity, as indicated by the associated numerals.)

10.6 Electrochromes based on electroactive conducting polymer composites

The oxidative polymerisation of monomers in the presence of selected additives has been a popular approach to the preparation of electroactive conducting polymers with tailored properties.[12]

10.6.1 Novel routes to castable poly(aniline) films

While electropolymerisation is a suitable method for preparing relatively low-surface-area electrochromic conducting polymer films, it may not be suitable for fabricating large-area coatings. As noted above for PEDOT materials, significant effort has gone into synthesising soluble poly(aniline) conducting polymers, such as poly(o-methoxyaniline), which can then be deposited as a thin film by casting from solution. In a novel approach, large-area electrochromic coatings have been prepared by incorporating poly(aniline) into poly(acrylate)–silica hybrid sol–gel networks generated from suspended particles or solutions, and then spraying or brush coating onto ITO surfaces.[87] Silane functional groups on the poly(acrylate) chain act as coupling and cross-linking agents to improve surface adhesion and mechanical properties of the resulting composite coatings.

A water-soluble poly(styrenesulfonic acid)-doped poly(aniline) has been prepared both by persulfate oxidative coupling and by anodic oxidation of aniline in aqueous dialysed poly(styrene sulfonic acid) solution.[88] Composites of poly(aniline) and cellulose acetate have been prepared both by casting of films from a suspension of poly(aniline) in a cellulose acetate solution, and by depositing cellulose acetate films onto electrochemically prepared poly(aniline) films.[89] The electrochromic properties of the latter films were studied by *in situ* spectroelectrochemistry, where the presence of the cellulose acetate was found not to impede the redox processes of the poly(aniline). The electroactivity and electrochromism of the graft copolymer of poly(aniline) and nitrilic rubber have been studied using stress–strain measurements, cyclic voltammetry, frequency response analysis (i.e. impedance spectroscopy) and visible-range spectroelectrochemistry.[90] The results indicated that the graft co-polymer exhibits mechanical properties similar to a cross-linked elastomer having the electrochromic and electrochemical properties typical of poly(aniline).

10.6.2 Encapsulation of dyes into electroactive conducting polymers

An example of a case where the additive itself is electrochromic is the encapsulation of the redox indicator dye Indigo Carmine within a poly(pyrrole) matrix.[91,92] The enhancement and modulation of the colour change on Indigo Carmine insertion into polypyrrole or poly(pyrrole)–dodecylsulfonate films was established.[93] As expected, the use of Indigo Carmine as dopant improves the electrochromic contrast ratio of the film.

10.7 ECDs using both electroactive conducting polymers and inorganic electrochromes

As noted in Chapter 8, numerous workers[94,95,96,97,98,99,100,101] have combined a poly(aniline) electrode with an electrode covered with the inorganic mixed valence complex, Prussian blue – PB, iron(III) hexacyanoferrate(II) – or with WO_3, in complementary ECDs that exhibit deep-blue to light-green electrochromism. Electrochromic compatibility is obtained by combining the coloured oxidised state of the polymer with the blue of PB, versus the (bleached) reduced state of the polymer coincident with the lightly coloured Prussian green (PG). An electrochromic window for solar modulation using PB, poly(aniline) and WO_3 has been developed,[97,98,100,101] where the symbiotic relationship between poly(aniline) and PB was exploited in a complete solid-state electrochromic 'window'. Compared to earlier results with a poly(aniline)–WO_3 window, much more light was blocked off by including PB within the poly(aniline) as matrix, while still retaining approximately the same transparency in the bleached state of the window.

A new complementary ECD has recently been described,[102] based on the assembly of PEDOT on ITO glass and PB on ITO glass substrates with a poly(methyl methacrylate) – PMMA-based gel polymer electrolyte. The colour states of the PEDOT (blue-to-colourless) and PB (colourless-to-blue) films fulfil the requirement of complementarity.

10.8 Conclusions and outlook

Intense interest continues to drive the highly novel research into the electrochromic properties of electroactive conducting polymers outlined here. Through the skills of organic chemists in the synthesis of novel monomers and soluble polymers, the possibilities in colour choice and performance characteristics seem endless and await further exploitation, particularly in the field of display applications. Tailoring the colour of electroactive conducting polymers remains a particularly active research area. Although not described in this chapter, in addition to the synthesis of novel functionalised monomers and use of composites, other chemical and physical methods are investigated for the control of the perceived colour of electrochromic polymers. Methods include the use of polymer blends, laminates and patterning using screen and ink-jet printing.[103] Furthermore, as described in Chapter 4, analysis of electrochrome and ECD colour changes are now routinely measured by *in situ* colour analysis, using Commission Internationale de l'Eclairage (CIE) (x,y)-chromaticity coordinates. This method is useful for the comparison of the electrochemical and optical properties of electroactive conducting polymers, and for gaining control of the colour of dual-polymer electrochromic devices.[104,105] As an example, by controlling the electron density and steric interactions along conjugated polymer backbones, a set of electrochromic polymers that provide colours through the full range of colour space has been developed through the study of twelve electrochromic polymers.[104]

References

1. Letheby, H. XXIX. On the production of a blue substance by the electrolysis of sulphate of aniline. *J. Chem. Soc.*, **15**, 1862, 161–3.
2. Shirakawa, H., Louis, E. J., MacDiarmid, A. G., Chiang, C. K. and Heeger, A. J. Synthesis of electrically conducting organic polymers: halogen derivatives of polyacetylene, $(CH)_x$. *J. Chem. Soc., Chem. Commun.*, 1977, 578–80.
3. Chiang, C. K., Druy, M. A., Gau, S. C., Heeger, A. J., Louis, E. J., MacDiarmid, A. G., Park, Y. W. and Shirakawa, H. Synthesis of highly conducting films of derivatives of polyacetylene, $(CH)_x$. *J. Am. Chem. Soc.*, **100**, 1978, 1013–15.
4. Chiang, C. K., Fincher, C. R., jr, Park, Y. W., Heeger, A. J., Shirakawa, H., Louis, E. J., Gan, S. C. and MacDiarmid, A. G. Electrical conductivity in doped polyacetylene. *Phys. Rev. Lett.*, **39**, 1977, 1098–101.
5. MacDiarmid, A. G. Nobel lecture: synthetic metals: a novel route for organic polymers. *Rev. Mod. Phys.*, **73**, 2001, 701–12.
6. Shirakawa, H., MacDiarmid, A. G. and Heeger, A. J. Focus article. Twenty-five years of conducting polymers. *Chem. Commun.*, 2003, 1–4.
7. Burroughes, J. H., Bradley, D. D. C., Brown, A. R., Marks, R. N., Mackay, K., Friend, R. H., Burns, P. L. and Holmes, A. B. Light-emitting diodes based on conjugated polymers. *Nature (London)*, **347**, 1990, 539–41.
8. Kraft, A., Grimsdale, A. C. and Holmes, A. B. Electroluminescent conjugated polymers – seeing polymers in a new light. *Angew, Chem., Int. Ed. Engl.*, **37**, 1998, 403–28.
9. Brabec, C. J., Sariciftci, N. S. and Hummelen, J. C. Plastic solar cells. *Adv. Funct. Mater.*, **1**, 2001, 15–26.
10. McQuade, D. Tyler, Pullen, A. E. and Swager, T. M. Conjugated polymer-based chemical sensors. *Chem. Rev.*, **100**, 2000, 2537–74.
11. Knobloch, A., Manuelli, A., Bernds, A. and Clemens, W. Fully printed integrated circuits from solution processable polymers. *J. Appl. Phys.*, **96**, 2004, 2286–91.
12. Monk, P. M. S., Mortimer, R. J. and Rosseinsky, D. R. *Electrochromism: Fundamentals and Applications*, Weinheim, VCH, 1995, ch. 9.
13. Heinze, J. Electronically conducting polymers. *Top. Curr. Chem.*, **152**, 1990, 1–47.
14. Evans, G. P. In Gerischer, H. and Tobias, C. W. (eds.), *Advances in Electrochemical Science and Engineering*, Weinheim, VCH, 1990, vol. 1, pp. 1–74.
15. Mastragostino, M. In Scrosati, B. (ed.), *Applications of Electroactive Polymers*, London, Chapman and Hall, 1993, ch. 7.
16. Roncali, J. Conjugated poly(thiophenes): synthesis, functionalization, and applications, *Chem. Rev.*, **92**, 1992, 711–38.
17. Higgins, S. J. Conjugated polymers incorporating pendant functional groups – synthesis and characterisation. *Chem. Soc. Rev.*, **26**, 1997, 247–57.
18. Skotheim, T. A., Elsebaumer, R. L. and Reynolds, J. R. (eds.), *Handbook of Conducting Polymers*, 2nd edn, New York, Marcel Dekker, 1998; Skotheim, T. A. and Reynolds, J. R. (eds.), *Handbook of Conducting Polymers* (3rd edn.) CRC Press, Taylor & Francis Group, Boca Raton, 2007.
19. Roncali, J. Electrogenerated functional conjugated polymers as advanced electrode materials. *J. Mater. Chem.*, **9**, 1999, 1875–93.
20. Bard, A. J. and Faulkner, L. R. Electrode reactions with coupled homogeneous chemical reactions. In *Electrochemical Methods: Fundamentals and Applications*, 2nd edn, New York, Wiley, 2001, ch. 12, pp. 471–533.

21. Street, G. B., Clarke, T. C., Geiss, R. H., Lee, V. Y., Nazzal, A. I., Pfluger, P. and Scott, J. C. Characterization of polypyrrole. *J. Phys.*, **44(C3)**, 1983, 599–606.
22. Ferraris, J. P., Henderson, C., Torres, D. and Meeker, D. Synthesis, spectroelectrochemistry and application in electrochromic devices of n-dopable and p-dopable conducting polymer. *Synth. Met.*, **72**, 1995, 147–52.
23. Wegner, G. The state of order and the relevance of phase transitions in conducting polymers. *Mol. Cryst. Liq. Cryst.*, **106**, 1984, 269–88.
24. Gazard, M. Application of polyheterocycles to electrochromic display devices. In Skotheim, T. A. (ed.), *Handbook of Conducting Polymers*, New York, Marcel Dekker, 1986, vol. 1, ch. 19.
25. Mastragostino, M. Electrochromic devices. In Scrosati, B. (ed.), *Applications of Electroactive Polymers*, London, Chapman and Hall, 1993, ch. 7.
26. Hyodo, K. Electrochromism of conducting polymers. *Electrochim. Acta*, **39**, 1994, 265–72.
27. Mortimer, R. J. Organic electrochromic materials. *Electrochim. Acta*, **44**, 1999, 2971–81.
28. Mortimer, R. J. Electrochromic polymers. In Kroschwitz, J. I. (ed.), *Encyclopedia of Polymer Science & Technology*, 3rd edn, New York, John Wiley & Sons, 2004, vol. 9, pp. 591–614.
29. Sonmez, G. Polymeric electrochromics. *Chem. Commun.*, 2005, 5251–9.
30. Mortimer, R. J., Dyer, A. L. and Reynolds, J. R. Electrochromic organic and polymeric materials for display applications. *Displays*, **27**, 2006, 2–18.
31. Barbarella, G., Melucci, M. and Sotgiu, G. The versatile thiophene: an overview of recent research on thiophene-based materials. *Adv. Mater.*, **17**, 2005, 1581–93.
32. Mastragostino, M., Arbizzani, C., Bongini, A., Barbarella, G. and Zambianchi, M. Polymer-based electrochromic devices, 1: poly(3-methylthiophenes). *Electrochim. Acta*, **38**, 1993, 135–40.
33. Kirchmeyer, S. and Reuter, K. Scientific importance, properties and growing applications of poly(3,4-ethylenedioxythiophene). *J. Mater. Chem.*, **15**, 2005, 2077–88.
34. Giglioti, M., Trivinho-Strixino, F., Matsushima, J. T., Bulhões, L. O. S. and Pereira, E. C. Electrochemical and electrochromic response of poly(thiophene-3-acetic acid) films. *Sol. Energy Mater., Sol. Cells*, **82**, 2004, 413–420.
35. Galal, A., Cunningham, D. D., Karagözler, A. E., Lewis, E. T., Nkansah, A., Burkhardt, A., Ataman, O. Y., Zimmer, H. and Mark, H. B. Electrochemical synthesis, characterization and spectroelectrochemical studies of some conducting poly(heterolene) films. *Proc. Electrochem. Soc.*, **90–2**, 1990, 179–91.
36. Mastragostino, M., Arbizzani, C., Ferloni, P. and Marinangeli, A. Polymer-based electrochromic devices, *Solid State Ionics*, **53–56**, 1992, 471–8.
37. Groenendaal, L., Jonas, F., Freitag, D., Pielartzik, H. and Reynolds, J. R. Poly(3,4-ethylenedioxythiophene) and its derivatives: past, present, and future. *Adv. Mater.*, **12**, 2000, 481–94.
38. Groenendaal, L., Zotti, G., Aubert, P.-H., Waybright, S. M. and Reynolds, J. R. Electrochemistry of poly(3,4-alkylenedioxythiophene) derivatives. *Adv. Mater.*, **15**, 2003, 855–79.
39. Jonas, F. and Schrader, L. Conductive modifications of polymers with polypyrroles and polythiophenes. *Synth. Met.*, **41–3**, 1991, 831–6.
40. Heywang, G. and Jonas, F. Poly(alkylenedioxythiophene)s – new, very stable conducting polymers. *Adv. Mater.*, **4**, 1992, 116–18.
41. Roncali, J., Blanchard, P. and Frère, P. 3,4-Ethylenedioxythiophene (EDOT) as a versatile building block for advanced functional π-conjugated systems. *J. Mater. Chem.*, **15**, 2005, 1598–610.

42. Sapp, S. A., Sotzing, G. A. and Reynolds, J. R. High contrast ratio and fast-switching dual polymer electrochromic devices. *Chem. Mater.*, **10**, 1998, 2101–8.
43. Gaupp, C. L., Welsh, D. M. and Reynolds, J. R. Poly(ProDOT-Et-2): a high-contrast, high-coloration efficiency electrochromic polymer. *Macromol. Rapid Commun.*, **23**, 2002, 885–9.
44. Welsh, D. M., Kumar, A., Meijer, E. W. and Reynolds, J. R. Enhanced contrast ratios and rapid switching in electrochromics based on poly(3,4-propylenedioxythiophene) derivatives. *Adv. Mater.*, **11**, 1999, 1379–82.
45. Kumar, A., Welsh, D. M., Morvant, M. C., Piroux, F., Abboud, K. A. and Reynolds, J. R. Conducting poly(3,4-alkylenedioxythiophene) derivatives as fast electrochromics with high-contrast ratios. *Chem. Mater.*, **10**, 1998, 896–902.
46. Sankaran, B. and Reynolds, J. R. High-contrast electrochromic polymers from alkyl-derivatised poly(3,4-ethylenedioxythiophenes). *Macromolecules*, **30**, 1997, 2582–8.
47. Welsh, D. M., Kloeppner, L. J., Madrigal, L., Pinto, M. R., Thompson, B. C., Schanze, K. S., Abboud, K. A., Powell, D. and Reynolds, J. R. Regiosymmetric dibutyl-substituted poly(3,4-propylenedioxythiophene)s as highly electron-rich electroactive and luminescent polymers. *Macromolecules*, **35**, 2002, 6517–25.
48. Kumar, A. and Reynolds, J. R. Soluble alkyl-substituted poly(ethylenedioxythiophene)s as electrochromic materials. *Macromolecules*, **29**, 1996, 7629–30.
49. Reeves, B. D., Grenier, C. R. G., Argun, A. A., Cirpan, A., McCarley, T. D. and Reynolds, J. R. Spray coatable electrochromic dioxythiophene polymers with high coloration efficiencies. *Macromolecules*, **37**, 2004, 7559–69.
50. Cirpan, A., Argun, A. A., Grenier, C. R. G., Reeves, B. D. and Reynolds, J. R. Electrochromic devices based on soluble and processable dioxythiophene polymers. *J. Mater. Chem.*, **13**, 2003, 2422–8.
51. Sotzing, G. A., Reynolds, J. R. and Steel, P. J. Electrochromic conducting polymers via electrochemical polymerization of bis(2-(3,4-ethylenedioxy)thienyl) monomers. *Chem. Mater.*, **8**, 1996, 882–9.
52. Sotzing, G. A., Reddinger, J. L., Katritzky, A. R., Soloducho, J., Musgrave, R. and Reynolds, J. R. Multiply colored electrochromic carbazole-based polymers. *Chem. Mater.*, **9**, 1997, 1578–87.
53. Irvin, J. A., Schwendeman, I., Lee, Y., Abboud, K. A. and Reynolds, J. R. Low-oxidation-potential conducting polymers derived from 3,4-ethylenedioxythiophene and dialkoxybenzenes. *J. Polym. Sci. Polym. Chem.*, **39**, 2001, 2164–78.
54. Gaupp, C. L. and Reynolds, J. R. Multichromic copolymers based on 3,6-bis(2-(3,4-ethylenedioxythiophene))-N-alkylcarbazole derivatives. *Macromolecules*, **36**, 2003, 6305–15.
55. Dubois, C. J., Abboud, K. A. and Reynolds, J. R. Electrolyte-controlled redox conductivity in n-type doping in poly(bis-EDOT-pyridine)s. *J. Phys. Chem. B*, **108**, 2004, 8550–7.
56. Dubois, C. J., Larmat, F., Irvin, D. J. and Reynolds, J. R. Multi-colored electrochromic polymers based on BEDOT-pyridines. *Synth. Met.*, **119**, 2001, 321–2.
57. Sonmez, G., Shen, C. K. F., Rubin, Y. and Wudl, F. A red, green, and blue (RGB) polymeric electrochromic device (PECD): the dawning of the PECD era. *Angew. Chem. Int. Ed. Eng.*, **43**, 2004, 1498–502.
58. Sonmez, G., Sonmez, H. B., Shen, C. K. F. and Wudl, F. Red, green and blue colors in polymeric electrochromics. *Adv. Mater.*, **16**, 2004, 1905–8.

59. Sonmez, G. and Wudl, F. Completion of the three primary colours: the final step toward plastic displays. *J. Mater. Chem.*, **15**, 2005, 20–2.
60. Sonmez, G., Sonmez, H. B., Shen, C. K. F., Jost, R. W., Rubin, Y. and Wudl, F. A processable green polymeric electrochromic. *Macromolecules*, **38**, 2005, 669–75.
61. Rauh, R. D., Peramunage, D. and Wang, F. Electrochemistry and electrochromism in star conductive polymers. *Proc. Electrochem. Soc.*, **2003–17**, 2003, 176–81.
62. Rauh, R. D., Wang, F., Reynolds, J. R. and Meeker, D. L. High coloration efficiency electrochromics and their application to multi-color devices. *Electrochim. Acta*, **46**, 2001, 2023–9.
63. Wang, F., Wilson, M. S., Rauh, R. D., Schottland, P., Thompson, B. C. and Reynolds, J. R. Electrochromic linear and star branched poly(3,4-ethylenedioxythiophene-didodecyloxybenzene) polymers. *Macromolecules*, **33**, 2000, 2083–91.
64. Wang, F., Wilson, M. S., Rauh, R. D., Schottland, P. and Reynolds, J. R. Electroactive and conducting star-branched poly(3-hexylthiophene)s with a conjugated core. *Macromolecules*, **32**, 1999, 4272–8.
65. Genies, E. M., Bidan, G. and Diaz, A. F. Spectroelectrochemical study of polypyrrole films. *J. Electroanal. Chem.*, **149**, 1983, 103–13.
66. Diaz, A. F., Castillo, J. I., Logan, J. A. and Lee, W. I. Electrochemistry of conducting polypyrrole films. *J. Electroanal. Chem.*, **129**, 1981, 115–32.
67. Wong, J. Y., Langer, R. and Ingber, D. E. Electrically conducting polymers can noninvasively control the shape and growth of mammalian cells. *Proc. Natl. Acad. Sci. USA*, **91**, 1994, 3201–4.
68. Schottland, P., Zong, K., Gaupp, C. L., Thompson, B. C., Thomas, C. A., Giurgiu, I., Hickman, R., Abboud, K. A. and Reynolds, J. R. Poly(3,4-alkylenedioxypyrrole)s: highly stable electronically conducting and electrochromic polymers. *Macromolecules*, **33**, 2000, 7051–61.
69. Gaupp, C. L., Zong, K. W., Schottland, P., Thompson, J. R., Thomas, C. A. and Reynolds, J. R. Poly(3,4-ethylenedioxypyrrole): organic electrochemistry of a highly stable electrochromic polymer. *Macromolecules*, **33**, 2000, 1132–3.
70. Sonmez, G., Schwendeman, I., Schottland, P., Zong, K. W. and Reynolds, J. R. N-substituted poly(3,4-propylenedioxypyrrole)s: high gap and low redox potential switching electroactive and electrochromic polymers. *Macromolecules*, **36**, 2003, 639–47.
71. Schwendeman, I., Hickman, R., Sonmez, G., Schottland, P., Zong, K., Welsh, D. M. and Reynolds, J. R. Enhanced contrast dual polymer electrochromic devices. *Chem. Mater.*, **14**, 2002, 3118–22.
72. Diaz, A. F. and Logan, J. A. Electroactive polyaniline films. *J. Electroanal. Chem.*, **111**, 1980, 111–14.
73. Kobayashi, T., Yoneyama, H. and Tamura, H. Polyaniline film-coated electrodes as electrochromic display devices. *J. Electroanal. Chem.*, **161**, 1984, 419–23.
74. MacDiarmid, A. G. and Epstein, A. J. Polyanilines – a novel class of conducting polymers. *Faraday Discuss. Chem. Soc.*, **88**, 1989, 317–32.
75. Ray, A., Richter, A. F., MacDiarmid, A. G. and Epstein, A. J. Polyaniline – protonation deprotonation of amine and imine sites. *Synth. Met.*, **29**, 1989, 151–6.
76. Rourke, F. and Crayston, J. A. Cyclic voltammetry and morphology of polyaniline-coated electrodes containing [Fe(CN)$_6$]$^{3-/4-}$ ions. *J. Chem. Soc., Faraday Trans.*, **89**, 1993, 295–302.

77. Hjertberg, T., Salaneck, W. R., Lundstrom, I., Somasiri, N. L. D. and MacDiarmid, A. G. A C-13 CP-MAS NMR investigation of polyaniline. *J. Polym. Sci., Polym. Lett.*, **23**, 1985, 503–8.
78. Watanabe, A., Mori, K., Iwasaki, Y., Nakamura, Y. and Niizuma, S. Electrochromism of polyaniline film prepared by electrochemical polymerization. *Macromolecules*, **20**, 1987, 1793–6.
79. Mortimer, R. J. Spectroelectrochemistry of electrochromic poly(o-toluidine) and poly(m-toluidine) films. *J. Mater. Chem.*, **5**, 1995, 969–73.
80. Ramirez, S. and Hillman, A. R. Electrochemical quartz crystal microbalance studies of poly(ortho-toluidine) films exposed to aqueous perchloric acid solutions. *J. Electrochem. Soc.*, **145**, 1998, 2640–7.
81. Wang, L., Wang, Q. Q. and Cammarata, V. Electro-oxidative polymerization and spectroscopic characterization of novel amide polymers using diphenylamine coupling. *J. Electrochem. Soc.*, **145**, 1998, 2648–54.
82. Stepp, J. and Schlenoff, J. B. Electrochromism and electrocatalysis in viologen polyelectrolyte multilayers. *J. Electrochem. Soc.*, **144**, 1997, L155–7.
83. DeLongchamp, D. and Hammond, P. T. Layer-by-layer assembly of PEDOT/polyaniline electrochromic devices. *Adv. Mater.*, **13**, 2001, 1455–9.
84. Cutler, C. A., Bouguettaya, M. and Reynolds, J. R. PEDOT polyelectrolyte based electrochromic films via electrostatic adsorption. *Adv. Mater.*, **14**, 2002, 684–8.
85. Argun, A. A., Cirpan, A. and Reynolds, J. R. The first truly all-polymer electrochromic devices. *Adv. Mater.*, **15**, 2003, 1338–41.
86. Mecerreyes, D., Marcilla, R., Ochoteco, E., Grande, H., Pomposo, J. A., Vergaz, R. and Sánchez Pena, J. M. A simplified all-polymer flexible electrochromic device. *Electrochim. Acta*, **49**, 2004, 3555–9.
87. Jang, G.-W., Chen, C. C., Gumbs, R. W., Wei, Y. and Yeh, J.-M. Large-area electrochromic coatings – composites of polyaniline and polyacrylate-silica hybrid set gel materials. *J. Electrochem. Soc.*, **143**, 1996, 2591–6.
88. Shannon, K. and Fernandez, J. E. Preparation and properties of water-soluble, poly(styrene-sulfonic acid)-doped polyaniline. *J. Chem. Soc., Chem. Commun.*, 1994, 643–4.
89. De Paoli, M. A., Duek, E. R. and Rodrigues, M. A. Poly(aniline) cellulose-acetate composites – conductivity and electrochromic properties. *Synth. Met.*, **41**, 1991, 973–8.
90. Tassi, E. L., De Paoli, M. A., Panero, S. and Scrosati, B. Electrochemical, electrochromic and mechanical-properties of the graft copolymer of polyaniline and nitrilic rubber. *Polymer*, **35**, 1994, 565–72.
91. Gao, Z., Bobacka, J., Lewenstam, A. and Ivaska, A. Electrochemical-behavior of polypyrrole film polymerized in indigo carmine solution. *Electrochim. Acta*, **39**, 1994, 755–62.
92. Li, Y. and Dong, S. Indigo-carmine-modified polypyrrole film electrode. *J. Electroanal. Chem.*, **348**, 1993, 181–8.
93. Girotto, E. M. and De Paoli, M. A. Polypyrrole color modulation and electrochromic contrast enhancement by doping with a dye. *Adv. Mater.*, **10**, 1998, 790–3.
94. Duek, E. A. R., De Paoli, M.-A. and Mastragostino, M. An electrochromic device based on polyaniline and Prussian blue. *Adv. Mater.*, **4**, 1992, 287–91.
95. Duek, E. A. R., De Paoli, M.-A. and Mastragostino, M. A solid-state electrochromic device based on polyaniline, Prussian blue and an elastomeric electrolyte. *Adv. Mater.*, **5**, 1993, 650–2.

96. Morita, M. Electrochromic behavior and stability of polyaniline composite films combined with Prussian blue. *J. Appl. Poly. Sci.*, **52**, 1994, 711–19.
97. Jelle, B. P., Hagen, G. and Nodland, S. Transmission spectra of an electrochromic window consisting of polyaniline, Prussian blue and tungsten oxide. *Electrochim. Acta*, **38**, 1993, 1497–500.
98. Jelle, B. P. and Hagen, G. Transmission spectra of an electrochromic window based on polyaniline, Prussian blue and tungsten oxide. *J. Electrochem. Soc.*, **140**, 1993, 3560–4.
99. Leventis, N. and Chung, Y. C. Polyaniline–Prussian blue novel composite-material for electrochromic applications. *J. Electrochem. Soc.*, **137**, 1990, 3321–2.
100. Jelle, B. P. and Hagen, G. Correlation between light absorption and electric charge in solid state electrochromic windows. *J. Appl. Electrochem.*, **29**, 1999, 1103–10.
101. Jelle, B. P. and Hagen, G. Performance of an electrochromic window based on polyaniline, Prussian blue and tungsten oxide. *Sol. Energy Mater. Sol. Cells*, **58**, 1999, 277–86.
102. Tung, T.-C. and Ho, K. C. A complementary electrochromic device containing 3,4-ethylenedioxythiophene and Prussian blue. *Proc. Electrochem. Soc.*, **2003–17**, 2003, 254–65.
103. Ferraris, J. P., Mudiginda, D. S. K., Meeker, D. L., Boehme, J., Loveday, D. C., Dan, T. M. and Brotherston, I. D. Color tailoring techniques for electroactive polymer-based electrochromic devices. Meeting Abstracts, volume 2003–01, Electrochromics Materials and Applications Symposium, at the 203rd Electrochemical Society Meeting, Paris, France, 27 April–2 May, 2003, Abstract No. 1329.
104. Thompson, B. C., Schottland, P., Zong, K. and Reynolds, J. R. In situ colorimetric analysis of electrochromic polymers and devices. *Chem. Mater.*, **12**, 2000, 1563–71.
105. Thompson, B. C., Schottland, P., Sonmez, G. and Reynolds, J. R. In situ colorimetric analysis of electrochromic polymer films and devices. *Synth. Met.*, **119**, 2001, 333–4.

11

The viologens

11.1 Introduction

The next major group of electrochromes are the bipyridilium species formed by the diquaternisation of 4,4'-bipyridyl to form 1,1'-disubstituted-4,4'-bipyridilium salts (Scheme 11.1). The positive charge shown localised on N is better viewed as being delocalised over the rings. The compounds are formally named as 1,1'-di-substituent-4,4'-bipyridilium if the two substituents at nitrogen are the same, and as 1-substitituent-1'-substituent'-4,4'-bipyridilium should they differ. The anion X^- in Scheme 11.1 need not be monovalent and can be part of a polymer. The molecules are zwitterionic (i.e. bearing plus and minus charge concentrations at different molecular regions or sites) when a substituent at one nitrogen bears a negative charge.[1,2]

A convenient abbreviation for any bipyridyl unit regardless of its redox state is 'bipm', with its charge indicated. The literature of these compounds contains several trivial names. The most common is 'viologen' following Michaelis,[3,4] who noted the violet colour formed when 1,1'-dimethyl-4,4'-bipyridilium undergoes a one-electron reduction to form a radical cation. 1,1'-Dimethyl-4,4'-bipyridilium is therefore called 'methyl viologen' (MV) in this nomenclature. Another extensively used name is 'paraquat', PQ, after the ICI brand name for methyl viologen, which they developed for herbicidal use. In this latter style, bipyridilium species other than the dimethyl are called '*substituent paraquat*'.

There are several reviews of this field extant. The most substantial is *The Viologens: Physicochemical Properties, Synthesis, and Applications of the Salts of 4,4'-Bipyridine* (1998) by Monk.[5] Other works are dated, but some still incorporate valuable bibliographic data, including 'Bipyridilium systems' (1995) by Monk *et al.*;[6] 'The bipyridines' (1984), by Summers,[7] deals at length with syntheses and properties of 4,4'-bipyridine, and Summers' 1980 book *The*

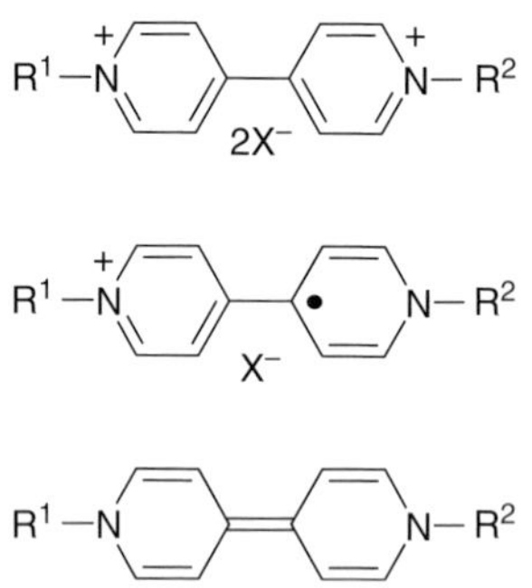

Scheme 11.1 The three common bipyridyl redox states. Different substituents as R^1 and R^2 may be attached to form unsymmetrical species. X^- is a singly charged anion.

Bipyridinium Herbicides[8] comprises copious detail. Although dated, the review entitled 'The Electrochemistry of the viologens' (1981) by Bird and Kuhn[9] is particularly relevant to this chapter. 'Formation, properties and reactions of cation radicals in solution' (1976) by Bard *et al.*[10] has a section on bipyridilium radical cations. Finally, the review, 'Chemistry of viologens' (1991) by Sliwa *et al.*[11] also alludes to electrochromism.

11.2 Bipyridilium redox chemistry

There are three common bipyridilium redox states: a dication (bipm^{2+}), a radical cation (bipm$^{+\bullet}$) and a di-reduced neutral compound (bipm^0). The dicationic salt is the most stable of the three and is the species purchased or first prepared in the laboratory. It is colourless when pure unless exhibiting optical charge transfer with the counter anion, or other charge-donating species. Such absorbances are feeble for anions like chloride, but are stronger for CT-interactive anions like iodide;[12] MV^{2+} 2I$^-$ is brilliant scarlet.

Reductive electron transfer to the dication forms a radical cation:

$$\text{bipm}^{2+} + e^- \rightarrow \text{bipm}^{+\bullet}. \quad (11.1)$$
$$\text{colourless} \qquad \text{intense colour}$$

Bipyridilium radical cations are amongst the most stable organic radicals, and may be prepared as air-stable solid salts.[13,14] In solution the colour of the radical will persist almost indefinitely[15] in the absence of oxidising agents like periodate or ferricyanide;[a] its reaction with molecular oxygen is particularly rapid.[16] The stability of the radical cation is attributable to the delocalisation

---

[a] 'Ferricyanide' is better termed hexacyanoferrate(III), but we stick to the usage in this field. Likewise, 'ferrocyanide' is properly hexacyanoferrate(II).

of the radical electron throughout the π-framework of the bipyridyl nucleus, the 1-and 1′-substituents commonly bearing some of the charge.

The potential needed to effect the reduction reaction in Eq. (11.1) depends on both the substituents at nitrogen and on the bipyridyl core – so-called 'nuclear substituted' compounds. For example, Hünig and co-workers have correlated the polarographic value of $E_{1/2}$, values of λ_{max} from electronic spectra, and the results of theoretical calculations, with informative parameters like σ and $\sigma*$[17,18,19] that relate empirically to electron densities and electronic shifts, as derived from the widely used linear free-energy relationships of physical organic chemistry.

Electrochromism occurs in bipyridilium species because, in contrast to the bipyridilium dications, radical cations are intensely coloured owing to optical charge transfer between the (formally) +1 and (formally) zero-charge nitrogens, in a simplified view of the phenomenon; however, because of the delocalisation already mentioned, the source of the colour is probably better viewed as an intramolecular photo-effected electronic excitation. The colours of radical cations depend on the substituents on the nitrogen.[5] Simple alkyl groups, for example, promote a blue-violet colour whereas aryl groups generally impart a variety of colours to the radical cation, the exact choice depending on the substituents. Manipulation of the substituents at N or the bipyridyl 'nucleus' to attain the appropriate molecular-orbital energy levels can also, in principle, tailor the colour as desired. The colour will also depend on the solvent.[b] Figure 11.1 shows the UV-visible spectrum of methyl viologen.

The molar absorptivity ε for the methyl viologen radical cation is large; for example, in water $\varepsilon = 13\,700\,\mathrm{dm^3\,mol^{-1}\,cm^{-1}}$ when extrapolated to zero concentration.[21] The value of ε is usually somewhat solvent dependent.[22] A few values of wavelength maxima and ε are listed in Table 11.1. The data refer to monomeric radical-cation species unless stated otherwise.

Comparatively little is known about the third redox form of the bipyridilium series, the di-reduced or so-called 'di-hydro'[32] compounds formed by one-electron reduction of the respective radical cation, Eq. (11.2):

$$\mathrm{bipm^{+\bullet} + e^- \rightarrow bipm^0}. \qquad (11.2)$$
intense colour   weak colour

---

[b] Kosower's solvent Z values (optical CT energies for the denoted solute with a variety of solvents) in ref. 20 were determined using the different but related system comprising 4-carboethoxy-1-methylpyridinium iodide. The Z values correlate well with many solvent–solute interactions. Other, comparable, CT scales have also been set up.

Table 11.1. *Optical data for some bipyridilium radical cations.*

R	Anion	Solvent	λ_{max}/nm	ε/dm^3 mol^{-1} cm^{-1}	Ref.
Methyl	Cl$^-$	H$_2$O	605	13 700	22
Methyl	I$^-$	H$_2$O–MeCN	605[a]	10 060	23,24
Methyl	Cl$^-$	H$_2$O	606	13 700	21
Methyl	Cl$^-$	MeCN	607	13 900	22
Methyl	Cl$^-$	MeOH	609	13 800	22
Methyl	Cl$^-$	EtOH	611	13 800	22
Methyl	Cl$^-$	H$_2$O	604	16 900	25
Ethyl	ClO$_4^-$	DMF	603	12 200	26
Heptyl	Br$^-$	H$_2$O	545[b,c]	26 000	27
Octyl	Br$^-$	H$_2$O	543[c]	28 900	28
Benzyl	Cl$^-$	H$_2$O	604	17 200	29
p-CN-Ph	BF$_4^-$	PC	674	83 300	30
p-CN-Ph	Cl$^-$	H$_2$O	535[b,c]	–	31

[a] Estimated from reported spectra. [b] Solid on OTE. [c] Solution-phase radical-cation dimer.

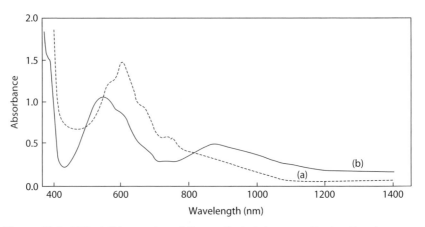

Figure 11.1 UV-visible spectra of the methyl viologen radical cation in aqueous solution. (a) ——— Monomeric (blue) radical cation and (b) – – – Red radical-cation dimer, the sample also containing a trace of monomer. (Figure reproduced from Monk, P. M. S., Fairweather, R. D., Duffy, J. A. and Ingram, M. D. 'Evidence for the product of viologen comproportionation being a spin-paired radical cation dimer'. *J. Chem. Soc., Perkin Trans. II*, 1992, 2039–41, by permission of The Royal Society of Chemistry.)

11.2 Bipyridilium redox chemistry

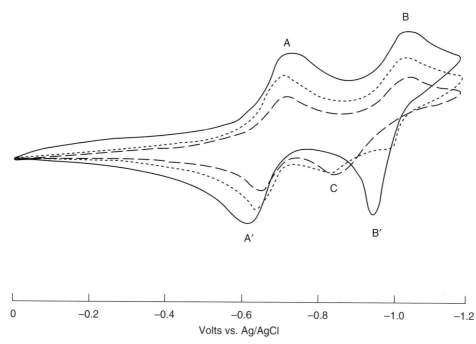

Figure 11.2 Cyclic voltammograms on glassy carbon of aqueous methyl viologen dichloride (1 mmol dm^{-3}) in KCl (0.1 mol dm^{-3}). Scan-rate dependence. Note the evidence of comproportionation – Eq. (11.7): the oxidation peak for spin-paired radical-cation dimer (**C**) is prominent while the peak for re-oxidation of bipm^0 (**B'**) is greatly diminished at slow scan rates. The outermost trace is fastest. (Figure reproduced from Datta, M., Jansson, R. E. and Freeman, J. J. '*In situ* resonance Raman spectroscopic characterisation of electrogenerated methyl viologen radical cation on carbon electrode'. *Appl. Spectrosc.*, **40**, 1986, 251–8, with permission of the Society of Applied Spectroscopy.)

This product may also be formed by direct two-electron reduction of the dication:

$$\text{bipm}^{2+} + 2\,e^- \rightarrow \text{bipm}^0. \tag{11.3}$$

Di-reduced compounds are often termed 'bi-radicals'[33] because of their extreme reactivity, but magnetic susceptibility measurements have shown such species to be diamagnetic[34] in the solid state, indicating that spins are paired. In fact, di-reduced bipm^0 compounds are simply reactive amines.[35] The intensity of the colour exhibited by bipm^0 species is often low since no obvious optical charge transfer or internal transition corresponding to visible wavelengths is accessible. Figure 11.2 shows cyclic voltammograms depicting these processes.

11.3 Bipyridilium species for inclusion within ECDs

The most extensive literature on a bipyridilium compound is that for 1,1′-dimethyl-4,4′-bipyridilium. The write–erase efficiency of an ECD with aqueous MV as electrochrome is low on a moderate time scale, its being type I as both dication and radical cation states are very soluble in polar solvents. The write–erase efficiency of such ECDs may be improved by retarding the rate at which the radical-cation product of electron transfer diffuses away from the electrode and into the solution bulk either by tethering the dication to the surface of an electrode, so forming a chemically modified ('derivatised') electrode (Section 1.4), or by immobilising the viologen species within a semi-solid electrolyte. These approaches, with the methyl viologen behaving as a pseudo-solid electrochrome, are described in Section 11.3.1.

The solubility–diffusion problem can also be avoided by the use of viologens having long alkyl-chain substituents at nitrogen, for which the coloured radical-cation product of Eq. (11.1) is insoluble, so here the viologen is a solution-to-solid type II electrochrome, as discussed in Section 11.3.3.

Effecting a large improvement in CR (60:1) and response times ($\tau_{colour} = 1$ ms, $\tau_{bleach} = 10$ ms) while employing light-scattering by a limited amount of HV^{2+} (deposited by 1 mC cm^{-2}), a complex optical system has been devised for display applications.[36,37]

11.3.1 Electrodes derivatised with viologens for ECD inclusion

Wrighton and co-workers[38,39] have often derivatised electrodes with bipyridilium species, initially using substituents at N consisting of a short alkyl chain terminating in the trimethoxysilyl group, which can bond to the oxide lattice on the surface of an optically transparent electrode (OTE). With chemical tethering of this type, Wrighton and co-workers attached the viologen (**I**)[38] and a benzyl viologen[40] species to electrode surfaces.

$$(MeO)_3Si(CH_2)_3\text{—}\overset{+}{N}\underset{}{\bigcirc}\text{—}\underset{}{\bigcirc}\overset{+}{N}\text{—}(CH_2)_3Si(OMe)_3$$

I

Wrighton and co-workers also diquaternised a bipyridilium nucleus with a short alkyl chain terminating in pyrrole (which was bonded to the alkyl chain at nitrogen[39]) – see **II**; anodic polymerisation of the pyrrole allowed an

11.3 Bipyridilium species for inclusion within ECDs

adherent film of the linked poly(pyrrole) to derivatise the electrode surface,[39] thereby attaching the bipyridilium units.

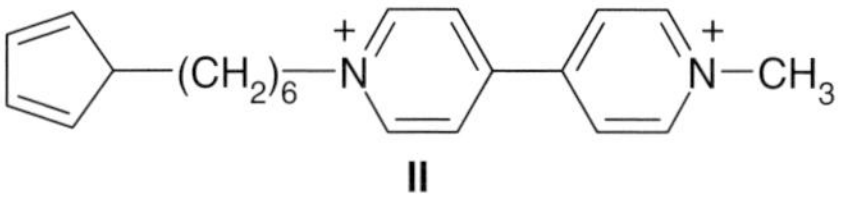

II

An identical analogue has been prepared with thiophene as the polymerisable heterocycle.[41] The electroactivity of the poly(thiophene) backbone in this latter polymer degraded rapidly after only a few doping/de-doping cycles, but the electroactivity of the viologen moiety remained high.

Itaya and co-workers[42] used polymeric electrolytes, but with an electrochromic salt bonded electrostatically to a poly(styrene sulfonate) electrolyte. A bipyridilium salt of poly(p- or m-xylyl)-4,4'-bipyridilium bromide (**III**, shown here as the p form) was employed in this manner: the interaction between the cationic bipyridilium nucleus and the sulfonyl group is coulombic. The electrode was prepared by dipping the conducting substrate into solutions of electrochrome-containing polymer which, after drying, is insoluble in aqueous solution.[42] Polymeric bipyridilium salts have also been prepared by Berlin et al.,[43] Factor and Heisolm,[44] Leider and Schlapfer,[45] Sato and Tamamura[46] and Willman and Murray.[47]

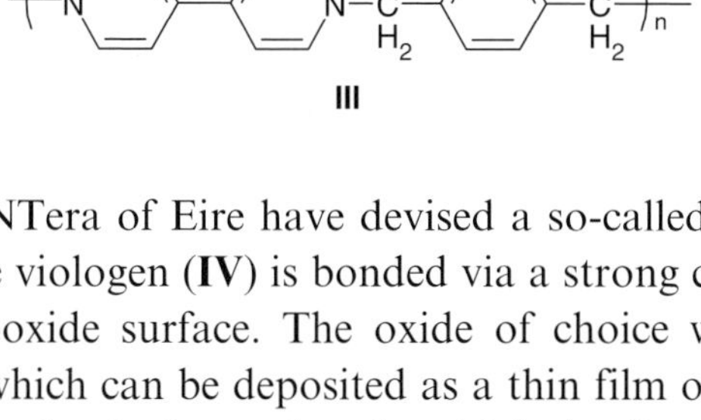

III

More recently, NTera of Eire have devised a so-called NanoChromics™ device in which the viologen (**IV**) is bonded via a strong chemisorptive interaction to a metal-oxide surface. The oxide of choice was nanostructured titanium dioxide, which can be deposited as a thin film of high surface area. The amount of **IV** adsorbed was therefore high, leading to a good contrast ratio. Fitzmaurice and co-workers[48] in 1994 were probably the first to use a viologen adsorbed onto such layers.

IV

The electrochromism of **IV** is discussed in Section 11.4 below. Corr et al.[49] of NTera also studied the electrochromic properties of an analogue of **IV**, in which the phosphonate substituent is replaced with a simple alkyl chain.

11.3.2 Immobilised viologen electrochromes for ECD inclusion

A different method of ensuring a high 'write–erase efficiency' is to embed the bipyridilium salt within a polymeric electrolyte. Thus, Sammells and Pujare[50,51] suspended heptyl viologen in poly(2-acrylamido-2-methylpropanesulfonic acid) – 'poly(AMPS)' – while Calvert et al.[52] used methyl viologen also in poly(AMPS). Both groups report an excellent long-term write–erase efficiency, and a good electrochromic memory. The response times of such devices are, as expected, intrinsically extremely slow.

Another means to a similar end is to employ a normally liquid solvent containing a gelling agent (silica, for example[53]) which is just as effective in immobilising viologens, the concentration of which can be[53] as high as 4 mol dm^{-3}.

11.3.3 Soluble-to-insoluble viologen electrochromes for ECD inclusion

As noted above, in aqueous solution, it is usual for the final product of reduction of a type-II dicationic viologen to be a solid film of radical-cation salt. The process of forming such a salt is usually termed 'electrodeposition'. Strictly, the term 'electrodeposition' implies that the solid product is the immediate product of electron transfer. Most workers now consider the formation of viologen radical-cation salts to be a three-step process, radical cation being formed at the electrode, Eq. (11.1), followed by acquisition of an anion X$^-$ in solution and thence precipitation of the salt from solution:

$$\text{bipm}^{+\bullet}\,(\text{aq}) + \text{X}^-(\text{aq}) \rightarrow [\text{bipm}^{+\bullet}\,\text{X}^-]\,(\text{s}). \tag{11.4}$$

Equation (11.4) represents the chemical step of an 'EC' type process in which the product of electron transfer – 'E' – undergoes a chemical reaction – 'C', Eq. (11.4). Such an overall EC reaction is strictly 'electroprecipitation' but commonly termed 'electrodeposition'. (If electroprecipitation occurs by two steps that are in effect instantaneously sequential, 'electrodeposition' *is* an adequate description.)

11.3.4 Applications of bipyridilium systems in electrochromic devices

The first ECD using bipyridilium salts was reported by Schoot *et al.*[54] (of Philips in the Netherlands) in 1973. Philips submitted Dutch patents in 1970[55] for heptyl viologen (HV = 1,1′-diheptyl-4,4′-bipyridilium) as the dibromide salt. The HV^{2+} dication is soluble in water, but forms an insoluble film of crimson-coloured radical-cation salt that adheres strongly to the electrode surface following a one-electron reduction, as in Eq. (11.4). The Philips ECD had a contrast ratio of 20:1, an erase time of 10 to 50 ms,[54] and cycle life of more than 10^5 cycles. Philips chose heptyl viologen for their ECD rather than a viologen with a shorter-chain, because reduction of the HV^{2+} dication formed a durable film on the electrode, whereas shorter alkyl chains yield somewhat soluble radical-cation salts. The Philips device was never marketed.

In 1971, ICI first submitted a patent for the use of the aryl-substituted viologen 1,1′-bis(*p*-cyanophenyl)-4,4′-bipyridilium ('cyanophenyl paraquat' or 'CPQ'),[56] which electroprecipitates according to Eq. (11.4) to form a green electrochrome with a superior colour and resistance to aerial oxidation. ICI preferred CPQ to HV owing to its greater extinction coefficient (and hence higher η) and therefore its faster response time per inserted charge. Figure 11.3 shows a schematic of the ECD cell, which is extremely simple. The conducting layer of the ITO (of fairly high resistance, ~80 Ω per square) acts as the working electrode that displays the colour. A strip of insulating cellulose acetate is placed near opposing edges of the base, and a stripe of conducting silver paint is applied to its upper surface to facilitate an ohmic contact with the electrode surface. The electrolyte layer was gelled with agar (5%) to improve its stability, and containing the electrochrome in a concentration of 10^{-3} mol dm^{-3} in sulfuric acid or potassium chloride (either of concentration 0.1 mol dm^{-3}). The layer is applied over the platinum-wire counter electrode, itself positioned over the insulating layer. The device is completed by encapsulating the electrolyte layer, so a sheet of plain non-conducting glass covers the device. Electrical connection is made to the counter electrode and the exposed end of the silver paint.

A potential of −0.2 V (relative to a small, internal silver|silver chloride electrode) is applied to the silver paint to effect electrochromic coloration – *cf.* Eq. (1.1) – to form a thin, even layer of insoluble, green radical-cation salt, Eq. (11.5):

$$CPQ^{2+}(\text{soln.}) + e^- + X^- \rightarrow [CPQ^{+} \bullet X^-](s). \quad (11.5)$$

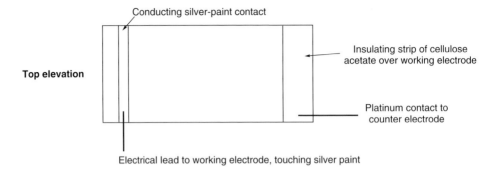
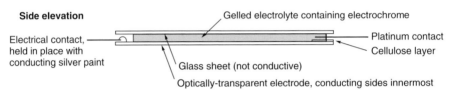

Figure 11.3 Schematic representation of an ECD operating by a type-II electrocoloration mechanism, with colourless CPQ^{2+} in solution being electro-reduced to form a coloured film of radical cation salt. (Figure reproduced from J. G. Kenworthy, ICI Ltd. British Patent, 1,314,049, 1973, with permission of ICI.)

It is best to prevent the formation of a further reduction product, the pale-red species CPQ^{0} (oxidation of which is slow), so the reducing potential should not exceed -0.4 V.

The intense green colour of the $CPQ^{+\bullet}$ radical is stable on open circuit, the colour persisting for many tens of hours. Reversing the polarity and applying a potential of $+1.0$ V (measured vs. silver–silver chloride electrode) oxidatively removes the electrogenerated colour in a bleaching time of ca. 1 minute.

The Pt counter electrode in Figure 11.3 is pre-coated with solid $CPQ^{+\bullet}$ and undergoes the reverse of Eq. (11.5) during coloration on the ITO. Then for bleaching at the ITO – i.e. the reverse of reaction (11.5) – the reaction (11.5) takes place at the Pt counter electrode in a confined, invisible volume. This pre-coating procedure represents an ingenious resolution of the often problematic choice of counter electrode. (In demonstration devices, often electrolysis of solvent is allowed to take place at the counter electrode, which in progressively destroying solvent will of course not serve in long-term use.)

A less-anodic potential of $+0.4$ V (vs. AgCl–Ag) can be used if the electrolyte is gelled and also contains sodium ferrocyanide (0.1 mol dm^{-3}) as an electron mediator to facilitate electro-bleaching; see p. 358.

Following extensive and successful field trials, this ECD was first marketed in the early 1970s as a data display device, but liquid-crystal displays (LCDs) entered the market at about the same time, and had faster response times τ; LCDs rapidly captured an unassailable market share. The slow kinetics of the ICI type-II cell were the result of including agar to gel the electrochrome-containing electrolyte. Removal of the agar allows for considerable improvements in device response times (to seconds or tenths of seconds), but the electrochromic image is usually streaky and uneven. Ultimately, a yellow-brown oil stains the electrode surface. Yasuda et al.[57] (of Sony Corporation) added encapsulating sugars such as β-cyclodextrin to aqueous heptyl viologen to circumvent the problem of 'oil' formation; but ICI believed that this molecular encapsulant would not improve the long-term write–erase efficiency of the different viologen CPQ.[58] The origin of 'oiling' as a result of dimer formation by the radicals is considered in more detail in Section 11.3.10 below.

11.3.5 The effect of the bipm N substituents

van Dam and Ponjeé[59] examined the effect that variations in the length of the alkyl chain have on the film-forming properties of the radical cation as the bromide salt (Table 11.2), and redox potentials have been added to this table from ref. 9. As the length of the alkyl chain is increased, the pentyl chain produces the first truly insoluble viologen radical-cation salt. The heptyl is the first salt for which the solubility product is small enough for realistic device usage.

Table 11.2 shows that an effective chain length in excess of four CH_2 units is necessary for stable solid films to form. The radical-cation salt of cyanophenyl paraquat (CPQ) is more insoluble in water than is $HV^{+\bullet}$, yet the dicationic salt is very soluble. The solubility product K_{sp} of $HV^{+\bullet}$ Br^- in water is[59] $3.9 \times 10^{-7}\,mol^2\,dm^{-6}$.

The radical cations of viologen species containing short alkyl chains have a blue colour becoming blue–purple when concentrated.[24] The colour of the radical cation tends towards crimson as the length of the alkyl chain increases, largely owing to increasing incidence of radical-cation dimerisation; the dimer of alkyl-substituted radical cations is red.[24]

By comparison, aryl-substituted viologens generally form green or dark-red radical-cation salts. Also, dication solubility and radical-cation stability (in thin films) are both greatly improved by using aryl substituents. This underlay ICI's use, presented in detail above, of the aryl-substituted viologens, particularly p-cyanophenyl CPQ in their ECD since the electrochromic colour of the heptyl

Table 11.2. *Symmetrical viologens: the effect of varying the alkyl chain length on radical-cation film stability (refs. 9 and 59). The $E^{\ominus}$ values are quoted against the SCE, and refer to viologen salts with the parenthesised anion.*

Substituent R	Effective length (units of CH_2)	Solid bromide salt film on Pt?	Colour	$E^{\ominus}$/mV
Methyl	1	No	Blue	−688 (Cl$^-$)
Ethyl	2	No	Blue	−691 (Cl$^-$)
Propyl	3	No	Blue	−690 (Br$^-$)
Butyl	4	No	Blue	−686 (Br$^-$)
Pentyl	5	Yes	Purple	−686 (Br$^-$)
Hexyl	6	Yes	Purple	−710 (Br$^-$)
Heptyl	7	Yes	Mauve	−600 (Br$^-$)
Octyl	8	Yes	Crimson	−705 (Br$^-$)
iso-Pentyl	4	Yes	Purple	−696 (Br$^-$)
Benzyl	4–5	Yes	Mauve	−573 (Cl$^-$)
$CH_3(Cl)CH_2OCH_2-$	4	No	–	
$CH_3-CH=CH-CH_2-$	4	No	–	
$H-CH=CH-(CH_2)_3-$	4–5	No	–	
$NC-C_3H_6-$	4–5	No	–	−362^a (Cl$^-$)

^{a} Polarographic $E_{1/2}$ value.

viologen radical cation was deemed insufficiently intense: the molar absorptivity (and therefore the *CR*) of aryl-substituted viologens is always greater than that of alkyl-substituted viologens (Table 11.1). Furthermore, the green radical cation of CPQ apparently[56] is more stable than the other aryl viologen radical cations.

11.3.6 *The effect of the counter anion*

The counter anion in the viologen salt may crucially affect the ECD performance. Different counter ions yield solid radical-cation products of electrodeposition having a wide range of solubilities and chemical stabilities.[30] For example, CPQ$^{+\bullet}$ is oxidised chemically by the nitrate ion via a rapid but complicated mechanism.[30] Studies of counter-ion effects may be performed using cyclic voltammetry (e.g. ref. 60) or by observing the time dependence of an ESR trace, which demonstrates the bipm$^{+\bullet}$ concentration.[30] The ICI group used the SO_4^{2-} salt of CPQ^{2+} in their prototype ECDs.[56]

The properties of heptyl viologen radical-cation films also depend on the anion as shown by van Dam and Ponjeé.[59] Jasinski[60] (Texas Instruments) found the optimum anion in water to be dihydrogen phosphate. Anions found

Table 11.3. *The effect of supporting electrolyte anion, and of electrode substrate, on the reduction potentials^a of heptyl viologen. Values of peak potential E_{pc} are cited against the SCE. (Table reproduced from Jasinski, R. J. 'The electrochemistry of some n-heptyl viologen salt solutions'. J. Electrochem. Soc., **124**, 1977, 637–41, with permission of The Electrochemical Society, Inc.)*

Anion	$E_{pc(1)}/V$ on Au	$E_{pc(2)}/V$ on Au	$E_{pc(1)}/V$ on Pt	$E_{pc(2)}/V$ on Pt	$E_{pc(1)}/V$ on Ag	$E_{pc(2)}/V$ on Ag
Bromide (0.3 mol dm^{-3})	−0.698	−1.008	−0.708	(< −0.818)	−0.708	−0.978
H$_2$PO$_4^-$ (2 mol dm^{-3})	−0.668	−1.048	−0.668	(< −0.818)	−0.668	
Formate (0.4 mol dm^{-3})	−0.848	−0.928			−0.828	−0.948
HCO$_3^-$ (1 mol dm^{-3})	−0.768	−0.958			−0.778	−0.948
Acetate (0.5 mol dm^{-3})	−0.828	−0.928				
Fluoride (1 mol dm^{-3})	−0.818^c	−0.878^c		−0.868		−0.808
Sulfate (0.3 mol dm^{-3})	−0.818^c	−0.928^c	−0.848^b	−0.898	−0.798	−0.918

^{a} Reduction potentials determined at pH 5.5. ^{b} Millimolar viologen dication employed for measurement. ^{c} No colour formed.

useful for ECDs were dihydrogen phosphate, sulfate, fluoride, formate and acetate. Bromide, chloride, tetrafluoroborate and perchlorate also proved satisfactory (as also concluded by van Dam and Ponjeé[59]). Heptyl viologen salts of bicarbonate (at pH 5.5), thiocyanate, tetrahydroborate, hexafluorophosphate, tetrafluoroantimonate and tetrafluoroarsenate are all water insoluble. Like CPQ$^{+\bullet}$, HV$^{+\bullet}$ is also oxidised by the nitrate ion,[60] presumably by a similar mechanism.

Jasinski's values[60] of reduction potentials for aqueous HV^{2+} on various metals as electrode substrate, with a variety of anions, are given in Table 11.3. Many other redox potentials for mono-reduction of bipyridilium salts are quoted in the reviews by Monk[5] and by Bird and Kuhn.[9]

The choice of anion in the viologen-containing solution can be important since it often participates in charge-transfer type interactions with the viologen. Recent evidence suggests the CT complex must dissociate prior to reductive electron transfer.[61] Reduction is therefore a two-step process: ion-pair dissociation → reduction. Electron transfer may be thought of as a special type of second-order nucleophilic substitution ('S$_N$2') reaction in which the 'nucleophile' is the electron and the leaving group is an anion.

The rates at which the CT complexes of methyl viologen dissociate vary: the complex with iodide dissociates at a rate of 8.7×10^5 s^{-1}, while that with chloride dissociates at $26.3 \times 10^5 \, \text{s}^{-1}$. The anion may thus also influence the speed of electrochromic coloration, as discussed more fully below.[61]

11.3.7 The kinetics and mechanism of viologen electrocoloration

Kinetic aspects of electro-coloration of type-I electrochromes are discussed in Section 5.1 on p. 75ff. Exemplar viologen systems include viologens in non-aqueous solutions, or short-chain-length viologens in water.

The coloration of type-II systems is considerably more complicated, as follows. Bruinink and van Zanten[62] (of Philips) and Jasinski[63] studied the kinetics of HV^{2+} dibromide reduction in response to a potential step; both groups found the kinetics of mono-reduction to depend on the electrode history and mode of preparation. For the HV^{2+}–$(H_2PO_4^-)_2$ system, the data obtained do not allow any distinction between two possible but different reduction mechanisms. Jasinski prefers a two-stage process of electroprecipitation (also favoured by van Dam[59] and Schoot et al.[54]) in which solution-phase HV^{2+} is reduced to form the radical cation, Eq (11.1), followed by anion acquisition and precipitation of the salt, Eq (11.6) for the bromide anion:

$$HV^{+\bullet}(aq) + Br^-(aq) \rightarrow [HV^{+\bullet} \, Br^-](s). \tag{11.6}$$

Bruinink and Kregting[64] (cf. Jasinski[65]), while citing the two-step electro-precipitation mechanism, found the reduction process to be compatible also with a theoretical model of metal deposition derived by Berzins and Delahay.[66] The rate of film growth is controlled by instantaneous three-dimensional nucleation, as seen by a current–time relationship of I vs. $t^{1/2}$. Ultimately, these nuclei overlap, the commencement of which is shown by a transition in the current–time domain to the expected Cottrell relationship of I dependence on $t^{-1/2}$. The number of nucleation sites available are suggested[66] to depend on the potential.

Fletcher et al.[67] reduced HV^{2+} in solutions of bromide or biphthalate at a disc of SnO_2 on glass, and agree that the reduction process proceeds via a nucleation step. At low overpotentials of mono-reduction, the rate of reduction was controlled by electron transfer and, at high overpotentials, the nucleation process, once initiated, was sufficiently fast for the crystal-growth process to be controlled by mass transport. Hemispherical diffusion was inferred, creating diffusion zones that could overlap, after formation, and lead to semi-infinite planar diffusion. In summary, the process may be written as electron transfer → nucleation → hemispherical diffusion → linear diffusion, but the process is too complicated to allow precise mathematical models of deposition to be used.

By way of confirmation, in cyclic voltammetry, the current–potential curve (cyclic voltammogram, CV) associated with electro-coloration is unusually steep, which usually implies a catalytic or nucleation process; the usual shape of the CV before the current peak is exponential. (The presence of radical-cation dimer on an electrode also causes a steep CV peak,[68] which may imply that nucleation sites are comprised of dimer.)

The morphology of $HV^{+\bullet}$ films has been addressed by Barna[69] (Texas Instruments). Deposited films are partially crystalline but largely amorphous, acquiring a greater degree of crystallinity with time; this acquisition of crystallinity is probably associated with additional sharp peaks observed during cyclic voltammetry of heptyl viologen films.[9,59,60,63] The time-dependent change within deposits of $HV^{+\bullet}$ on an OTE has been observed by Goddard et al.[70] using UV-visible spectroscopy with a novel potential cycling technique (but now rendered obsolete by use of diode-array spectrophotometry).

Bewick et al.[71,72] have investigated HV^{2+} dibromide and many asymmetric bipyridilium salts (that is, with substituents at N and N' being different), by diode-array spectroscopy. The initial solid product of mono-reduction was considered to be $HV^{+\bullet}$ radical cation in a salt which incorporates some unreduced HV^{2+} dication.[72] Subsequent aging effects and previously inexplicable additional CV peaks are explained in terms of this composite form of solid deposit. A similar explanation for the complicated cyclic voltammetric behaviour observed during the formation of solid $CPQ^{+\bullet}$ radical-cation salt has also been advanced.[13,73]

11.3.8 Micellar species

Association of bipyridilium species to form π-dimers is a well-documented phenomenon[23,24] for the viologen radical cation, but is not so well attested for the dication (although see references 13, 37, 74 and 75). A particular problem with aqueous solutions of viologen can be the formation of micelles of dication, particularly if the substituents are large aryl groups or long alkyl chains that are hydrophobic. In this latter case, the analogy between such viologens and quaternary ammonium cationic surfactants is clear.[76] Barclay et al.[37] (of IBM) quote a critical micelle concentration cmc for the HV^{2+} dication of 10^{-2} mol dm^{-3} in aqueous bromide solution.

Electrochemistry at these micelles is envisaged to proceed in discrete steps, with dication on the micelle periphery being reduced preferentially.[77] If the concentration of HV^{2+} lies above the cmc, interaction of such a micelle with a cathodically biased electrode causes reduction of the outside of the micelle to form $bipm^{+\bullet}$, yet the inside of the micelle remains fully oxidised dication.[13]

A similar explanation for the complicated cyclic voltammetric behaviour observed during the formation of solid CPQ$^{+\bullet}$ radical-cation salt has also been advanced.[13,73] Heyrovský has also postulated the existence of solution-phase mixed-valence species of methyl viologen in water.[78,79,80]

Although such mixed-valence viologens are not involved often, and comprise very small amounts of material, they are capable of greatly complicating the electrochemistry of solutions containing them.

In an effort to mimic the properties of bipyridilium species within micelle environments, Kaifer and Bard[74] investigated the electrochemistry of methyl viologen in the presence of various surfactants (anionic, cationic and non-ionic), finding that the properties of methyl viologen were largely unaffected by the presence of surfactant when below the *cmc*, but above it, the properties of the methyl viologen were markedly different, e.g. the EPR spectrum of MV$^{+\bullet}$ lost all hyperfine coupling due to rapid inter-radical spin swapping.

Engleman and Evans[81,82] also investigated the electrochemical reduction of MV^{2+} in the presence of micellar anions. In the presence of anionic surfactant, the position of the monomer–dimer equilibrium was displaced significantly in favour of the monomeric form when above the *cmc*, whereas cationic and non-ionic surfactant did not affect the equilibrium either way. Cyclic voltammetry also differed above and below the *cmc*.

Given the complexity of the overall electroprecipitation mechanism, and the different speciations in solution during each of the steps during electroprecipitation, it is again worth remarking that comparison of results from different authors is difficult since each step is defined by the number and nature of each anion in solution, and different experimental conditions (where known) have been employed by each author.

11.3.9 The write–erase efficiency

The write–erase efficiency of a viologen electrochrome is very high if the solvent is non-aqueous and rigorously dried. For example, the archetypal type-I Gentex mirror described in Section 12.1 contains a viologen (which is probably zwitterionic) as the primary electrochrome. Byker, formerly of Gentex,[83] cites a cycle life of '$> 4 \times 10^4$ cycles' for the related system benzyl viologen (as the BF$_4^-$ salt) in propylene carbonate (PC). Ho et al.[84,85] have modelled the electrochemical behaviour and cycle life of similar electrochromic devices, particularly one with heptyl viologen as the primary electrochrome and tetramethylphenylenediamine (TMPD) as the secondary. The cycle lives are high, but the response times are quite slow, for the reasons discussed above.

11.3 Bipyridilium species for inclusion within ECDs

The mechanism of deposition was examined at length since here the nature of the solid deposit is important. For example, firstly a fresh film of $HV^{+\bullet}$ is amorphous, even[60] and smooth,[86] yet soon after deposition ($<10\,s^{70}$) the film appears patchy as an aging process occurs, which probably involves ordering (crystallisation) of radical moieties. Re-oxidation of the film to bleach the colour is rapid for fresh $HV^{+\bullet}$ films, but patchy films that show signs of aging are more difficult to oxidise, requiring a higher potential or a longer re-oxidation time. Dimerisation of radicals could participate in this complication.

Secondly, after prolonged cycling between the coloured and bleached states, bipyridilium ECD devices form an unsightly yellow–brown stain on the electrode. Some evidence now suggests that this stain is a form of crystalline radical-cation salt[30] containing spin-paired radical-cation dimer, or intervalence species comprising both $bipm^{2+}$ and $bipm^{+\bullet}$. Spectroscopic studies[87] (a surface-enhanced resonance Raman analysis of a disulfide-containing dimeric viologen adsorbed on rough silver) strongly suggest the presence of a 'liquid-like' environment at the electrode surface following reduction to form dimerised radical $(bipm)_2^{2+}$.

It is also likely that reordering of radical species ('recrystallisation') occurs within the electroprecipitated viologen deposit soon after it forms. In order to understand these processes, thin films of $HV^{+\bullet}$ salt have been studied by many techniques including UV-visible spectroelectrochemistry,[72,88,89,90] EPR,[91] Raman spectroscopy,[92,93,94] photoacoustic spectroscopy,[95,96] photothermal spectroscopy[97] and the electrochemical quartz-crystal microbalance (EQCM).[86]

Scharifker and Wehrmann[98] investigated phase changes within radical-cation salt deposits of $HV^{+\bullet}$ and benzyl viologen radical cation, and Gołden and Przyłuski[99] looked at $HV^{+\bullet}$. Both groups found the aging effect to be due, in part, to the dimerisation of radical cation in solution. Belinko[33] suggested that device failure is also due to production of di-reduced bipyridilium ($bipm^0$) as a minor electrode product. The formation of diamagnetic HV^0 (at large negative potentials) should be avoided since it is only electrochemically quasi-reversible electrochemically, i.e. slow, in aqueous solution.[9] Belinko[33] investigated the write–erase efficiency of $HV^{+\bullet}$ films by cyclic voltammetry, making the lower scanning limit progressively more negative deliberately to generate $bipm^0$. After $bipm^0$ is formed, it may react with $bipm^{2+}$ from the solution in the comproportionation:

$$bipm^{2+} + bipm^0 \rightarrow (bipm^{+\bullet})_2 \rightarrow 2\,bipm^{+\bullet}. \qquad (11.7)$$

The immediate product of Eq. (11.7) is the radical-cation dimer. In solution, subsequent dimer dissociation yields monomeric radical cation[100] but often solid deposits of 'bipm$^{+\bullet}$' exhibit spectroscopic IR bands attributable to the spin-paired (bipm$^{+\bullet}$)$_2$ dimer.[101] In effect, spin pairing is 'locked into' solid deposits of viologen radical cation.

Recent work has shown that the radical-cation dimer is electrochemically only quasi-reversible, that is, its electro-oxidation is slow,[100,102] hence the observed failure of ECDs containing traces of dimer.

The 1998 review by Monk[5] demonstrates how widely comproportionations occur in viologen redox chemistry. Its fast rate constant and moderate equilibrium constant make it almost certain that comproportionation processes always occur whenever bipm^0 is formed electrochemically. Invoking the participation of Eq. (11.7) can greatly simplify Belinko's otherwise complicated mechanistic observations. (In this context, see also the way comproportionation can simplify mechanistic observations in ref. 103.)

Engelmann and Evans have also published[104] studies of potentiostatic deposited MV^0, EtV^0, BzV^0 and HV^0; each was formed at a glassy-carbon rotated ring-disc electrode (RRDE) from solutions containing the respective dications. The reductions are concerted two-electron reactions. To summarise their findings, deposition is initiated by nucleation of supersaturated bipm^0 close to the electrode; the rate of deposition decreases as the bulk of the deposit increases, i.e. as the surface of the disc becomes blocked. Comproportionation of bipm^{2+} (aq) from the solution with bipm^0 (s) on the disc becomes increasingly important with time, so that the total amount of bipm^0 on the disc decreases until the amount reaches a steady state. That comproportionation occurs in the solid state has been confirmed for CPQ^0 and CPQ^{2+} from aqueous electrolytes.[30,105]

The mechanism of comproportionation differs when ferrocyanide is involved (see footnote a on p. 342). This result may be important because this ion is a popular choice of electron mediator.[106] Generally, the bipm^{2+} and bipm^0 species approach and thence form a sandwich-like structure with their π-orbitals overlapping. Comproportionation occurs when the electron transfers through these orbitals. However, when ferrocyanide is involved, the ferrocyanide ion is believed to lie between the two viologen species, in a structure reminiscent of a metallocene. Equation (11.7) could thus occur by electron transfer through the ferrocyanide possibly by a concerted double-exchange mechanism.[106] Hence the two radical-cation moieties produced by reaction (11.7) are never in contact; after reaction, they separate from the ferrocyanide to form individual ions.

Benzyl viologen has also been extensively investigated since it will also form an insoluble film of radical-cation salt following one-electron reduction.[70,89,98,107]

To summarise, the speciation of the viologen dication is complicated prior to the transfer of an electron: the rate of anion–dication separation prior to (or during) electron transfer follows[61] the rate k_{et} and may in fact dictate its magnitude; the rates of anion–radical cation association following the electron transfer is completely unknown; the way the length of the substituents at nitrogen dictates the solubility constants of radical cation–anion pairs is fairly well understood; and the way the solubility index dictates the rate of precipitation has been investigated extensively.

11.3.10 Attempts to improve the write–erase efficiency

The first and most effective method of improving the write–erase efficiency is to employ non-aqueous solutions, although the coloration time will necessarily be slow, and the bleaching time slower still.

The second method used to prevent the non-erasure of films of $HV^{+\bullet}$ salt is to add an auxiliary redox couple (that is, an electron mediator) to the dication-containing electrolyte solution. The mediators used include hydroquinones,[55] ferrous ion,[56] ferrocyanide,[56,57,92] or cerous ion,[50,51] and ferrocene in acetonitrile has been used in a type-II device.[108] During electro-coloration, $bipm^{2+}$ ion is reduced to $bipm^{+\bullet}$ but, during re-oxidation at a positive potential, it is the mediator (e.g. ferrocyanide) that is oxidised at the electrode. The oxidised form of the mediator – in this example, ferricyanide, i.e. hexacyanoferrate(III) – allows for *chemical* oxidation of the radical-cation film, to reform the dication. Such oxidation is very rapid.[15] Mediators facilitate the electro-oxidation of the radical cations of type-II species, such as heptyl viologen. For aryl viologens in aqueous solution, a mediator is always necessary to ensure complete colour removal on re-oxidation.[5] As ferrocyanide is known to form a charge-transfer complex with methyl viologen dication[109,110] and also with the dications of $CPQ^{5,12,111}$ and HV,[57] it will be the free-anion equilibrium fraction of the species that can be assumed to act.

The unsightly yellow–brown stains still persist, however, even with the HV^{2+} and CPQ^{2+} systems that contain $K_4Fe(CN)_6$.[56,92] In a notable advance, addition of the sugar β-cyclodextrin to the voltammetry solution has been found to impede the formation of yellow–brown stains,[57,91] probably by encapsulating the dication within the cavity of the cyclodextrin in a guest–host relationship. Because close contact between bipyridilium dications is greatly impeded in such a guest–host relationship, association of $bipm^{2+}$ cations in solution[57] is largely thereby prevented, so alignment of $bipm^{+\bullet}$ species in the solid deposit is impossible. However, such 'oiling' is claimed still to occur 'ultimately' with $CPQ^{+\bullet}$.[58]

Other attempts to stop the ageing phenomenon have used different, modified bipyridilium compounds.[112,113,114] For example, Bruinink et al.[112] prepared the compound **V** in which the two pyridinium rings are separated by methylene linkages.

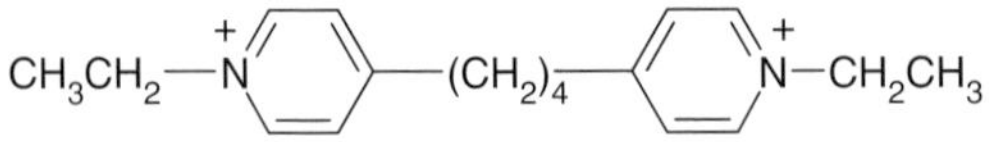

$$CH_3CH_2-\overset{+}{N}\underset{}{\bigcirc}-(CH_2)_4-\underset{}{\bigcirc}\overset{+}{N}-CH_2CH_3$$

V

To a similar end, Barna and Fish[113] prepared asymmetric bipyridilium salts, that is species in which $R^1 \neq R^2$ (Scheme 11.1), thereby inhibiting the crystallisation process: for example, a compound was made having $R^1 = C_7H_{15}$ and $R^2 = C_{18}H_{37}$. Barltrop and Jackson[114] have prepared similar asymmetric viologens, and a diquaternised (that is, made cationic by alkyl or aryl addition) 3,8-phenanthroline salt (**VI**), together with a series of nuclear-substituted bipyridyls (species in which substituents are directly bonded to carbon in the pyridine rings). Again, films with superior write–erase properties were formed.

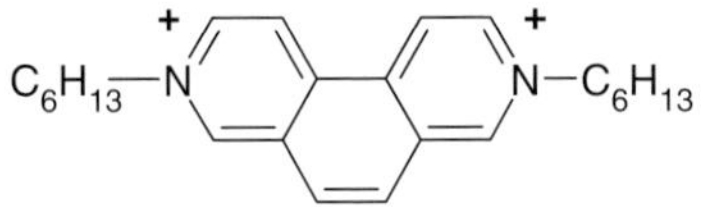

$$C_6H_{13}-\overset{+}{N}\underset{}{\bigcirc\bigcirc}\overset{+}{N}-C_6H_{13}$$

VI

Despite the many drawbacks recounted above, a large number of prototype viologen ECD devices have been made.[36,37,55,56,115] For example, an impressive device from the IBM laboratories utilised a 64 × 64 pixel integrated ECD device with eight levels of grey tone of heptyl viologen[115] on a 1 inch square silicon chip, to give quite detailed images (Figure 11.4). These devices were not exploited further owing to competition from LCD systems, though they may still have a size advantage in large devices.

11.4 Recent elaborations

The majority of the new developments reported here aim to enhance the rate of coloration in bipyridilium-based ECDs.

11.4.1 Displays based on viologens adsorbed on nanostructured titania

Nanostructured electrodes are easily prepared by spreading a concentrated colloidal suspension on a conducting substrate and firing the resulting gel film

Figure 11.4 Reproduction of an IBM electrochromic image displayed on a 64 × 64 pixel integrated ECD device with eight levels of 'grey tone' of heptyl viologen. The original is clearer. (Figure reproduced from Barclay, D. J. and Martin, D. H. 'Electrochromic displays'. In Howells, E. R. (ed.), *Technology of Chemicals and Materials for the Electronics Industry*, Chichester, Ellis Horwood, 1984, 266–76, by permission of Ellis Horwood.)

at 450 °C.[116] Such electrodes have been widely investigated for use in dye-sensitised photoelectrochemical cells.[116,117] The rough surface of the porous titanium dioxide film consists of a network of interconnected semi-conducting metal oxide nanocrystals. Because the oxide crystals are so small, such films have an extraordinarily high internal surface area.

The ratio between the internal surface area and the smooth geometrical area of the electrode (the 'roughness factor') approaches 1000 for a film that is only 4 μm thick.[117] This means that a high number of electrochromic viologen molecules can occupy a relatively small area, leading to a high coloration efficiency η. Furthermore, as they are surface-confined, the viologen molecules need not diffuse to the electrode surface, which leads to shorter switching times. Nanostructured titanium dioxide in its anatase form can be deposited as a thin film of high surface area. Viologens are strongly adsorbed on its surface owing to their electron deficiency. Such systems have long been investigated in research on dye-sensitised solar cells, for example Grätzel's work on his photoelectrochemical cell.[117,118]

Originally developing a spin-off from the Grätzel cell, Fitzmaurice and co-workers at the Dublin-based NTera Ltd[119] (founded in 1997, having manufacturing facilities in Ireland and Taiwan) have developed a 'next generation display technology' called NanoChromics™ displays that are based on these principles.[49,120,121,122] NTera also describe their ECD as a 'paper quality' electrochromic display, that is, an ECD of very high definition. An assembled NanoChromics™ electrochromic device uses two metal-oxide films – one at the negative electrode and, unusually, one at the positive electrode. In a typical device[122] (borrowed from Grätzel[123]) the negative F-doped tin oxide conducting glass electrode (the cathode on coloration) is coated with the wide bandgap titanium dioxide film 4 μm thick, followed by a monolayer of self-assembled, chemisorbed phosphonated viologen molecules. The positive F-doped tin oxide conducting glass counter electrode (i.e. the anode on coloration) carries a film of heavily doped antimony tin oxide (SnO_2:Sb) 3 μm thick, followed by a monolayer of self-assembled, chemisorbed phosphonated phenothiazine molecules. The TiO_2 film is further modified with an adsorbed monolayer of viologen (**IV**), bis(2-phosphonoethyl)-4,4'-bipyridilium dichloride. The electrolyte was γ-butyrolactone containing $LiClO_4$ (0.2 mol dm^{-3}) and ferrocene (0.05 mol dm^{-3}).[120,121] In trials, their device had a coloration efficiency η of 170 cm^2 C^{-1} at 608 nm,[121] and was said to be stable over 10 000 'standard' test cycles.

The counter electrode is viewed as having a high capacitance, which assists charge storage during coloration. The ECD is sealed with a thermoplastic gasket and a UV-curable epoxy resin. Application of a potential of 1.2 V reduces the dicationic viologen to its blue radical cation, and oxidises the phenothiazine from its weak yellow colour to red. The overall colour change is therefore from virtually colourless to a blue-red purple.

Placing a diffuse reflector between the electrodes, e.g. a layer of an ion-permeable nanostructured solid film of titanium dioxide, gives on coloration the visual effect of ink on pure white paper. Without the intermediate TiO_2 layer the display is transparent while retaining readability. Different colours can be achieved in ECDs depending on the nature of the substituent(s) on the viologen molecule.[120] In such devices, many thousands of switches are possible before there is significant degradation of performance. Some open-circuit memory persists, the colour remaining for more than 10 min after the voltage is switched off, but readily regenerated. Electrodes can be micro-patterned for display applications.

Fitzmaurice's display is said to be 'ultra fast',[122] although the criterion for this claim is unclear, since the switching time is 1 s for a change in absorbance of 0.60.[120] However, this is certainly faster than most of the other viologen-based devices, since the anchored viologen electrochrome avoids the diffusion delay before electron transfer. Fitzmaurice notes that charge compensation within the viologen layer is also fast because many counter ions are also adsorbed on the TiO_2 layer.

If the counter electrode is covered with a secondary electrochrome such as a phenothiazine, the value of η increases to about 270 $cm^2 C^{-1}$ and the response time is decreased to 250 ms.[122] Published spectra suggest an optical density (OD) change of about 0.55, again at 608 nm.

The NTera group state that they are working with a number of market-leading strategic partners for access to the market. Recently, NTera have demonstrated a NanoChromics™ display operating in a converted iPod (the portable digital audio players from Apple Computer Corporation).[124] The NTera website[119] provides 'consumer product reference designs' for digital clocks and an eight-digit calculator. That NanoChromics™ displays can be manufactured by existing LCD manufacturing processes will clearly enhance the likely success in the commercial development of this technology.

NTera also state that their flexible display prototype can, in principle, be applied to all the product types: displays, windows and mirrors, giving rise to products such as 'smart card' displays, dimmable window laminates, applications in toys and games and ultimately flexible electronic paper displays. The company notes that signs using NanoChromics™ display technology are ideal for sports player-substitution boards. They claim that the current LED boards can become bleached out and difficult to read in bright daylight in sports stadia, and that NanoChromics™ display signs are perfect for this application as they are easy to read in bright daylight and at all angles.

Several workers have adapted these ideas. Grätzel et al.,[125,126] for example, have prepared such devices with a series of viologens, with aryl as well as alkyl substituents. In each case, the anchored group attaching the viologen to the titania was benzoate, salicylate or phosphonate (as in **IV**). Electrochromic devices they have constructed include shutters and displays. The cell OTE|TiO_2-poly(viologen)|glutaronitrile–$LiN(SO_2CF_3)_2$|Prussian-blue|OTE exhibited an optical density change of about 2; the colour changes on reduction were transparent to blue, or yellowish to green, and (at higher potentials) to red–brown. They report switching times in the range of 1–3 s. Higher optical density changes are possible if the switching times are slower.[126] Grätzel and co-workers also made a variety of cell geometries for ECDs operating on

364 *The viologens*

reflectors. The viologens in such devices were generally oligomers rather than polymers.

In a similar way, Boehlen et al.[127] prepared a salt of 2,2′-bipyridine (**VII**) calling it a 'viologen'; they generated a pink colour on reduction (which could indicate that a proportion of the viologen exists as radical-cation dimer).

Edwards et al.[128,129,130,131,132] have prepared many similar systems for devices, with viologen electrochromes adsorbed on titania, naming such devices 'electric paint'. They generally employed the viologen **IV** to produce amazing clarity. For example, Figure 11.5 shows a prototype, demonstrating clarity capable of high-definition patterning. The response time is about 0.5 s.

Figure 11.5 Prototype electrochromic display showing an 'electric paint' display: the primary electrochrome was viologen (**IV**) adsorbed on nanocrystalline TiO_2. (Figure reproduced from Pettersson, H., Gruszecki, T., Johansson, L.-H., Edwards, M.O.M., Hagfeldt, A. and Matuszczyk, T. 'Direct-driven electrochromic displays based on nanocrystalline electrodes'. *Displays*, **25** 2004, 223–30, with permission of Elsevier Science Ltd.)

11.4.2 The use of pulsed potentials

Pulses of current have been shown to enhance the rate at which electrochromic colour is formed, relative to coloration with a continuous potential.[133] The procedure relies on the solution-phase redox reaction between bipm^{2+} (from the bulk solution) and bipm^0 electrogenerated during the current pulse. The reaction is comproportionation, Eq. (11.7), so a sufficiently cathodic potential must be applied at the working electrode.

The amounts of bipm^{2+} and bipm^0 at the electrode and in the region around the electrode depleted of bipm^{2+} will govern the rate of comproportionation and hence the rate of product colour formation. Thus for a given concentration of bipm^{2+} and bipm^0 in such a region, the most intense colour will ensue when the two species are in equal concentration. It is envisaged that the pulse procedure possibly favours this equality.

11.4.3 Electropolychromism

Bipyridilium salts may typically possess three colours, one for each oxidation state in Scheme 11.1, although the dication in solution is essentially colourless. Viologen electrochromes comprising n bipyridilium units could thus, in principle, exhibit $2n + 1$ colours. This maximal number is not achieved however when delocalisation allows simultaneous coloration of two or more of the bipyridiliums.[134] Several approaches have employed a number of bipyridilium units connected either with alkyl linkages[135,136] or benzylic moieties.[134]

A different, highly promising, combination, the complementary use of a bipyridilium with a Prussian blue electrochrome, allows the fabrication of a five-colour ECD.[137,143]

11.4.4 Viologens incorporated within paper

Viologen electrochromes have been incorporated within paper, to effect electrochromic writing. These include methyl viologen,[138,139,140,141] heptyl viologen,[141] and the asymmetric system, methyl–benzyl paraquat (**VIII**).[139] The adsorption of methyl viologen onto the carbohydrate structures of paper follows Langmuir adsorption isotherms that imply chemisorptive behaviour.[138]

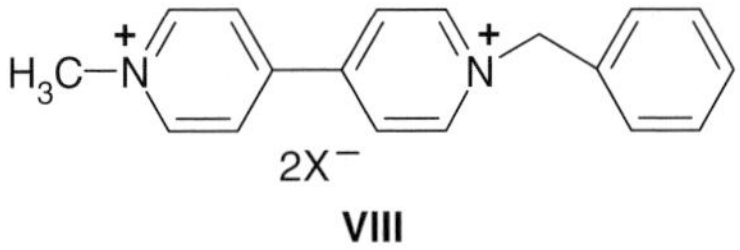

2X$^-$

VIII

While methyl viologen in paper is electrochromic,[138,140] its response time is prohibitively slow. The speed is faster if the paper is layered with the polyelectrolyte poly(AMPS), presumably because it provides an additional source of ions. With MV^{2+}, the speed of response depends critically on the paper's relative moisture. The results can be summarised as showing that in paper of marginal moistness, the solution-phase electrochemistry of both Prussian blue and viologens can be reproduced as though in a standard electrochemical cell.

Alternatively, incorporation within Nafion™ has been shown to produce good results. Several viologen electrochromes have been incorporated into Nafion™ as a host matrix[142,143] in which the viologen cation is immobilised by electrostatic interactions. Coloration is faster then bleaching. The five-colour bipm/Nafion/PB system could find application here.[137,143]

However, commercial utilisation of the processes just outlined seems at present somewhat questionable, as colour printing in say newsprint is now commonplace.

References

1. Kamagawa, H. and Suzuki, T. Organic solid photochromism *via* a photoreduction mechanism: photochromism of viologen crystals. *J. Chem. Soc., Chem. Commun.*, 1985, 525–6.
2. Sariciftci, N. S., Mehring, M. and Neugebauer, N. *In situ* studies on the structural mechanism of zwitter-viologen system during electrochemical charge-transfer reactions. *Synth. Met.*, **41–43**, 1991, 2971–4.
3. Michaelis, L. Semiquinones, the intermediate steps of reversible organic oxidation–reduction. *Chem. Rev.*, **16**, 1935, 243–86.
4. Michaelis, L. and Hill, E. S. The viologen indicators. *J. Gen. Physiol.*, **16**, 1933, 859–73.
5. Monk, P. M. S. *The Viologens: Physicochemical Properties, Synthesis and Applications of the Salts of 4,4'-Bipyridine*. Chichester, Wiley, 1998.
6. Monk, P. M. S., Mortimer, R. J. and Rosseinsky, D. R. *Electrochromism: Fundamentals and Applications*, Weinheim, VCH, 1995, ch. 8.
7. Summers, L. A. The bipyridines. *Adv. Heterocyc. Chem.*, **35**, 1984, 281–394.
8. Summers, L. A. *The Bipyridinium Herbicides*, London, Academic Press, 1980.
9. Bird, C. L. and Kuhn, A. T. The electrochemistry of the viologens. *Chem. Soc. Rev.*, **10**, 1981, 49–82.
10. Bard, A. J., Ledwith, A. and Shine, H. J. Formation, properties and reactions of cation radicals in solution. *Adv. Phys. Org. Chem.*, **13**, 1976, 155–278.
11. Sliwa, W., Bachowska, B. and Zelichowicz, N. Chemistry of viologens. *Heterocycles*, **32**, 1991, 2241–73.
12. Rosseinsky, D. R. and Monk, P. M. S. Comproportionation in propylene carbonate of substituted bipyridiliums. *J. Chem. Soc., Faraday Trans.*, **89**, 1993, 219–22.

13. Rosseinsky, D. R. and Monk, P. M. S. Solid-state conductivities of CPQ [1,1′-bis(*p*-cyanophenyl)-4,4′-bipyridilium] salts, redox-state mixtures and a new intervalence adduct. *J. Chem. Soc., Faraday Trans.*, **90**, 1994, 1127–31.
14. Emmert, B. and Varenkamp, O. Über chinhydronartige Verbindungen der *N, N′*dialkyl-[dihydro-γγ′dipyridyle]. *Berichte*, **56**, 1923, 490–501.
15. Levey, G. and Emmertson, T. W. Methyl viologen radical reactions with several oxidizing reagents. *J. Phys. Chem.*, **87**, 1983, 829–32.
16. Leest, R. E. V. D. The coulometric determination of oxygen with the electrochemically generated viologen radical-cation. *J. Electroanal. Chem.*, **43**, 1973, 251–5.
17. Hünig, S. and Schenk, W. Einfluß von N-substitutenten in 4,4′-bipyridylen auf das Redox-verhalten, die Radikalstabilität und die Elektronenspektren. *Liebigs Ann. Chem.*, 1979, 1523–33.
18. Čársky, P., Hünig, S., Scheutzow, D. and Zahradník, R. Theoretical study of redox equilibria. *Tetrahedron*, **25**, 1969, 4781–96.
19. Hünig, S. and Groß, J. Reversible Redoxsysteme vom Weitz-typ: eine polarographische Studie. *Tetrahedron Lett.*, **21**, 1968, 2599–604.
20. Kosower, E. M. *An Introduction to Physical Organic Chemistry*, New York, Wiley, 1968.
21. Thorneley, R. N. F. A convenient electrochemical preparation of reduced methyl viologen and kinetic study of the reaction with oxygen using an anaerobic stopped-flow apparatus. *Biochim. Biophys. Acta*, **333**, 1974, 487–96.
22. Watanabe, T. and Hondo, K. Measurement of the extinction coefficient of the methyl viologen cation radical and the efficiency of its formation by semiconductor photocatalysis. *J. Phys. Chem.*, **86**, 1982, 2617–19.
23. Kosower, E. and Cotter, J. L. Stable free radicals, II: the reduction of 1-methyl-4-cyanopyridinium ion to methylviologen radical cation. *J. Am. Chem. Soc.*, **86**, 1964, 5524–7.
24. Schwarz W., Jr. Ph.D thesis, 1962, University of Wisconsin, as cited in ref. 23.
25. Stargardt, J. F. and Hawkridge, F. M. Computer decomposition of the ultraviolet–visible adsorption spectrum of the methyl viologen cation radical and its dimer in solution. *Anal. Chim. Acta*, **146**, 1983, 1–8.
26. Imabayashi, S.-I., Kitamura, N., Tazuke, S. and Tokuda, K. The role of intramolecular association in the electrochemical reduction of viologen dimers and trimers. *J. Electroanal. Chem.*, **243**, 1988, 143–60.
27. *New Electronics*, **7**, 1986, 66 (editorial).
28. Monk, P. M. S., Hodgkinson, N. M. and Ramzan, S. K. Spin pairing ('dimerisation') of the viologen radical cation: kinetics and equilibria. *Dyes Pigm.*, **43**, 1999, 207–17.
29. Müller, F. and Mayhew, S. G. Dimerisation of the radical cation of Benzyl Viologen in aqueous solution. *Biochem. Soc. Trans.*, **10**, 1982, 176–7.
30. Compton, R. G., Waller, A. M., Monk, P. M. S. and Rosseinsky, D. R. Electron paramagnetic resonance spectroscopy of electrodeposited species from solutions of 1,1′-bis(*p*-cyanophenyl)-4,4′-bipyridilium (cyanophenyl paraquat, CPQ). *J. Chem. Soc., Faraday Trans.*, **86**, 1990, 2583–6.
31. Mori, H. and Mizuguchi, J. Green electrochromism in the system *p*-cyanophenyl viologen and potassium ferrocyanide. *Jpn. J. Appl. Phys.*, **26**, 1987, 1356–60.
32. Emmert, B. Ein Radikal mit vierwertigem Stickstoff. *Ber.*, **53**, 1920, 370–7.
33. Belinko, K. Electrochemical studies of the viologen system for display applications. *Appl. Phys. Lett.*, **29**, 1976, 363.

34. Müller, E. and Bruhn, K. A. Über das Merichinoide, N, N'-Dibenzyl-γ, γ'dipyridinium-subchlorid. *Chem. Ber.*, **86**, 1953, 1122–32.
35. Carey, J. E., Cairns, J. E. and Colchester, J. E. Reduction of 1,1'-dimethyl-4,4'-bipyridilium dichloride to 1,1'-dimethyl-1,1'-dihydro-4,4'-bipyridilyl. *J. Chem. Soc., Chem. Commun.*, 1969, 1290–1.
36. Barclay, D. J., Bird, C. L., Kirkman, D. H., Martin, D. H. and Moth, F. T. An integrated electrochromic data display. *SID Digest*, 1980, 124–5.
37. Barclay, D. J., Dowden, A. C., Lowe, A. C. and Wood, J. C. Viologen-based electrochromic light scattering display. *Appl. Phys. Lett.*, **42**, 1983, 911–13.
38. Bookbinder, D. C. and Wrighton, M. S. Electrochromic polymers covalently anchored to electrode surfaces: optical and electrochromic properties of a viologen-based polymer. *J. Electrochem. Soc.*, **130**, 1983, 1081–7.
39. Dominey, R. N., Lewis, T. J. and Wrighton, M. S. Synthesis and characterization of a benzylviologen surface-derivatizing reagent: N, N'-bis [p-(trimethoxysilyl)benzyl]-4,4'-bipyridilium dichloride. *J. Phys. Chem.*, **87**, 1983, 5345–54.
40. Shu, C.-F. and Wrighton, M. S. Synthesis and charge-transport properties of polymers derived from the oxidation of 1-hydro-1'-(6-(pyrrol-1-yl)hexyl)-4,4'-bipyridinium bis(hexafluorophosphate) and demonstration of a pH-sensitive microelectrochemical transistor derived from the redox properties of a conventional redox center. *J. Phys. Chem.*, **92**, 1988, 5221–9.
41. Ko, H. C., Park, S.-A., Paik, W.-K. and Lee, H. Electrochemistry and electrochromism of the polythiophene derivative with viologen pendant. *Synth. Met.*, **132**, 2002, 15–20.
42. Akahoshi, H., Toshima, S. and Itaya, K. Electrochemical and spectroelectrochemical properties of polyviologen complex modified electrodes. *J. Phys. Chem.*, **85**, 1981, 818–22.
43. Berlin, A. A., Zherebtsova, L. V. and Rabazobovskii, Y. F. Polymers with a conjugated system, XXXVII: synthesis of polymers with charged heteroatoms in the macromolecular chain (onium polymerization). *Polym. Sci. (USSR)*, **6**, 1964, 67–74.
44. Factor, A. and Heisolm, G. E. Polyviologens – a novel class of cationic polyelectrolyte redox polymers. *Polym. Lett.*, **9**, 1971, 289–95.
45. Lieder, M. and Schlapfer, C. W. Synthesis and electrochemical properties of new viologen polymers. *J. Appl. Electrochem.*, **27**, 1997, 235–9.
46. Sato, H. and Tamamura, T. Polymer effect in electrochromic behavior of oligomeric viologens. *J. Appl. Polym. Sci.*, **24**, 1979, 2075–85.
47. Willman, K. W. and Murray, R. W. Viologen homopolymer, polymer mixture and polymer bilayer films on electrodes: electropolymerization, electrolysis, spectroelectrochemistry, trace analysis and photoreduction. *J. Electroanal. Chem.*, **133**, 1982, 211–31.
48. Marguerettaz, X., O' Neill, R. and Fitzmaurice, D. Heterodyads: electron transfer at a semiconductor electrode–liquid electrolyte interface modified by an adsorbed spacer–acceptor complex. *J. Am. Chem. Soc.*, **116**, 1994, 2629–30.
49. Corr, D., Bach, U., Fay, D. *et al.* Coloured electrochromic 'paper-quality' displays based on modified mesoporous electrodes. *Solid State Ionics*, **165**, 2003, 315–21.
50. Sammells, A. F. Semi conductor/solid electrolyte junctions for optical information storage. *US Government Reports and Announcements Index*, **87**, 1987, Abstract no. 703, 869, as cited in *Chem. Abs.* **107**: 86,064m.

51. Sammells, A. F. and Pujare, N. U. Electrochromic effects on heptylviologen incorporated within a solid polymer electrolyte cell. *J. Electrochem. Soc.*, **133**, 1986, 1270–1.
52. Calvert, J. M., Manuccia, T. J. and Nowak, R. J. A polymeric solid-state electrochromic cell. *J. Electrochem. Soc.*, **133**, 1986, 951–3.
53. Byker, H. J. Electrochromics and polymers. *Electrochim. Acta*, **46**, 2001, 2015–22.
54. Schoot, C. J., Ponjeé, J. J., van Dam, H. T., van Doorn, R. A. and Bolwijn, P. J. New electrochromic memory device. *Appl. Phys. Lett.*, **23**, 1973, 64–5.
55. As cited in Philips Ltd. Image display apparatus. British Patent 1,302,000, 1971.
56. Kenworthy, J. G. ICI Ltd. Variable light transmission device. British Patent, 1,314,049, 1973.
57. Yasuda, A., Mori, H., Takehana, Y. and Ohkoshi, A. Electrochromic properties of n-heptyl viologen–ferrocyanate system. *J. Appl. Electrochem.*, **14**, 1984, 323–8.
58. J. G. Allen, personal communication, 1987.
59. van Dam, H. T. and Ponjeé, J. J. Electrochemically generated colored films of insoluble viologen radical compounds. *J. Electrochem. Soc.*, **121**, 1974, 1555–8.
60. Jasinski, R. J. The electrochemistry of some n-heptyl viologen salt solutions. *J. Electrochem. Soc.*, **124**, 1977, 637–41.
61. Monk, P. M. S. and Hodgkinson, N. M. Charge-transfer complexes of the viologens: effects of complexation on the rate of electron transfer to methyl viologen. *Electrochim. Acta*, **43**, 1998, 245–55.
62. Bruinink, J. and van Zanten, P. The response of an electrochromic display with viologens on a potential step. *J. Electrochem. Soc.*, **124**, 1977, 1232–3.
63. Jasinski, R. On the cathodic growth of n-heptylviologen radical cation films. *J. Electrochem. Soc.*, **126**, 1979, 167–70.
64. Bruinink, J. and Kregting, C. G. A. The voltammetric behaviour of viologens at SnO_2 electrodes. *J. Electrochem. Soc.*, **125**, 1978, 1397–401.
65. Jasinski, R. n-Heptylviologen radical cation films on transparent oxide electrodes. *J. Electrochem. Soc.*, **125**, 1978, 1619–23.
66. Berzins, T. and Delahay, P. Oscillographic polarographic waves for the reversible deposition of metals on solid electrodes. *J. Am. Chem. Soc.*, **75**, 1953, 555–9.
67. Fletcher, S., Duff, L. and Barradas, R. G. Nucleation and charge-transfer kinetics at the viologen/SnO_2 interface in electrochromic device applications. *J. Electroanal. Chem.*, **100**, 1979, 759–70.
68. Tang, X. Y., Schneider, T. W., Walker, J. W. and Buttry, D. A. Dimerized-complexes in self-assembled monolayers containing viologens: an origin of unusual wave shapes in the voltammetry of monolayers. *Langmuir*, **12**, 1996, 5921–33.
69. Barna, G. G. The morphology of viologen films on transparent oxide electrodes. *J. Electrochem. Soc.*, **127**, 1980, 1317–19.
70. Goddard, N. J., Jackson, A. C. and Thomas, M. G. Spectroelectrochemical studies of some viologens used in electrochromic display devices. *J. Electroanal. Chem.*, **159**, 1983, 323–35.
71. Bewick, A., Lowe, A. C. and Wederell, C. W. Recrystallisation processes in viologen-based electrochromic deposits: voltammetry coupled with rapid time-resolved spectroscopy. *Electrochim. Acta*, **28**, 1983, 1899–902.
72. Bewick, A., Cunningham, D. W. and Lowe, A. C. Electrochemical and spectroscopic characterisation of structural reorganisation in N, N'-dipyridinium cation radical deposits. *Macromol. Chem. Macromol. Symp.*, **8**, 1987, 355–60.

73. Rosseinsky, D. R., Monk, P. M. S. and Hann, R. A. Anion-dependent aqueous electrodeposition of electrochromic 1,1'-*bis*-cyanophenyl-4,4'-bipyridilium (cyanophenylparaquat) radical cation by cyclic voltammetry and spectroelectrochemical studies. *Electrochim. Acta*, **35**, 1990, 1113–23.
74. Kaifer, A. E. and Bard, A. J. Micellar effects on the reductive electrochemistry of methylviologen. *J. Phys. Chem. B*, **89**, 1985, 4876–80.
75. Rosseinsky, D. R. and Monk, P. M. S. Electrochromic cyanophenylparaquat (CPQ: 1,1'-bis-cyanophenyl-4,4'-bipyridilium) studied voltammetrically, spectroelectrochemically and by ESR. *Sol. Energy Mater. Sol. Cells*, **25**, 1992, 201–10.
76. Shaw, D. J. *Introduction to Colloid and Surface Chemistry*, 3rd edn, London, Butterworths, 1980, p. 74.
77. Hoshino, K. and Saji, T. Electrochemical formation of thin film of viologen by disruption of micelles. *Chem. Lett.*, 1987, 1439–42.
78. Heyrovský, M. Catalytic and photocatalytic reduction of water by the reduced forms of methylviologen. *J. Chem. Soc., Chem. Commun.*, 1983, 1137–8.
79. Heyrovský, M. Effect of light upon electroreduction of 4,4'-bipyridyl and methyl viologen in aqueous solutions. *J. Chem. Soc., Faraday Trans. 1*, **82**, 1986, 585–96.
80. Heyrovský, M. The electroreduction of methyl viologen. *J. Chem. Soc., Chem. Commun.*, 1987, 1856–7.
81. Engelman, E. E. and Evans, D. H. Explicit finite-difference digital simulation of the effects of rate-controlled product adsorption or deposition in double-potential-step chronocoulometry. *J. Electroanal. Chem.*, **331**, 1992, 739–49.
82. Engelman, E. E. and Evans, D. H. Treatment of the electrodeposition of alkyl sulfate salts of viologen radical cations as an equilibrium process governed by solubility product. *Anal. Chem.*, **66**, 1994, 1530–4.
83. Byker, H. J., Gentex Corporation. 1990, Single-compartment, self-erasing, solution-phase electrochromic devices, solutions for use therein and uses thereof. US Patent No. 4,902,108, 1990.
84. Ho, K.-C., Fang, J. G., Hsu, Y.-C. and Yu, F.-C. A study on the electro-optical properties of HV and TMPD with their application in electrochromic devices. *Proc. Electrochem. Soc.*, **2003-17**, 2003, 266–78.
85. Ho, K.-C., Fang, Y.-W., Hsu, Y.-C. and Chen, L.-C. The influences of operating voltage and cell gap on the performance of a solution-phase electrochromic device containing HV and TMPD. *Solid State Ionics*, **165**, 2003, 279–87.
86. Ostrom, G. S. and Buttry, D. A. Quartz crystal microbalance studies of deposition and dissolution mechanisms of electrochromic films of diheptylviologen bromide. *J. Electroanal. Chem.*, **256**, 1988, 411–31.
87. Tang, X. Y., Schneider, T. and Buttry, D. A. A vibrational spectroscopic study of the structure of electroactive self-assembled monolayers of viologen derivatives. *Langmuir*, **10**, 1994, 2235–40.
88. Beden, B., Enea, O., Hahn, F. and Lamy, C. Investigation of the adsorption of methyl viologen on a platinum electrode by voltammetry coupled with 'in situ' UV-visible reflectance spectroscopy. *J. Electroanal. Chem.*, **170**, 1984, 357–61.
89. Crouigneau, P., Enea, O. and Beden, B. 'In situ' investigation by simultaneous voltammetry and UV-visible reflectance spectroscopy of some viologen radicals absorbed on a platinum electrode. *J. Electroanal. Chem.*, **218**, 1987, 307–17.
90. Reichman, B., Fan, F.-R. F. and Bard, A. J. Semiconductor electrodes, XXV: the p-GaAs / heptyl viologen system: photoelectrochemical cells and photoelectrochromic cells. *J. Electrochem. Soc.*, **127**, 1980, 333–8.

91. Crouigneau, P., Enea, O. and Lamy, C. A comparative electron spin resonance study of adsorbed cation-radicals generated 'in situ' by electrochemical and photoelechemical reduction of some viologen derivatives. *Nouv. J. Chem.*, **10**, 1986, 539–43.
92. Yasuda, A., Kondo, H., Itabashi, M. and Seto, J. Structure changes of viologen + β-cyclodextrin inclusion complex corresponding to the redox state of viologen. *J. Electroanal. Chem.*, **210**, 1986, 265–75.
93. Lu, T. and Cotton, T. M. *In situ* Raman spectra of the three redox forms of heptylviologen at platinum and silver electrodes: counterion effects. *J. Phys. Chem.*, **91**, 1987, 5978–85.
94. Osawa, M. and Suëtaka, W. Electrochemical reduction of heptyl viologen at platinum studied by time-resolved resonance Raman spectroscopy. *J. Electroanal. Chem.*, **270**, 1989, 261–72.
95. Sawada, T. and Bard, A. J. Laser-photoelectric-photoacoustic observation of the electrode surface. *J. Photoacoustics*, **1**, 1982/3, 317–27.
96. Malpas, R. E. and Bard, A. J. *In situ* monitoring of electrochromic systems by piezoelectric detector photoacoustic spectroscopy of electrodes. *Anal. Chem.*, **52**, 1980, 109–12.
97. Brilmyer, G. H. and Bard, A. J. Application of photothermal spectroscopy to in-situ studies of films on metals and electrodes. *Anal. Chem.*, **52**, 1980, 685–91.
98. Scharifker, B. and Wehrmann, C. Phase formation phenomena during electrodeposition of benzyl and heptyl viologen bromides. *J. Electroanal. Chem.*, **185**, 1985, 93–108.
99. Gołden, A. and Przyłuski, J. Studies of electrochemical properties of N-heptylviologen bromide films. *Electrochim. Acta*, **30**, 1985, 1231–5.
100. Monk, P. M. S., Fairweather, R. D., Duffy, J. A. and Ingram, M. D. Evidence for the product of viologen comproportionation being a spin-paired radical cation dimer. *J. Chem. Soc., Perkin Trans. II*, 1992, 2039–41.
101. Poizat, O., Sourisseau, C. and Corset, J. Vibrational and electronic study of the methyl viologen radical cation $MV^{+\bullet}$ in the solid state. *J. Mol. Struct.*, **143**, 1986, 203–6.
102. Rosseinsky, D. R. and Monk, P. M. S. Kinetics of the comproportionation of the bipyridilium salt p-cyanophenyl paraquat in propylene carbonate studied by rotating ring-disc electrodes. *J. Chem. Soc., Faraday Trans.*, **86**, 1990, 3597–601.
103. Monk, P. M. S. Comment on: 'Dimer formation of viologen derivatives and their electrochromic properties', *Dyes Pigm.*, **39**, 1998, 125–8.
104. Engelman, E. E. and Evans, D. H. Investigation of the nature of electrodeposited neutral viologens formed by reduction of the dications. *J. Electroanal. Chem.*, **349**, 1992, 141–58.
105. Compton, R. G., Monk, P. M. S., Rosseinsky, D. R. and Waller, A. M. An ESR study of the comproportionation of 1,1'-bis(p-cyanophenyl)-4,4'-bipyridilium (cyanophenyl paraquat) in propylene carbonate. *J. Electroanal. Chem.*, **267**, 1989, 309–12.
106. Monk, P. M. S. The effect of ferrocyanide on the performance of heptyl viologen-based electrochromic display devices. *J. Electroanal. Chem.*, **432**, 1997, 175–9.
107. Rosseinsky, D. R., Slocombe, J. D., Soutar, A., Monk, P. M. S. and Glidle, A. Simple diffuse reflectance monitoring of emerging surface-attached species. *J. Electroanal. Chem.*, **259**, 1989, 233–9.
108. Yasuda, A. and Mori, H. Sony Corp. Electrochromic display devices. Jpn. Kokai Tokkyo Koho JP 60,198,521, 1985, as cited in *Chem. Abs.* **104**: P99,571n.

109. Nakahara, A. and Wang, J. H. Charge-transfer complexes of methylviologen. *J. Phys. Chem.*, **67**, 1963, 496–8.
110. Murthy, A. S. N. and Bhardwaj, A. P. Electronic absorption spectroscopic studies on charge-transfer interactions in a biologically important molecule: N, N'-dimethyl-4,4'-bipyridylium chloride (paraquat or methyl viologen) as an electron acceptor. *Spectrochim. Acta*, **38A**, 1982, 207–12.
111. Kramarenko, S. F., Krainov, I. P., Pretsenko, E. G. and Vargalyuk, B. F. Electrochemical chromism of 1,1'-diaryl-4,4'-bipyridilium perchlorates. *Ukr. Khim. Zh.*, **51**, 1985, 501–4 [in Russian], cited in *Chem. Abs.* **103**: 112,232.
112. Bruinink, J., Ponjeé, J. J. and Kregting, C. G. A. Modified viologens with improved electrochemical properties for display applications. *J. Electrochem. Soc.*, **124**, 1977, 1854–8.
113. Barna, G. G. and Fish, J. G. An improved electrochromic display using a symmetric viologen. *J. Electrochem. Soc.*, **128**, 1981, 1290–2.
114. Barltrop, J. A. and Jackson, A. C. The synthesis and electrochemical study of new electrochromic viologen-based materials. *J. Chem. Soc., Perkin Trans. II*, 1984, 367–71.
115. Barclay, D. J. and Martin, D. H. Electrochromic displays. In Howells, E. R. (ed.), *Technology of Chemicals and Materials for the Electronics Industry*, Chichester, Ellis Horwood, 1984, 266–76,
116. Nazeeruddin, M. K. and Grätzel, M. Conversion and storage of solar energy using dye-sensitized nanocrystalline TiO_2 cells. In McCleverty, J. A. and Meyer, T. J. (eds.), *Comprehensive Coordination Chemistry II: From Biology to Nanotechnology*, Oxford, Elsevier, 2004, vol. 9, pp. 719–58.
117. O'Regan, B. and Grätzel, M. A low-cost, high-efficiency solar-cell based on dye-sensitized colloidal TiO_2 films. *Nature (London)*, **353**, 1991, 737–40.
118. [Online] at http://www.fki.uu.se/research/nano/textfile2/integration3.html, (accessed 27 January 2006).
119. [Online] at www.ntera.com (accessed 27 January 2006).
120. Cinnsealach, R., Boschloo, G., Nagaraja Rao, S. and Fitzmaurice, D. Coloured electrochromic windows based on nanostructured TiO_2 films modified by adsorbed redox chromophores. *Sol. Energy Mater. Sol. Cells*, **57**, 1999, 107–25.
121. Cinnsealach, R., Boschloo, G., Nagaraja Rao, S. and Fitzmaurice, D. Electrochromic windows based on viologen-modified nanostructured TiO_2 films. *Sol. Energy Mater. Sol. Cells*, **55**, 1998, 215–23.
122. Cummins, D., Boschloo, G., Ryan, M., Corr, D., Rao, S. N. and Fitzmaurice, D. Ultrafast electrochromic windows based on redox-chromophore modified nanostructured semiconducting and conducting films. *J. Phys. Chem. B*, **104**, 2000, 11449–59.
123. Grätzel, M. Ultrafast colour displays. *Nature (London)*, **409**, 2001, 575–6.
124. Graham-Rowe, D. Mirror trick leads chase for electronic paper. *New Scientist*, 26 February 2005, p. 27.
125. Bonhôte, P., Gogniat, E., Campus, F., Walder, L. and Grätzel, M. Nanocrystalline electrochromic displays. *Displays*, **20**, 1999, 137–44.
126. Campus, F., Bonhôte, P., Grätzel, M., Heinen, S. and Walder, L. Electrochromic devices based on surface-modified nanocrystalline TiO_2 thin-film electrodes. *Sol. Energy Mater. Sol. Cells*, **56**, 1999, 281–97.
127. Boehlen, R., Felderhoff, M., Michalek, R. and Walder, L. A new 2,2'-bipyridinium salt with pink electrochromism for the modification of nanocrystalline TiO_2-electrodes. *Chem. Lett.*, **27**, 1998, 815–16.

128. Edwards, M. O. M., Andersson, M., Gruszecki, T., Pettersson, H., Thunman R., Thuraisingham, G., Vestling, L. and Hagfeldt, A. Charge–discharge kinetics of electric-paint displays. *J. Electroanal. Chem.*, **565**, 2004, 175–84.
129. Edwards, M. O. M., Boschloo, G., Gruszecki, T., Pettersson, H., Sohlberg, R. and Hagfeldt. 'Electric-paint displays' with carbon counter electrodes. *Electrochim. Acta*, **46**, 2001, 2187–93.
130. Edwards, M. O. M., Gruszecki, T., Pettersson, H., Thuraisingham, G. and Hagfeldt, A. A semi-empirical model for the charging and discharging of electric-paint displays. *Electrochem. Commun.*, **4**, 2002, 963–7.
131. Edwards, M. O. M. and Hagfeldt, A. Switch-speed considerations for viologen–metal oxide displays. *Proc. Electrochem. Soc.*, **2003–17**, 2003, 305–9.
132. Pettersson, H., Gruszecki, T., Johansson, L.-H., Edwards, M. O. M., Hagfeldt, A. and Matuszczyk, T. Direct-driven electrochromic displays based on nanocrystalline electrodes. *Displays*, **25**, 2004, 223–30.
133. Monk, P. M. S., Fairweather, R. D., Ingram, M. D. and Duffy, J. A. Pulsed electrolysis enhancement of electrochromism in viologen systems: influence of comproportionation reactions. *J. Electroanal. Chem.*, **359**, 1993, 301–6.
134. Rosseinsky, D. R. and Monk, P. M. S. Studies of tetra-(bipyridilium) salts as possible polyelectrochromic materials. *J. Appl. Electrochem.*, **24**, 1994, 1213–21.
135. Porat, Z., Tricot, Y.-M., Rubinstein, I. and Zinger, B. New multi-charged viologen derivatives, 1: electrochemical behaviour in Nafion films. *J. Electroanal. Chem.*, **315**, 1991, 217–23.
136. Porat, Z., Tricot, Y.-M., Rubinstein, I. and Zinger, B. New multi-charged viologen derivatives, 2: unusual electrochemical behaviour in solution. *J. Electroanal. Chem.*, **315**, 1991, 225–43.
137. Mortimer, R. J. Five color electrochromicity using Prussian blue and Nafion / methyl viologen layered films. *J. Electrochem. Soc.*, **138**, 1991, 633–4.
138. Monk, P. M. S., Delage, F. and Costa Vieira, S. M. Electrochromic paper: utility of electrochromes incorporated in paper. *Electrochim. Acta*, **46**, 2001, 2195–202.
139. Rosseinsky, D. R. and Monk, J. L. Thin layer electrochemistry in a paper matrix: electrochromography of Prussian blue and two bipyridilium systems. *J. Electroanal. Chem.*, **270**, 1989, 473–8.
140. Monk, P. M. S., Turner, C. and Akhtar, S. P. Electrochemical behaviour of methyl viologen in a matrix of paper. *Electrochim. Acta*, **44**, 1999, 4817–26.
141. Mortimer, R. J. and Warren, C. P. Cyclic voltammetric studies of Prussian blue and viologens within a paper matrix for electrochromic printing applications. *J. Electroanal. Chem.*, **460**, 1999, 263–6.
142. John, S. A. and Ramaraj, R. Role of acidity on the electrochemistry of Prussian blue at plain and Nafion film-coated electrodes. *Proc. Ind. Acad. Sci.*, **107**, 1995, 371–383.
143. Mortimer, R. J. and Dillingham, J. L. Electrochromic 1,1'-dialkyl-4,4'-bipyridilium-incorporated Nafion electrodes. *J. Electrochem. Soc.*, **144**, 1997, 1549–53.

12

Miscellaneous organic electrochromes

12.1 Monomeric electrochromes

A large number of organic compounds that are molecular aromatics form a coloured species on electron transfer. Indeed, most redox indicators are by definition electrochromic, and are thus straightforward candidates for exploitation as electrochromes. Most standard texts on quantitative and qualitative analytical chemistry cite many examples of redox indicators, like the compendia in ref. 1. There is a severe lack of any systematic survey of the electrochromic properties of such species for ECD application.

Most of the aromatic species in this chapter form either a molecular radical cation or radical anion following electron transfer. All the organic species in this chapter, and the viologen species in the previous chapter, are 'violenes' – a conceptual classification pioneered by Hünig.[2] A violene is a conjugated molecular fragment of the form $X\text{-}(CH=CH)_n\text{-}CH\text{-}X$, where $X = O$, N or S. The conjugated $(CH=CH)_n$ portion is normally part of an aromatic ring or series of rings. As a direct consequence of their structure, all violenes typically possess three stable redox states: an uncharged species, a species with a double charge, and a species of intermediate redox state that is either a radical cation or radical anion. The conjugation within the violene that allows extensive delocalisation is the ultimate cause of the extraordinary stability of many such radicals.

12.1.1 Aromatic amine electrochromes

Aromatic amines are generally colourless unless they undergo some form of charge-transfer interaction with an electron-deficient acceptor species. By contrast, the product of one-electron oxidation yields a radical cation which, in organic solution, possesses a brilliant colour. Aromatic amines are thus candidate electrochromes.

12.1 Monomeric electrochromes

If both the neutral and radical-cation redox states are soluble, such amines show type-I electrochromism. In relatively non-polar solvents, the radical cation together with an electrolyte anion may deposit as a salt.[3,4] In such solvent systems, the aromatic amines are type-II electrochromes.

These species fall into two categories:[2] (i) the nitrogen is incorporated into an aromatic ring and is a derivative of the pyridine ring C_5H_5N, for example; (ii) the amine is attached to an aromatic (e.g. C_6H_5-) ring, like the NH_2 group of aniline, $C_6H_5-NH_2$. The nitrogens of all the amine groups need to be fully substituted, to preclude polymerisation reactions: the radical cations of aromatic secondary amines retaining an N–H functionality readily form an inherently conducting polymer, for example poly(aniline), as described in Section 10.4.

The monomeric aromatic amine that has probably received the most attention for its electrochromic prospects is tetramethylphenylenediamine (TMPD) as the *p*- (**I**) and *o*-isomers.[5,6,7,8,9,10,11,12] The radical cation of **I** is stable and is brilliant blue-green. Ho *et al.*[10,11] have modelled the electrochemical behaviour and cycle lives of electrochromic devices in which **I** was the secondary electrochrome, against heptyl viologen (see Chapter 11) as the primary. The cycle lives of such devices are high, but the response times are slow owing to the requirement for all solution-phase electrochromes to diffuse toward the electrode–solution interphase prior to electron transfer; bleaching times likewise are also long.

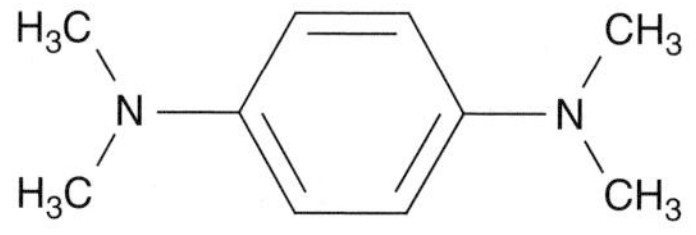

I

A more bulky electrochrome is the triphenylamine derivative **II**. When a solution of **II** is injected between two ITO-coated electrodes (with *n*-tetra-butylammonium perchlorate as an inert electrolyte) and a voltage of 2.2 V applied, the initially colourless neutral compound forms a brilliant bluish-red radical with λ_{max} of 530 nm. Changes to the substituent causes a blue shift in the colour of the radical.

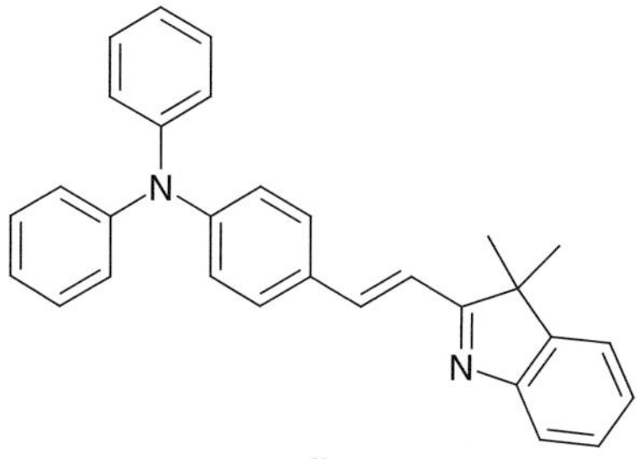

II

Several aromatic amines show electrochromic activity in the near infrared (NIR).[13] Table 12.1 contains data for several such species, as prepared and studied by the US Gentex Corporation. In each case, the amine was the secondary electrochrome, and an aryl-substituted viologen was the primary. All amines colour anodically and are envisaged for use within solution-phase (type-I) devices. Some of the changes in optical transmission are marked. For example, most compounds in Table 12.1 show a contrast ratio of 5:1 on coloration.[13] Thus, compound **III** has a visible transmission of 75% in its clear state and only 9% in its coloured state; compound **VIII** ('Crystal violet') has a NIR transmission of 46% in its clear state and 14% in its coloured state.

12.1.2 Carbazole electrochromes

The monomeric, substituted carbazole species **X** readily undergoes a one-electron oxidation to form a radical-cation salt. In their neutral form, carbazoles are soluble and essentially colourless, whereas films of radical cation generated oxidatively according to Eq. (12.1) form a highly coloured, solid precipitate on the electrode:

$$\text{carbazole (soln)} + X^- \rightarrow [\text{carbazole}^{+\bullet} : X^-]^0 \text{ (s)} + e^-. \quad (12.1)$$
  colourless   strongly coloured

The carbazoles therefore represent an example of type-II electrochromism. Table 12.2 summarises results obtained by Dubois and co-workers for a few carbazole electrochromes.[14]

X

12.1.3 Cyanine electrochromes

Spiropyrans[15,16,17] such as **XI** are both electrochromic and photoelectrochromic. The initial product of electroreducing **XI** is a radical anion, while further reduction yields a ring-opened merocyanine species. In the absence of an electrode, photolysis of **XI** yields the merocyanine directly. A plot of

Table 12.1. *Aromatic amine electrochromes that modulate NIR radiation. (Table reproduced with permission from Theiste, D., Baumann, K. and Giri, P. 'Solution phase electrochromic devices with near infrared attenuation'. Proc. Electrochem. Soc., **2003–17**, 2003, 199–207, with permission of The Electrochemical Society.)*

Electrochrome	λ_{max}/nm	ε/dm^3 mol^{-1} cm^{-1}
III	727	1500
IV	1694	15 000
V	912	19 000
VI	1198	38 500

Table 12.1. *(cont.)*

Electrochrome	λ_{max}/nm	ε/dm^3 mol^{-1} cm^{-1}
VII	954	47 000
VIII	968	46 000
IX	907	

absorbance *Abs* against Q for species **XI** is essentially linear, and allows $\eta = 21$ cm^2 C^{-1} to be calculated.[17]

XI

Table 12.2. *Colours and electrode potentials of oligomers derived from various carbazole electrochromes in MeCN solution.* (*Table reproduced from: Desbène-Monvernay, A., Lacaze, P.-C. and Dubois, J.-E. 'Polaromicrotribometric (PMT) and IR, ESCA, EPR spectroscopic study of colored radical films formed by the electrochemical oxidation of carbazoles, part I: carbazole and N-ethyl, N-phenyl and N-carbazyl derivatives'. J. Electroanal. Chem., 129, 1981, 229–41, with permission of Elsevier Science Ltd.*)

Monomer	Colour of radical cation	$E^\ominus$/V
Carbazole	Dark green	+0.9
N-ethylcarbazole	Green	+1.3
N-phenylcarbazole	'Iridescent'	+1.2
N-carbazylcarbazole	Yellow–brown	+1.1

Another recently discovered series of electrochromes are the squarylium dyes,[18] such as **XII**, which have some structural elements in common with **XI**. The reduced form of the dyes are blue, while the radical species formed on oxidation are green.

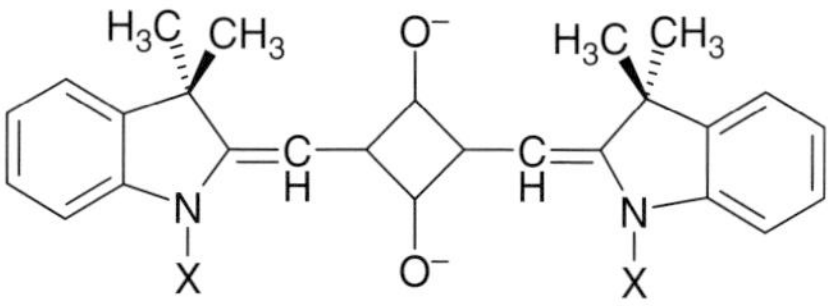

(X = H, Me or Et)

XII

12.1.4 Methoxybiphenyl electrochromes

The next class of compounds are the violenes based on a core of polymethoxybiphenyl. The uncharged parent compounds are essentially colourless, while electro-oxidation yields a thin, solid film of brilliantly coloured radical-cation salt.

Methoxybiphenyl compounds have been studied by the groups of Parker[19,20] and Grant,[21] although Parker mentions neither electrochromism nor electrochemical applications. Many of these species should more correctly be called 'biphenyls', 'fluorenes' or 'phenanthrenes' according to the nature of the bridging group (if any) connecting the two aromatic rings.

The stability of the radical cation formed by one-electron oxidation of the neutral species is a function of molecular planarity, as demonstrated by the stability series **XIII** ≪ **XIV** < **XV**: compound **XIII** is forced out of planarity by the steric repulsion induced by the two *o*-methoxy groups, whereas **XV**, by necessity, is always planar owing to the methylene bridge.

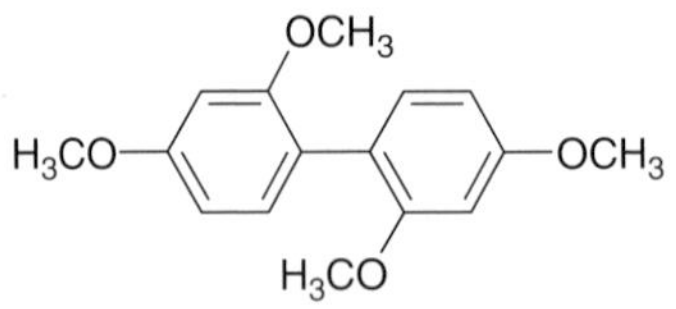

XIII

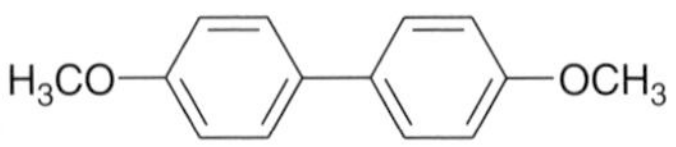

XIV

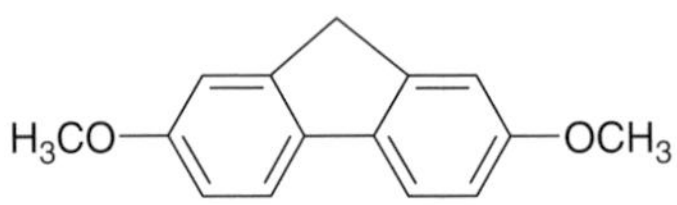

XV

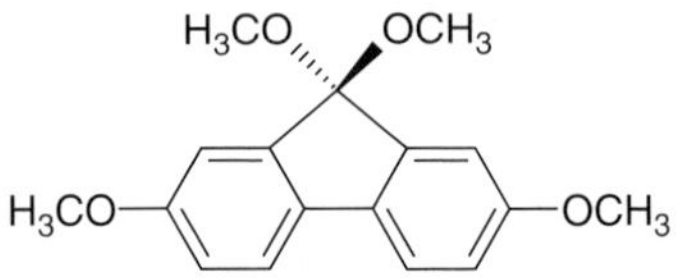

XVI

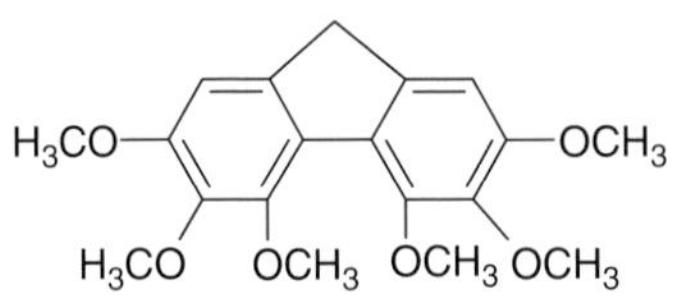

XVII

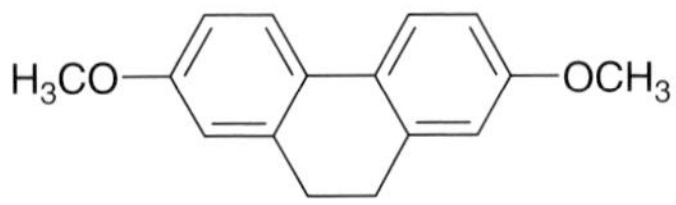

XVIII

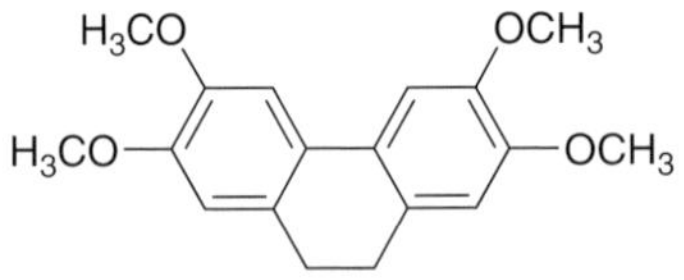

XIX

As a crude generalisation,[21] fluorenes with a single methoxy group are oxidised irreversibly, but the electrochemistry of compounds with two or more methoxy groups is much more reversible, i.e. monomethoxy species are not truly violenes. *Ortho*- and *meta*-methoxy substituents engender lower redox potentials than *para* groups.[21] The fluorene compounds that appear most suitable for ECD inclusion, that is, those yielding the most stable films of radical-cation salt, are listed in Table 12.3.

Other fluorene compounds investigated did not form radical cations of sufficient stability for viable use as electrochromes, or evinced irreversible electrochemistry. For example, Table 12.4 lists some biphenyl compounds of interest, within which compound **XVIII** aromatises slowly by deprotonating to form 2,7-dimethoxyphenanthrene.

Table 12.3. *Colours, CV peak potentials, and spectral properties for methoxybiphenyl species forming a solid radical-cation film on reduction in MeCN solutions. (Table reproduced from Grant, B., Clecak, N. J. and Oxsen, M. 'Study of the electrochromism of methoxyfluorene compounds'. J. Org. Chem.,* **45**, *1980, 702–5, with permission of The American Chemical Society.)*

Compound	Colour of radical	E_{pa}/V	E_{pc}/V	λ_{max}/nm	ε/dm^3 mol^{-1} cm^{-1}
XV	Blue	+0.91	+0.79	411	40 400
XVI	–	+0.96	+0.84	385	32 800
XVII	Blue	+0.87	+0.81	415	44 300

Table 12.4. *Colours, CV peak potentials and spectral properties for methoxybiphenyl species forming only a soluble radical cation on reduction in dichloromethane–TFA (5:1 v:v) solution. (Table reproduced from Ronlán, A., Coleman, J., Hammerich, O. and Parker, V. D. 'Anodic oxidation of methoxybiphenyls: effect of the biphenyl linkage on aromatic cation radical and dication stability'. J. Am. Chem. Soc.,* **96**, *1974, 845–9, with permission of The American Chemical Society.)*

Compound	Colour of radical	$E_{pa(1)}$/V	$E_{pc(1)}$/V	λ_{max}/nm	ε/dm^3 mol^{-1} cm^{-1}
XIV	–	+1.28	+1.22	417	29 512
XVIII	Green	+1.14	+1.07	386	20 420
XIX	Green	+0.94	+0.88		

12.1.5 Quinone electrochromes

Many quinone species are soluble, stable, and only moderately coloured as neutral molecules but on one-electron reduction form brightly coloured, stable, solid films of radical anion on the electrode surface.[22,23,24,25,26,27] For example, the electrochromism of several benzoquinones has been studied such as the *ortho* (**XX**) and *para* (**XXI**) isomers. The most comprehensive study of (**XXI**) involved the electrochrome dissolved in a solution of propylene carbonate containing LiClO$_4$ as supporting electrolyte.[24]

The quinone to have received the most attention is probably p-2,3,5,6-tetrachlorobenzoquinone ('p-chloranil' **XXII**),[22] which forms a pink radical cation; see Eq. (12.2):

$$M^+ + (\mathbf{XXII})^0 \text{ (soln)} + e^- \rightarrow [M^+(\mathbf{XXII})^{\bullet-}] \text{ (s)}, \qquad (12.2)$$

where the alkali or alkaline-earth cation M is needed to co-deposit with the radical anion when forming an insoluble salt. Desbène-Monvernay et al.[27] say the best results are obtained if the cation forms a 'visible light-forming charge transfer complex between [the] o-chloranil$^{\bullet-}$ and the counter ion M^+.' This is doubtful for they also say the best results are obtained when M = Na, but as the sodium cation does not undergo colour-forming charge-transfer interactions, merely undergoing co-deposition with the quinone radical cation, the source of the quinone radical-cation colour is best conceived as an *internal* charge-transfer transition modified by M^+.

The colour of the radical cation depends on the substituents around the quinone: the tetrafluoro analogue of (**XXII**) 'fluoranil' forms a yellow radical anion, and the radical anion of p-2,3-dicyano-5,6-dichloroquinone is pink. Table 12.5 lists a few sample quinone species together with electrochemical and optical data. Figure 12.1 shows the absorbance spectrum of a film of p-chloranil radical anion on ITO polarised to −0.6 V vs. SCE.

In CH$_3$CN solution, only p-benzoquinone, o-chloranil (**XXII**) and o-bromoanil form films that are both stable and adherent.[22] Desbène-Monvernay and

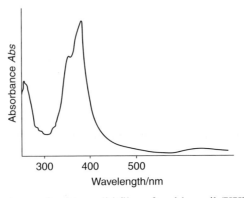

Figure 12.1. Spectrum of a thin, solid film of p-chloranil (**XXII**) as the radical anion salt on an ITO electrode polarised to −0.6 V. The spectrum baseline was that of the uncharged, colourless p-chloranil prior to charge passage. (Figure reproduced from Desbène-Monvernay, A., Lacaze, P. C. and Cherigui, A. 'UV-visible spectroelectrochemical study of some *para-* and *ortho-*benzoquinoid compounds: comparative evaluation of their electrochromic properties'. *J. Electroanal. Chem.*, **260**, 1989, 75–90, by permission of Elsevier Science.)

Table 12.5. *Quinone systems: film-forming properties, colours, wavelength maxima, and reduction potentials. Values of E_{pc} were obtained from CVs, or standard electrode potentials $E^{\ominus}$; all solutions in MeCN with tetraethylammonium perchlorate (0.1 mol dm^{-3}).*

Quinone (R–Q)	Solid film?	Colour of R–Q$^{-\bullet}$	λ_{max}/nm	$E_{pc(1)}$/V	$E_{pc(2)}$/V	Ref.
o-3,4,5,6-tetrachloro-benzoquinone	Yes	Intense blue		−0.170	+0.210	22,23
o-3,4,5,6-tetrabromo-benzoquinone	Yes	Blue		−0.190	+0.140	22
p-benzoquinone	Yes	Light blue		−0.720	−0.430	22,24
p-2,3,5,6-tetrafluoro-benzoquinone	No	Yellow		−0.430	−0.100	22
p-2,3,5,6-tetrachloro-benzoquinone	No	Yellow		−0.420	−0.060	22
p-2,3-dicyano-5,6-dichlorobenzo-quinone	No	Pink		+0.070	+0.330	22
5-aminonaphtho-quinone	Yes	Purple–blue	410	$E^{\ominus}_{(1)} = -0.83$		25
1-aminoanthra-quinone	Yes	–	–	$E^{\ominus}_{(1)} = -1.03$		25
2-aminoanthra-quinone	Yes	–	–	$E^{\ominus}_{(1)} = -0.99$		25
1,5-diaminoanthra-quinone	Yes	Purple	570	$E^{\ominus}_{(1)} = -1.10$		25

co-workers[22] say that *o*-bromanil forms a superior radical-cation film to any of the other *para*-substituted quinones, from its low solubility product and good adherence.

In general, electrochromes based on *ortho* quinones are superior to the *para* analogues: they are more electrochemically stable,[22] and the solubility constants K_s are lower. The values of K_s for the *para* isomers are generally too high, sometimes allowing soluble radical cation to diffuse back into the solution bulk, which therefore represents type-I response rather than the perhaps more desirable type-II electrochromism.[22]

The quinone evincing the highest electrochemical stability is *o*-chloranil (the *ortho* analogue of **XXII**). Its electrochromic properties are 'outstanding',[22,28] with a cycle life exceeding 10^5 write–erase cycles.[22]

While the electrochromism of most quinones requires the formation of radical species, i.e. a transition from pale to intense colour, a recent example operates differently: the red quinone species 1-amino-4-bromoanthraquinone-2-sulfonate may be electroreduced in aqueous solution to form a colourless dihydroxy compound, rather than a coloured quinone radical,[29] *cf.* the so-called 'quinhydrone electrode', a 1:1 compound of *p*-benzoquinone **XX** and dihydroxybenzene (both depicted in Eq. (12.3)).

$$\text{benzoquinone} \xrightarrow{2e^- + 2H^+} \text{hydroquinone} \tag{12.3}$$

Molecular naphthaquinone and anthraquinone species are also type-I electrochromes. Exemplar species include 1,4-naphthaquinone (**XXIII**) and anthra-9,10-quinone (**XXIV**). Aminoanthraquinones show a more complicated electrochemical behaviour than the naphthaquinone: at moderate potentials, two redox couples are exhibited during cyclic voltammetry, representing first Eq. (12.4):

$$\text{quinone}^0 + e^- \longrightarrow \text{quinone}^{\bullet -}, \tag{12.4}$$

followed at more negative potentials by a second reduction reaction, Eq. (12.5):

$$\text{quinone}^{\bullet -} + e^- \longrightarrow \text{quinone}^{2-}. \tag{12.5}$$

In addition to this behaviour, polymerisation of the amine moiety occurs when the electrode is made very positive, *cf.* the formation of poly(aniline) in Section 10.4.

XXIII   **XXIV**

More advanced again is a trichromic ECD[30] with the capacity to form the colours red, green and green-blue, which has been developed using 2-ethylanthraquinone in PC together with 4,4′-bis(dimethylamino)diphenylamine. The electrolyte is gelled with a white 'filler' to enhance the contrast ratio. In this way, the anthraquinone compound produces the red colour when reduced ($CR = 2:1$ at $\lambda_{max} = 545$ nm), while the other colours derive from the diphenylamine, which yields two different oxidation states: its first oxidation product is a green radical cation ($CR = 2:1$) and a subsequent oxidation product is a green-blue dication ($CR = 3.5:1$ at $\lambda_{max} \approx 500$ nm). Because the electrochromes are not encapsulated in separate pixels, the various redox states formed will diffuse back into the solution bulk and undergo radical-annihilation reactions. Furthermore, being violenes, it is also likely that the 2+ and 0 redox states will undergo comproportionation thus: $(2+) + (0) \rightarrow 2(+\bullet)$.

12.1.6 Thiazine electrochromes

Thiazine compounds contain a heterocyclic ring comprising both nitrogen and sulfur moieties. Methylene Blue (**XXV**), the common dye and biological stain, is the archetypal thiazine. The Greek descriptor *Leucos* ('white') is used in organic chemistry and in the dyestuffs industry to describe the colourless form of a redox dye, so **XXV** is blue when oxidised and colourless following reduction to form the neutral radical, so called *leuco*-Methylene Blue.

XXV

The thiazine **XXV** is soluble in a wide range of solvents, but has been occasionally considered for ECD usage when immobilised in a semi-solid polymer matrix, as described in Section 12.3.

The world's best-selling electrochromic device is undoubtedly the Night Vision System (NVS©) produced by the US Gentex Corporation,[31,32] a self-darkening rear-view mirror that is a standard feature in many millions of expensive high-performance cars.[33] The Gentex device comprises two electrochromes, a viologen species (see Chapter 11) and a phenothiazine.[31,34,35] In MeCN solution, thiazines such as **XXV** are used. At heart, each NVS© mirror incorporates a front electrode of ITO-coated glass and a metallic rear electrode having a highly reflective surface. These two parallel electrodes separated by a sub-millimetre gap form the basis of the cell. (In a similar device containing heptyl viologen and tetramethylphenylenediamine in PC, the cell would only function when the gap was narrower than 0.28 mm.[12]) The dual-electrochrome solution is injected into the cavity between the electrodes.

The exact composition of the Gentex NVS© mirror is obscured within densely worded patents, but it is possible to infer some details of the operation: a substituted viologen species 'bipm' (see Section 11.3), undoubtedly cationic as 'bipm^{2+}', serves as the cathodic electrochrome. When the mirror is switched on, mass transport occurs as the positive charge of the uncoloured precursor propels it toward the cathode in response to ohmic migration (the electrolyte in the Gentex mirror is free of additional swamping electrolyte). Reductive coloration then occurs at the cathode, Eq. (12.6):

$$\text{bipm}^{2+}(\text{soln}) \rightarrow \text{bipm}^{+\bullet}(\text{soln}). \tag{12.6}$$

The other electrochrome (which is initially in its reduced form) is probably a molecular thiazine 'TA' (or perhaps a phenylenediamine species, see p. 375 above). The TA is uncharged and depletion by oxidation at the anode ensures that mass transport of TA ensues by diffusion alone. Oxidation of TA evokes colour, Eq. (12.7):

$$\text{TA (soln)} \rightarrow \text{TA}^{+\bullet}(\text{soln}) + e^-. \tag{12.7}$$

In operation, the colour in a commercially-available NVS© mirror is an intense blue–green. The colour-forming reduction process, $\text{bipm}^{2+} + e^- \rightarrow \text{bipm}^{+\bullet}$, and the complementary oxidation reaction, $\text{TA} \rightarrow \text{TA}^{+\bullet} + e^-$ occur in dual electro-coloration processes in tandem. The coloured species diffuse away from the respective electrodes and meet in the intervening solution where their mutual reaction ('radical annihilation') ensues, Eq. (12.8)

$$\text{TA}^{+\bullet}(\text{soln}) + \text{bipm}^{+\bullet}(\text{soln}) \rightarrow \text{TA (soln)} + \text{bipm}^{2+}(\text{soln}), \tag{12.8}$$

that regenerates the original uncoloured species. These reactions are depicted schematically in Figure 12.2.

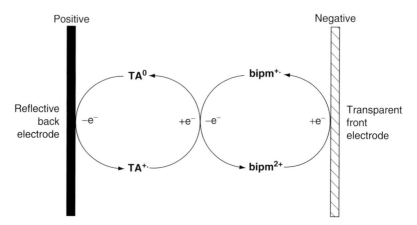

Figure 12.2. Schematic representation of the redox cycles occurring within the Gentex Night Vision System©. Coloration occurs electrochemically at both electrodes; bleaching occurs chemically at the centre of the cell by radical annihilation.

The radical annihilation in Eq. (12.7) represents a divergence from one of the benefits of electrochromism since the 'memory effect' is lost. Thus, maintenance of coloration requires the passage of a continuous (albeit minute) current to replenish the coloured electrochromes lost by the annihilation. Reaction (12.7) obviates any need to electro-bleach the Gentex NVS© mirror, since colour fades spontaneously on switch-off. For this reason, the Gentex NVS© is sometimes termed the *self-erasing* mirror. United States law requires the 'failure mode', on loss of current, to be the clear condition, with which these mirrors comply.

12.1.7 Miscellaneous monomeric electrochromes

A trichromic ECD has been fabricated including 2,4,5,7-tetranitro-9-fluorenone (**XXVI**) as the red-forming material, 2,4,7-trinitro-9-fluorenylidene malononitrile (**XXVII**) as the green and tetracyanoquinodimethane (TCNQ) as the blue electrochrome.[36]

Finally, a Japanese group has prepared a TCNQ derivative and studied its spectra as a function of applied potential.[37] While their study was not concerned with electrochromic activity, their results may facilitate the preparation of new electrochromes.

12.2 Tethered electrochromic species

12.2.1 Pyrazoline electrochromes

A tethered organic system that has received some attention is that based on the oxidation of the pyrazolines **XXVIII** and **XXIX**, spectral details for which are listed in Table 12.6.

Kaufman *et al.*[38,39] have published most of the current work on tethered pyrazolines. Such species are more intensely absorbing than the tetrathiafulvalene (TTF) species below, and have faster response times τ.[39] Pure pyrazoline monomers are readily prepared, and are soluble in many solvents prior to polymerisation.[39]

Table 12.6. *Half-wave potentials $E_{1/2}$, colours, and response times τ for tethered pyrazoline species bound covalently to an electrode substrate, immersed in MeCN solution containing TEAP electrolyte (0.1 mol dm^{-3}). (Table reproduced from Kaufman, F. B. and Engler, E. M. Solid-state spectroelectrochemistry of crosslinked donor bound polymer films. J. Am. Chem. Soc., **101**, 1979, 547–9, with permission of The American Chemical Society.)*

Compound	$E_{1/2}$/V	Colour change	λ_{max}/nm	τ/ms
XXVII	+0.55	Yellow-to-green	510	50
XXVIII	+0.45	Yellow-to-red	554	100

A solid-state ECD which incorporates such polymeric pyrazolines has been constructed,[40] and has a response time of 10 ms and a *CR* of 10:1.

12.2.2 *Tetracyanoquinodimethane (TCNQ) electrochromes*

Neutrally charged TCNQ is a stable, colourless molecule that forms a blue-green coloured radical anion following one-electron reduction.[36,37] The stability of the tetracyanoquinonedimethanide radical is ascribed to appreciable delocalisation of the single negative charge over the four CN groups.

Since TCNQ and its radical anion are both soluble in most common solvents, Chambers *et al.*[41,42,43] have improved the electrochromic write–erase efficiency by chemically tethering the TCNQ species **XXX** to an electrode surface by means of polymerisation. The oligomer **XXX** is estimated[41] to have a molecular weight of about 2200 g mol^{-1}, i.e. a chain comprising an average chain length of 6.3 electrochrome units.

Table 12.7. *Spectroscopic data for a modified electrode bearing a thin film of the TCNQ-based polymer* **XXX**, *immersed in MeCN solution. (Reproduced from Inzelt, G., Day, R. W., Kinstle, J. F. and Chambers, J. Q. 'Spectroelectrochemistry of tetracyanoquinodimethane modified electrodes'. J. Electroanal. Chem.,* **161**, *1984, 147–61 with permission of Elsevier Science.)*

Species	λ_{max}/nm	$\ln(\varepsilon/dm^3 \, mol^{-1} \, cm^{-1})$
TCNQ^0	408	5.06
	430	5.06
TCNQ$^{-\bullet}$	445	4.30
	660	3.38
	728	3.92
	812	4.20

XXX

Electrodes modified with **XXX** are electrochemically reversible.[41] Spectroscopic data for TCNQ and TCNQ$^{-\bullet}$ are listed in Table 12.7.

In solution, additional species to those in Table 12.7 have also been identified, including a dianion $(TCNQ)^{2-}$, and (in aqueous solution only) a dianion $(TCNQ)_2^{2-}$ dimer.[42]

12.2.3 Tetrathiafulvalene (TTF) electrochromes

Like TCNQ, TTF has been used in ECDs chemically tethered to an electrode surface. In this way Kaufman and co-workers[44,45] used the two species **XXXI** and **XXXII** to modify electrodes. In early trials, a TTF device underwent $>10^4$ cycles without visible deterioration.[44] The electrochromic TTF colouration accompanies oxidation of neutral TTF to form a radical cation. Spectral characteristics of **XXXI** and **XXXII** are listed in Table 12.8.

Table 12.8. *Half-wave potentials $E_{1/2}$, colours, wavelength maxima and response times τ for tethered TTF species.(Data reproduced from Kaufman, F. B., Schroeder, A. H., Engler, E. M. and Patel, V. V. 'Polymer-modified electrodes: a new class of electrochromic materials'. Appl. Phys. Lett., 36, 1980, 422–5, with permission of American Institute of Physics.)*

Compound	$E_{1/2}$/V	Colour change	λ_{max}/nm	τ/ms[a]
XXXI	+0.45	Orange-to-brown	515	200
XXXII	+0.35	Yellow-to-green	650	150

[a] Time required for a charge injection of 1 mC cm^{-2} into a film of thickness 5 μm.

Table 12.9. *Spectroscopic data for TTF redox species in MeCN solution. (Data reproduced from Kaufman, F. B. 'New organic materials for use as transducers in electrochromic display devices', Conference Record of the IEEE, Biennial Display Research Conference, 1978, New York, p. 23–5, with permission of The IEEE.)*

Species	λ_{max}/nm
TTF$^{+\bullet}$	393, 653
(TTF$^{+\bullet}$)$_2$	1800
(TTF)$^{+\bullet}{}_2$	820
TTF^{2+}	533

XXXI X = —O—C(=O)—

XXXII X = O—C$_6$H$_4$—

Electrochemical studies show the rate-determining step during coloration is ion movement into and through the film;[39,46] furthermore, electron transport through the film proceeds via hopping or tunnelling between TTF sites.

In addition to TTF$^{+\bullet}$, the other TTF species listed in Table 12.9 will also form in the layer around the electrode; their spectral characteristics are reproduced in Table 12.9. Although the minor species in Table 12.9 do not contribute much to the colouration of a TTF device, they greatly complicate any electrochemical interpretation.

Recent TTF displays comprise solid-state devices with polymeric electrolytes.[40]

Table 12.10. *Electrochromes dispersed within semi-solid polymer 'matrices'.*

Electrochrome	Polarity to yield colour	Polymer	Colour	Ref.
p-Diacetylbenzene	Cathodic	PVPD	Green	50,51
Diethyl terephthalate	Cathodic	PVPD	Red	51
Dimethyl terephthalate	Cathodic	PVPD	Red	50,51
Methylene Blue (**XXV**)	Anodic	poly(AMPS)	Blue	48,49

12.3 Electrochromes immobilised within viscous solvents

The write–erase efficiency can be enhanced by dissolving or dispersing an electrochrome in a semi-solid electrolyte of high viscosity. Such immobilised species are essentially type-III electrochromes. The usual matrix for entrapment is an electrolyte gel of high viscosity,[47] such as the polyelectrolytes or polymeric electrolytes described in Chapter 14. In this context, the host polymers of choice are semi-solid poly(AMPS),[48] poly(aniline),[49] and poly(1-vinyl-2-pyrrolidinone-co-*N,N'*-methylenebisacrylamide) PVPD.[50] Table 12.10 lists a few electrochromes which have been immobilised in this way.

Clearly, only a small proportion of the electrochrome dispersed in viscous electrolyte will ever be juxtaposed with the electrode, or can reach the electrode within a tolerable time lag. For this reason, the majority of the electrochrome must be considered to be 'passive', with most remaining in its colourless form. Methylene Blue (**XXV**) is thus unpromising as an electrochrome as its colourless *leuco* form reverts back to the coloured form quite rapidly, especially when exposed to oxygen.

In the studies by Tsutseumi *et al.*,[50,51] the electrochromes dispersed in PVPD were all ester-based. In each case, the colour formed after the potential had been applied for a few seconds, but a rapid self-bleaching process occurred under open circuit. Such gel films therefore lack any optical memory effect.

Carbazoles (*cf.* Section 12.1 above) have similarly been immobilised in a 'matrix' of poly(siloxane) to yield viable ECDs.[52,53,54,55,56,57]

References

1. Kodama, K. *Methods of Quantitative Inorganic Analysis*, New York, Interscience, 1963, ch. 15.
2. Hünig, S. Stable radical ions. *Pure Appl. Chem.*, **15**, 1967, 109–22.
3. Michaelis, L., Schubert, M. P. and Gramick, S. The free radicals of the type of Wurster's salts. *J. Am. Chem. Soc.*, **61**, 1939, 1981–92.
4. Michaelis, L. Semiquinones, the intermediate steps of reversible organic oxidation–reduction. *Chem. Rev.*, **16**, 1935, 243–86.

5. Ling-Ling, W., Jin, L. and Zhong-Hua, L. Spectroelectrochemical studies of poly-*o*-phenylenediamine, part 1: in situ resonance Raman spectroscopy. *J. Electroanal. Chem.*, **417**, 1996, 53–8.
6. Long, J. W., Rhodes, C. P., Young, A. L. and Rolison, D. R. Ultrathin, protective coatings of poly(*o*-phenylenediamine) as electrochemical porous gates: making mesoporous MnO_2 nanoarchitectures stable in acid electrolytes. *Nano Lett.*, **3**, 2003, 1155–61.
7. Nishikitani, Y., Kobayashi, M., Uchida, S. and Kubo, T. Electrochemical properties of non-conjugated electrochromic polymers derived from aromatic amine derivatives. *Electrochim. Acta*, **46**, 2001, 2035–40.
8. Yano, J. and Yamasaki, S. Three-color electrochromism of an aramid film containing polyaniline and poly(*o*-phenylenediamine). *Synth. Met.*, **102**, 1999, 1157.
9. Zhang, A. Q., Cui, C. Q., Chen, Y. Z. and Lee, J. Y. Synthesis and electrochromic properties of poly-*o*-aminophenol. *J. Electroanal. Chem.*, **373**, 1994, 115–21.
10. Ho, K.-C., Fang, J. G., Hsu, Y.-C. and Yu, F.-C. A study on the electro-optical properties of HV and TMPD with their application in electrochromic devices. *Proc. Electrochem. Soc.*, **2003–17**, 2003, 266–78.
11. Ho, K.-C., Fang, Y.-W., Hsu, Y.-C. and Chen, L.-C. The influences of operating voltage and cell gap on the performance of a solution-phase electrochromic device containing HV and TMPD. *Solid State Ionics*, **165**, 2003, 279–87.
12. Leventis, N., Muquo, C., Liapis, A. I., Johnson, J. W. and Jain, A. Characterisation of 3×3 matrix arrays of solution-phase electrochromic cells. *J. Electrochem. Soc.*, **145**, 1998, L55–8.
13. Theiste, D., Baumann, K. and Giri, P. Solution phase electrochromic devices with near infrared attenuation. *Proc. Electrochem. Soc.*, **2003–17**, 2003, 199–207.
14. Desbène-Monvernay, A., Lacaze, P.-C. and Dubois, J.-E. Polaromicrotribometric (PMT) and IR, ESCA, EPR spectroscopic study of colored radical films formed by the electrochemical oxidation of carbazoles, part I: carbazole and *N*-ethyl, *N*-phenyl and *N*-carbazyl derivatives. *J. Electroanal. Chem.*, **129**, 1981, 229–41.
15. Zhi, J. F., Baba, R. and Fujishima, A. An electrochemical study of some spirobenzopyran derivatives in dimethylformamide. *Ber. Bunsen.-Ges. Phys. Chem.*, **100**, 1996, 1802–7.
16. Zhi, J. F., Baba, R., Hashimoto, K. and Fujishima, A. A multifunctional electrooptical molecular device: the photoelectrochemical behavior of spirobenzopyrans in dimethylformamide. *Ber. Bunsen.-Ges. Phys. Chem.*, **99**, 1995, 32–9.
17. Zhi, J. F., Baba, R., Hashimoto, K. and Fujishima, A. Photoelectrochromic properties of a spirobenzopyran derivative. *J. Photochem. Photobiol.*, **A92**, 1995, 91–7.
18. Kim, S. H. and Huang, S. H. Electrochromic properties of functional squarylium dyes. *Dyes Pigm.*, **36**, 1998, 139–48.
19. Ronlán, A., Hammerich, O. and Parker, V. D. Anodic oxidation of methoxybibenzyls: products and mechanism of the intramolecular cyclization. *J. Am. Chem. Soc.*, **95**, 1973, 7132–8.
20. Ronlán, A., Coleman, J., Hammerich, O. and Parker, V. D. Anodic oxidation of methoxybiphenyls: effect of the biphenyl linkage on aromatic cation radical and dication stability. *J. Am. Chem. Soc.*, **96**, 1974, 845–9.
21. Grant, B., Clecak, N. J. and Oxsen, M. Study of the electrochromism of methoxyfluorene compounds. *J. Org. Chem.*, **45**, 1980, 702–5.

22. Desbène-Monvernay, A., Lacaze, P. C. and Cherigui, A. UV-visible spectroelectrochemical study of some *para-* and *ortho-*benzoquinoid compounds: comparative evaluation of their electrochromic properties. *J. Electroanal. Chem.*, **260**, 1989, 75–90.
23. Dubois, J. E., Desbène-Monvernay, A., Cherigui, A. and Lacaze, P. C. *Ortho-*chloranil – a new electrochromic material. *J. Electroanal. Chem.*, **169**, 1984, 157–66.
24. Yashiro, M. and Sato, K. A new electrochromic material: 1,4-benzoquinone in a non-aqueous solution. *Jpn. J. Appl. Phys.*, **20**, 1981, 1319–20.
25. Gater, V. K., Liu, M. D., Love, M. D. and Leidner, C. R. Quinone molecular films derived from aminoquinones. *J. Electroanal. Chem.*, **257**, 1988, 133–46.
26. Gater, V. K., Love, M. D., Liu, M. D. and Leidner, C. R. Quinone molecular films derived from 1,5-diaminoanthraquinone. *J. Electroanal. Chem.*, **235**, 1987, 381–5.
27. Desbène-Monvernay, A., Lacaze, P. C., Dubois, J. E. and Cherigui, A. Ion-pair effects on the electroreduction and electrochromic properties of *ortho-*chloranil in dipolar aprotic solvents. *J. Electroanal. Chem.*, **216**, 1987, 203–12.
28. Cherigui, A., Desbène-Monvernay, A. and Lacaze, P.-C. Electrochromism of the *o*-CA/*o*-CA$^-$ system in display cells. *J. Electroanal. Chem.*, **240**, 1988, 321–4.
29. Yano, J. Electrochromism of polyaniline film incorporating a red quinone 1-amino-4-bromoanthraquinone-2-sulfonate. *J. Electrochem. Soc.*, **144**, 1997, 477–81.
30. Ueno, T., Hirai, Y. and Tani, C. Three color switching electrochromic display using organic redox-pair dyes. *Jpn. J. Appl. Phys.*, **24**, 1985, L178–80.
31. Byker, H. J., Gentex Corporation. Single-compartment, self-erasing, solution-phase electrochromic devices, solutions for use therein and uses thereof. US Patent No. 4,902,108, 1990.
32. [online] at www.gentex.com/auto_how_nvs_work.html (accessed 6 September 2005).
33. [online] at http://uk.cars.yahoo.com/010419/4/749k.html (accessed 25 July 2003).
34. Byker, H. J. Electrochromics and polymers. *Electrochim. Acta*, **46**, 2001, 2015–22.
35. Byker, H. J. Commercial developments in electrochromics. *Proc. Electrochem. Soc.*, **94–2**, 1994, 1–13.
36. Yasuda, A. and Seto, J. Electrochromic properties of vacuum-evaporated organic thin films, part 4: the case of 2,4,7-trinitro-9-fluorenylidene malononitrile. *J. Electroanal. Chem.*, **303**, 1991, 161–9.
37. Higuchi, H., Ichioka, K., Kawai, H., Fujiwara, K., Ohkita, M., Tsuji, T. and Suzuki, T. Three-way-output response system by electric potential: UV-vis, CD, and fluorescence spectral changes upon electrolysis of the chiral ester of tetracyanoanthraquinodimethane. *Tetrahedron Lett.*, **45**, 2004, 3027–30.
38. Kaufman, F. B. and Engler, E. M. Solid-state spectroelectrochemistry of crosslinked donor bound polymer films. *J. Am. Chem. Soc.*, **101**, 1979, 547–9.
39. Kaufman, F. B., Schroeder, A. H., Engler, E. M. and Patel, V. V. Polymer-modified electrodes: a new class of electrochromic materials. *Appl. Phys. Lett.*, **36**, 1980, 422–5.
40. Hirai, Y. and Tani, C. Electrochromism for organic materials in polymeric all-solid-state systems. *Appl. Phys. Lett*, **43**, 1983, 704–5.
41. Day, R. W., Inzelt, G., Kinstle, J. F. and Chambers, J. Q. Tetracyanoquinodimethane-modified electrodes. *J. Am. Chem. Soc.*, **104**, 1982, 6804–5.
42. Inzelt, G., Day, R. W., Kinstle, J. F. and Chambers, J. Q. Spectroelectrochemistry of tetracyanoquinodimethane modified electrodes. *J. Electroanal. Chem.*, **161**, 1984, 147–61.

43. Inzelt, G., Day, R. W., Kinstle, J. F. and Chambers, J. Q. Electrochemistry and electron spin resonance of tetracyanoquinodimethane modified electrodes: evidence for mixed-valence radical anions in the reduction process. *J. Phys. Chem.*, **87**, 1983, 4592–8.
44. Kaufman, F. B. New organic materials for use as transducers in electrochromic display devices. Conference Record of the IEEE, Biennial Display Research Conference, New York, 1978, p. 23–5.
45. Torrance, J. B., Scott, B. A., Welber, B., Kaufman, F. B. and Seiden, P. E. Optical properties of the radical cation tetrathiafulvalenium (TTF^+) in its mixed-valence and monovalence halide salts. *Phys. Rev.*, **B19**, 1979, 730–41.
46. Kaufman, F. B., Schroeder, A. H., Engler, E. M., Kramer, S. R. and Chambers, J. Q. Ion and electron transport in stable, electroactive tetrathiafulvalene polymer coated electrodes. *J. Am. Chem. Soc.*, **102**, 1980, 483–8.
47. Tsutsumi, H., Nakagawa, Y., Miyazaki, K., Morita, M. and Matsuda, Y. Polymer gel films with simple organic electrochromics for single-film electrochromic devices. *J. Polym. Chem.*, **30**, 1992, 1725–9.
48. Calvert, J. M., Manuccia, T. J. and Nowak, R. J. A polymeric solid-state electrochromic cell. *J. Electrochem. Soc.*, **133**, 1986, 951–3.
49. Kuwabata, S., Mitsui, K. and Yoneyama, H. Preparation of polyaniline films doped with methylene blue-bound Nafion and the electrochromic properties of the resulting films. *J. Electroanal. Chem.*, **281**, 1990, 97–107.
50. Tsutsumi, H., Nakagawa, Y. and Tamura, K. Single-film electrochromic devices with polymer gel films containing aromatic electrochromics. *Sol. Energy Mater. Sol. Cells*, **39**, 1995, 341–8.
51. Tsutsumi, H., Nakagawa, Y., Miyazaki, K., Morita, M. and Matsuda, Y. Single polymer gel film electrochromic device. *Electrochim. Acta*, **37**, 1992, 369–70.
52. Goldie, D. M., Hepburn, A. R., Maud, J. M. and Marshall, J. M. Carrier mobility studies of carbazole modified polysiloxanes. *Mol. Cryst. Liq. Cryst.*, **234**, 1993, 777–82.
53. Goldie, D. M., Hepburn, A. R., Maud, J. M. and Marshall, J. M. Dynamics of colouration and bleaching in cross-linked carbazole modified polysiloxane thin films. *Synth. Met.*, **55**, 1993, 1650–5.
54. Goldie, D. M., Hepburn, A. R., Maud, J. M. and Marshall, J. M. Characterisation and application of carbazole modified polysiloxanes in electrochemical displays. *Mol. Cryst. Liq. Cryst.*, **234**, 1993, 627–34.
55. Maud, J. M., Vlahov, A., Goldie, D. M., Hepburn, A. R. and Marshall, J. M. Carbazolylalkyl substituted cyclosiloxanes: synthesis and properties. *Synth. Met.*, **55**, 1993, 890–5.
56. Bartlett, I. D., Marshall, J. M. and Maud, J. M. Characterization and application of carbazole modified polysiloxanes to electrochromic displays. *J. Non-Cryst. Solids*, **198–200**, 1996, 665–8.
57. Hepburn, A. R., Marshall, J. M. and Maud, J. M. Novel electrochromic films via anodic oxidation of carbazolyl substituted polysiloxanes. *Synth. Met.*, **43**, 1991, 2935–8.

13

Applications of electrochromic devices

13.1 Introduction

While the applications of electrochromism are ever growing, all devices utilising electrochromic colour modulation fall within two broad, overlapping categories according to the mode of operation: electrochromic devices (ECDs) operating by transmission (see schematic in Figure 13.1) or by reflection (see the schematic representation in Figure 13.2).

Several thousand patents have been filed to describe various electrochromic species and devices deemed worthy of commercial exploitation, so the field is vast. Much duplication is certain in such patents, but it is clear how large scale are the investments directed toward implementing electrochromism as viable in displays or light modulation. In this field, vital details of compositions are often well hidden, as these comprise the valued intellectual property rights on which substantial financial considerations rest.

The most common applications are electrochromic mirrors and windows, as below. These and other applications are reviewed at length by Lampert[1] (1998), who cites all the principal manufacturers of electrochromic goods worldwide, and also several novel applications.

13.2 Reflective electrochromic devices: electrochromic car mirrors

Mirrors, which obviously operate in a reflectance mode, illustrate the first application of electrochromism (*cf.* Figure 13.2). Self-darkening electrochromic mirrors, for automotive use at night, disallow the lights of following vehicles to dazzle by reflection from the driver's or the door mirror. Here an optically absorbing electrochromic colour is evoked over the reflecting surface, reducing reflection intensity and thereby alleviating driver discomfort. However, total opacity is to be avoided as muted reflection must persist in the darkened state. The back electrode is a reflective material

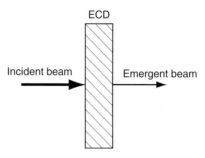

Figure 13.1 Schematic diagram of an ECD operating in transmittance mode. Both the front and back electrodes are optically transparent. The respective widths of the arrows indicate the relative magnitudes of the light intensities.

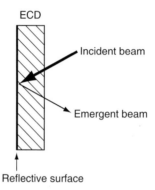

Figure 13.2 Schematic diagram of an ECD operating in reflectance mode. The front electrode is optically transparent and the back electrode is made of polished platinum or platinum-based alloy. The respective widths of the arrows indicate the relative magnitudes of the light intensities.

allowing customary mirror reflection in the bleached state. Reference 2 has some nice graphics that illustrate the necessary components.

The best-selling electrochromic mirror is the Gentex Night-Vision System[3,4] (NVS©), of which many millions have been sold, probably 90 to 95% of all self-darkening mirror sales.[5] Its operation employing type-I electrochromes is described in detail in Chapter 12, Section 12.1 under 'Thiazine electrochromes', and the mechanism is illustrated in Figure 12.2 (None of the accounts available reveal the dramatic events at the onset of Gentex's first big auto contract, when a small adventurous inventive company that had some impressive demo devices had suddenly to tool up for mass production. The Gentex Corporation we refer to here are based in Zeeland, Michigan, and are not to be confused with an identically named but independent firm in Pennsylvania that supplies amongst other things protective clothing, fireproofing and the like for aeronauts and astronauts.)

An example of all-solid-state mirror is the SchottDonnelly[6] solid polymer matrix (SPM^{TM}) mirror for lorries and trucks, which relies on WO_3 and NiO, and is thus a type-III system. A different solid-state electrochromic mirror is based on WO_3.[7,8,9,10,11]

Electrochromic mirrors are fitted on luxury cars made by, among others, Audi, Bentley, BMW, Daewoo, DaimlerChrysler, Fiat, Ford, General Motors, Hyundai, Infiniti, Kia Motors, Lexus, Mitsubishi, Nissan, Opel, Porsche, Rolls Royce and Toyota.[12]

The likely thermal and other stresses resulting from mounting ECDs in or on cars require particularly stringent tests of ECD design and fabrication. The durability of ECDs is discussed in Chapter 16.

Gesheva *et al.*[13] have developed an electrochromic mirror that is, apparently, reflective to X-rays. Films of WO_3, MoO_3 or mixed W–Mo oxide, were deposited on wafers of silica by plasma-enhanced CVD from a metal-carbonyl precursor. Electrochromic modulation changes the X-reflectivity of the underlying silica.

13.3 Transmissive ECD windows for buildings and aircraft

13.3.1 Buildings

Svensson and Granqvist coined the term 'smart window' in 1985 to describe windows that electrochromically change in transmittance.[14] The term has since been augmented with 'smart windows' and 'self-darkening windows' to describe novel fenestrative applications. The British Fenestration Rating Council describes electrochromic windows as 'Chromogenic glazing'.[15] However, the terms 'electrochromic window' and 'smart glass' are now widespread and attract attention, particularly in popular-science articles.[16]

The construction of electrochromic windows has often been reviewed, for example, 'Toward the smart window: progress in electrochromics' by Granqvist *et al.* (in 1999),[17] 'Electrochromic windows: an overview' by Rauh (also in 1999),[18] 'Windows' by Bell *et al.* (2002),[19] and 'Electrochromic smart windows: energy efficiency' by Azens and Granqvist (in 2003).[20] Smart windows for automotive usage have not been reviewed so often: one of the few reviews to mention this application explicitly is 'Angular selective window coatings: theory and experiments' by Granqvist *et al.*[21] in 1997. Although dated (1991), 'A review on electrochromic devices for automotive glazing' by Demiryont[22] is still relevant.

The appeal of smart windows is both economic and environmental: if successful, they preclude much solar radiation from a room or a car. The exact cost of air conditioning in summer is unknown, but is surely greater than losses through windows in winter, which in 2003 cost $25 billion in the USA alone.[23] Smart windows might thus both improve working environments and alleviate costs. Nevertheless, Lee and DiBartolomeo[24] suggest that electrochromic windows 'may not be able to fulfil both energy-efficiency and visual comfort objectives when low winter direct sun is present'.

The rush to develop smart windows is also a response to pressure from environmental campaigners of the 'Green' lobby.[25,26] Many 'green' considerations are assessed by Griffiths et al.[27] and Syrrakou et al.[28]

Architectural applications are at present the subject of intense research activity: the web page from the National Renewable Energy Laboratories (NREL) in ref. 29 aims to cite all the present-day producers of electrochromic windows; the number of manufacturers appears to be expanding rapidly. However, many of these products are poorly described in the associated publicity, so the identities of the electrochromes are unclear. This is scarcely informative, or a boost for electrochromic applications. For example, from their the web site SAGE Electrochromics Inc.[30] of Minnesota clearly produce two products, one of which is said to be 'organic' and the other 'inorganic', without identifying either electrochrome.

Many websites show video clips of electrochromic windows: the short sequences available in ref. 31 show dramatic colour changes of organic films of PEDOT-based polymers (see Section 10.2). Reference 32 contains a short video clip of an electrochromic window measuring 3 ft × 6 ft, made by Research Frontiers Inc. (though it is not clear whether this is an ECD or an SPD[33]), and ref. 34 contains several longer .mpeg clips of varying clarity. Colour Plate 4 shows a window made by Gentex.

In the smart-window application, individual panes of glass or whole windows can be coloured electrochromically to darken sunlight intensity in rooms or offices. Similar electrochromic applications are planned for car sun-roofs,[22,35] and the motorcycle helmets and ski goggles developed by the Granqvist group in Sweden;[36,37,38] see Colour Plate 5. Recently, Zinzi[39] published a full study describing the preferences of office workers, as follows. He made a full mock-up room, illuminated internally with conventional fluorescent and incandescent bulbs, and externally by solar radiation that entered the room through electrochromic windows. The results were interesting and not always as expected. Most workers preferred to control the external lighting via electrochromic windows rather than blinds or other mechanical forms of shutter. When the transparency of the windows was changed automatically, via a photocell connected to a microprocessor, the alteration of the visual environment was sufficiently smooth and slow that few workers actually perceived the changes; those who did were not unhappy with its effects. Nevertheless, most preferred a manually operated transparency control, presumably to 'personalise' their own working space. Some workers wanted electrochromic windows to adjust more rapidly, to accommodate fluctuating ambient illumination.

Many manufacturers of electrochromic windows prefer so-called 'neutral' colours, i.e. shades of grey, to the richer blue colour of (for example) H_xWO_3 alone. Office workers are said to favour such grey hues, because other colours can induce nausea.[40] Thus there is now a considerable research effort to optimise the hue for the working environment, with many researchers seeking to effect subtle changes in optical bands, for example by mixing various metal oxides in precise relative amounts. In this regard, a promising electrochrome is a mixture of vanadium and tungsten oxides, which evinces an electrochromic colour that is more grey than that of either constituent oxide alone, because the absorption spectrum comprises several broad and overlapping optical bands.[41,42,43] The quest for 'neutral colour' is presented in refs. 42,43,44,45, 46,47,48, while mixtures of metal oxide are considered in greater detail in Section 6.5.

Electrochromes commonly show colour in the visible spectrum. While most WO_3-based ECDs develop a band peaking in the near infrared (NIR), it is sufficiently broad for much visible light also to be absorbed. Some smart windows, however, develop a band almost wholly in the NIR. While not altering the *perceived colour* of an electrochromic window, such electrochromic windows do further regulate the transmission of the thermal components of sunlight. This desirable property is found also in two unexpected electrochromes, namely the fullerene[49] Li_xC_{60}, the reduced form having a band maximum at 1060–80 nm, and electrodeposited diamond,[50] which is yellow but becomes brown following reduction due to a band with a maximum in the UV (see Section 9.2).

While light transmittances may be modulated between, say, 85% when bleached, to 15% when coloured, *complete* blocking of sunlight would need the dissipation of much absorbed heat, unless the solar radiation could be reflected metallically by the electrochrome, requiring a material with metallic, specular, reflectivity. Some few electrochromes show specular reflectivity, the most remarkable being yttrium hydride, which can be cycled between highly transparent and mirror-like conditions (see Section 9.4). Other electrochromes indicating specular reflectance are the inorganic systems copper oxide,[51] iridium oxide,[52] lithium pnictide,[51] tungsten oxyfluoride[53] and tungsten trioxide;[54] and organic systems such as poly(pyrrole) composites,[55] poly(diphenylamine)[56] and PEDOT.[56]

The reflectance of iridium oxide in ref. 52 arose from a thin layer of electrochromic IrO_2 deposited on opaque Ir metal, so the reflectance may be that of the metallic under-layer. Other systems in which a metallic layer is electrodeposited are outlined in Section 9.3.

While in theory there is no absolute upper limit to the contrast ratio CR in ordinary electrochromism, in practice the values are never particularly high. Thus, unusually large values of CR are assumed to indicate reflective effects, and a sputtered film of WO_3, with a reported[57] CR of 1000:1, probably comprises WO_3 particles that reflect some of the incident light.

Schott Glass has shown demonstration models of their 'Ucolite' room-illumination system at Schott Glass Singapore (June 2000). The 40 or 50 cm diameter circular electrochromic window was no doubt WO_3-based, but darker, so possibly comprising a nickel hydroxy-oxide counter electrode. Designed to be fitted into the ceiling of an interior room, sunlight was to be funnelled to it down a tube from the outer roof, with a clear glass roof-lid, the intensity to be electrochromically controlled from within. The inside placing would protect from solar photodegradation, but it could be of use only for top-floor or single-storey illumination. Furthermore it would have found use only in tropical or equatorial sunshine intensities as a light source, when other windows could be permanently darkened against solar heating. Apart from the undisclosed cost of the window itself, the funnel-tube and roof-installation expense has probably vitiated any commercial appeal.

The Stadtsparkasse Bank in Dresden however has operating electrochromic external windows supplied by Flabeg Gmbh, who acquired Pilkington Glass's ECD technology based on WO_3 with FTO substrates.[58] Asahi Glass in Japan have an electrochromic window of small panes (*ca.* 30 × 30 cm), also WO_3, which they claim to have been operating in a building for some years. The stage is poised for a wider use of ECD windows in buildings, but the period between 'possible' (it works) and 'commercial' (it will pay its way) can be appreciable.

13.3.2 Aircraft – the first ubiquitous ECD window application: Gentex and Boeing

In December 2005 Gentex Corporation, PPG Aerospace and Boeing signed agreements to install electrochromic windows in the new long-range Boeing (B787) aircraft, the 'Dreamliner' (an artistic name).[59,60] The windows are said to be 25% greater in area than the usual. Several hundred of the B787 aircraft, due to operate in 2008, are already on order. The Gentex–PPG systems will allow passengers to set the windows from clear to five increasing levels of darkening up to virtual opacity. The electrochromic system employed has not been revealed. The screen is said to be sited between the external cabin window and the plastic dust shield; whether the outermost glass layer is part of the ECD system is not disclosed.

The company PPG Aerospace is an experienced aircraft-window manufacturer and an ideally imaginative collaboration with Gentex has been created, to effect the first mass-produced application of ECDs of appreciable size. Clearly a specialist, niche, application is involved here, but it represents a substantial advance on the only other mass-produced device, the car mirror, that has thoroughly proved its worth on the smaller scale.

The costs to Boeing are reported as being $50 million (of which the larger part goes to Gentex). The B787 has 100 windows, for the 221 seats. One can do a little simple arithmetic to arrive at a pricey sum per window, but perhaps only about 10 times that of an ECD car-mirror installation. As at present constructed, an avenue to mass-produced architectural applications is not yet open, but this substantial growth in window production can only lead to advances towards accommodating the requirements of buildings.

Airbus are reported to be considering electrochromic windows for the A380, their new aircraft undergoing development, as are no doubt other aircraft manufacturers.

13.3.3 Capital screening: sunglasses and visors

The Swedish invention of motorcyclists' ECD visors is referred to below (see p. 422). Electrochromic sunglasses, also necessarily operating in a transmittance mode (*cf.* Figure 13.1), have been produced that may be darkened at will, contrasting with the automatic operation of the now widely available photochromic lenses that darken automatically. Nikon were the first to market electrochromic sunglasses in 1981, calling them a 'variable-opacity lens filter'.[61] Subsequently Nikon marketed WO_3-based sunglasses in 1993, but these are no longer available. Donnelly have also produced electrochromic sunglasses that apparently operate via a different mechanism.[62]

13.4 Electrochromic displays for displaying images and data

Electrochromic devices operating as displays can act in either reflectance or transmissive modes, the majority being of the reflectance type.

The two reviews in 1986 by Agnihotry *et al.* covering both physicochemical properties[63] and device technology[64] delineate the historical development of such devices. Additionally, Faughnan and Crandall's still useful 1980 review[65]

'Electrochromic devices based on WO_3' helps justify the claim that these workers introduced the concept of electrochromic displays. Although dated, the extensive review by Bowonder *et al.*[66] in 1994 helps establish the place of electrochromism within the wide varieties of display device. Byker's two reviews of ECDs, 'Commercial developments in electrochromics' (in 1994)[5] and 'Electrochromics and polymers' (in 2001),[67] provide much detail.

Electrochromic devices are often termed 'passive' since they do not emit light and hence require external illumination, a possible disadvantage: light-emitting diodes (LEDs) and cathode-ray tubes (CRTs) are emissive, but liquid crystal displays (LCDs) and almost all mechanical displays are also non-emissive. A newer emissive competitor is the 'plasma' screen, in televisions and large advertising displays, which comprise individual pixels of three (for tri-colour emission) minute, gas-filled, fluorescent light-emitting units. The construction costs are high, which, however, users seem prepared to bear.

The global display market is expanding rapidly. For example, the total global display market was $11.6 billion in 1994 and will top $100 billion in 2007.[66] The market for 'flat-panel' information displays was worth approximately $18 billion in 2003, and is growing very fast. Since 1996, LCD devices have formed a larger proportion than CRTs. They now dominate with about 90% of the market share, and are superseding CRTs in applications such as television screens and visual-display units (VDUs) for computers and instruments requiring monitors.

Electrochromic devices have been proposed for flat-panel displays for applications such as television and VDU screens (but note possible disqualifications spelt out below), data boards at transport terminuses, advertising boards,[68] and even as an electrochromic 'indicator' (based on WO_3) on a cash card.[69] The range of applications for flat-panel displays increases rapidly, and are incorporated into a wide array of electronic devices both large and small, from calculators and watches to, perhaps, mobile phones and screens on lap-, palm- or desk-top computers. For example, at the 'DEMO 2005' show, NTera of Eire demonstrated an iPod with a Nanochromics™ screen (as below). One commentator thought the new electrochromic screen 'definitely exceeded the original iPod [screen] in crisp- and brightness'.[70] There are many other nascent applications of ECDs, so when technological barriers are overcome, these materials are likely to play an increasing role in such uses.

The first application suggested for ECDs was in watch faces.[71] A modern variant is the face of the so-called *Moonwatch*;[72] here the face does not tell the time but represents a display with fourteen separate areas, which darken progressively to indicate the phases of the moon. Other specialist ECDs designed for use as watch faces are cited in refs. 73,74,75.

13.4 Displaying images and data

Liquid crystal displays can be fabricated extremely cheaply, sometimes so cheaply as to be disposable. The main reason for their cheapness is the sheer volume of production worldwide, which decreases the capital costs. Electrochromic devices must compete with LCDs for commercial viability, and therefore possess economic advantages over them. The claimed advantages are as follows: firstly, ECDs consume little power in producing images which, once formed, remain with little or no additional input of power – the so-called 'memory effect' outlined on p. 53. Secondly, in principle, there is no limit to the size an ECD can take, so a device may be constructed having a larger electrode expanse or a greater number of small electrodes. Electrochromic devices may be either flat or curved for wide-angle viewing. By contrast, large-area LCDs are expensive, and large CRTs require a huge electron 'gun' behind the screen, which is both bulky and prohibitively expensive.

Realistically, however, ECDs have insufficiently fast response times τ to be considered for applications such as television and (most) VDU screens, and cycle lives are probably also somewhat low (see Section 16.1). Indicative response times can be roughly estimated from Eq. (3.16), $l \approx (Dt)^{1/2}$, for type-I and type-III electrochromes. Typical distances l to be traversed by a key species in a coloration step are between 10 and 100 nm, say ~50 nm intermediately. With D for type-I (solution-phase) species about 10^{-7} cm^2 s^{-1}, a response time of less than a millisecond is obtained, but the type-I coloration in solution will be mobile. For immobile coloration, as obtained with type III, D is typically 10^{-12} cm^2 s^{-1}, giving a response time of ~20 s. For televisions and VDUs, the image must be coloured at fixed points, so requiring responses from a type-III system, which our order-of-magnitude arithmetic shows to be slow. Displays of digits and alphanumeric displays could however comprise liquid-containing elements or solids with faster diffusion coefficients $D \approx 10^{-10}$ cm^2 s^{-1}, so responding within a range of a few milliseconds to a second or so. (Note that these estimates, while of illustrative value, ride roughshod over the detail of the mechanisms summarised in Chapter 5. Furthermore, tethered monolayer systems, with l but a few nm – see Section 11.3.1 – could be 10^2 to 10^3 times faster than these 'guesstimates'.) Accordingly, the most suitable roles envisaged at this stage involve displaying information more slowly, for long-term perusal, e.g. at transport terminuses as mentioned above, for re-useable price labels, or on advertising boards and frozen-food monitors.

To produce such an image, multiple electrodes – 'picture elements' or 'pixels' – allow text or images to be displayed rather than mere blocks of colour. The electrochromic '3' shown in Figure 4.1 is achieved with seven relatively large electrodes; the IBM Laboratories made an ECD with a 64 × 64 pixel image on a one inch square silicon chip[76] and the NTera NanoChromics™ display (see further detail in Section 11.4, p. 361) comprises an array of transparent electrodes, each about 0.25 mm square, or about 100 dots per inch.[77] Colour Plate 6 shows a reflective cell with nine pixels.

In such multi-pixel ECDs, tonal variation is achieved by stippling with dots as with LCD displays; alternatively, the image may be intensified by passing more charge into specified areas where more of the coloured substance is to be formed. There is however the technical problem with any large-area ECD. Areas of patchy colour may form when the current distribution is uneven across the electrode surface, since the electric field can be larger at the edges of the electrode substrate nearest the metallic leads, if the electrode substrate is semiconductive (like ITO). This allows a potential drop with distance towards the centre of the conducting area. Increasing the viscosity of the electrolyte, and subtle choice of potentials and dimensions, can more-or-less obviate this problem.[67]

13.5 ECD light modulators and shutters in message-laser applications

In addition to displays and windows, electrochromic systems find a novel application as optical shutters or light modulators where the ECD operates in a transmissive mode (Figure 13.1). It is often the case that in fibre-optic message-laser applications the transmitting front-end puts out too high an intensity for the fibre. This is best remedied by a permanent filter, which could be a once-for-all photochromically evoked colour filter for the particular laser wavelength (the photochemistry of this coloration being effected by a pulse from a laser of different wavelength from that of the message laser). However, at the receiving end a variety of detectors are in use, with an associated variety of sensitivities, not always commensurate with the incoming signal. To match the output laser intensity to the detector sensitivity, an adjustable ECD is inserted in the optical path before the detector, that needs particular circuitry to evoke the most fitting coloration intensity. This task requires that the ECD remains almost constant for any one transmission. As this is a preliminary setting preceding message reception, instant (i.e. nanosecond) responses are not required. As receivers get messages from a number of sources with varying intensities, automatic adjustment preceding reception is desirable; this takes place during the communication-linking protocol. A patent describes the circuitry detail required for this purpose.[78]

However, for *operation* of fibre-optic message transmissions (or in optical computer action), a response time of sub-nanoseconds is necessary, so no redox ECDs are sufficiently fast to act in this particular role as on–off shutters. Possibly for more leisurely optical data storage, pixels need only represent either 'off' or 'on', as in Figure 13.1 when coloured or bleached respectively, which thus totally interrupts (or not) a light beam, without regard to gradations of intensity. Electrochromic data storage is thus not precluded.

13.6 Electrochromic paper

The impetus behind developing electrochromic paper is environmental: electrochromes embedded within a sheet of paper can in principle be switched reversibly between coloured and bleached, thereby allowing the paper to be re-used, rather than recycled.

Relatively little work has yet been done on electrochromic materials impregnated into paper. Talmay[79,80] patented an idea for electrochromic printing in 1942 with 'electrolytic writing paper' consisting of paper pre-impregnated with particulate MoO_3 and WO_3 that formed an image following reduction at an inert-metal electrode acting as a pen.

The electroformation of Prussian blue within the fibres of the paper has also been suggested: cf. the comments in Chapter 2 concerning 'blue prints'.

Several recent patents have been issued for elaborations of electrochromic printing systems usually based on organic electrochromic dyes, as cited in ref. 5. In 1989 Rosseinsky and Monk[82] investigated whether voltammetry in paper was possible, revealing marginal problems associated with *IR* drop across the paper and variations in its internal humidity. Moist paper was impregnated with a variety of viologens or Prussian blue precursors, together with an ionic electrolyte in sufficient concentration. In paper of marginal moistness, the electrochemistry of both Prussian blue and viologen electrochromes was quite well reproduced as though in a laboratory electrochemical cell, establishing electrochromic reactions to occur within the paper, as described in Section 11.4. Details of the study have been improved on.[81]

In 1989, IBM[83] prepared a form of electrochromic paper capable of multiple coloration, but the complexity of their system precluded economic commercialisation.

Investigations on electrochromes impregnated into paper included viologens,[82,84,85] Prussian blue[82,84] and the metal oxides MoO_3 and WO_3.[84] Incorporation of an electrochrome within a thin layer of Nafion® as a host matrix has also been shown to produce good results: the electrochromes included viologen,[86] Methylene Blue[87] and phenolsafranine dyes.[87]

Printable electrochromic paper has not been further pursued. However, NTera of Eire have developed a product called 'electrochromic paper' (and marketed as NanoChromics™), which is based on the viologen **I**.[88] The display, not based on paper, uses phosphonate groups bound by chemisorption to a metal-oxide surface such as a titanium dioxide film deposited on FTO. The oxidised form of **I** is colourless while the reduced form is blue–mauve.

$$\text{HO-P(=O)(OH)-CH}_2\text{CH}_2\text{-N}^+\text{C}_5\text{H}_4\text{-C}_5\text{H}_4\text{N}^+\text{-CH}_2\text{CH}_2\text{-P(=O)(OH)-OH} \quad 2\text{Cl}^-$$

I

NTera call their ECD a nanochromic display (NCD), claiming their technology has more than four times the reflectivity and contrast of a liquid crystal display (LCD).

13.7 Electrochromes applied in quasi-electrochromic or non-electrochromic processes: sensors and analysis

It is of interest to consider the substances that can be electrochromic when they are used in another, analytical, context. 'Gasochromic' coloration outlined in Section 1.2 involves a mechanism of the kind further contemplated here. Only the first example of Co_3O_4 that we cite below has some electrochromic basis to its operation; we then suggest an extension of this principle. In the solely analytical applications presented below, these normally electrochromic substances acquire or lose electrons from (or to) solution or gaseous species, rather than from (or to) electrode substrates when in electrochromic mode. The analytical relevance arises from the ensuing colour changes: a direct relationship exists between the absorbance of an electrochrome and the amount of charge passed, and thus the amount of test substance present.

In a quasi-electrochromic application, Shimizu et al.[89,90] made a sensor based on cobalt oxide Co_3O_4, which, electrically polarised, colorises in the presence of phosphate ion: a thin Co_3O_4 film changed transmittance T in the range 550–800 nm, becoming coloured when polarised to 0.4 V vs. SCE, but only in the presence of sufficient phosphate ion. No colour ensued in the absence of $[HPO_4]^{2-}$. The sensor transmittance T depends on the logarithm of $[HPO_4]^{2-}$ concentration in the range 10^{-6} to 10^{-2} mol dm^{-3}, via a mechanism depending on the redox reaction of Eq. (6.19) in Chapter 6, but with the electrons now coming from a chemical reductant. (So it is not truly electrochromic.)

Other electrochromic sensors have been fabricated in which the optical absorbance relates to pH,[91] or NO_3^-, or Cl^- concentrations.[89]

Table 13.1. *Electrochromes utilised in gasochromic sensing devices, responding to gaseous analyte.*

Electrochrome	Analyte	Refs.
Chromium oxide	Ozone	99
Metalloporphyrins	Chlorine	100
Nickel oxide	Ozone	99
Phthalocyanine	Chlorine	101
Phthalocyanine	NO_2, toluene	92,93,94,95,96,97,98
Tungsten trioxide	Hydrogen	102,103,104,105,106,107,108,109
Tungsten trioxide	Oxygen	109
Tungsten trioxide	$CH_4/NH_3/CO$	110
Tungsten trioxide	H_2S	111,112,113,114
Tungsten trioxide	NO	115,116,117

The term 'gasochromic', Section 1.2, describes devices that operate with a gas-phase reductant or oxidant providing or accepting the electrons that would be necessary were these electrochromic redox processes. Thus, in a non-electrochromic analytical application, Cook and co-workers used a variety of phthalocyanines, e.g. in the form of a Langmuir–Blodgett film, to test for such diverse gases as NO_2 and toluene.[92,93,94,95,96,97,98]

Many examples of gasochromic sensing devices are listed in Table 13.1, all electrochromes remaining in the solid state during coloration. While not strictly electrochromic, they are cited here because the chemical compositions and device geometry could readily be transformed into reversibly electrochromic systems, with the possibility of re-use for testing. In several cases, the device changes transmittance chemically following contact with gaseous analysis sample, but can be refreshed electrochemically for re-use.

In an interesting gasochromic–analytical application, Khatko *et al.* show that doping a solid layer of WO_3 with different metals increases the sensitivity and selectivity to different gases.[118] Thin films of tungsten trioxide respond readily and rapidly to gaseous hydrogen. Many of the WO_3-based gasochromic devices cited in Table 13.1 incorporate tungsten trioxide bearing a thin layer of platinum coated on the outer surface. In such cases, the WO_3 is responding to atomic hydrogen formed by a 'spillover' process catalysed by Pt, as described by Wittwer *et al.*[109]

13.8 Miscellaneous electrochromic applications

Portable identification cards for membership or security purposes can all bear an electrochromic fragment. Obvious applications include cash-point machines and credit cards, etc., for which patents have already been filed.[69] Other security-related applications possible with an electrochrome impregnated into (solid) paper include security devices such as vouchers, tokens and tickets – even bank notes – where fraudulent copying is likely.

The only extant review of electrochromic printing is ours[119] in 1995.

Some applications rely on a *thermo*-electrochromic system, in which the speed of electrochromic coloration depends on temperature. In such applications, the device is usually so slow when cold that it is effectively switched 'off', even when a suitable potential is applied. As the temperature rises, so the speed of operation increases until a threshold is attained, above which the device will colour and bleach quite normally.

The temperature dependence of a thermoelectrochromic device is best achieved by incorporating an ionic electrolyte for which the movement of counter ions has a high activation energy, E_a. The magnitude of E_a ensures that a relatively small change in temperature causes a substantial increase in ionic conductivity, and hence in device operation. Scrosati et al.[120] were probably the first to make such a device: the electrochrome was WO_3 and the electrolyte comprised poly(ethylene oxide) containing dissolved $LiClO_4$.

More recently, Owen and co-workers[121] developed a thermo-electrochromic device for displaying the safety of food, and is to be positioned above shop refrigerators. The electrolyte is again poly(ethylene oxide) containing dissolved $LiClO_4$,. The rate of coloration followed an Arrhenius-type expression at temperatures in the range 30 to –25 °C, provided the electrolytes remained amorphous (achieved by adding a high concentration of $LiClO_4$ and also a small amount of ZnI_2). So long as the rate of electro-coloration is essentially the same as the rate at which harmful bacteria multiply in the food, then the food is safe to eat while the device has not formed any colour. Conversely, the refrigerated food may be unsafe when the thermoelectrochromic ECD *has* changed its colour, because bacteria in the food will have had time to multiply.

The Eveready Battery Company have produced a long, narrow electrochromic strip to indicate the state of charge, for use with dry-cell batteries.[122] During use, the two ends of the 'charge indicator' strip are attached to the two termini of a battery: the level of charge within the battery is indicated via the intensity of the strip's colour and the proportion of the strip's length that has become coloured. The identity of the electrochrome is obscured by the prose of the patent. (The strip on Duracell batteries is based on liquid-crystal technology, and is not electrochromic.)

Kojimo and Terao[123] have developed an electrochromic system as a component within a DVD. Here an electrochromic layer serves as the multi-information-layer for an optical disk system. The active electrochrome is PEDOT (see Section 10.2). The claimed advantages of the electrochromic layer disk are in its large capacity, high sensitivity in recording, and the relative simplicity of the attendant hardware.

The military in the USA are investigating fitting electrochromic panels as camouflage. The organic electrochromes are being developed by EIC Laboratories in conjunction with the Reynolds group in Florida.[124]

13.9 Combinatorial monitoring of multiples of varied electrode materials

A hugely ingenious application of electrochromism, a major aid to multiple monitorings of electrode processes, has just been announced.[125] It matches the 'combinatorial' methods of organic chemistry in which mixtures of products from concurrently occurring organic reactions in one pot are simultaneously analysed at the conclusion of reaction.

As illustration, using a sheet of WO_3 deposited onto a FTO on glass of surface resistance 50 ohm per square, the electro-oxidation of methanol by a variety of Pt catalysts was employed. The 56 electrodes undergoing tests comprised various masses (groups of 6, 12, 18 or 24 μg) of Pt-containing electrode catalysts, each of similar diameter, 3 mm. These were deposited on vitreous carbon electrodes mounted on a non-conducting poly(tetrafluoroethylene (PTFE) planar support in a 7×8 matrix. The counter electrode, placed only 1 mm apart from the matrix, was the single WO_3-coated sheet. The methanol reactant was at 1 mol dm^{-3} while the electrolyte was very dilute (H_2SO_4, 1 mmol dm^{-3}), but the otherwise high resistance engendered is totally mitigated by the closeness of the two electrode sheets. The several millimetre lateral spacing between the Pt 'dots' confers high inter-dot resistances and thereby 'focusses' currents onto WO_3 areas directly opposite the Pt electrodes.

For a suitable fixed duration, with the same potential simultaneously applied versus the WO_3 electrode to all the Pt electrodes, the relative effectiveness of each Pt electrode, as measured by the current or charge passed by each, is recorded as a small disc of blue coloration on the WO_3, in a matrix corresponding to the geometry of the Pt electrodes. The intensity of coloration of each dot is directly proportional to the charge or current passed by each Pt catalyst. The simple photometric measurement of the colour intensity of each, from say a CCD camera image, bypasses separate or seriatim monitorings by voltammetry or galvanometry of each Pt electrode, by this simple and convenient quantitative method. For rapid comparative purposes, viewing by eye provides an instant estimate, if the quantity or quality of the catalyst in the monitored electrodes are arranged in sequence in the electrode mountings.

A filter paper interposed between the electrodes acted both as a cell separator and a diffuse reflector aiding the optical monitoring by CCD camera. In the experiments reported in the paper, but not essential in application, separate currents were individually monitored for comparison with the optical imprints on the WO_3, providing very satisfactory evidence of the quantitative precision of the method. (This current monitoring, being expensive of apparatus or time, would not of course be needed except perhaps introductorily once-off in actual test applications.) Several tests on smaller groups of electrodes confirmed the satisfactory operation.

The initially clear WO_3 was preconditioned by being cycled from 0 to -200 mV with respect to an SCE, and finally pre-set at -50 mV before use, which ensured linearity of coloration intensity with current passed. The actual test was initiated by stepping the voltage across the multiplex cell from 0 to 0.4 V (the Pt being positive), which set the electro-oxidation reaction going. The size of the WO_3 electrode allowed its use as a quasi-reference electrode, its potential in separate tests remaining adequately constant.

While it may be critically argued that such tests are limited by intercalation into the WO_3 only of such cations as H^+ or Li^+, it is just these cations that are important players in catalysis: by the former in fuel cells, and by the latter in lithium battery material. Further redox and electrocatalytic scenarios employing the ingenious new geometry might also be envisaged, possibly involving test-bed materials other than WO_3.

References

1. Lampert, C. M. Smart switchable glazing for solar energy and daylight control. *Sol. Energy Mater. Sol. Cells*, **52**, 1998, 207–21.
2. Bange, K. and Gambke, T. Electrochromic materials for optical switching devices. *Adv. Mater.*, **2**, 1992, 10–16.
3. Byker, H. J., Gentex Corporation. Single-compartment, self-erasing, solution-phase electrochromic devices, solutions for use therein and uses thereof. US Patent 4,902,108, 1990.
4. [Online] at www.gentex.com/auto_how_nvs_work.html (accessed 6 September 2005).
5. Byker, H. J. Commercial developments in electrochromics. *Proc. Electrochem. Soc.*, **94-2**, 1994, 1–13.
6. Schierbeck, K. L., Donnelly Corporation. Digital electrochromic mirror system. US Patent 06089721, 2000.
7. Baucke, F. G. K. Electrochromic mirrors with variable reflectance. *Rivista della Staz. Sper. Vetro*, **6**, 1986, 119–22.
8. Baucke, F. G. K. Electrochromic mirrors with variable reflectance. *Sol. Energy Mater*, **16**, 1987, 67–77.
9. Baucke, F. G. K. Reflecting electrochromic devices – construction, operation and application. *Proc. Electrochem. Soc.*, **20-4**, 1990, 298–311.

10. Baucke, F. G. K., Bange, K. and Gambke, T. Reflecting electrochromic devices. *Displays*, **9**, 1988, 179–87.
11. Baucke, F. G. K. Beat the dazzlers. *Schott Information*, **1**, 1983, 11–13.
12. Gentex announces new Intelligent high-beam headlamp control technology: miniature camera to control vehicle high beams. Machine Vision Online, 2004.
13. Gesheva, K., Ivanova, T. and Hamelmann, F. Optical coatings of CVD-transition metal oxides as functional layers in 'smart windows' and X-ray mirrors. *J. Optoelectronics Adv. Mater.*, **7**, 2005, 1243–52.
14. Svensson, J. S. E. M. and Granqvist, C. G. Electrochromic coatings for 'smart windows'. *Sol. Energy Mater.*, **12**, 1985, 391–402.
15. [Online] at www.bfrc.org/Technical_Publications-Thermal_definitions.htm (accessed 6 September 2005).
16. [Online] at home.howstuffworks.com/smart-window.htm and home.howstuffworks.com/smart-window2.htm (accessed 6 September 2005).
17. Granqvist, C. G., Azens, A., Isidorsson, J., Kharrazi, M., Kullman, L., Lindstrom, T., Niklasson, G. A., Ribbing, C.-G., Rönnow, D., Strømme Mattson, M. and Veszelei, M. Towards the smart window: progress in electrochromics. *J. Non-Cryst. Solids*, **218**, 1997, 273–9.
18. Rauh, R. D. Electrochromic windows: an overview. *Electrochim. Acta*, **44**, 1999, 3165–76.
19. Bell, J. M., Skryabin, I. L., Matthews, J. P. and Matthews, J. P. Windows. In Schwartz, M. (ed.), *Encyclopedia of Smart Materials*, New York, Wiley, 2002, vol. 2, pp. 1134–45.
20. Azens, A. and Granqvist, C. G. Electrochromic smart windows: energy efficiency. *J. Solid State Electrochem.*, **7**, 2003, 64–8.
21. Mbise, G. W., Le Bellac, D., Niklasson, G. A. and Granqvist, C. G. Angular selective window coatings: theory and experiments. *J. Phys. D.*, **30**, 1997, 2103–22.
22. Demiryont, H. A review on electrochromic devices for automotive glazing. *Proc. SPIE*, **1536**, 1991, 2–28.
23. [Online] at eetd.lbl.gov/EA/mills/Lab2Mkt/Windows.html (accessed 6 September 2005).
24. Lee, E. S. and DiBartolomeo, D. L. Application issues for large-area electrochromic windows in commercial buildings. *Sol. Energy Mater. Sol. Cells*, **71**, 2002, 465–91.
25. Harary, J. M. Automated window shading, available [online] at www.earthtoys.com/emagazine.php?issue_number=02.09.01&article=harary (accessed 6 September 2005).
26. [Online] at www.consumerenergycenter.org/homeandwork/homes/inside/windows/future.html (accessed 6 September 2005).
27. Griffiths, P., Eames, P., Lo, S. and Norton, B. Energy and environmental life-cycle analysis of advanced windows. *Renewable Energy*, **8**, 1996, 219–22.
28. Syrrakou, E., Papaefthimiou, S. and Yianoulis, P. Environmental assessment of electrochromic glazing production. *Sol. Energy Mater. Sol. Cells*, **85**, 2005, 205–40.
29. [Online] at www.nrel.gov/buildings/windows/producers.html (accessed 6 September 2005).
30. [Online] at www.sage-ec.com/pages/technol.html (accessed 6 September 2005).
31. [Online] at www.chem.ufl.edu/~reynolds (accessed 19 June 2007).
32. [Online] at www.nrel.gov/buildings/windows.html (accessed 6 September 2005).
33. [Online] at www.rjfalkner.com/page.cfm?pageid = 2241 (accessed 2 April 2006).
34. [Online] at http://windows.lbl.gov/materials/Chromogenics/ec_radiance/simulations.html (accessed 6 September 2005).

35. [Online] at www.saint-gobain-recherche.com/anglais/index.htm (accessed 6 September 2005).
36. [Online] at www.chromogenics.se/index_eng.htm (accessed 5 September 2005).
37. Azens, A., Gustavsson, G., Karmhag, R. and Granqvist, C. G. Electrochromic devices on polyester foil. *Solid State Ionics*, **165**, 2003, 1–5.
38. Buyan, M., Brühwiler, P. A., Azens, A., Gustavsson, G., Karmhag, R. and Granqvist, C. G. Facial warming and tinted helmet visors. *Int. J. Ind. Ergonomics*, **36**, 2006, 11–16.
39. Zinzi, M. Office worker preferences of electrochromic windows: a pilot study. *Buildings and Environment*, **41**, 2005, 1262–73.
40. Siddle, J., Pilkington PLC, personal communication, 1991.
41. Munro, B., Kramer, S., Zapp, P., Krug, H. and Schmidt, H. All sol–gel electrochromic system for plate glass. *J. Non-Cryst. Solids*, **218**, 1997, 185–8.
42. von Rottkay, K., Ozer, N., Rubin, M. and Richardson, T. Analysis of binary electrochromic tungsten oxides with effective medium theory. *Thin Solid Films*, **308–309**, 1997, 50–5.
43. Fang, G. J., Yao, K.-L. and Liu, Z.-L. Fabrication and electrochromic properties of double layer $WO_3(V)/V_2O_5(Ti)$ thin films prepared by pulsed laser ablation technique. *Thin Solid Films*, **394**, 2001, 63–70.
44. Mathew, J. G. H., Sapers, S. P., Cumbo, M. J., O'Brien, N. A., Sargent, R. B., Raksha, V. P., Lahaderne, R. B. and Hichwa, B. P. Large area electrochromics for architectural applications. *J. Non-Cryst. Solids*, **218**, 1997, 342–6.
45. Rougier, A., Blyr, A., Garcia, J., Zhang, Q. and Impey, S. A. Electrochromic W–M–O (M = V, Nb) sol–gel thin films: a way to neutral colour. *Sol. Energy Mater. Sol. Cells*, **71**, 2002, 343–57.
46. Bell, J. M., Barczynska, J., Evans, L. A., MacDonald, K. A., Wang, J., Green, D. C. and Smith G. B. Electrochromism in sol–gel deposited TiO_2 films. *Proc. SPIE*, **2255**, 1994, 324–31.
47. Gao, W., Lee, S.-H., Benson, D. K. and Branz, H. M. Novel electrochromic projection and writing device incorporating an amorphous silicon carbide photodiode. *J. Non-Cryst. Solids*, **266–9**, 2000, 1233–7.
48. Impey, S. A., Garcia-Miguel, J. L., Allen, S., Blyr, A., Bouessay, I. and Rougier, A. Colour neutrality for thin oxide films from pulsed laser deposition and sol–gel. *Proc. Electrochem. Soc.*, **2003–17**, 2003, 103–18.
49. Klein, J. D., Yen, A., Rauh, R. D. and Causon, S. L. Near-infrared electrochromism in Li_xC_{60} films. *Appl. Phys. Lett.*, **63**, 1993, 599–601.
50. Kulak, A. I., Kokorin, A. I., Meissner, D., Ralchenko, V. G., Vlasou, I. I., Kondratyuk, A. V. and Kulak, T. I. Electrodeposition of nanostructured diamond-like films by oxidation of lithium acetylide. *Electrochem. Commun.*, **5**, 2003, 301–5.
51. Richardson, T. J. New electrochromic mirror systems. *Solid State Ionics*, **165**, 2003, 305–8.
52. Manevich, R. M. L., Shamritskaya, I. G., Sokolova, L. A. and Kolotyrkin, Y. M. The electroreflection spectra of anodically oxidized iridium and adsorption of water. *Russ. J. Electrochem.*, **32**, 1996, 1237–44.
53. Rönnow, D., Kullman, L. and Granqvist, C. G. Spectroscopic light scattering from electrochromic tungsten-oxide-based films. *J. Appl. Phys.*, **80**, 1996, 423–30.
54. Goldner, R. B., Mendelsohn, D. H., Alexander, J., Henderson, W. R., Fitzpatrick, D., Haas, T. E., Sample, H. H., Rauh, R. D., Parker, M. A. and Rose, T. L. High near-infrared reflectivity modulation with polycrystalline electrochromic WO_3 films. *Appl. Phys. Lett.*, **43**, 1983, 1093–5.

55. Otero, T. F. and Bengoechea, M. *In situ* absorption-reflection study of polypyrrole composites – switching stability. *Electrochim. Acta*, **41**, 1996, 1871–6.
56. Pages, H., Topart, P. and Lemordant, D. Wide band electrochromic displays based on thin conducting polymer films. *Electrochim. Acta*, **46**, 2001, 2137–43.
57. Schlotter, P. High contrast electrochromic tungsten oxide layers. *Sol. Energy Mater. Sol. Cells*, **16**, 1987, 39–46.
58. [Online] at www.chemsoc.org/chembytes/ezine/2002/ashton_jun02.htm (accessed 16 March 2006).
59. [Online] at www.Gentex.com (accessed 29 March 2006).
60. [Online] at www.ppg.com/gls_ppgglass/aircraft/22779.pdf (accessed 29 March 2006).
61. [Online] at www.nikon.co.jp/main/eng/portfolio/about/history/corporate_history.htm (accessed 6 September 2005).
62. Taylor, D. J., Cronin, J. P., Allard, L. F. and Birnie, D. P. Microstructure of laser-fired, sol–gel-derived tungsten oxide films. *Chem, Mater.*, **8**, 1996, 1396–401.
63. Agnihotry, S. A., Saini, K. K. and Chandra, S. Physics and technology of thin film electrochromic displays, part I: physicochemical properties. *Ind. J. Pure Appl. Phys.*, **24**, 1986, 19–33.
64. Agnihotry, S. A., Saini, K. K. and Chandra, S. Physics and technology of thin film electrochromic displays, part II: device technology. *Ind. J. Pure Appl. Phys.*, **24**, 1986, 34–40.
65. Faughnan, B. W. and Crandall, R. S. Electrochromic devices based on WO_3. In Pankove, J. L. (ed.), *Display Devices*, Berlin, Springer-Verlag, 1980, pp. 181–211.
66. [Online] at www.elecdesign.com/Articles/ArticleID/15783/15783.html (accessed 19 June 2007).
67. Byker, H. J. Electrochromics and polymers. *Electrochim. Acta*, **46**, 2001, 2015–22.
68. [Online] at www.napa.ufl. edu/2001news/colors.htm (accessed 6 September 2005).
69. Tadashi, N. Cash card having electrochromic indicator. Japanese Patent, JP 59,197,980, 1984.
70. [Online] at www.mobileread.com/forums/showthread.php?threadid=3375 (accessed 27 January 2006).
71. Schoot, C. J., Ponjeé, J. J., van Dam, H. T., van Doorn, R. A. and Bolwijn, P. J. New electrochromic memory device. *Appl. Phys. Lett.*, **23**, 1973, 64–5.
72. [Online] at www.moonwatch.com/article.html (accessed 6 September 2005. The webpage comprises a journalistic account entitled 'The Moonwatch story'.).
73. Ando, E., Kawakami, K., Matsuhiro, K. and Masuda, Y. Performance of a-WO_3/$LiClO_4$–PC electrochromic displays. *Displays*, **6**, 1985, 3–10.
74. Kaneko, N., Tabata, J. and Miyoshi, T. Electrochromic device watch display. *SID Int. Symp. Digest*, **12**, 1981, 74–5.
75. Schoot, C. J., Bolwijn, P. T., van Dam, H. T., van Doorn, R. A., Ponjeé, J. J. and van Houten, G. Elektrochrome Anzeige mit Speichereigenschaften (Electrochrome displays with storage properties: construction and functioning of storage-type electrochrome cell), *Elektronikpraxis*, **10**, 1975, 11–14 [in German].
76. Barclay, D. J. and Martin, D. H. Electrochromic displays. in Howells, E. R. (ed.), *Technology of Chemicals and Materials for the Electronics Industry*, Chichester, Ellis Horwood, 1984, 266–76.
77. Advanced electrochromic displays find markets. *Printed Electronics Review*, 2005; available [online] at www.idtechex.com/printelecreview/en/articles/00000149.asp (accessed 14 September 2005).
78. Freeman, W., Rosseinsky, D., Jiang, H. and Soutar, A., Finisar Corporation. Control systems for electrochromic devices. US Patent 6,940,627 B2, 2005.

79. Talmay, P. US Patent 2,319,765, 1943; as cited in Granqvist, C. G., *Handbook of Inorganic Electrochromic Materials*, Amsterdam, Elsevier, 1995.
80. Talmay, P. US Patent 2,281,013, 1942; as cited in Granqvist, C. G., *Handbook of Inorganic Electrochromic Materials*, Amsterdam, Elsevier, 1995.
81. Mortimer, R. J. and Warren, C. P. Cyclic voltammetric studies of Prussian blue and viologens within a paper matrix for electrochromic printing applications. *J. Electroanal. Chem.*, **460**, 1999, 263–6.
82. Rosseinsky, D. R. and Monk, J. L. Thin layer electrochemistry in a paper matrix: electrochromography of Prussian blue and two bipyridilium systems. *J. Electroanal. Chem.*, **270**, 1989, 473–8.
83. Balanson, R. D., Corker, G. A. and Grant, B. D. *IBM Technical Disclosure Bulletin*, **26**, 1983, 2930, as cited in ref. 75.
84. Monk, P. M. S., Delage, F. and Costa Vieira, S. M. Electrochromic paper: utility of electrochromes incorporated in paper. *Electrochim. Acta*, **46**, 2001, 2195–202.
85. Monk, P. M. S., Turner, C. and Akhtar, S. P. Electrochemical behaviour of methyl viologen in a matrix of paper. *Electrochim. Acta*, **44**, 1999, 4817–26.
86. John, S. A. and Ramaraj, R. Electrochemical, *in situ* spectrocyclic voltammetric and electrochromic studies of phenosafranine in Nafion® film. *J. Electroanal. Chem.*, **424**, 1997, 49–59.
87. Ganesan, V., John, S. A. and Ramaraj, R. Multielectrochromic properties of methylene blue and phenosafranine dyes incorporated into Nafion® film. *J. Electroanal. Chem.*, **502**, 2001, 167–73.
88. [Online] at www.ntera.ie/nano.pdf (accessed 27 January 2006).
89. Shimizu, Y. and Furuta, Y. An opto-electrochemical phosphate-ion sensor using a cobalt-oxide thin-film electrode. *Solid State Ionics*, **113–15**, 1998, 241–5.
90. Shimizu, Y., Furuta, Y. and Yamashita, T. Optical phosphate-ion sensor based on electrochromism of metal-oxide thin-film electrode. *Trans. Inst. Elect. Eng. Jpn.*, **119**, 1999, 285–9.
91. Talaie, A., Lee, J. Y., Lee, Y. K., Jang, J., Romagnoli, J. A., Taguchi, T. and Maeder, E. Dynamic sensing using intelligent composite: an investigation to development of new pH sensors and electrochromic devices. *Thin Solid Films*, **363**, 2000, 163–6.
92. James, S. A., Ray, A. K., Thorpe, S. C. and Cook, M. J. Thermopower of copper tetra(4-tert-butyl)phthalocyanine Langmuir–Blodgett films. *Thin Solid Films*, **226**, 1993, 3–5.
93. Wright, J. D., Roisin, P., Rigby, G. R., Nolte, R. J. M., Cook, M. J. and Thorpe, S. C. Crowned and liquid-crystalline phthalocyanines as gas-sensor materials. *Sens. Actuators*, **B13**, 1993, 276–80.
94. Cole, A., McIlroy, R. J., Thorpe, S. C., Cook, M. J., McMurdo, J. and Ray, A. K. Substituted phthalocyanine gas sensors. *Sens. Actuators*, **B13–14**, 1993, 416–19.
95. Ray, A. K., Mukhopadhyay, S. and Cook, M. J. Hopping conduction in Langmuir–Blodgett films of amphiphilic phthalocyanine molecules. *Thin Solid Films*, **229**, 1993, 8–10.
96. Crouch, D., Thorpe, S. C., Cook, M. J., Chambrier, I. and Ray, A. K. Langmuir–Blodgett films of an asymmetrically substituted phthalocyanine: improved gas-sensing properties. *Sens. Actuators*, **B18–19**, 1994, 411–14.
97. Lukas, B., Silver, J., Lovett, D. R. and Cook, M. J. Electrochromism in the octapentyloxy nickel phthalocyanines and related phthalocyanines. *Chem. Phys. Lett.*, **241**, 1995, 351–4.

98. Baker, P. S., Petty, M. C., Monkman, A. P., McMurdo, J., Cook, M. J. and Pride, R. A hybrid phthalocyanine/silicon field-effect transistor sensor for NO_2. *Thin Solid Films*, **285**, 1996, 94–7.
99. Azens, A., Kullman, L. and Granqvist, C. G. Ozone coloration of Ni and Cr oxide films. *Sol. Energy Mater. Sol. Cells*, **76**, 2003, 147–53.
100. Yahaya, M. B., Salleh, M. M. and Yusoff, N. Y. N. Electrochromic sensor using porphyrin thin films to detect chlorine. *Proc. SPIE*, **5276**, 2004, 422–7.
101. Schiffrin, D. J. New Applications of Electrochromism: Displays, Light Modulation and Printing Meeting, Scientific Societies Lecture Hall, London, 3 April 1991, presentation.
102. Schweiger, D., Georg, A., Graf, W. and Wittwer, V. Examination of the kinetics and performance of a catalytically switching (gasochromic) device. *Sol. Energy Mater. Sol. Cells*, **54**, 1998, 99–108.
103. Georg, A., Graf, W., Neumann, R. and Wittwer, V. The role of water in gasochromic WO_3 films. *Thin Solid Films*, **384**, 2001, 269–75.
104. Georg, A., Graf, W., Neumann, R. and Wittwer, V. Stability of gasochromic WO_3 films. *Sol. Energy Mater. Sol. Cells*, **63**, 2000, 165–76.
105. Opara Krašovec, U., Orel, B., Georg, A. and Wittwer, V. The gasochromic properties of sol–gel WO_3 films with sputtered Pt catalyst. *Sol. Energy*, **68**, 2000, 541–51.
106. Shanak, H., Schmitt, H., Nowoczin, J. and Ziebert, C. Effect of Pt-catalyst on gasochromic WO_3 films: optical, electrical and AFM investigations. *Solid State Ionics*, **171**, 2004, 99–106.
107. Georg, A., Graf, W., Neumann, R. and Wittwer, V. Mechanism of the gasochromic coloration of porous WO_3 films. *Solid State Ionics*, **127**, 2000, 319–28.
108. Salinga, C., Weis, H. and Wuttig, M. Gasochromic switching of tungsten oxide films: a correlation between film properties and coloration kinetics. *Thin Solid Films*, **414**, 2002, 288–95.
109. Wittwer, V., Datz, M., Ell, J., Georg, A., Graf, W. and Walze, G. Gasochromic windows. *Sol. Energy Mater. Sol. Cells*, **84**, 2004, 305–14.
110. Shaver, P. Activated tungsten oxide gas detectors. *Appl. Phys. Lett*, **11**, 1967, 255–7.
111. Dwyer, D. G. Surface chemistry of gas sensors: H_2S on WO_3 films. *Sens. Actuators*, **B5**, 1991, 155–9.
112. Solis, J. L., Saukko, S., Kish, L., Granqvist, C. G. and Lantto, V. Semiconductor gas sensors based on nanostructured tungsten oxide. *Thin Solid Films*, **391**, 2001, 255–60.
113. Solis, J. L., Saukko, S., Kish, L. B., Granqvist, C. G. and Lantto, V. Nanocrystalline tungsten oxide thick-films with high sensitivity to H_2S at room temperature. *Sens. Actuators*, **B77**, 2001, 316–21.
114. Heszler, P., Reyes, L. F., Hoel, A., Landstrome, L., Lantto, V. and Granqvist, C. G. Nanoparticle films made by gas phase synthesis: comparison of various techniques and sensor applications. *Proc. SPIE*, **5055**, 2003, 106–19.
115. Tomchenko, A. A., Emelianov, I. L. and Khatko, V. V. Tungsten trioxide-based thick-film NO sensor: design and investigation. *Sens. Actuators*, **B57**, 1999, 166–70.
116. Tomchenko, A. A., Khatko, V. V. and Emelianov, I. L. WO_3 thick-film gas sensors. *Sens. Actuators*, **B46**, 1998, 8–14.
117. Ho, J.-J. Novel nitrogen monoxide (NO) gas sensors integrated with tungsten trioxide (WO_3)/pin structure for room temperature operation. *Solid State Electronics*, **47**, 2003, 827–30.

118. Khatko, V., Guirado, F., Hubalek, J., Llobet, E. and Correig, Z. X-Ray investigation of nanopowder WO_3 thick films. *Physica Status Solidi*, **202**, 2005, 1973–9.
119. Monk, P. M. S., Mortimer, R. J. and Rosseinsky, D. R. *Electrochromism: Fundamentals and Applications*, Weinheim, VCH, 1995.
120. Pantaloni, S., Passerini, S. and Scrosati, B. Solid state thermoelectrochromic device. *J. Electrochem. Soc.*, **134**, 1987, 753–75.
121. Colley, R. A., Budd, P. M., Owen, J. R. and Balderson, S. Poly[oxymethylene-oligo(oxyethylene)] for use in subambient temperature electrochromic devices. *Polym. Int.*, **49**, 2000, 371–6.
122. Bailey, J. C. Eveready Battery Company. Electrochromic thin film state-of-charge detector for on-the-cell application. US Patent 05458992, 1995.
123. Kojima, K. and Terao, M. Proposal of a multi-information-layer electrically selectable optical disk (ESD) using the same optics as DVD. *Proc. SPIE*, **5069**, 2003, 300–5.
124. [Online] at www.nttc.edu/resources/funding/awards/dod/1998sbir/982army.asp (accessed 6 September 2005).
125. Brace, K., Hayden, B. E., Russell, K. E. and Owen, J. R. A parallel optical screen for the rapid combinatorial analysis of electrochemical materials. *Adv. Mater.*, **18**, 2006, 3253–70.

14

Fundamentals of device construction

14.1 Fundamentals of ECD construction

All electrochromic devices are electrochemical cells, so each contains a minimum of two electrodes separated by an ion-containing electrolyte. Since the colour and optical-intensity changes occurring within the electrochromic cell define its utility, the compositional changes within the ECD must be readily seen under workplace illumination. In practice, high visibility is usually achieved by fabricating the cell with one or more optically transparent electrodes (OTEs), as below.

Electrochromic operation of the ECD is effected via an external power supply, either by manipulation of current or potential. Applying a constant potential in 'potentiostatic coloration' is referred to in Chapter 3, while imposing a constant current is said to be 'galvanostatic'. Galvanostatic coloration requires only two electrodes, but a true potentiostatic measurement requires three electrodes (Chapter 3), so an approximation to potentiostatic control, with two electrodes, is common.

The electrolyte between the electrodes is normally of high ionic conductivity (although see p. 386). In ECDs of types I and II, the electrolyte viscosity can be minimised to aid a rapid response. For example, a liquid electrolyte (that actually comprises the electrochromes) is employed in the world's best-selling ECD, the Gentex rear-view mirror described in Section 13.2. The electrolyte in a type-III cell is normally solid or at least viscoelastic, e.g. a semi-solid or polymer, as below.

In fact, virtually all the type-III cells in the literature are designed to remain solid during operation, as 'all-solid-state devices', or 'ASSDs'. Such solid-state ECDs have multilayer structures, and a wide range of device geometries has been contemplated,[1,2,3,4,5,6,7,8,9,10] involving variations in the positions of the counter and working electrodes. Figure 14.1 shows schematically one such solid-state device. Layer (**i**) is an optical electrode comprising a glass slide coated with ITO,

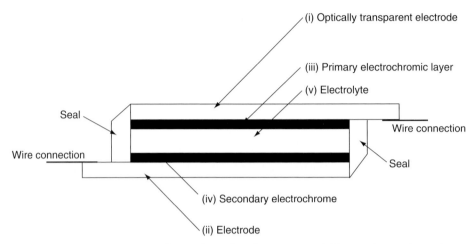

Figure 14.1 Schematic of a typical all-solid-state, multi layer electrochromic cell. Layer (**i**) is an optically transparent electrode, OTE. The second electrode (**ii**) could be another inward-facing OTE. Layer (**iii**) is the primary electrochrome and layer, (**iv**) is the secondary. Layer (**v**) is the electrolyte.

the conductive side innermost. The second electrode (**ii**) could be another inward-facing OTE if the device is to operate in a transmittance mode. Alternatively, devices operating in a reflectance mode generally require the second electrode to be made of polished metal, the metal being chosen both for its electronic conductivity and its aesthetic qualities, including its ability to act as a reflector, as described in Section 14.3 below. However, the colour and reflectivity of the second electrode are unimportant if it is positioned *behind* a layer of electrolyte containing an opaque filler; see p. 421.

The other layers of an all-solid-state ECD lie parallel and between the two electrodes. At least one of the 'ion-insertion layers' will be electrochromic. The primary electrochromic layer (**iii**) is juxtaposed with the front OTE; the secondary electrochrome (**iv**) is deposited on the rear, counter electrode. Finally, an electrolyte layer (**v**) separates the two ion-insertion layers, as described in Section 14.2.

Since the primary electrochrome is oxidised concurrently with reduction of the secondary (and *vice versa* when switching off), it is sometimes necessary to construct an all-solid-state ECD with one of the layers precharged with mobile ions. In practice, this is rarely a simple procedure. To effect this with say WO_3, lithium metal can be evaporated in vacuo onto the surface of one electrochrome film before device assembly – so-called 'dry lithiation'.[11,12,13,14,15,16] Elemental lithium is a powerful reducing agent, so gaseous lithium diffuses into the solid layer to effect chemical reduction, as in Eq. (14.1):

$$WO_3 \text{ (s)} + x \text{ Li}^0 \text{ (g)} \rightarrow \text{Li}_x WO_3 \text{ (s)}; \qquad (14.1)$$

x should not exceed about 0.3, since subsequent electrochemical extraction of Li$^+$ in attempted re-oxidation is irreversible (see p. 142).

Somewhat similarly, nickel oxide, in some commercial prototypes, is pre-charged using ozone;[17,18] in practice, films were irradiated with UV light in the presence of gaseous oxygen.

14.2 Electrolyte layers for ECDs

Reviews of electrolyte layers for ECD usage include 'Electrical and electrochemical properties of ion conducting polymers' by Linford[19] (in 1993), 'Sol–gel electrochromic coatings and devices: a review' by Livage and Ganguli[20] (in 2001) and 'Electrochromics and polymers' by Byker[21] (in 2001).

The layer of electrolyte between the two electrodes must be ionically conductive but electronically an insulator. In type-I and type-II ECDs, the electrochrome is dissolved in a liquid electrolyte, which can be either aqueous or a polar organic solvent such as acetonitrile or a variety of other nitriles, dimethylformamide, propylene carbonate or γ-butyrolactone. The electrochrome approaches the working electrode through this milieu during electrochromic coloration. Solutions may also contain a dissolved supporting electrolyte in high concentration to suppress migration effects (see Sections 3.3.2 and 3.3.3).

A thickener, such as acrylic polymer, poly(vinylbutyral) or colloidal silica,[21] may be added to the solution to increase its viscosity. This practice improves the appearance of an ECD because the coloration develops at different rates in different areas in a fast device (see end of Section 13.4), hence artificially slowing the rate of coloration helps ensure an even coloration intensity. Thickening also improves the safety of a device should breakage occur, and helps minimise mass transport by convection (Section 3.3). Gelling the electrolyte, e.g. by adding a polyether such as PEO, is claimed to enhance the electrochemical stability.[22]

In type-III systems, while the electrolyte holds no soluble electrochrome, it now enacts two roles (see Chapter 3). Firstly, during coloration and bleaching, for electroneutrality it supplies the mobile counter ions that enter and leave the facing solid-electrochrome layers. However, secondly, the electrolyte still effects the accompanying conduction between the electrodes. Quite neglecting the latter, however, the electrolyte layer is called by some an 'ion-storage (IS) layer', which represents only the former action. Thus an 'ion-storage layer' and an 'electrolyte layer' are by no means equivalent terms. Better (but possibly too late and too long) is an inclusive term such as 'ionogenic electrolyte layer'; or – shorter – 'ion-supplying layer', which at least allows of both roles.

Type-III ECDs operating with protons as the mobile ions can contain aqueous acids. In Deb's ECD,[23] for example, the electrolyte was aqueous sulfuric acid of concentration $0.1\,\mathrm{mol\,dm^{-3}}$. Liquid acids are rarely used today owing to their tendency to degrade or dissolve electrochromes, and from safety considerations should the device leak. A majority of type-III ECDs now employ inorganic solids or viscoelastic organic polymers, the latter being flexible and resistant to mechanical shock. Solid organic acids of amorphous structure might serve similarly, although considerably higher potentials would be needed to drive any such ECD. They are apparently untested in this role, their electrical connectivity with electrochromes being critical. Ionic liquids somewhat below their solidification temperature might also serve but their ion-insertion capability could be questionable.

14.2.1 Inorganic and mixed-composition electrolytes

Many ECDs contain as electrolyte a thin layer of solid inorganic oxide; thin-film Ta_2O_5 is becoming widely used. Such layers are generally evaporated or sputtered. However, they are mechanically weak and cannot endure bending or mechanical shock. There may be a role here for mixed organic/inorganic solids like tetraalkylammonium salts with small inorganic anions, or alkali-metal salts containing large organic anions (provided that insertions only of the smaller ion are required); these might evince greater mechanical robustness. Like organic acids – previous paragraph – these also appear not to have been tried.

14.2.2 Organic electrolytes

Semi-solid organic electrolytes fall within two general categories: polyelectrolytes and polymer electrolytes, as described below.

Polyelectrolytes

Polyelectrolytes are polymers containing ion-labile moieties at regular intervals along the backbone. A popular example is poly(2-acrylamido-2-methylpropanesulfonic acid), 'poly(AMPS)', in which the proton-donor moiety is an acid. The molar ionic conductivity Λ of polymers such as poly(AMPS) depends critically on the extent of water incorporation; wholly dehydrated poly(AMPS) is not conductive, but Λ increases rapidly as the water content increases. Table 14.1 lists some polyelectrolytes used in solid-state ECDs.

14.2 Electrolyte layers for ECDs

Table 14.1. *Solid ion-conducting electrolytes for use in ECDs.*

Electrolyte	Refs.
Inorganic electrolytes	
$LiAlF_4$	24
$LiNbO_3$	25,26,27,28
Sb_2O_5 (inc. $HSbO_3$)	29,30,31
$HSbO_3$ based polymer	32
Ta_2O_5 (including 'TaO_x')	33,34,35,36,37,38,39,40
TiO_2 (including 'TiO_x')	40
$H_3UO_3(PO_4) \cdot 3H_2O$ ('HUP')	41
ZrO_2	42,43,44,45,46
Organic polymers	
Nafion^{TM}	47,48,49
Poly(acrylic acid)	50,51,52
Poly(AMPS)	47,53,54,55
Poly(methyl methacrylate), PMMA ('Perspex')	56,57,58,59,60,61,62,63,64, 65,66,67,68,69,70
Poly(2-hydroxyethyl methacrylate)	42,56,71,72
Poly(ethylene oxide), PEO	73,74,75,76,77,78,79,80,81,82,83
Poly(vinyl chloride), PVC	84,85

Polymer electrolytes

Polymer electrolytes contain, as solvent, neutral macromolecules such as poly(ethylene oxide) – PEO, poly(propylene glycol) – PPG, or poly(vinyl alcohol) – PVA. Added inert salt acts to form an inorganic electrolyte layer. Common examples include $LiClO_4$, triflic acid CF_3SO_3H, or H_3PO_4.

The viscosity of such polymers increases with increasing molecular weight, so polymers range from liquid, at low molecular weight, through to longer polymers which behave as rigid solids. Table 14.1 lists a selection of polymer electrolytes and polyelectrolytes used in solid-state ECDs.

It is quite common for polymeric electrolytes to have an opaque white 'filler' powder added, such as TiO_2 to enhance the contrast ratio in displays. A white layer also dispenses with any need to tailor the optical properties of the secondary layer. Thus, Duffy and co-workers[9] have described a device in which WO_3 forms both the primary and secondary electrodes, a device which could not show any observable change in colour unless the rear electrode was screened from view by incorporating such an opaque filler in the intervening electrolyte. The inclusion of particulate TiO_2 does not seem to affect the response times of such 'filled' ECDs, but the photocatalytic activity of TiO_2 may accelerate photolytic deterioration of organic materials such as the electrolyte.

The stability of electrolyte layers is discussed in Section 16.3.

14.3 Electrodes for ECD construction

All ECD devices require at least one transparent electrode. Devices operating in a transmissive mode, such as spectacles, goggles, visors or whole windows, must of course operate with a second OTE as the rear electrode, whereas devices operating in a reflective sense, as in information displays, do not. It is common but expensive for polished platinum to act as both mirror and supporting electrode in a reflecting ECD. Otherwise, the electrolyte-with-filler ploy (previous paragraph) is used.

Reviews of materials for OTE construction for electrochromic devices include 'Transparent conductors: a status review' by Chopra et al.[86] (in 1983), 'Transparent electronic conductors' by Lynam[87] (in 1990), 'Transparent conductive electrodes for electrochromic devices – a review' by Granqvist[88] (in 1993), 'Transparent and conducting ITO films: new developments and applications' by Granqvist and Hultåker[89] (in 2002), and 'Frontier of transparent oxide semiconductors' by Ohta et al.[90] (in 2003).

14.3.1 Transparent conductors

The most common choice of OTE is indium–tin oxide as a thin film sputtered onto glass. Another common choice is fluorine-doped tin oxide (FTO), an example being so-called 'K-glass™' from Pilkington, which comprises FTO on glass.[57,71,91,92,93] Its UV-visible absorption is less than 2% and its thermal infrared reflectance exceeds 90%.

Indium–tin oxide is electrically semiconducting rather than metallic. The relatively high innate resistance of semiconducting ITO (or other OTEs) can cause complications such as IR drop[94] and the so-called 'terminal effect'. As a consequence of IR drop, a gradient of potential forms across the electrode surface: the potential near the external contact is higher than elsewhere, so the electrochromic coloration or image formed during coloration is generated at different speeds across the electrode surface, and the intensity of colour will often be more intense near the external electrical contact, leading to a non-uniform image. Ho et al.[95] discuss such 'terminal effects' in ECDs.

The best conductivity of ITO is about 20 Ω per square; substrates of higher electronic conductivity are attainable, but are slightly yellow. Thus the conductivity of OTEs is relatively poor, considerably affecting ECD response times;[96] see p. 349.

Ways of combating *IR* drop and terminal effects involve increasing the electronic conductivity. Methods adopted include incorporating an ultra-thin layer of metallic nickel between the electrochrome and ITO,[97] or depositing an ultra-thin layer of precious metal on the electrolyte-facing side of the electrochrome.[98,99,100,101] Thin films of Cr_2O_3[102] or MgF_2[103,104] can also fulfil this goal.

The idea of flexible ECDs is attractive for lightweight, temporary electrochromic window coverings and the like.[105,106,107,108,109,110,111,112,113] Azens et al.[114] describe the fabrication and applications of such electrochromic 'foils'. Clearly any such device will need to be enclosed within thin sheets of an appropriate polymer. Furthermore, all the layers, including the conductive ITO and both electrochromes, must be durable, since any cracks formed by bending cause irreversible insulating discontinuities that lead to certain device failure. A review (1995) that addressed the use of polymeric substrates for electrochromic purposes is the short work by Antinucci et al.[78]

The deposition conditions must be milder when ITO is to be deposited onto polymeric substrates rather than on glass. Such deposition is now relatively easy but, nevertheless, the differing deposition conditions result in ITO layers with poorer electrical conductivity to that made on glass. Bertran et al.[115] overcame this problem by incorporating small amounts of silver within their ITO films, which is known to lower the electrical resistivity[116] albeit with a slight decrease in optical transmittance. The highest electrical conductivities were achieved in depositions using low Ar pressures of 0.4 Pa (without oxygen) and the relatively high power density of 2×10^4 W m^{-2}. Glass and polyester substrates exhibited different growth rates and samples deposited onto glass substrates showed better film-to-polymer adhesion. Nevertheless, ITO for counter-electrode use has been deposited on sheet plastics such as Mylar,[117] poly(ethyleneterephthalate) or PET[78,105,106,107,112,118] and polyester.[111,115,119] (Such flexible displays could also be *photo*-electrochromic.[110]) Several all-polymer ECDs have also been fabricated; see Section 10.5.

The stability of ITO electrodes is discussed in Section 16.2.

14.3.2 Opaque and metallic conductors

The most common choice of rear electrode is platinum or Pt-based alloys.[3,5,6,10] Other materials have also been advocated: Liu and Richardson[120] suggest an alloy of antimony and copper.

The second electrode need not bear a separate layer of electrochrome: redox-active counter electrodes can themselves 'absorb charge' with the

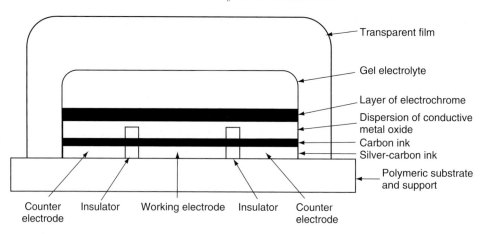

Figure 14.2 'Side-by-side' design of a screen-printed electrochromic display device: schematic representation illustrating the arrangement of the electrodes. (Figure redrawn from Liu, J. and Coleman, J. P. 'Nanostructured metal oxides for printed electrochromic displays'. *Mater. Sci. Eng. A*, **286**, 2000, 144–8, by permission of Elsevier Science.)

accompaniment of counter-ion intercalation. For example, ECDs have been constructed in which charge is intercalated into a counter electrode of carbon: examples of such counter electrodes include 'carbon'[29,121,122,123] or 'carbon-based' materials,[79,124] screen-printed carbon black,[125] and graphite.[126] All these counter electrodes remain black during electrochromic operation, and need therefore to remain hidden behind a layer of electrolyte containing an opaque white filler.

14.3.3 ECDs requiring no transparent conductor

Transparent conductors are not always needed. A novel design by Liu and Coleman[113] has recently been described which employs a 'side-by-side' structure. Ultrafine electrodes are screen-printed onto a non-conductive glass substrate, with electrochrome deposited above and between them; see Figure 14.2.

14.4 Device encapsulation

The process of assembling the components of a commercial device, and the mounting materials, are clearly as important as (in some views more important than) the operation of the parts taken individually.

In devices containing a liquid or semi-solid electrolyte, the separation between the two electrodes can be maintained by introducing flat or spherical 'spacers', acting in a similar manner to the minute spherical beads of constant diameter employed in fabricating an LCD, to maintain the precisely defined distance between the two parallel electrodes. For example, PPG Industries used this approach.[127,128]

Finally, the device must be sealed. In fact, the fabrication of a robust, leak-proof seal to encapsulate a type-I or -II ECD is not a trivial problem: Byker (at that time, of Gentex Corporation) recently stated, 'polymer sealant materials are often crucial to the life of an EC device, and may represent as big a R&D challenge as the EC system itself',[21] in bringing a device to commercial viability. One of the principal problems is chemical durability; a second is the hydrostatic pressures that form in large devices containing liquid electrolytes, since the weight of liquid causes the bottom of the device to swell, yet can push the top of the panes together till they break. Byker believes that all-solid-state systems also require an elastomeric polymer seal.[21] He discusses the use of polymers as electrolytes within ECDs in ref. 21. To these ends, PPG employed an adhesive layer to coat the edges of their devices,[127,128] and Gentex designed a complicated type of clip,[129] to withstand hydrostatic pressures.

The sealant around a device must be chemically stable. It is regrettable – but perhaps inevitable in view of industrial competitiveness – how many reports of actual devices (prototype and in production) fail to divulge details of device encapsulation. Of the few mentioned in the literature, Syrrakou et al.[57] employed an acetate silicone material; and the 'electric paint' displays made by Edwards and co-workers[130] at Uppsala University are encapsulated with the DuPont thermoplastic, Surlyn. This latter polymer performs the role 'reasonably well'.[121]

References

1. Baucke, F. G. K. Electrochromic applications. *Mater. Sci. Eng. B*, **10**, 1991, 285–92.
2. Baucke, F. G. K. Electrochromic mirrors with variable reflectance. *Rivista della Staz. Sper. Vetro*, **6**, 1986, 119–22.
3. Baucke, F. G. K. Electrochromic mirrors with variable reflectance. *Sol. Energy Mater.*, **16**, 1987, 67–77.
4. Baucke, F. G. K. Reflecting electrochromic devices – construction, operation and application. *Proc. Electrochem. Soc.*, **20–4**, 1990, 298–311.
5. Baucke, F. G. K., Bange, K. and Gambke, T. Reflecting electrochromic devices. *Displays*, **9**, 1988, 179–87.
6. Baucke, F. G. K. Beat the dazzlers. *Schott Information*, **1**, 1983, 11–13.
7. Baucke, F. G. K. Reflectance control of automotive mirrors. *Proc. SPIE*, **IS4**, 1990, 518–38.
8. Baucke, F. G. K. and Duffy, J. A. Darkening glass by electricity. *Chem. Br.*, **21**, 1985, 643–6 and 653.

9. Baucke, F. G. K., Duffy, J. A. and Smith, R. I. Optical absorption of tungsten bronze thin films for electrochromic applications. *Thin Solid Films*, **186**, 1990, 47–51.
10. Baucke, F. G. K. and Gambke, T. Electrochromic materials for optical switching devices. *Adv. Mater.*, **2**, 1990, 10–16.
11. Ashrit, P. V. Dry lithiation study of nanocrystalline, polycrystalline and amorphous tungsten trioxide thin-films. *Thin Solid Films*, **385**, 2001, 81–8.
12. Ashrit, P. V., Benaissa, K., Bader, G., Girouard, F. E. and Truong, V.-V. Lithiation studies on some transition metal oxides for an all-solid thin film electrochromic system. *Solid State Ionics*, **59**, 1993, 47–57.
13. Yonghong, Y., Jiayu, Z., Peifu, G. and Jinfa, T. Study on the WO_3 dry lithiation for all-solid-state electrochromic devices. *Sol. Energy Mater. Sol. Cells*, **46**, 1997, 349–55.
14. Yonghong, Y., Jiayu, Z., Peifu, G. and Jinfa, T. Study on the dry lithiation of WO_3 films. *Acta Energiae Solaris Sinica*, **19**, 1998, 371–375 [in Chinese]; as cited at www.engineering village 372.org (accessed 16 December 2004).
15. Ashrit, P. V. Structure dependent electrochromic behaviour of WO_3 thin films under dry lithiation. *Proc. SPIE*, **3789**, 1999, 158–69.
16. Taj, A. and Ashrit, P. V. Dry lithiation of nanostructured sputter deposited molybdenum oxide thin films. *J. Mater. Sci.*, **39**, 2004, 3541–4.
17. Azens, A. and Granqvist, C. G. Electrochromic smart windows: energy efficiency. *J. Solid State Electrochem.*, **7**, 2003, 64–8.
18. Azens, A., Kullman, L. and Granqvist, C. G. Ozone coloration of Ni and Cr oxide films. *Sol. Energy Mater. Sol. Cells*, **76**, 2003, 147–53.
19. Linford, R. G. Electrical and electrochemical properties of ion conducting polymers. In Scrosati, B. (ed.), *Applications of Electroactive Polymers*, London, Chapman and Hall, 1993, pp. 1–28.
20. Livage, J. and Ganguli, D. Sol–gel electrochromic coatings and devices: a review. *Sol. Energy Mater. Sol. Cells*, **68**, 2001, 365–81.
21. Byker, H. J. Electrochromics and polymers. *Electrochim. Acta*, **46**, 2001, 2015–22.
22. Mitsui Chemicals Inc. Ion conductive macromolecular gel electrolyte and solid battery using ion-conductive macromolecular gel electrolyte. Japanese Patent 2000-207934-A, 2000.
23. Deb, S. K. Optical and photoelectric properties and colour centres in thin films of tungsten oxide. *Philos. Mag.*, **27**, 1973, 801–22.
24. Oi, T., Miyake, K. and Uehara, K. Electrochromism of $WO_3/LiAlF_4/LiIn$ thin-film overlayers. *J. Appl. Phys.*, **53**, 1982, 1823.
25. Goldner, R. B., Haas, T., Seward, G., Wong, G., Norton, P., Foley, G., Berera, G., Wei, G., Schulz, S. and Chapman, R. Thin film solid state ionic materials for electrochromic smart window^{TM} glass. *Solid State Ionics*, **28–30**, 1988, 1715–21.
26. Goldner, R. B. Electrochromic smart window^{TM} glass. In Chowdari, B. V. R. and Radhakrishna, S. (eds.), *Proceedings of the International Seminar on Solid State Ionic Devices*, Singapore, World Publishing Co., 1988, pp. 379–89.
27. Goldner, R. B., Arntz, F. O., Berera, G., Haas, T. E., Wei, G., Wong, K. K. and Yu, P. C. A monolithic thin-film electrochromic window. *Solid State Ionics*, **53–6**, 1992, 617–27.
28. Goldner, R. B., Arntz, F. O., Dickson, K., Goldner, M. A., Haas, T. E., Liu, T. Y., Slaven, S., Wei, G., Wong, K. K. and Zerigian, P. Some lessons learned from research on a thin film electrochromic window. *Solid State Ionics*, **70–1**, 1994, 613–18.
29. Kuwabara, K. and Noda, Y. Potential wave-form measurements of an electrochromic device, $WO_3/Sb_2O_5/C$, at coloration–bleaching processes using a new quasi-reference electrode. *Solid State Ionics*, **61**, 1993, 303–8.

30. Vaivars, G., Kleperis, J. and Lusis, A. Antimonic acid hydrate xerogels as proton electrolytes. *Solid State Ionics*, **61**, 1993, 317–21.
31. Lusis, A. Solid state ionics and optical materials technology for energy efficiency, solar energy conversion and environmental control. *Proc. SPIE*, **1536**, 1991, 116–24.
32. Granqvist, C. G., Azens, A., Hjelm, A., Kullman, L., Niklasson, G. A., Rönnow, D., Strømme Mattson, M., Veszelei, M. and Vaivars, G. Recent advances in electrochromics for smart windows applications. *Sol. Energy*, **63**, 1998, 199–216.
33. Corbella, C., Vives, M., Pinyol, A. *et al.* Influence of the porosity of RF sputtered Ta_2O_5 thin films on their optical properties for electrochromic applications. *Solid State Ionics*, **165**, 2003, 15–22.
34. Hutchins, M. G., Butt, N. S., Topping, A. J., Porqueras, I., Person, C. and Bertran, E. Tantalum oxide thin film ionic conductors for monolithic electrochromic devices. *Proc. SPIE*, **4458**, 2001, 120–7.
35. Kitao, M., Akram, H., Machida, H. and Urabe, K. Ta_2O_5 electrolyte films and solid-state EC cells. *Proc. SPIE*, **1728**, 1992, 165–72.
36. Kitao, M., Akram, H., Urabe, K. and Yamada, S. Properties of solid-state electrochromic cells using Ta_2O_5 electrolyte. *J. Electron. Mater.*, **21**, 1992, 419–22.
37. Klingler, M., Chu, W. F. and Weppner, W. Three-layer electrochromic system. *Sol. Energy Mater. Sol. Cells*, **39**, 1995, 247–55.
38. Özer, N., He, Y. and Lampert, C. M. Ionic conductivity of tantalum oxide films prepared by sol–gel process for electrochromic devices. *Proc. SPIE*, **2255**, 1994, 456–66.
39. Sone, Y., Kishimoto, A. and Kudo, T. Amorphous tantalum oxide proton conductor derived from peroxo-polyacid and its application for EC device. *Solid State Ionics*, **70–1**, 1994, 316–20.
40. Cantao, M. P., Laurenco, A., Gorenstein, A., Córdoba de Torresi, S. I. and Torresi, R. M. Inorganic oxide solid state electrochromic devices. *Mater. Sci. Eng. B*, **26**, 1994, 157–61.
41. Howe, A. T., Sheffield, S. H., Childs, P. E. and Shilton, M. G. Fabrication of films of hydrogen uranyl phosphate tetrahydrate and their use as solid electrolytes in electrochromic displays. *Thin Solid Films*, **67**, 1980, 365–70.
42. Azens, A., Kullman, L., Vaivars, G., Nordborg, H. and Granqvist, C. G. Sputter-deposited nickel oxide for electrochromic applications. *Solid State Ionics*, **113–15**, 1998, 449–56.
43. Larsson, A.-L. and Niklasson, G. A. Infrared emittance modulation of all-thin-film electrochromic devices. *Mater. Lett.*, **58**, 2004, 2517–20.
44. Larsson, A.-L. and Niklasson, G. A. Optical properties of electrochromic all-solid-state devices. *Sol. Energy Mater. Sol. Cells*, **84**, 2004, 351–60.
45. van der Sluis, P. and Mercier, V. M. M. Solid state Gd–Mg electrochromic devices with ZrO_2H_x electrolyte. *Electrochim. Acta*, **46**, 2001, 2167–71.
46. Mercier, V. M. M. and van der Sluis, P. Toward solid-state switchable mirrors using a zirconium oxide proton conductor. *Solid State Ionics*, **145**, 2001, 17–24.
47. Randin, J.-P. Ion-containing polymers as semisolid electrolytes in WO_3-based electrochromic devices. *J. Electrochem. Soc.*, **129**, 1982, 1215–20.
48. Kim, E., Rhee, S. B., Shin, J.-S., Lee, K.-Y. and Lee, M.-H. All solid-state electrochromic window based on poly(aniline N-butylsulfonate)s. *Synth. Met.*, **85**, 1997, 1367–8.
49. Pennisi, A. and Simone, F. An electrochromic device working in absence of ion storage counter-electrode. *Sol. Energy Mater. Sol. Cells*, **39**, 1995, 333–40.

50. Choy, J.-H., Kim, Y.-I., Kim, B.-W., Campet, G., Portier, J. and Huong, P. V. Grafting mechanism of electrochromic PAA–WO$_3$ composite film. *J. Solid State Chem.*, **142**, 1999, 368–73.
51. Choy, J.-H., Kim, Y.-I., Park, N.-G., Campet, G. and Grenier, J.-C. New solution route to poly(acrylic acid)/WO$_3$ hybrid film. *Chem. Mater.*, **12**, 2000, 2950–6.
52. Ohno, H. and Yamazaki, H. Preparation and characteristics of all solid-state electrochromic display with cation-conductive polymer electrolytes. *Solid State Ionics*, **59**, 1993, 217–22.
53. Randin, J.-P. Chemical and electrochemical stability of WO$_3$ electrochromic films in liquid electrolytes. *J. Electron. Mater.*, **7**, 1978, 47–63.
54. Monk, P. M. S., Turner, C. and Akhtar, S. P. Electrochemical behaviour of methyl viologen in a matrix of paper. *Electrochim. Acta*, **44**, 1999, 4817–26.
55. Zukowska, G., Williams, J., Stevens, J. R., Jeffrey, K. R., Lewera, A. and Kulesza, P. J. The application of acrylic monomers with acidic groups to the synthesis of proton-conducting polymer gels. *Solid State Ionics*, **167**, 2004, 123–30.
56. Inaba, H., Iwaku, M., Nakase, K., Yasukawa, H., Seo, I. and Oyama, N. Electrochromic display device of tungsten trioxide and Prussian blue films using polymer gel electrolyte of methacrylate. *Electrochim. Acta*, **40**, 1995, 227–32.
57. Syrrakou, E., Papaefthimiou, S. and Yianoulis, P. Environmental assessment of electrochromic glazing production. *Sol. Energy Mater. Sol. Cells*, **85**, 2005, 205–40.
58. Nishikawa, M., Ohno, H., Kobayashi, T., Tsuchida, E. and Hirohashi, R. All solid-state electrochromic device containing poly[oligo(oxyethylene) methylmethacrylate]/LiClO$_4$ hybrid polymer ion conductor. *J. Soc. Photogr. Sci. Technol. Jpn.*, **81**, 1988, 184–90 [in Japanese].
59. Bohnke, O., Frand, G., Rezrazi, M., Rousselot, C. and Truche, C. Fast ion transport in new lithium electrolytes gelled with PMMA, 1: influence of polymer concentration. *Solid State Ionics*, **66**, 1993, 97–104.
60. Deepa, M., Sharma, N., Agnihotry, S. A., Singh, S., Lal, T. and Chandra, R. Conductivity and viscosity of liquid and gel electrolytes based on LiClO$_4$, LiN(CF$_3$SO$_2$)$_2$ and PMMA. *Solid State Ionics*, **152–3**, 2002, 253–8.
61. Stevens, J. R., Such, K., Cho, N. and Wieczorek, W. Polyether-PMMA adhesive electrolytes for electrochromic applications. *Sol. Energy Mater. Sol. Cells*, **39**, 1995, 223–37.
62. Su, L., Fang, J., Xiao, Z. and Lu, Z. An all-solid-state electrochromic display device of Prussian blue and WO$_3$ particulate film with a PMMA gel electrolyte. *Thin Solid Films*, **306**, 1997, 133–6.
63. Su, L., Lu, Z. and Xiao, Z. All solid-state electrochromic device with PMMA gel electrolyte. *Mater. Chem. Phys.*, **52**, 1998, 180–3.
64. Tsutsumi, N., Ueda, Y. and Kiyotsukuri, T. Measurement of the internal electric field in a poly(vinylidene fluoride)/poly(methyl methacrylate) blend. *Polymer*, **33**, 1992, 3305–7.
65. Vondrak, J., Reiter, J., Velicka, J. and Sedlarikova, M. PMMA-based aprotic gel electrolytes. *Solid State Ionics*, **170**, 2004, 79–82.
66. Rauh, R. D., Wang, F., Reynolds, J. R. and Meeker, D. L. High coloration efficiency electrochromics and their application to multi-color devices. *Electrochim. Acta*, **46**, 2001, 2023–9.
67. Reynolds, J. R., Kumar, A., Reddinger, J. L., Sankaran, B., Sapp, S. A. and Sotzing, G. A. Unique variable-gap polyheterocycles for high-contrast dual polymer electrochromic devices. *Synth. Met.*, **85**, 1997, 1295–8.

68. Sönmez, G., Schwendeman, I., Schottland, P., Zong, K. and Reynolds, J. R. N-Substituted poly(3,4-propylenedioxypyrrole)s: high gap and low redox potential switching electroactive and electrochromic polymers. *Macromolecules*, **36**, 2003, 639–47.
69. Sotzing, G. A., Reddinger, J. L., Reynolds, J. R. and Steel, P. J. Redox active electrochromic polymers from low oxidation monomers containing 3,4-ethylenedioxythiophene (EDOT). *Synth. Met.*, **84**, 1997, 199–201.
70. Welsh, D. M., Kumar, A., Morvant, M. C. and Reynolds, J. R. Fast electrochromic polymers based on new poly(3,4-alkylenedioxythiophene) derivatives. *Synth. Met.*, **102**, 1999, 967–8.
71. Pennisi, A., Simone, F., Barletta, G., Di Marco, G. and Lanza, M. Preliminary test of a large electrochromic window. *Electrochim. Acta*, **44**, 1999, 3237–43.
72. Varshney, P., Deepa, M., Agnihotry, S. A. and Ho, K. C. Photo-polymerized films of lithium ion conducting solid polymer electrolyte for electrochromic windows (ECWs). *Sol. Energy Mater. Sol. Cells*, **79**, 2003, 449–58.
73. Pedone, P., Armand, M. and Deroo, D. Voltammetric and potentiostatic studies of the interface WO_3/polyethylene oxide–H_3PO_4. *Solid State Ionics*, **28–30**, 1988, 1729–32.
74. Agnihotry, S. A., Ahmad, S., Gupta, D. and Ahmad, S. Composite gel electrolytes based on poly(methylmethacrylate) and hydrophilic fumed silica. *Electrochim. Acta*, **49**, 2004, 2343–9.
75. Agnihotry, S. A., Nidhi, P. and Sekhon, S. S. Li^+ conducting gel electrolyte for electrochromic windows. *Solid State Ionics*, **136–7**, 2000, 573–6.
76. Aliev, A. E. and Shin, H. W. Image diffusion and cross-talk in passive matrix electrochromic displays. *Displays*, **23**, 2002, 239–47.
77. Andrei, M., Roggero, A., Marchese, L. and Passerini, S. Highly conductive solid polymer electrolyte for smart windows. *Polymer*, **35**, 1994, 3592–7.
78. Antinucci, M., Chevalier, B. and Ferriolo, A. Development and characterisation of electrochromic devices on polymeric substrates. *Sol. Energy Mater. Sol. Cells*, **39**, 1995, 271–87.
79. Asano, T., Kubo, T. and Nishikitani, Y. Durability of electrochromic windows fabricated with carbon-based counterelectrode. *Proc. SPIE*, **3788**, 1999, 84–92.
80. Kuwabara, K., Sugiyama, K. and Ohno, M. All-solid-state electrochromic device, 1: electrophoretic deposition film of proton conductive solid electrolyte. *Solid State Ionics*, **44**, 1991, 313–18.
81. Kuwabara, K., Ohno, M. and Sugiyama, K. All-solid-state electrochromic device, 2: characterization of transition-metal oxide thin films for counter electrode. *Solid State Ionics*, **44**, 1991, 319–23.
82. Nishio, K. and Tsuchiya, T. Electrochromic thin films prepared by sol–gel process. *Sol. Energy Mater. Sol. Cells*, **68**, 2001, 279–93.
83. Scrosati, B. Ion conducting polymers and related electrochromic devices. *Mol. Cryst. Liq. Cryst.*, **190**, 1990, 161–70.
84. Lianyong, S., Hong, W. and Zuhong, L. All solid-state electrochromic smart window of electrodeposited WO_3 and Prussian blue film with PVC gel electrolyte. *Supramol. Sci.*, **5**, 1998, 657–9.
85. Su, L., Xiao, Z. and Lu, Z. All solid-state electrochromic window of electrodeposited WO_3 and prussian blue film with PVC gel electrolyte. *Thin Solid Films*, **320**, 1998, 285–9.
86. Chopra, K. L., Major, S. and Pandya, D. K. Transparent conductors: a status review. *Thin Solid Films*, **102**, 1983, 1–46.

87. Lynam, N. R. Transparent electronic conductors. *Proc. Electrochem. Soc.*, **90–2**, 1990, 201–31.
88. Granqvist, C. G. Transparent conductive electrodes for electrochromic devices – a review. *Appl. Phys. A*, **57**, 1993, 19–24.
89. Granqvist, C. G. and Hultåker, A. Transparent and conducting ITO films: new developments and applications. *Thin Solid Films*, **411**, 2002, 1–5.
90. Ohta, H., Nomura, K., Hiramatsu, H., Ueda, K., Kamiya, T., Hirano, M. and Hosono, H. Frontier of transparent oxide semiconductors. *Solid-State Electron.*, **47**, 2003, 2261–7.
91. Di Marco, G., Lanza, M., Pennisi, A. and Simone, F. Solid state electrochromic device: behaviour of different salts on its performance, *Solid State Ionics*, **127**, 2000, 23–9.
92. Papaefthimiou, S., Leftheriotis, G. and Yianoulis, P. Study of WO_3 films with textured surfaces for improved electrochromic performance. *Solid State Ionics*, **139**, 2001, 135–44.
93. Vroon, Z. A. E. P. and Spee, C. I. M. A. Sol–gel coatings on large area glass sheets for electrochromic devices. *J. Non-Cryst. Solids*, **218**, 1997, 189–95.
94. Michalak, F. M. and Owen, J. R. Parasitic currents in electrochromic devices. *Solid State Ionics*, **86–8**, 1996, 965–70.
95. Ho, K.-C., Singleton, D. E. and Greenberg, C. B. Effect of cell size on the performance of electrochromic windows. *Proc. Electrochem. Soc.*, **90–2**, 1990, 349–64.
96. Nagai, J., Kamimori, T. and Mizuhashi, M. Transmissive electrochromic device. *Proc. SPIE*, **562**, 1985, 39–45.
97. Jeong, D. J., Kim, W.-S. and Sung, Y. E. Improved electrochromic response time of nickel hydroxide thin films by ultra-thin nickel metal underlayer. *Jpn. J. Appl. Phys.*, **40**, 2001, L708–10.
98. He, T., Ma, Y., Cao, Y., Yang, W. and Yao, J. Enhanced electrochromism of WO_3 thin film by gold nanoparticles. *J. Electroanal. Chem.*, **514**, 2001, 129–32.
99. Yao, J. N., Yang, Y. A. and Loo, B. H. Enhancement of photochromism and electrochromism in MoO_3/Au and MoO_3/Pt thin films. *J. Phys. Chem. B*, **102**, 1998, 1856–60.
100. Haranahalli, A. R. and Holloway, P. H. The influence of metal overlayers on electrochromic behavior of tungsten trioxide films. *J. Electronic Mater.*, **10**, 1981, 141–72.
101. Haranahalli, A. R. and Dove, D. B. Influence of a thin gold surface layer on the electrochromic behavior of WO_3 films. *Appl. Phys. Lett.*, **36**, 1980, 791–3.
102. Inoue, E., Kawaziri, K. and Izawa, A. Deposited Cr_2O_3 as a barrier in a solid-state WO_3 electrochromic cell. *Jpn. J. Appl. Phys.*, **16**, 1977, 2065–6.
103. Stocker, R. J., Singh, S., van Uitert, L. G. and Zydzik, G. J. Efficiency and humidity dependence of WO_3–insulator electrochromic display structures. *J. Appl. Phys.*, **50**, 1979, 2993–4.
104. Yoshimura, T., Watanabe, M., Kiyota, K. and Tanaka, M. Electrolysis in electrochromic device consisting of WO_3 and MgF_2 thin films. *Jpn. J. Appl. Phys.*, **21**, 1982, 128–32.
105. Michalak, F. and Aldebert, P. A flexible electrochromic device based on colloidal tungsten oxide and polyaniline. *Solid State Ionics*, **85**, 1996, 265–72.
106. Bessière, A., Badot, J.-C., Certiat, M.-C., Livage, J., Lucas, V. and Baffier, N. Sol–gel deposition of electrochromic WO_3 thin film on flexible ITO/PET substrate. *Electrochim. Acta*, **46**, 2001, 2251–6.

107. Bessière, A., Duhamel, C., Badot, J.-C., Lucas, V. and Certiat, M.-C. Study and optimization of a flexible electrochromic device based on polyaniline. *Electrochim. Acta*, **49**, 2004, 2051–5.
108. Coleman, J. P., Lynch, A. T., Madhukar, P. and Wagenknecht, J. H. Printed, flexible electrochromic displays using interdigitated electrodes. *Sol. Energy Mater. Sol. Cells*, **56**, 1999, 395–418.
109. Mecerreyes, D., Marcilla, R., Ochoteco, E., Grande, H., Pomposo, J. A., Vergaz, R. and Sarchez Pena, J. M. A simplified all-polymer flexible electrochromic device. *Electrochim. Acta*, **49**, 2004, 3555–9.
110. Pichot, F., Ferrere, S., Pitts, J. R. and Gregg, B. A. Flexible photoelectrochromic windows. *J. Electrochem. Soc.*, **146**, 1999, 4324–6.
111. Azens, A., Gustavsson, G., Karmhag, R. and Granqvist, C. G. Electrochromic devices on polyester foil. *Solid State Ionics*, **165**, 2003, 1–5.
112. De Paoli, M.-A., Nogueira, A. F., Machado, D. A. and Longo, C. All-polymeric electrochromic and photoelectrochemical devices: new advances. *Electrochim. Acta*, **46**, 2001, 4243–9.
113. Liu, J. and Coleman, J. P. Nanostructured metal oxides for printed electrochromic displays. *Mater. Sci. Eng. A*, **286**, 2000, 144–8.
114. Azens, A., Avendaño, E., Backholm, J., Berggren, L., Gustavsson, G., Karmhag, R., Niklasson, G. A., Roos, A. and Granqvist, C. G. Flexible foils with electrochromic coatings: science, technology and applications. *Sol. Energy Mater. Sol. Cells*, **119**, 2005, 214–23.
115. Bertran, E., Corbella, C., Vives, M., Pinyol, A., Person, C. and Porqueras, I. RF sputtering deposition of Ag/ITO coatings at room temperature. *Solid State Ionics*, **165**, 2003, 139–48.
116. Hultåker, A., Jarrendahl, K., Lu, J., Granqvist, C. G. and Niklasson, G. A. Electrical and optical properties of sputter deposited tin doped indium oxide thin films with silver additive. *Thin Solid Films*, **392**, 2001, 305–10.
117. Brotherston, I. D., Mudigonda, D. S. K., Osborn, J. M., Belk, J., Chen, J., Loveday, D. C., Boehme, J. L., Ferraris, J. P. and Meeker, D. L. Tailoring the electrochromic properties of devices via polymer blends, copolymers, laminates and patterns. *Electrochim. Acta*, **44**, 1999, 2993–3004.
118. Yu, P. C., Backfisch, D. L., Slobodnik, J. B. and Rukavina, T. G., PPG Industries Ohio, Inc. Fabrication of electrochromic device with plastic substrates. US Patent 06136161, 2000.
119. Rousselot, C., Gillet, P. A. and Bohnke, O. Electrochromic thin films deposited onto polyester substrates. *Thin Solid Films*, **204**, 1991, 123–31.
120. Liu, G. and Richardson, T. J. Sb–Cu–Li electrochromic mirrors. *Sol. Energy Mater. Sol. Cells*, **86**, 2005, 113–21.
121. Edwards, M. O. M., Andersson, M., Gruszecki, T., Petterson, H., Thunman, R., Thuraisingham, G., Vestling, L. and Hagfeldt, A. Charge–discharge kinetics of electric-paint displays. *J. Electroanal. Chem.*, **565**, 2004, 175–84.
122. Edwards, M. O. M., Boschloo, G., Gruszecki, T., Petterson, H., Sohlberg, R. and Hagfeldt, A. 'Electric-paint displays' with carbon counter electrodes. *Electrochim. Acta*, **46**, 2001, 2187–93.
123. Edwards, M. O. M., Gruszecki, T., Pettersson, H., Thuraisingham, G. and Hagfeldt, A. A semi-empirical model for the charging and discharging of electric-paint displays. *Electrochem. Commun.*, **4**, 2002, 963–7.

124. Nishikitani, Y., Asano, T., Uchida, S. and Kubo, T. Thermal and optical behavior of electrochromic windows fabricated with carbon-based counterelectrode. *Electrochim. Acta*, **44**, 1999, 3211–17.
125. Wang, J., Tian, B. M., Nascomento, V. B. and Angnes, L. Performance of screen-printed carbon electrodes fabricated from different carbon inks. *Electrochim. Acta*, **43**, 1998, 3459–65.
126. Yu, P., Popov, B. N., Ritter, J. A. and White, R. E. Determination of the lithium ion diffusion coefficient in graphite. *J. Electrochem. Soc.*, **146**, 1999, 8–14.
127. Backfisch, D. L., PPG Industries Ohio, Inc. Method for laminating a composite device. US Patent 06033518, 2000.
128. Backfisch, D. L., PPG Industries Ohio, Inc. Method for sealing a laminated electrochromic device edge. US Patent 05969847, 2000.
129. Tonar, W. L., Bauer, F. T., Bostwick, D. J. and Stray, J. A., Gentex Corporation. Clip for use with transparent conductive electrodes in electrochromic devices. US Patent 06064509, 2000.
130. Pettersson, H., Gruszecki, T., Johansson, L.-H., Edwards, M. O. M., Hagfeldt, A. and Matuszczyk, T. Direct-driven electrochromic displays based on nanocrystalline electrodes. *Displays*, **25**, 2004, 223–30.

15
Photoelectrochromism

15.1 Introduction

Systems that change colour electrochemically, but only on being illuminated, are termed *photoelectrochromic* (*cf. electrochromic* or *photochromic* when only one of these stimuli is applied). Relatively few photoelectrochromic systems have been examined as such, although in some studies of photoelectrochemistry, colour changes are mentioned; see refs. 1,2,3. One study calls such devices 'user controllable photochromic devices'.[4]

Few reviews of the topic are extant: the chapter on photoelectrochromism in our 1995 book[5] is dated, but still the most comprehensive. Others include 'Photoelectrochromic cells and their applications' by Gregg (of NREL in Colorado)[6] in 1997, and 'All-polymeric electrochromic and photoelectrochemical devices: new advances' by De Paoli *et al.*[7] in 2001.

Two bases of photoelectrochromic operation are available. In the first, the potential required to evoke electrochromism is already applied but can act only through a photo-activated switch, filter or trigger. A separate photoconductor or other photocell serves as a switch, or the actual electrochromic electrode surface itself could be a photoconductor, or sandwiched together with a photoconductor. Such *photo-activated* systems contrast with *photo-driven* devices, in which illumination of one or other part of the circuit produces the photovoltaic potential required to drive the electrochromic current.

15.2 Direction of beam

The direction of illumination during cell operation is important. If the incident beam traverses a (minimum) distance in the cell prior to striking the photo-active layer, then illumination is said to be 'front-wall',[8] as shown by arrow (a) in Figure 15.1. Conversely 'back-wall' illumination, arrow (b), Figure 15.1,

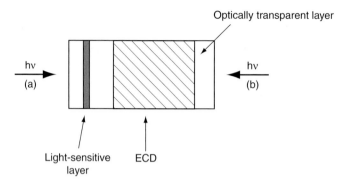

Figure 15.1 Schematic representation of a photoelectrochromic cell. Illumination from direction (a) represents 'front-wall' illumination and (b) 'back-wall' illumination.

operates with the beam directed from behind the cell, so traversing more cell material before reaching the photosensitive layer. Front-wall illumination generally yields superior results since additional absorptions by other layers within the ECD are minimised. Back-wall illumination is used only if undesirable photolytic processes occur with front-wall illumination of the cell.

15.3 Device types

15.3.1 Devices acting in tandem with a photocell

The simplest circuits for photoelectrochromic device operation comprise a conventional *electrically* driven ECD together with a photo-operated switch. The switch operates by illumination of a suitable photocell, be it photovoltaic or photoconductive, which triggers a microprocessor or similar element which in turn switches on the already 'poised' cell.

Such an arrangement is not intrinsically photoelectrochromic but is switched on by photocontrolled circuitry: the cell itself could be any straightforward electrochromic system.

15.3.2 Photoconductive layers

Photoconductive materials are insulators in the absence of light but become conductive when illuminated. Such photoconductors were traditionally semiconductors like amorphous silicon but, in recent years, many organic photoconductors have become candidates, as below. The mechanism of photoconduction involves the photo-excitation of charge carriers (electrons or holes) from localised sites, or from bonds in the valence band, into the delocalised energy levels

15.3 Device types

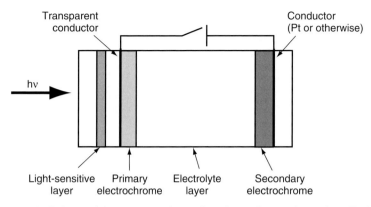

Figure 15.2 Schematic representation of a photoelectrochromic cell: front-wall illumination of an ECD containing a photoconductive layer between the transparent conductor and the primary electrochrome layer.

forming the conduction band. The mobilised charges can be driven by an externally applied potential,[9] yielding a current that can effect electrochromism.

Electrochromic cells may employ a layer of photoconductive material in one of two ways.[10,11] In the first, a photoconductive component is positioned *outside* the ECD and acts as a photocell switch: illumination of the photoconductor completes the circuit, allowing for electrochromic coloration. Current ceases in the dark, so coloration stops. In the second arrangement, a photoconductive layer is incorporated *within* the electrochromic cell. Figure 15.2 shows an ECD with a photoconductor (light-sensitive layer) positioned between an optically conducting substrate and a film of electrochrome. During electrochromic coloration or bleaching, ions from the electrolyte enter the electrochromic layer as in normal operation (see Section 1.4 on page 11), but electrons enter via the photoconductor. This arrangement has the difficulty that, since most photoconductors are somewhat opaque, ECDs operating with a photoconductor will probably have to operate in a reflective mode. Back-wall illumination of the ECD in Figure 15.2 would allow for strong, metallic electrodes to be employed as the photoconductor support. A few photoelectrochromic devices have been fabricated with *semi-transparent* photoconductors.[12,13]

In a variation of this latter arrangement, the photoconductor might conceivably be located *between* the electrochrome and the electrolyte layers (Figure 15.3).[10,14] Here the photoconductor would need to be completely ion-permeable, although note that the attendant physical stresses of continual ion movement through the photoconductor could lead to its eventual disintegration. Accordingly, the arrangement in Figure 15.2 is preferred.

Several workers[12,15,16,17] of the NREL laboratories in Colorado, made a photoelectrochromic device in which the photoconductor was a thin,

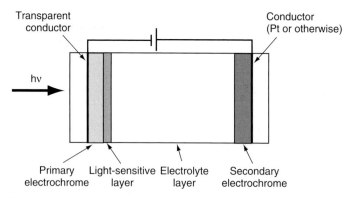

Figure 15.3 Schematic representation of a photoelectrochromic cell: front-wall illumination of an ECD containing a photoconductive layer between the primary electrochrome layer and the electrolyte.

semi-transparent layer of hydrogenated amorphous silicon. It yielded a photocurrent of 3.9 mA cm^{-2}, and an open-circuit potential V_{oc} of 0.92 V, which is deemed adequate to colour a lithium-based device with a response time τ of less than one minute. Their window covering could be produced on a flexible polymer substrate, allowing it to be affixed to the inside surface of a window, i.e. this represents a *photo*-electrochromic 'smart-glass' window (*cf.* Section 13.3); NREL called the device a 'stand-alone photovoltaic-powered electrochromic window'. The primary electrochrome was WO_3.

Photoelectrochromic 'writing' has been suggested by several authors; NREL made a photoelectrochromic prototype that could be bleached with a light pen:[18] they envisaged use in light-on-dark viewgraph projection or possibly within children's toys. The writing appeared light yellow on a black background. The photoconductor within the display was hydrogenated amorphous silicon carbide. The primary electrochrome was WO_3, with ion-conducting $LiAlF_4$ as the electrolyte, and Ni–W oxide as the counter electrode. Similarly, Yoneyama[19] labelled his device 'a photo-rewritable ... image'.

In fact, many intrinsically conducting polymers are photoconductive:[20] photoelectrochromic devices employing poly(aniline) as a photoconductor have been made by Fitzmaurice,[21] Hagen,[22] Ileperuma,[23] Kobayashi[24,25,26,27] and their co-workers. The electrochrome in Kobayashi's cell was methyl viologen[27] (*cf.* Chapter 11), with a variant of ruthenium tris(bipyridyl) as a photosensitiser.

Other polymer electrochromes than poly(aniline) have been used as photoconductive layers within photoelectrochromic devices: poly(pyrrole),[28] poly(*o*-methoxyaniline)[7] and the thiophene-based polymers poly(3-methylthiophene)[29] and PEDOT.[7]

Titanium dioxide (in its anatase allotrope) is one of the most intensely studied photo-active materials, and has been incorporated into many photo-electrochromic devices. For example, Hagen's et al.'s[22] photoelectrochromic device employed a nanocrystalline layer of TiO_2 as a photoconductor, in addition to poly(aniline), as above. The coloration process was photosensitised using a dye based on ruthenium tris(2,2'-bipyridine). Their 'self-powered' cell was able to modulate its transmission over the whole visible spectral region. (The illuminating lamps simulated solar spectral intensities.)

The photoactive TiO_2 need not be a continuous layer: in the device fabricated by Liao and Ho,[30] *particulate* titanium dioxide was the photoactive material; a ruthenium complex acted as a photosensitiser, and the I^-/I_3^- redox couple was incorporated as the electron mediator. The electrochrome was a thin layer of PEDOT polymer, yielding a device with an overall coloration efficiency η of $280 \, cm^2 \, C^{-1}$.

15.3.3 Photovoltaic materials

A photovoltaic material produces a potential when illuminated, from a process similar to the excitation of electrons within a photoconductor but with an internal rectifying field to provide a driving force on the electrons. The ionic charges needed to accompany the electrochromic transition enter the film from juxtaposed electrolyte or an electron mediator. The photovoltaic layer is not consumed in this process.

The photovoltage produced need not be large; indeed, its actual magnitude is not a problem because an external bias can be applied until the cell is 'poised'. Illumination of such a poised cell generates a photovoltage which, when supplementing the external bias, is sufficient to enable the coloration process to proceed, even if the photovoltage is itself too small to effect the required redox chemistry. For example, a cell comprising tungsten trioxide deposited on TiO_2 requires a bias[31] since the photovoltage generated is insufficient.

Prussian blue (PB) has also been used as the electrochrome in photoelectrochromic devices, with a photovoltage coming from polycrystalline n-type $SrTiO_3$,[32,33] TiO_2[34,35] or CdS[36] as the photolayer. (Indeed, PB has been used with WO_3 to make a photorechargeable battery.[36]) Other photoelectrochromic cells operating via photovoltaism include WO_3 on CdS,[14,37] GaAs,[38] GaP,[39] or on TiO_2,[40,41] Films of indium hexacyanometallate grown in a bath containing colloidal TiO_2 are also photoelectrochromic.[42,43]

Few monomeric organic systems claim photoelectrochromism, perhaps owing to their tendency to photodegrade. Among the few in the literature are Methylene

Blue[44] (**I**) and the spirobenzopyran[45] (**II**), both of which undergo reversible photoelectrochromic transitions at TiO_2 electrodes.

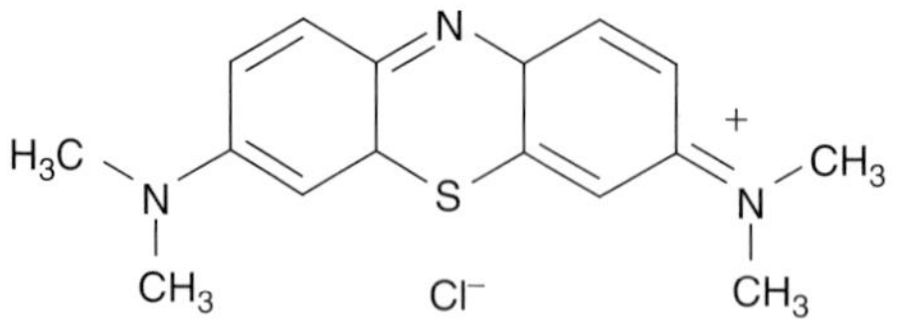

I

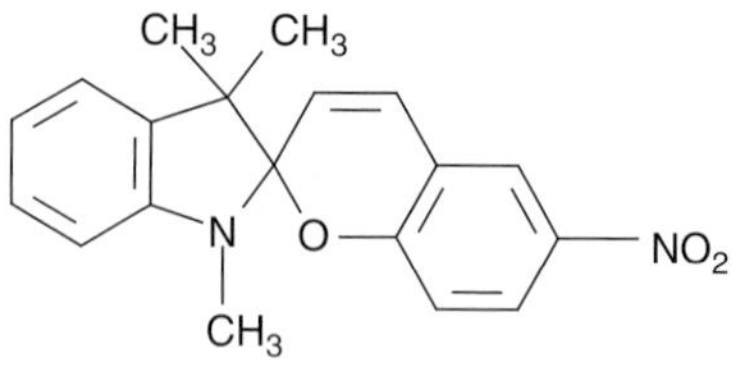

II

15.3.4 Photogalvanic materials

Photogalvanic materials generate current when illuminated. The photogalvanic material is generally consumed during the photoreaction[14] which inevitably causes the (photo-operated) write–erase efficiency to be poor.

Photoelectrochromism in the cell $WO_3|PEO, H_3PO_4$ (MeCN)$|V_2O_5$ is believed to operate in a photogalvanic sense[14] since the brown colour of the V_2O_5 layer disappears gradually during illumination. Curiously, the cell is still photoelectrochromic even after the colour of the V_2O_5 has gone and an alternative cathodic reaction (possibly catalysed consumption of oxygen, or reduction of VO_2?) must be envisaged.

15.4 Photochromic–electrochromic systems

Some systems are not photoelectrochromic in the sense defined above, yet do not function as electrochromic or photochromic alone. For example, De Filpo and co-workers devised 'photoelectrochromic systems' comprising either ethyl viologen[46] or Methylene Blue (**I**) in solution,[46,47] together with a suitable electron donor such as an amine. Irradiation e.g. with a He–Ne laser induces an electron-transfer process with concomitant formation of colour. The colour-forming process is straightforwardly photochromic. The colour may be erased *electro* chromically. We adopt the compound adjective 'photochromic–electrochromic' for those systems that colour and bleach via the alternate use of photochromism and electrochromism.

Yoneyama *et al.*[19,48] developed a photochromic–electrochromic cell functioning in the opposite sense to that of De Filpo's, so the colour bleached *photo* chromically and was regenerated *electro* chromically. Yoneyama's photo chromic–electrochromic device employed poly(aniline) as the colour-changing material. The polymer film contained entrapped particles of TiO_2, enabling the poly(aniline) to act as both photoconductor and colour-changing

15.4 Photochromic–electrochromic systems

material. The device was assembled with the polymer as one layer in a multi-layer 'sandwich'. Illumination effected photoreduction of the poly(aniline) with concomitant bleaching of the polymer's dark-blue colour. During illumination, the film was immersed in aqueous methanol, the methanol acting as a sacrificial electron donor. In this example, the dark blue colour of the poly(aniline) was subsequently recoloured *electro* chromically.

The colour of the poly(aniline) did not bleach completely during illumination, presumably because the photoconducting properties of poly(aniline) decrease in proportion to the extent of the bleaching; it is the oxidised form of the polymer that photoconducts.

The poly(aniline) film can only photoconduct through those areas that are illuminated, so *images*, rather than uniform blocks of tone, may be formed if the light source passes through a patterned mask or photographic negative. To this end, Yoneyama et al.[19] illuminated their photochromic–electrochromic poly(aniline) film through a photographic negative to form the notable image in Figure 15.4. Kobayashi et al.[49] have also generated impressive images by illuminating a film of poly(aniline) through a photographic negative.

Figure 15.4 Photoelectrochromic image generated on a thin film of poly(aniline)–TiO_2: the film was immersed in a solution of phosphate buffer (0.5 mol dm^{-3} at pH 7) containing 20 wt% methanol as a sacrificial electron donor. The film was illuminated through a photographic negative with a 500 W xenon lamp for 1 min. (Figure reproduced from Yoneyama, H., Takahashi, N. and Kuwabata, S. Formation of a light image in a polyaniline film containing titanium(IV) oxide particles. *J. Chem. Soc., Chem. Commun.*, 1992, 716–17, with permission of The Royal Society of Chemistry.)

References

1. Hirochi, K., Kitabatake, M. and Yamazaki, O. Electrochromic effects of Li–W–O films under ultraviolet light exposure. *J. Electrochem. Soc.*, **133**, 1986, 1973–4.
2. Buttner, W., Rieke, P. and Armstrong, N. R. Photoelectrochemical response of GaPc-Cl thin film electrode using two photon sources and two illumination devices. *J. Electrochem. Soc.*, **131**, 1984, 226–8.
3. Stilkans, M. P., Purans, Y. Y. and Klyavin, Y. K. Integral photoelectrical properties of thin-film systems based on photosensitive conductor and photochromium material. *Zh. Tekh. Fiz.*, **61**, 1991, 91–7 [in Russian]. The title and abstract are cited in *Curr. Contents*, **31**, 1991, 1969.
4. Teowee, G., Gudgel, T., McCarthy, K., Agrawal, A., Allemand, P. and Cronin, J. User controllable photochromic (UCPC) devices. *Electrochim. Acta*, **44**, 1999, 3017–26.
5. Monk, P. M. S., Mortimer, R. J. and Rosseinsky, D. R. *Electrochromism: Fundamentals and Applications*, Weinheim, VCH, 1995, ch. 12.
6. Gregg, B. A. Photoelectrochromic cells and their applications. *Endeavour*, **21**, 1997, 52–5.
7. De Paoli, M.-A., Nogueira, A. F., Machado, D. A. and Longo, C. All-polymeric electrochromic and photoelectrochemical devices: new advances. *Electrochim. Acta*, **46**, 2001, 4243–9.
8. Rauh, R. D. Cadmium chalcogenides. *Stud. Phys. Chem.*, **55**, 1988, 277–327.
9. Duffy, J. A. *Bonding, Energy Levels and Inorganic Solids*, London, Longmans, 1990.
10. Shizukuishi, M., Shimizu, S. and Enoue, E. Application of amorphous silicon to WO_3 photoelectrochromic device. *Jpn. J. Appl. Phys.*, **20**, 1981, 2359–63.
11. Yoneyama, H., Wakamoto, K. and Tamura, H. Photoelectrochromic properties of polypyrrole-coated silicon electrodes. *J. Electrochem. Soc.*, **132**, 1985, 2414–17.
12. Bullock, J. N., Bechinger, C., Benson, D. K. and Branz, H. M. Semi-transparent *a*-SiC:H solar cells for self-powered photovoltaic-electrochromic devices. *J. Non-Cryst. Solids*, **198–200**, 1996, 1163–7.
13. Bechinger, C. and Gregg, B. A. Development of a new self-powered electrochromic device for light modulation without external power supply. *Sol. Energy Mater. Sol. Cells*, **54**, 1998, 405–10.
14. Monk, P. M. S., Duffy, J. A. and Ingram, M. D. Electrochromic display devices of tungstic oxide containing vanadium oxide or cadmium sulphide as a light-sensitive layer. *Electrochim. Acta*, **38**, 1993, 2759–64.
15. Deb, S. K., Lee, S.-H., Tracy, C. E., Pitts, J. R., Gregg, B. A. and Branz, H. M. Stand-alone photovoltaic-powered electrochromic smart window. *Electrochim. Acta*, **46**, 2001, 2125–30.
16. Benson, D. K. and Branz, H. M. Design goals for a photovoltaic-powered electrochromic window covering. *Sol. Energy Mater. Sol. Cells*, **39**, 1995, 203–11.
17. Deb, S. K. Recent developments in high efficiency photovoltaic cells. *Renewable Energy*, **15**, 1998, 467–72.
18. Gao, W., Lee, S.-H., Benson, D. K. and Branz, H. M. Novel electrochromic projection and writing device incorporating an amorphous silicon carbide photodiode. *J. Non-Cryst. Solids*, **266–9**, 2000, 1233–7.
19. Yoneyama, H., Takahashi, N. and Kuwabata, S. Formation of a light image in a polyaniline film containing titanium(IV) oxide particles. *J. Chem. Soc., Chem. Commun.*, 1992, 716–17.

20. Inganäs, O., Carlberg, C. and Yohannes, T. Polymer electrolytes in optical devices. *Electrochim. Acta*, **43**, 1998, 1615–21.
21. Sotomayor, J., Will, G. and Fitzmaurice, D. Photoelectrochromic heterosupramolecular assemblies. *J. Mater. Chem.*, **10**, 2000, 685–92.
22. Li, Y., Hagen, J. and Haarer, D. Novel photoelectrochromic cells containing a polyaniline layer and a dye-sensitized nanocrystalline TiO_2 photovoltaic cell. *Synth. Met.*, **94**, 1998, 273–7.
23. Ileperuma, O., Dissanayake, M., Somusunseram, S. and Bandara, L. Photoelectrochemical solar cells with polyacrylonitrile-based and polyethylene oxide-based polymer electrolytes. *Sol. Energy Mater. Sol. Cells*, **84**, 2004, 117–24.
24. Kobayashi, N., Yano, T., Teshima, K. and Hirohashi, R. Photoelectrochromism of poly(aniline) derivatives in a Ru complex–methylviologen system containing a polymer electrolyte. *Electrochim. Acta*, **43**, 1998, 1645–9.
25. Kobayashi, N., Hirohashi, R., Kim, Y. and Teshima, K. Photorewritable conducting polyaniline image formation with photoinduced electron transfer. *Synth. Met.*, **101**, 1999, 699–700.
26. Kim, Y., Teshima, K. and Kobayashi, N. Improvement of reversible photoelectrochromic reaction of polyaniline in polyelectrolyte composite film with the dichloroethane solution system. *Electrochim. Acta*, **45**, 2000, 1549–53.
27. Kobayashi, N., Fukuda, N. and Kim, Y. Photoelectrochromism and photohydrolysis of sulfonated polyaniline containing $Ru(bpy)_3^{2+}$ film for negative and positive image formation. *J. Electroanal. Chem.*, **498**, 2001, 216–22.
28. Inganäs, O. and Lundström, I. Some potential applications for conducting polymers. *Synth. Met.*, **21**, 1987, 13–19.
29. De Saja, J. A. and Tanaka, K. Photoelectrochemical cells with *p*-type poly(3-methylthiophene). *Phys. Stat. Sol. A*, **108**, 1988, K109–14.
30. Liao, J. and Ho, K.-C. A photoelectrochromic device using a PEDOT thin film. *J. New Mater. Electrochem. Syst.*, **8**, 2005, 37–47.
31. Ohtani, B., Atsumi, T., Nishimoto, S. and Kagiya, T. Multiple responsive device: photo- and electrochromic composite thin film of tungsten trioxide with titanium oxide. *Chem. Lett.*, 1988, 295–8.
32. Ziegler, J. P., Lesniewski, E. K. and Hemminger, J. C. Polycrystalline *n*-$SrTiO_3$ as an electrode for the photoelectrochromic switching of Prussian blue films. *J. Appl. Phys.*, **61**, 1987, 3099–104.
33. Ziegler, J. P. and Hemminger, J. C. Spectroscopic and electrochemical characterization of the photochromic behaviour of Prussian blue on *n*-$SrTiO_3$. *J. Electrochem. Soc.*, **134**, 1987, 358–63.
34. DeBerry, D. W. and Viehbeck, A. Photoelectrochromic behaviour of Prussian blue-modified TiO_2 electrodes. *J. Electrochem. Soc.*, **130**, 1983, 249–51.
35. Itaya, K., Uchida, I., Toshima, S. and De La Rue, R. M. Photoelectrochemical studies of Prussian blue on *n*-type semiconductor (*n*-TiO_2). *J. Electrochem. Soc.*, **131**, 1984, 2086–91.
36. Kaneko, M., Okada, T., Minoura, H., Sugiura, T. and Ueno, Y. Photochargeable multilayer membrane device composed of CdS film and Prussian blue battery. *J. Electrochem. Soc.*, **35**, 1990, 291–3.
37. Stikans, M., Kleparis, J. and Klevins, E. J. Photoelectric characterization of solid-state photochromic system. *Latv. P. S. R. Zinat. Akad. Vestis. Fiz. Tekh. Zinat. Ser.*, **4**, 1988, 43.

38. Reichman, B., Fan, F.-R. F. and Bard, A. J. Semiconductor electrodes, XXV: the p-GaAs / heptyl viologen system: photoelectrochemical cells and photoelectrochromic cells. *J. Electrochem. Soc.*, **127**, 1980, 333–8.
39. Butler, M. A. Photoelectrochemical imaging. *J. Electrochem. Soc.*, **131**, 1984, 2185–90.
40. Opara Krašovec, U., Georg, A., Georg, A., Wittwer, V., Luther, J. and Topic, M. Performance of a solid-state photoelectrochromic device. *Sol. Energy Mater. Sol. Cells*, **84**, 2004, 369–80.
41. Hauch, A., Georg, A., Abumgärtner, S., Opara Krašovec, U. and Orel, B. New photoelectrochromic device. *Electrochim. Acta*, **46**, 2001, 2131–6.
42. de Tacconi, N. R., Rajeshwar, K. and Lezna, R. O. Photoelectrochemistry of indium hexacyanoferrate–titania composite films. *J. Electroanal. Chem.*, **500**, 2001, 270–8.
43. de Tacconi, N. R., Rajeshwar, K. and Lezna, R. O. Preparation, photoelectrochemical characterization, and photoelectrochromic behavior of metal hexacyanoferrate–titanium dioxide composite films. *Electrochim. Acta*, **45**, 2000, 3403–11.
44. de Tacconi, N. R., Carmona, J. and Rajeshwar, K. Reversibility of photoelectrochromism at the TiO_2/methylene blue interface. *J. Electrochem. Soc.*, **144**, 1997, 2486–90.
45. Zhi, J. F., Baba, R., Hashimoto, K. and Fujishima, A. Photoelectrochromic properties of spirobenzopyran derivative. *J. Photochem. Photobiol.*, **A92**, 1995, 91–7.
46. Macchione, M., De Filpo, G., Nicoletta, F. and Chidichimo, G. Improvement of response times in photoelectrochromic organic film. *Chem. Mater.*, **16**, 2004, 1400–1.
47. Macchione, M., De Filpo, G., Mashin, A., Nicoletta, F. and Chidichimo, G. Laser-writable electrically erasable photoelectrochromic organic film. *Adv. Mater.*, **15**, 2003, 327–9.
48. Yoneyama, H. Writing with light on polyaniline films. *Adv. Mater.*, **5**, 1993, 394–6.
49. Kobayashi, M., Hashimoto, K. and Kim, Y. Photoinduced electrochromism of conducting polymers and its application. *Proc. Electrochem. Soc.*, **2003–17**, 2003, 157–65.

16
Device durability

16.1 Introduction

Like all other types of display device, mechanical or electronic, no electrochromic device will continue to function indefinitely. For this reason, cycle lives are reported. The definition of cycle life has not been conclusively settled. Even by the definition in Section 1.4, reported lives vary enormously: some workers suggest their devices will degrade and thereby preclude realistic use after a few cycles while others claim a device surviving several million cycles. Table 16.1 contains a few examples; in each case the cycle life cited represents 'deep' cycles, as defined on p. 12. Some of these longer cycle lives were obtained via methods of accelerated testing, as outlined below.

It is important to appreciate that results obtained with a typical three-electrode cell in conjunction with a potentiostat can yield profoundly different results from the same components assembled as a device: most devices operate with only two electrodes.

The results of Biswas *et al.*,[9] who potentiostatically cycled a thin film of WO_3 immersed in electrolyte, are typical insofar as the electrochemical reversibility of the cycle remained quite good with little deterioration. Their films retained their physical integrity, but the intensity of the coloration decreased with the number of cycles.

Some devices are intended for once-only use, such as the freezer indicator of Owen and co-workers;[10] other applications envisage at most a few cycles, like the Eveready battery-charge indicator.[11] Clearly, degradation can be allowed to occur after no more than a few cycles with applications like the latter. Conversely, applications such as a watch display will need to withstand many billions of cycles without significant deterioration – a stringent requirement.

Devices can fail for one or more of three related reasons: failure of the conductive electrodes; failure of the electrolyte layer; and failure of the

Table 16.1. *A selection of cycle lives of electrochromic devices, reported as number of cycles survived.*

Primary electrochrome	Secondary electrochrome	Cycle life	Ref.
WO_3	Poly(aniline)	20 000	1
WO_3	Prussian blue	20 000	2,3
WO_3	'VO_xH_y'	30 000	4
WO_3	$(CeO_2)_x(TiO_2)_{1-x}$	50 000	5
WO_3	Nickel oxide	100 000	6
WO_3	Iridium oxide	10 000 000	7
Electrodeposited bismuth	Prussian blue	50 000 000	8

electrochromes. The durability of individual electrochromes is discussed in their respective chapters.

Here the durability of transparent electrodes is discussed in Section 16.2; that of electrolyte layers is discussed in Section 16.3; and general methods of enhancing electrochrome durability are outlined in Section 16.4. Finally, Section 16.5 contains details of how cycle lives are assessed for complete, assembled, devices.

16.2 Durability of transparent electrodes

The first reason for device failure is breakdown of an optically transparent electrode, OTE. The most common cause of OTE degradation is decomposition of ITO, which occurs readily in acidic solutions: the oxides within the ITO layer are themselves reduced when not in contact with solution. Such reduction both decreases its chemical stability and increases its electrical resistance:[12,13,14] while the oxidised form of ITO is chemically stable, reduced ITO is unstable and rarely bears the strains of repeated redox cycling because it dissolves readily in aqueous acids.[15] Indeed, in aqueous solution, the subsequent reaction of *over*-reduction to form metallic tin is difficult to stop.[16,17,18] For this reason, some workers tentatively suggest that *all* moisture must be excluded rigorously from the electrolyte of an ECD.[19,20,21]

In the study by Bressers and Meulenkamp[22] it was shown that a thin layer of metallic indium forms on the surface of the ITO during reduction, possibly facilitating the observed dissolution in water-containing electrolytes, which is faster if the ITO is partially reduced.[14,23]

Even ITO in contact with semi-solid poly(ethylene) oxide (PEO) electrolyte can deteriorate: Radhakrishnan *et al.*[15] show how ITO electrodes in contact with PEO deteriorate after repeated cycling, both in terms of their conductivity and transparency. Their XPS studies of ITO electrodes clearly show the metallic impurities being expelled into the PEO. The change of composition leads to eventual diminution of the ITO conductivity, with concomitant decrease in ECD cycle life.

16.3 Durability of the electrolyte layers

The second reason for device failure is electrolyte breakdown. Most organic polymers have relatively poor photolytic stability, particularly when in solution or intimately mixed with an ionic salt, as is typical for ECD usage.[24] Hence long-term solar irradiation will inevitably cause ECD breakdown. In an ECD operating in a reflectance mode, such as a mirror, a particularly photo-unstable primary electrochrome can be placed adjacent to the reflective back electrode rather than situated on the front OTE, i.e. *behind* the electrolyte and secondary electrochrome layers (provided both have a high optical transparency for all wavelengths).

It is quite common for polymeric electrolytes to be 'filled' with an opaque white powder such as TiO_2, to enhance the contrast ratio of the primary electrochrome. While the inclusion of particulate TiO_2 does not affect the response times of an ECD, its photo-activity (particularly if the TiO_2 is in its anatase form) will significantly accelerate photolytic deterioration of organic polymers.[25,26,27]

A further danger associated with devices operating via proton conduction is underpotential (catalysed) generation of molecular hydrogen gas, formed according to Eq. (16.1), which both removes protonic charges and also forms insulating bubbles of gas inside the ECD:

$$2H^+ (soln) + 2e^- \rightarrow H_2 (g). \tag{16.1}$$

Areas of the electrode adjacent to such a bubble are insulated, thereby disabling the device.

16.4 Enhancing the durability of electrochrome layers

Great care is needed when the electrolyte in an ECD is a layer of rigid inorganic solid, since most type-III electrochromes change volume during redox changes, owing to chemical volume changes and the volume decrease of a dielectric in a field, *electrostriction*. Thin-layer WO_3, for example, expands by about 6% during reduction[28] from $H_{\rightarrow 0}WO_3$ to H_1WO_3 (see pp. 87, 129). Most type-III devices comprise two solid layers of electrochrome. Therefore the extent of chemical-volume change in either layer could be approximately the same, changing in a complementary sense with one expanding while the other contracts; however, electrostriction acts only by contracting. Placing an elastomeric (semi-solid polymer) electrolyte between the two ECD electrochromic layers considerably cushions the strains engendered by expansion and contraction in a two-layer ECD. To confirm the scope for cushioning the effects of electrostriction, Scrosati and co-workers[29] note how the stresses engendered by ion insertion/egress within the cell WO_3|electrolyte|NiO are similar in both the WO_3 and NiO layers. (Methods of quantifying the stresses induced during electrochromic activity are discussed on p. 130).

Many electrochromes dissolve in, or are damaged by prolonged contact with, the electrolyte layer. To protect the interphase between the electrochrome and electrolyte, several studies suggest depositing a thin, protective film over the electrochrome film. Enhancement of chemical stability will obviously extend the cycle life of an ECD. There are a number of examples of this practice. Thus for example, Haranahalli and Dove[30] deposited a thin semi-transparent layer of gold on their WO_3, so protecting it from chemical attack, and incidentally also accelerating the speed of device operation. Similarly, in one study, Granqvist and co-workers[31] deposited a thin film of tungsten oxyfluoride on solid WO_3, and in another, deposited a thin protective layer of electron-bombarded WO_3 onto a layer of metal oxyfluoride.[32,33] Yoo et al.[34] coated WO_3 with lithium phosphorus oxynitride. Deb and co-workers[35] coated V_2O_5 with a protective layer of $LiAlF_4$, which exhibited improved durability and electrochemical charge capacity during 800 write–erase cycles. Long et al.[36] electrodeposited poly(o-phenylenediamine) onto porous MnO_2. He et al.[37] accelerated the operation of a WO_3-based device with a surface layer of gold nanoparticles. In the same way, the perfluorinated polymer Nafion® has been coated on Prussian blue,[38] tantalum pentoxide[39] and tungsten oxide,[40,41] in each case improving the stability and enhancing the electrochromic characteristics of these electrochromes.

While such barrier films protect the electrochrome from chemical degradation, they also hinder the motion of the counter ions needed for charge balance. Movement across the electrolyte–electrochrome interphase will therefore increase the ECD response time. However, the acceleration noted by Haranahalli and Dove[30] and He et al.[37] follows because a potential was applied to the gold layer, itself conductive, covering the respective electrode surfaces.

16.5 Durability of electrochromic devices after assembly

Studies describing the durability of assembled electrochromic devices are to be found in the following reports: 'Durability evaluation of electrochromic

16.5 Durability after assembly

devices – an industry perspective' by Lampert et al.[42] (in 1999), 'Failure modes of sol–gel deposited electrochromic devices' by Bell and Skryabin[43] (in 1999) and 'A feasibility study of electrochromic windows in vehicles' by Jaksic and Salahifar[44] (in 2003).

Many individual studies of device durability are extant. For example, Nishikitani and co-workers[45] of the Japanese Nippon Mitsubishi Oil Corporation employed a variety of weathering tests on electrochromic windows designed for automotive applications. Their two-year outdoor weathering tests suggest their ECDs are highly durable, but as expected, outdoor exposure ultimately causes device degradation.

Many workers consider it impractical to wait for results from such trials in real time, so considerable effort has been expended in the use of accelerated testing methods. A few exemplar studies below will suffice. Asahi Glass failed to detect deterioration in their lithium-based ECD windows stored during 1000 hours of testing at 70 °C and 90% humidity. Similarly, Deb and co-workers[46] of the National Renewable Energy Laboratory (NREL) in Colorado, USA, used accelerated testing conditions on several prototype ECDs. Deb and co-workers[47] have also described the way such devices were illuminated with a high-intensity UV lamp to mimic the effects of long-term exposure to solar light, and concluded that the effects of long-term exposure can indeed be mimicked readily within a considerably shortened time – even a few days – with concomitant savings in overheads. However, the applicability of the NREL results is limited since all devices were fabricated by anonymous US companies.

Sbar et al.[48] of SAGE Electrochromics in New Jersey, USA, tested electrochromic architectural windows during external exposure at test sites in New Jersey and the Arizona desert. Their accelerated testing methods included electrochemical cycling over a range of temperatures, with changes in illumination and/or humidity. They concluded that their windows showed 'good switching performance'.

Colour Plate 7 shows similar testing of a Gentex window.

Skryabin et al.[49] present a more fundamental, partly theoretical, assessment of testing and quality control criteria for large devices. The durability of electrochromic devices was assessed from three perspectives: mimicking the device behaviour with an equivalent circuit; arranging the external electrical connections; and optimizing the switching procedure. Their principal conclusion was that mimicking is difficult: ECDs are 'inherently complicated devices'.

Nagai et al.,[6] also using a programme of accelerated testing, concluded that their device, Glass|ITO|NiO|Ta_2O_5 (electrolyte)|WO_3|ITO|adhesive-film|Glass was capable of 10^5 cycles at 60 °C.

Mathew et al.[50] of The Optical Coating Laboratory, in Santa Rosa, USA, consider electrochromic devices for large-area architectural applications, viability requirements for minimum acceptable performance encompassing depth of colour, switching time, chromatism and durability. Within these criteria, windows were deemed acceptable if they coloured to a contrast ratio of 10:1 and were capable of 20 000 cycles.

The brief list above demonstrates the way criteria for study can differ considerably: many studies do not even state the criteria chosen. The report 'Evaluation criteria and test methods for electrochromic windows' (1990, but made widely available in 1999) by Czanderna and Lampert[51] was compiled to address this problem, and goes some way toward generating a template for reproducible testing of electrochromic devices. These authors elaborate the requirements in a subsequent paper,[47] but excessive use of unexplained abbreviations detracts from clarity.

Device durability is then defined in terms of the following five criteria:[47,51]

1. The environment for a specific application, which clearly dictates the speed at which the device must operate.
2. The upper and lower temperatures of operation (they chose −40 °C to 50 °C). Device operation was discussed in terms of likely variations in temperature in the USA; rather wider variations around the globe are to be expected.
3. Stresses induced in a device by 'thermal shock' as it cools and warms rapidly. In the authors' Californian climate, no drastic temperature changes occurred during electrochromic operation. They conclude that no major stresses born of thermal shock occur during clear, sunny days, nor when the sky is continually overcast; rapid temperature changes were only observed when the sun appeared from behind a cloud, or was partially obscured; and during thunderstorms. Such variations scarcely cover conditions in other countries, let alone other US states. Holidaymakers in Skegness, UK, for example, would need more assurance of ECD robustness against the weather.
4. The effect of deterioration owing to solar exposure, especially by UV light. The UV was provided by a xenon light source of output $0.55\,\text{W m}^{-2}$ at 340 nm, a severe test in view of the peak daylight intensity in Miami of $0.8\,\text{W m}^{-2}$.
5. The effect of additional stresses such as changes in humidity, and mechanical shock. Devices operating via proton movement may need water; the concentration of water needed for optimum performance needs to remain within a narrow, desirable range, so such devices are sealed to minimise changes in internal humidity levels. A robust seal also protects against oxygen ingress. Device encapsulation is described on p. 424. Strong frames are required for rough handling or percussive incidents.

Having noted that variations on the above test methodology will depend on many factors (the choice of electrochrome, device construction, customer specifications, the intended application, and so on), they conclude:[47]

Our major conclusions are that substantial R&D is [still] necessary to understand the factors that limit electrochromic windows [ECWs] durability, ...[but] that it is possible to predict the service lifetime of ECWs.

They add, 'The accelerated tests are reasonable for the evaluation of the lifetime of EC glazing but have not been verified with real time testing.'[47]

References

1. Bernard, M.-C., Hugot-Le Goff, A. and Zeng, W. Elaboration and study of a PANI/PAMPS/WO$_3$ all solid-state electrochromic device. *Electrochim. Acta*, **44**, 1998, 781–96.
2. Ho, K.-C. Cycling and at-rest stabilities of a complementary electrochromic device based on tungsten oxide and Prussian blue thin films. *Electrochim. Acta*, **44**, 1999, 3227–35.
3. Ho, K.-C. Cycling stability of an electrochromic system at room temperature. *J. Electrochem. Soc.*, **139**, 1992, 1099–104.
4. Lechner, R. and Thomas, L. K. All solid state electrochromic devices on glass and polymeric foils. *Sol. Energy Mater. Sol. Cells*, **54**, 1998, 139–46.
5. Sun, D. L., Heusing, S., Puetz, J. and Aegerter, M. A. Influence of water on the electrochemical properties of $(CeO_2)_x(TiO_2)_{1-x}$ and WO$_3$ sol–gel coatings and electrochromic devices. *Solid State Ionics*, **165**, 2003, 181–9.
6. Nagai, J., McMeeking, G. D. and Saitoh, Y. Durability of electrochromic glazing. *Sol. Energy Mater. Sol. Cells*, **56**, 1999, 309–19.
7. Beni, G. Recent advances in inorganic electrochromics. *Solid State Ionics*, **3–4**, 1981, 157–63.
8. Ziegler, J. P. and Howard, B. M. Applications of reversible electrodeposition electrochromic devices. *Sol. Energy Mater. Sol. Cells*, **39**, 1995, 317–31.
9. Biswas, P. K., Pramanik, N. C., Mahapatra, M. K., Ganguli, D. and Livage, J. Optical and electrochromic properties of sol–gel WO$_3$ films on conducting glass. *Mater. Lett.*, **57**, 2003, 4429–32.
10. Colley, R. A., Budd, P. M., Owen, J. R. and Balderson, S. Poly[oxymethylene]-oligo(oxyethylene) for use in subambient temperature electrochromic devices. *Polym. Int.*, **49**, 2000, 371–6.
11. Bailey, J. C., Eveready Battery Company. Electrochromic thin film state-of-charge detector for on-the-cell application. US Patent 05458992, 1995.
12. Armstrong, N. R., Liu, A. W. C., Fujihira, M. and Kuwana, T. Electrochemical and surface characteristics of tin oxide and indium oxide electrodes. *Anal. Chem.*, **48**, 1976, 741–50.
13. Scholten, M. and van den Meerakker, J. E. A. M. On the mechanism of ITO etching: the specificity of halogen acids. *J. Electrochem. Soc.*, **140**, 1993, 471–4.
14. van den Meerakker, J. E. A. M., Baarslag, P. C. and Scholten, M. On the mechanism of ITO etching in halogen acids: the influence of oxidizing agents. *J. Electrochem. Soc.*, **142**, 1995, 2321–6.
15. Radhakrisnan, S., Unde, S. and Mandale, A. B. Source of instability in solid state polymeric electrochromic cells: the deterioration of indium tin oxide electrodes. *Mater. Chem. Phys.*, **48**, 1997, 268–71.

16. Goldner, R. B., Foley, G., Goldner, E. L., Norton, P., Wong, K., Haas, T., Seward, G. and Chapman, R. Electrochromic behaviour in ITO and related oxides. *Appl. Optics*, **24**, 1985, 2283–4.
17. Svensson, J. S. E. M. and Granqvist, C. G. No visible electrochromism in high-quality e-beam evaporated In_2O_3:Sn films. *Appl. Opt.*, **24**, 1984, 2284–5.
18. Golden, S. J. and Steele, B. C. H. Thin-film tin-doped indium oxide counter electrode for electrochromic applications. *Solid State Ionics*, **28–30**, 1988, 1733–7.
19. Duffy, J. A., Ingram, M. D. and Monk, P. M. S. The effect of moisture on tungsten oxide electrochromism in polymer electrolyte devices. *Solid State Ionics*, **58**, 1992, 109–14.
20. Sharma, P. K., Fantini, M. C. A. and Gorenstein, A. Synthesis, characterization and electrochromic properties of NiO_xH_y thin film prepared by a sol–gel method. *Solid State Ionics*, **113–15**, 1998, 457–63.
21. Shiyanovskaya, I. V. Structure rearrangement and electrochromic properties of amorphous tungsten trioxide films. *J. Non-Cryst. Solids*, **187**, 1995, 420–4.
22. Bressers, P. M. M. C. and Meulenkamp, E. A. Electrochromic behavior of indium tin oxide in propylene carbonate. *J. Electrochem. Soc.*, **145**, 1998, 2225–30.
23. Monk, P. M. S. and Man, C. M. Reductive ion insertion into thin-film indium tin oxide (ITO) in aqueous acidic solutions: the effect of leaching of indium from the ITO. *J. Mater. Sci., Electron. Mater.*, **10**, 1999, 101–7.
24. Halim Hamid, S. (ed). *Handbook of Polymer Degradation*, 2nd edn, New York, Marcel Dekker, 2000.
25. Jin, C., Christensen, P. A., Egerton, T. A., Lawson, E. J. and White, J. R. Rapid measurement of polymer photo-degradation by FTIR spectrometry of evolved carbon dioxide. *Polymer Degradation and Stability*, **91**, 2006, 1086–96.
26. Gesenhues, U., Influence of titanium dioxide pigments on the photodegradation of poly(vinyl chloride), *Polym. Degrad. Stab.*, **68**, 2000, 185–96.
27. Su, C., Hong, B.-Y. and Tseng, C.-M. Sol–gel preparation and photocatalysis of titanium dioxide. *Catal. Today*, **96**, 2004, 119–26.
28. Green, M. Atom motion in tungsten bronze thin films. *Thin Solid Films*, **50**, 1978, 148–50.
29. Passerini, S., Scrosati, B., Hermann, V., Holmblad, C. and Bartlett, T. Laminated electrochromic windows based on nickel oxide, tungsten oxide, and gel electrolytes. *J. Electrochem. Soc.*, **141**, 1994, 1025–8.
30. Haranahalli, A. R. and Dove, D. B. Influence of a thin gold surface layer on the electrochromic behavior of WO_3 films. *Appl. Phys. Lett.*, **36**, 1980, 791–3.
31. Azens, A., Hjelm, A., Le Bellac, D., Granqvist, C. G., Barczynska, J., Pentjuss, E., Gabrusenoks, J. and Wills, J. M. Electrochromism of W-oxide-based films: some theoretical and experimental results. *Proc. SPIE*, **2531**, 1995, 92–104.
32. Azens, A. and Granqvist, C. G. Electrochromic films of tungsten oxyfluoride and electron bombarded tungsten oxide. *Sol. Energy Mater. Sol. Cells*, **44**, 1996, 333–40.
33. Azens, A., Granqvist, C. G., Pentjuss, E., Gabrusenoks, J. and Barczynska, J. Electrochromism of fluorinated and electron-bombarded tungsten oxide. *J. Appl. Phys.*, **78**, 1995, 1968–74.
34. Yoo, S. J., Lim, J. W. and Sung, Y.-E. Improved electrochromic devices with an inorganic solid electrolyte protective layer. *Sol. Energy Mater. Sol. Cells*, **90**, 2006, 477–84.
35. Lee, S.-H., Cheong, H. M., Liu, P., Tracy, C. E., Pitts, J. R. and Deb, S. K. Improving the durability of ion insertion materials in a liquid electrolyte. *Solid State Ionics*, **165**, 2003, 81–7.

36. Long, J. W., Rhodes, C. P., Young, A. L. and Rolison, D. R. Ultrathin, protective coatings of poly(o-phenylenediamine) as electrochemical porous gates: making mesoporous MnO_2 nanoarchitectures stable in acid electrolytes. *Nano. Lett.*, **3**, 2003, 1155–61.
37. He, T., Ma, Y., Cao, Y., Yang, W. and Yao, J. Enhanced electrochromism of WO_3 thin film by gold nanoparticles. *J. Electroanal. Chem.*, **514**, 2001, 129–32.
38. John, S. A. and Ramaraj, R. Role of acidity on the electrochemistry of Prussian blue at plain and Nafion film-coated electrodes. *Proc. Ind. Acad. Sci.*, **107**, 1995, 371–83.
39. Sone, Y., Kishimoto, A., Kudo, T. and Ikeda, K. Reversible electrochromic performance of Prussian blue coated with proton conductive $Ta_2O_5.nH_2O$ film. *Solid State Ionics*, **83**, 1996, 135–43.
40. Chen, K. Y. and Tseung, A. C. C. Effect of Nafion dispersion on the stability of Pt/WO_3 electrodes. *J. Electrochem. Soc.*, **143**, 1996, 2703–8.
41. Shen, P. K., Huang, H. and Tseung, A. C. C. Improvements in the life of WO_3 electrochromic films. *J. Mater. Chem.*, **2**, 1992, 497–9.
42. Lampert, C. M., Agrawal, A., Baertlien, C. and Nagai, J. Durability evaluation of electrochromic devices – an industry perspective. *Sol. Energy Mater. Sol. Cells*, **56**, 1999, 449–63.
43. Bell, J. M. and Skryabin, I. L. Failure modes of sol–gel deposited electrochromic devices. *Sol. Energy Mater. Sol. Cells*, **56**, 1999, 437–48.
44. Jaksic, N. I. and Salahifar, C. A feasibility study of electrochromic windows in vehicles. *Sol. Energy Mater. Sol. Cells*, **79**, 2003, 409–23.
45. Kubo, T., Shinada, T., Kobayashi, Y., Imafuku, H., Toya, T., Akita, S., Nishikitani, Y. and Watanabe, H. Current state of the art for NOC–AGC electrochromic windows for architectural and automotive applications. *Solid State Ionics*, **165**, 2003, 209–16.
46. Tracy, C. E., Zhang, J.-G., Benson, D. K., Czanderna, A. W. and Deb, S. K. Accelerated durability testing of electrochromic windows. *Electrochim. Acta*, **44**, 1999, 3195–202.
47. Czanderna, A. W., Benson, D. K., Jorgensen, G. J., Zhang, J.-G., Tracy, C. E. and Dep, S. K. Durability issues and service lifetime prediction of electrochromic windows for buildings applications. *Sol. Energy Mater. Sol. Cells*, **56**, 1999, 419–36.
48. Sbar, N., Badding, M., Budziak, R., Cortez, K., Laby, L., Michalski, L., Ngo, T., Schulz, S. and Urbanik, K. Progress toward durable, cost effective electrochromic window glazings. *Sol. Energy Mater. Sol. Cells*, **56**, 1999, 321–41.
49. Skryabin, I. L., Evans, G., Frost, D., Vogelman, G. and Bell, J. M. Testing and control issues in large area electrochromic films and devices. *Electrochim. Acta*, **44**, 1999, 3203–9.
50. Mathew, J. G. H., Sapers, S. P., Cumbo, M. J., O'Brien, N. A., Sargent, R. B., Raksha, V. P., Lahaderne, R. B. and Hichwa, B. P. Large area electrochromics for architectural applications. *J. Non-Cryst. Solids*, **218**, 1997, 342–6.
51. Czanderna, A. W. and Lampert, C. M. 1990: SERI/TP-255-3537, as cited in ref. 47.

Index

absorbance, optical, 43, 52
  change in, 53
  shape of bands, 53
accelerated ECD testing
  humidity, 447, 448
  weathering, 447
  xenon arc, 448
acetate silicone, ECD encapsulation, 425
acetonitrile, ECD electrolyte, 254, 262, 359, 385, 390T, 419, 438
achromatic, centre of colour diagram, 67
acidity constants K_a, 4
acrylic powder, ECD electrolyte thickener, 419
activation energy, 86, 87, 108, 111, 112
  in colouring metal oxides, 93
  in nickel oxide, 111T
  in tungsten trioxide, 111T
  to bacterial growth, 408
  to counter-ion movement, 408
  to diffusion, 83, 86
  to electron transfer, 47
activity, 36–7, 38, 84
  coefficient, 36, 40, 96
    of pure solids, 38
admittance, 50
AEIROF, 156, 158
agar, as gelling agent, 349, 351
AGFA, 332
air conditioning, ix, 398
Airbus, ECD, windows, 401
AIROF, 155, 156
  degradation of, 156
alkoxides
  CVD precursors, 131
    forming molybdenum trioxide, 152
    forming titanium dioxide, 184
alloy
  Inconel-600, oxide mixture on, 203
  nickel–aluminium, 200
all-polymer devices, 332
all-solid-state-devices, 417
alpha particles, 160
aluminium–cobalt oxide, 195, 196
aluminium–nickel alloy, 200
aluminium–nickel oxide, 200
aluminium–silicon–cobalt oxide, 204
amino-4-bromoanthaquinone-2-sulfonate, 384
aminonaphthaquinone, 384
amorphisation, tungsten–molybdenum oxide, 193
amorphous, oxides, 88
  made by vacuum evaporation, 81
amorphous silicon, 15
  as photoconductor, 436
anatase, see *titanium dioxide*
Anderson transition, 81, 99, 142, 149, 307
ANEEPS, 3
aniline, 313
aniline black, 312
aniline–polypyridyl complexes, 256
annealing
  endothermic process, 140
  to effect crystallisation, 88, 89
    cerium oxide, 166
    cobalt oxide, 168
    CVD product, 131
    iridium oxide, 158
    iron oxide, 173, 175
    molybdenum trioxide, 152, 154
    nickel oxide, 161
    niobium pentoxide, 177
    rhodium oxide, 181
    spin-coated products, 135
    spray pyrolysis product, 135
    tungsten trioxide, 88, 140, 141, 148
    vanadium pentoxide, 185
anodic coloration, coloration efficiency negative, 55
anodic reactions, definition, 46
antimony pentoxide
  as electrochromic host, 193
  as ECD electrolyte, 421T
antimony–copper alloy, substrate, 423
antimony-doped tin oxide, 193–5, 196, 274, 362
  optically passive, 193
applications, ECD
  battery charge indicator, 408, 443
  camouflage, 409

452

displays, 401–4
  advertising boards, 402
  bank notes, 408
  cash- and credit cards, 402, 408
  display, cashpoint machines, 407
  computer screens, 363, 402
    data display, 149, 265, 363
  electric paint, 364
  games, 363
  iPod, 363, 402
  laptop computer screens, 402
  mobile phone screens, 402
  NanoChromics, 347, 362, 363, 402, 406
  optical data storage, 265
  palmtop computer screens, 402
  smart cards, 363
  tickets, 408
  tokens, 408
  toys, 363, 436
  transport terminus screens, 402
  vouchers, 408
  watch faces, 149, 402, 443
electrochromic paper, 363, 405–6
eye wear, goggles, 398, 422
  motorcycle helmets, 398
  sunglasses, 401, 422
  visors, 401, 422
fibre-optics, 265
light modulation, 404–5
medicine, 265
mirrors, ix, 11, 44, 149, 307, 356, 363, 385–7, 395–7
optical attenuator, 270
shutters, 363, 404–5
solar-energy storage, 265, 266
temperature management, 265
windows, 149, 200, 363, 422
  Airbus, 401
  aircraft, 400–1
  Asahi Glass, 400
  Boeing 'Dreamliner', 400
  car sun roof, 398
  chromogenic glazing, 397
  dimmable laminates, 363
  Flabeg Gmbh, 400
  Gentex, 396, 398, 400
  optical attenuator, 270
  Pilkington Glass, 400
  PPG Aerospace, 400
  Schott Glass, 400
  shutters, 363
  Stadsparkasse Bank, 400
  X-ray reflector, 397
Aramid resin, 198
aromatic amines, 374–6, 377T–378T
  charge transfer, 374
  contrast ratio, 376
  near infrared absorption, 376
  response time, 375
  type-I electrochromism, 375
  type-II electrochromism, 375

Arrhenius equation, 83, 408
aryl viologens, 11, 28
Asahi Glass, 400, 447
asymmetric viologens, 355, 360
automotive mirrors, see *Applications, ECD mirror*
azulene, 313
Azure A, coloration efficiency, 57T
Azure B, coloration efficiency, 57T

back potential, 92, 93, 98, 102, 105, 106, 110–11, 115
bacteria
  growth, activation energy, 408
  reactions, 4
Bacteriorhodopsin, 3
band conduction, 81
band structure, poly(thiophene)s, 152
bandgap, 316
  of PEDOT, 322
Basic Blue 3, coloration efficiency, 57T
batteries, 14, 54, 167
  dry cell, 408
  ECD charge indicator, 408
  ECD like a secondary, 54
  photo-chargeable, of Prussian blue and tungsten trioxide, 437
  rechargeable, manganese oxide, 176
Bayer AG, 323
Baytron M, 323
Baytron P, 323
beam direction, photoelectrochromism, 433
BEDOT, 326
  BEDOT-NMeCz, 326
Beer–Lambert law, 53, 55, 146, 147, 148, 150, 151, 176
Bell Laboratories, 29
benzoquinones
  benzoquinone, *o*-, 381
  benzoquinone, *p*-, 381, 382
benzyl viologen, 8, 344T, 346, 352T, 356, 358
  di-reduced, 358
  radical, recrystallisation, 357
Berlin green, see *Prussian green*
betaines, 5
biological membrane potentials, 3
biphenyls, 379, 380
bipolaron
  in poly(thiophene)s, 320
  in tungsten trioxide, 147
bis(dimethylamino)diphenylamine, 4,4'-, 384
bismuth oxide, 166
  coloration efficiency, 56T, 166
  formation via evaporation, 166; rf sputtering, 166
  response time, 166
bismuth
  as secondary electrochrome, 444T
  electrodeposition of, 8, 27, 304, 305–6
    coloration efficiency, 306
    cycle life, 305
    ECD, 306

bismuth (cont.)
    electrochemistry, 304, 305
    electron mediation, 305
bithiophenes, 320
  conducting polymers, 316
bleaching
  chemical, viologens, 359
  models, 105–8
    Faughnan and Crandall, 105–8
    Green, 108, 109
  rate, 33
    electrochromes, for nickel oxide, 164; for vanadium pentoxide, 188
    potentiostatic, 105–8
  self, 15, 54, 150, 153
  types
    type-II electrochromes, 79–115
    type-III electrochromes, 79–115
blueprints, Prussian blue, 26, 405
Boeing 'Dreamliner', ECD, windows, 400
brightness, and colour analysis, 64
British Fenestration Rating Council, 397
bromoanil, o-, 382, 383
  solubility product, 383
bronze, 82, 103
  of lithium tungsten trioxide, electro-irreversibility of, 82
  of metal oxide, 61, 81, 82, 103
  of molybdenum trioxide, 103, 151
  of sodium tungsten trioxide, 27
  of tungsten trioxide, 81, 113, 144
Butler–Volmer equation, 29, 42, 46–8, 95
butyl viologen, 352T
γ-butyrolactone, ECD electrolyte, 167, 186, 303, 304, 362, 419

cadmium sulfide, 437
calomel reference electrode, see *saturated calomel electrode*
camouflage, ECD application, 409
capacitance effects, 11, 50
  electrolytic capacitors, 52
car mirrors, see *applications, ECD, mirrors*
car sun roof, ECD application, 398
carbazoles, 313, 376, 379T
  immobilised, 391
  N-carbazylcarbazole, 379T
  N-ethylcarbazole, 379T
  N-phenylcarbazole, 379T, 381T
  type-II electrochromes, 376
carbon electrochromes, 'carbon based', 303, 305
  screen printed carbon, 303, 305
  see also *diamond, fullerene and graphite*
carbon, substrate, 424
castable films, poly(aniline), 332–3
catalytic silver paint, depositing Prussian blue, 283
catechole, 271, 272
cathode ray tube
  power consumption, 15
  television, 402, 403

cathodic coloration, coloration efficiency positive, 55
cathodic-arc deposition, of vanadium pentoxide, 185
cathodic, definition, 46
CE, see *coloration efficiency*
cells, 34
  aqueous, 37–9
  electrochemical, 417
  electroneutrality in, 38
cellulose acetate, composite with poly(aniline), 333
cerianite, 166
cerium oxide, 166–7, 194
  annealing of, 166
  chemical diffusion coefficient, 85T
  electrochemistry of, 167
  electrochromic host, 193–5
  formation via
    dip coating, 135
    physical vapour deposition, 166
    spin coating, 135
    spray pyrolysis, 135, 166
  optical properties, 166
  optically passive, 166
cerium vanadate, 202
cerium–nickel oxide, 200
cerium–praseodymium oxide, 179
cerium–tin oxide, 201
cerium–titanium oxide, 194
  as secondary electrochrome, 444T
  chemical diffusion coefficient, 194
  coloration efficiency, 194
  EXAFS, of, 194
  optically passive, 194
  via dc magnetron sputtering, 194, 195
cerium–titanium–titanium oxide, 203
cerium–titanium–zirconium oxide, 203
cerium–tungsten oxide, see *tungsten–cerium oxide*
cerium–vanadium–titanium oxide, 203
cerium–zirconium oxide, 203
cerous ion, as electron mediator, 359
characteristic time, in Faughnan and Crandall model of coloration, 95, 111
charge
  electronic, 42
  faradaic, 52
charge capacity, 200
charge density, 55
charge dispersibility, 127
charge transfer, 42
  aromatic amines, 374
  complexation of
    cyanophenyl paraquat, 60, 359
    ferrocyanide, 359
    heptyl viologen, 359
    methyl viologen, 359
    viologens, 342–5, 353, 359
  intervalence, 60–1, 127, 145
  orbitals, 61
  oxides
    and cobalt ion, 169
    in iron–titanium oxide, 202

in oxide ion, 168
in permanganate, 60
in tungsten trioxide, 60
rate of, 95
resistance to, 105
charging, double-layer, 52
chemical diffusion coefficient, 46, 84T, 87, 88, 90, 96, 101, 102, 112, 190, 195T
   and diffusion coefficient, 84
   and insertion coefficient, 90–1
   definition of, 84
   electrode reactions when, 47
   ions through oxides, 85T
     cobalt oxide, 195T
     ions through WO_3, 87, 88
     Li^+ in Li_xWO_3, 91
     molybdenum trioxide, 153
     nickel oxide, 85T
     niobium pentoxide, 85T
     titanium dioxide, 184
     tungsten trioxide, 84T, 85T, 101, 195T
     vanadium pentoxide, 85T
   ions through oxide mixtures
     cerium–titanium oxide, 194
     cobalt–tungsten oxide, 195, 195T
     indium–tin oxide, 197
     tungsten–cobalt oxide, 195T
   ions through phthalocyanines
     lutetium phthalocyanine, 85T
     zinc phthalocyanine, 85T
   ions through conducting polymers
     poly(carbazole), 85T
     poly(isothianaphene), 85T
chemical potential, of H^+ in WO_3, 93, 94
chemical tethering, write–erase efficiency, 346
chemical vapour deposition
   annealing needed, 131
   of metal oxides, 131–2
     iron oxide, 174
     molybdenum trioxide, 151, 397
     nickel oxide, 161
     praseodymium oxide, 178, 179
     tantalum oxide, 182
     tungsten trioxide, 141, 148, 150, 397
   of mixtures of metal oxide,
     tungsten–molybdenum oxide, 397
   precursors
     alkoxides, 131
     hexacarbonyls, 131, 135–6, 397
   products impure, 132
   process is two-step, 131
chemically modified electrode, see *derivatised electrodes*
chloranil, 382
   o-, 382, 383
   cycle life, 383
   p-, 382
chloride ion, gasochromic, sensor for, 406
Chroma meter, 62
   and colour analysis, 63, 64–71

chromatic colour, and colour analysis, 62, 64
chromium oxide, 167
   and batteries, 167
   coloration efficiency, 167
   electrochemistry, 167
   formation via
     electron-beam evaporation, 167
     rf sputtering, 167
   gasochromic, 407T
   terminal effect suppressor, 423
chromium phthalocyanine, 261
chromium–iron–nickel oxide, 203
chromium–molybdenum oxide, 199
chromium–nickel oxide, 200
chromogenic glazing, 397; see also *ECD, windows*
chromophore, definition, 2
chronoabsorptometry, 57
chronoamperometry, 83
   peaks, 99–101
chronocoulometry, 57, 59
CIE, see *Commission internationale de l'eclairage*
circuit element, 50
clusters, c-WO_3 in a-WO_3, 88
cobalt acetylacetonate complex, 168
cobalt hydroxide, 169
cobalt oxide, 167–70, 195
   annealing, 168
   charge transfer in, 169
   chemical diffusion coefficient, 195T
   coloration efficiency, 56T, 169, 172T
   ECDs of, 170
   electrochemistry of, 168–9
   electrochromic host, 195–6
     incorporating gold, 204T
   formation via
     CVD, 172T
     dip coating, 168
     electrodeposition, 132, 172T
     evaporation, 172T
     oxidation of cobalt, 168, 169
     peroxo species, 168
     rf sputtering, 167
     sol–gel, 135, 168, 172T, 195
     sonication, 133, 134
     spin coating, 135
     spray pyrolysis, 135, 168, 169, 172T
   gasochromic applications, 406
   lithium deficient, 167
   optical properties, 168, 169–70
   secondary electrochrome, 170
cobalt oxyhydroxide, 81, 168
   formation via electrodeposition, 168
cobalt phthalocyanine, 261
cobalt tartrate complex, 168
cobalt–aluminium oxide, 195, 196
   coloration efficiency, 195
   via sol–gel, 195
cobalt–aluminium–silicon oxide, 195, 204
cobalt–nickel–iridium oxide, 203

cobalt–tungsten oxide, 195
  chemical diffusion coefficient, 195, 195T
colloid, via sol–gel, 134
coloration, 2
  and colour analysis, 66
  extrinsic, 52–3
  after potential stopped, 114
  chemical, 101–2
  galvanostatic 96–8, 104, 417
  iridium oxide, 157
    phase changes in, 157
  metal oxides
    involves counter ions, 80
    involves ionisation of water, 81
  potentiostatic, 99, 104, 358, 417
    three-electrode, 443
  potential step, 354–5
  pulsed, 87, 365
  pulsed current, titanium dioxide, 184
  tailoring, 334
  tungsten trioxide, 80
    hysteresis, 143
    involves water, 80
    two-electron process, 103
  type-II electrochromes, 79–115
coloration efficiency, 10, 15, 16, 42, 54–60, 88, 139
  and conjugation length, 60
  and extinction coefficient, 55
  anodic coloration, η is negative, 55
  cathodic coloration, η is positive, 55
  composite CCE, 55, 57–60
  definition, 15
  intrinsic, 54–60
  metal hexacyanoferrates
    Prussian blue, 59T
    Prussian white, 59T
  metal hydrides
    magnesium–samarium hydride, 308
    samarium–magnesium hydride, 308
  metal oxides, 56T
    bismuth oxide, 56T, 166
    chromium oxide, 167
    cobalt oxide, 56T, 169, 172T
    copper oxide, 172
    iridium oxide, 56T, 70, 158
    iron oxide, 56T, 70, 175, 175T, 201T
    manganese oxide, 176
    molybdenum trioxide, 56T, 154, 155T, 199, 199T
    nickel oxide, 56T, 70, 165T
    niobium pentoxide, 56T, 178, 181T, 199T–201T, 200
    niobium pentoxide, mixtures, 201
    rhodium oxide, 56T, 181
    tantalum oxide, 56T, 183
    titanium dioxide, 56T, 184, 185T
    tungsten trioxide, 56T, 146, 147, 148, 148T, 191, 193, 201
    vanadium pentoxide, 56T, 189, 190T
  metal oxyfluorides
    titanium oxyfluoride, 205
    tungsten oxyfluoride, 205
  metals, bismuth, 306
  mixtures of metal oxide
    cerium–titanium oxide, 194
    cobalt–aluminium oxide, 195
    indium–tin oxide, 197, 199, 199T
    iron oxide, mixtures, 198
    iron–niobium oxide, 201T
    molybdenum–tin oxide, 199T–201T
    nickel–titanium oxide, 202
    nickel–tungsten oxide, 200
    niobium–iron oxide, 201T
    niobium–tungsten oxide, 201
    samarium–vanadium oxide, 202
    titanium–molybdenum oxide, 199
    tungsten–niobium oxide, 201
    tungsten–molybdenum oxide, 56T, 192
    tungsten–vanadium oxide, 202
    vanadium–samarium oxide, 202
    zirconium–tantalum oxide, 203
  organic dyes
    Azure A, 57T
    Azure B, 57T
    Basic Blue ix, 57T
    Indigo Blue, 57T
    Methylene Blue, 57T
    Nile Blue, 57T
    Resazurin, 57T
    Resorufin, 57T
    Safranin O, 57T
    Toluylene Red, 57T
  organic electrochromes, 57T
    cyanines, 378
    fullerene, 304, 305–6
  organic polymers
    PEDOT, 59T, 437
    poly(3,4-ethylenedioxy thiophenedidode-cyloxybenzene), 57T
    poly(3,4-propylenedioxypyrrole), 57T
    poly(3,4-propylenedioxythiophene), 57T
  phthalocyanines
    lutetium phthalocyanine, 260
  quantum mechanical, 55
  sign of, 55
  viologens, 349, 361, 362, 363; methyl viologen, 57T
coloration models
  Bohnke, 101–2, 113, 115
  Faughnan and Crandall, 91–6, 99, 102, 110, 111, 113, 115
  Green, 96–8, 102, 113, 115
  Ingram, Duffy, Monk, 99–101, 102, 113, 115
  W^{IV} and W^{V}, 102–3
coloration rate, 33, 139, 149
  and flux, 75
  mixing oxides, enhances rate, 200
  nickel oxide, 163
  vanadium pentoxide, 188
colorimetric theory, 62
colour analysis, 62–71
  and light sources, 64

conducting polymers, 62
  Prussian blue, 62, 70
colour diagram, achromatic centre, 67
colour formed, amount of, 53
colour manipulation, metal-oxide mixtures, 190
colour space, 63, 64–71
colour tailoring, 399
combinatorial chemistry, 409
Commission internationale de l'eclairage (CIE), 62, 63, 334
complementarity, during cell operation, 41
complementary electrochromism, 290
complexes, see, *charge-transfer complexation; coordination complexes*
composite coloration efficiency CCE, 55, 57–60
  determined at wavelength maximum, 59
  determined with reflected light, 57
composites, conducting polymer, 332–3
comproportionation
  tungsten trioxide, 103
  viologens, 357–8, 365
computer screen, ECD applications, 363
concentration gradient, 45, 51, 93, 97, 98, 104, 110, 112, 114, 115, 303, 305
conducting polymers, 9, 62, 80, 312–34
  and electroluminescent organic light-emitting diodes, 312
  and field-effect transistors, 312
  and sensors, 312
  and solar-energy conversion, 312
  colour analysis of, 62
  composites, 332–3
  electrochromic, 57, 60
  high resistance, 11
  history, 312
  oxidative polymerisation, 313–14
  p-doping, 315
  type-III electrochromes, 317
conductivity, electronic, 113
  phthalocyanine complexes, 263
  silver paint, 349
  through bands, 81, 127, 147
conductivity
  indium–tin oxide, 422
  ionic, metal oxides, 89
  M_xWO_3, 81, 113, 142
  protons in tantalum oxide, 181, 183
  of amorphous and polycrystalline WO_3, 82
conjugation length, 314
  and coloration efficiency, 60
construction, ECD, 417
contact lithography, 258
contrast ratio, 9, 14, 104, 146, 156, 189, 197, 333, 346, 348, 349, 352, 376, 384, 385, 388, 400
  all-polymer ECD, 332
  and electrolyte fillers, 445
convection, 43, 44, 75, 76
  absent in solid-state ECDs, 44
coordination complexes, 253
  intervalence charge transfer, 253
  metal-to-ligand charge transfer, 253

copper ethoxide, 170
copper hexacyanoferrate, 294
copper oxide, 170–2
  as secondary electrochrome, 165
  coloration efficiency, 172
  electrochemistry, 172
  formation via
    copper ethoxide, 170
    electrodeposition, 171
    electron mediator, 305, 306
    sol–gel, 170
  specular reflectance, 407T
Cottrell equation, 76, 77, 354
Coulomb's law, 102
counter electrode, 41, 48
  electrochromic, see *secondary electrochrome*
counter ion
  activation energy, 408
  movement, 82–5, 188
    during coloration of metal oxides, 80
  rate of, 33
  size, 87–8
  swapping of, 87
  through solid film, 86
  viologens, effect of, 352–4
  Ag^+ through tungsten trioxide, 142, 146
  CN^- through iridium oxide, 157
  Cs^+ through tungsten trioxide, 146
  deuterons through tungsten trioxide, 87
  F^- through iridium oxide, 157
  K^+ through
    iron oxide, 174
    iron oxide mixtures, 198
    tin oxide, 205
    tungsten trioxide, 142, 146
  Li^+ through
    cerium oxide, 167
    cerium–titanium oxide, 194
    cobalt oxide, 169
    fullerene, 303, 304–5
    graphite, 303, 304
    iron oxide, 173, 174
    iron oxide mixtures, 198
    ITO, 197
    manganese oxide, 176
    molybdenum trioxide, 152
    nickel oxide, 163
    niobium pentoxide, 178
    praseodymium oxide, 179
    tin oxide, 197, 205
    titanium dioxide, 184
    tungsten trioxide, 88, 89, 90, 96, 113, 130, 142, 146, 148, 150, 151, 410, 419
    vanadium pentoxide, 186, 188
  Mg^{2+} through
    molybdenum oxide, 152
    tungsten trioxide, 146
  Na^+ through
    iron oxide, 174
    iron oxide mixtures, 198
    tin oxide, 205

counter ion (cont.)
    tungsten trioxide, 87, 90, 109, 113, 130, 142, 146, 147
    vanadium pentoxide, 186
    OH^- through
    anodic oxides, 87
    cobalt oxide, 195, 195T
    lutetium phthalocyanine, 259
CR, see *contrast ratio*
critical micelle concentration, heptyl viologen, 355
CRT, see *cathode ray tube*
crystal lattice
    changes during coloration, 86–7
    motion of is rate limiting, 87
    stresses in, 130
crystal violet, 376
crystallisation, by annealing, 89
CT, see *charge transfer*
current, 38, 41
    as rate, 38
    coloration, 93
    definition, 42, 77
    depends on rates, 75
    faradaic, 45, 76
    leakage, 52
    limiting, 76
    non-faradaic, 76
    parasitic, 52
CVD, see *chemical vapour deposition*
cyanines, 376
    coloration efficiency, 378
    electrochromic, 60
    merocyanines, 376
    spiropyrans, 376
    squarylium, 379
cyanophenyl paraquat, 8, 28, 60, 344T, 349, 350, 351, 352, 356, 358
    charge transfer complexation, 359
    diffusion coefficient, 77T
    optical charge transfer in, 60
cyanotype photography, of Prussian blue, 26
cycle life, 12–13, 172, 178, 179, 188, 197, 205, 269, 294, 303, 304, 305, 308, 362, 383, 389, 443, 444T, 447
    and kinetics, 11
    deep and shallow cycles, 12, 443
    enhanced by mixing oxides, 200
    measurement of, 12
cyclic voltammetry, 48–50, 83
    of conducting polymers, poly(aniline), 333
    of metal hexacyanoferrates
        copper hexacyanoferrate, 294
        Prussian blue, 286, 287
    of metal oxides
        iridium oxide, 156
        niobium pentoxide, 178
        rhodium oxide, 181
        tungsten trioxide, 93
        vanadium pentoxide, 187
    of viologens, 352, 355, 356, 357, 359
    schematic, 48
cyclodextrin, beta, 351, 359

Darken relation, 85
data display, ECD applications, 149, 363
dc magnetron sputtering, 136
    of mixtures of metal oxide
        cerium–titanium oxide, 194, 195
        indium–tin oxide, 136
        titanium–cerium oxide, 136
        tungsten–cerium oxide, 136
    of metal oxides
        molybdenum trioxide, 136, 141, 151
        nickel oxide, 136
        niobium pentoxide, 136, 177
        praseodymium oxide, 136, 178
        tantalum oxide, 136, 182
        tungsten trioxide, 136
        vanadium pentoxide, 136, 185
    of metal oxyfluorides
        titanium oxyfluoride, 205
        tungsten oxyfluoride, 205
    onto ITO, 136
DDTP, 326
decomposition, of electrochrome, 49
deep cycles, cycle life, 443
defect sites, 103, 127, 146
DEG, see *diethylene glycol*
degradation, 443
    acid, sulfuric, 420
    aquatic, 89
    mechanical stresses, 397
        caused by ion movement, 13
    fullerene electrochromes, 303, 305
    indium–tin oxide, 423, 444–5
    lutetium phthalocyanine, 260
    metal oxides, photolytic, 54, 125
    molybdenum trioxide, 153
    nickel oxide, 163
    tungsten trioxide, 149, 150
        via Cl^- ion, 150; yields tungstate, 89
    vanadium pentoxide in acid, 186
    viologens, 351, 357
DEMO 2005 show, 402
deposition *in vacuo*, 137–8
depth profiling, 89
derivatised electrodes, 7
    definition, 12
    pyrazolines, 387–8
        contrast ratio, 388
        ECD, 388
        response times, 387
        TCNQ species, 388–9
        reversibility, 389
        write–erase efficiency, 388
    TTF species, 387, 389–90
        cycle life, 389
        ion hopping, 390
        ion tunnelling, 390
    viologen ECDs, 346–8, 361
desolvation, during ion insertion, 89

deuteron, motion through WO_3, 87
diacetylbenzene, p-, 77, 78
  immobilised, 390T, 391T
diamond electrochromes, 303, 305
  absorption in near infrared, 399
dielectric properties, 50
diethyl terephthalate, immobilised, 391T
diethylene glycol, 331
diffusion, 43, 111, 386
  activation energy for, 83
  energetics of, 112
  fast track, 98
  length, 45, 101, 403
  linear, 76
  of electrochromes, 12
diffusion and migration, concurrent, 83
diffusion coefficient, 44, 45, 48, 77, 77T, 83, 90–1, 96, 97, 101, 102, 112, 403, 407T
  and chemical diffusion coefficient, 84
  and oxygen deficiency, 103
  includes migration effects, 83
  solution-phase species
    cyanophenyl paraquat, 77T
    ferric ion, 77T
    methyl viologen, 77T
diffusion rate, 33
digital video disc, 408
dihedral angle, 314, 320
  poly(thiophene)s, 323T
dihydro viologen, see *viologen, doubly reduced*
dimer, of W^V–W^V, 103, 145, 147
dimethoxyphenanthrene, 2,7-, 380
dimethylterephthalate, immobilised, 391T
dimmable window laminates, ECD applications, 363
dinuclear ruthenium complexes, mixed-valency, 268T
  near infrared electrochromism, 268T
diode-array spectroscopy, 355
dioxypyrrole, 327–8
dip coating
  of metal oxides, 135
    cerium oxide, 135
    cobalt oxide, 168
    iridium oxide, 135
    iron oxide, 135
    nickel oxide, 135, 161
    niobium pentoxide, 135
    tantalum oxide, 182
    titanium dioxide, 135, 184
    tungsten trioxide, 135, 141
    vanadium pentoxide, 135
  of mixed metal oxides, 135
    iron–titanium oxide, 202
    titanium–iron oxide, 201, 202
  substrates, ITO, 135
directed assembly, ECD, 157
  of Prussian blue, 285
di-reduced, viologens, 343, 357, 358
  ethyl viologen, 358
  heptyl viologen, 358

methyl viologen, 358
displays, see *applications, ECD*
dissolution, of WO_3, 89
dithiolene complexes, 266–7
DMF, as ECD electrolyte, 254, 419
DMSO, as ECD electrolyte, 150, 157, 261
dodecylsulfonate, within poly(pyrrole), 333
dominant wavelength, and colour analysis, 62
Donnelly mirror, 11
  as sunglasses, 401
double insertion, of ions and electrons, 138
double potential step and cycle life, 12
double-layer, charging, 11, 52
Dreamliner, Boeing, windows, 400
Drude theory, 101, 142
Drude–Zener theory, 142
dry-cell, battery, 408
dry lithiation, 418
  of tungsten trioxide, 418
dual insertion, of ions and electrons, 138
  during coloration and bleaching, 83
DuPont, 425
durability, 443–9
  accelerated tests
    humidity, 447, 448
    weathering, 447
    xenon arc, 448
  of ECD electrolyte, 445
  of ECDs during pulsing, 104
  of substrates, 444–5
Duracell, 408
DVD, see *digital video disc*
dyes, encapsulated within poly(aniline), 333
dynamic electrochemistry, 46–8
dysprosium–vanadium pentoxide, 202

$E_{(cell)}$, see *emf*
ECD, 53, 60, 76, 106, 108, 112
  all polymer, 330, 331–2
  applications, see *applications, ECD*
  assembly, 157, 417
  directed assembly, 157
  dual organic–inorganic, 333–4
  durability, 443–9
  electrodeposited bismuth, 306
  electrodes, 52, 419–24
  electrochromes
    conducting polymers,
      PEDOT, 409
      poly(aniline)s, 330, 331
      poly(pyrrole)s, 328
    inorganic electrochromes, oxo-molybdenum complexes, 269
    metal hexacyanoferrates, Prussian blue, 289–91
    metal hydrides, lanthanide hydride, 308
    metal oxides
      cobalt oxide, 170
      iridium oxide, 159
      manganese oxide, 176
      molybdenum trioxide, 154–5, 397
      nickel oxide, 164–5, 397

ECD (cont.)
   niobium pentoxide, 178
   of tungsten trioxide, 61, 82, 87, 104, 139, 397, 399, 402, 408, 409, 410
   vanadium pentoxide, 189–90
  mixtures of metal oxide
   indium–tin oxide, 197
   tungsten–molybdenum oxide, 397
  phthalocyanine complexes, 263
   lutetium phthalocyanine, 259, 260
  organic
   pyrazolines, 388
   quinones, 384
   thiazines, 385
  viologens, 346–8, 349, 352, 357, 362, 385
   heptyl viologen, 360
   viologens, paper quality, 362
electrolytes
  acetonitrile, 254, 262, 359, 385, 390T, 419, 438
  antimony pentoxide, 421T
  DMF, 254, 419
  DMSO, 150, 157, 261
  ethylene glycol, 259
  fillers (titanium dioxide), 421
  γ-butyrolactone, 167, 186, 303, 304, 362, 419
  gelled, 106, 305, 350, 384
  hydrogen uranyl phosphate, 421T
  inorganic, 420
  lead fluoride, 159
  lead tetrafluorostannate, 159
  lithium niobate, 421T
  lithium pentafluoroarsenate, 150
  lithium perchlorate, 82, 150, 151, 152, 163, 166, 167, 169, 173, 176, 184, 186, 188, 197, 199, 205, 362, 408, 421
  lithium phosphorous oxynitride, 363
  lithium tetrafluoroaluminate, 150, 152, 421T, 436
  Nafion, 421T
  organic, 420–2
  perchloric acid, 150, 157
  phosphoric acid, 167, 421, 438
  polyelectrolytes, 420–2
   Nafion, 421T
   poly(AMPS), 150, 260, 348, 366, 391, 391T, 395–410, 420
  polymer electrolytes, 421–2
   poly(acrylic acid), 150, 421T
   poly(ethylene oxide), 150, 290, 408, 421, 438
   poly(methyl methacrylate), 291, 334
   poly(propylene glycol), 421
   poly(vinyl alcohol), 421
   poly(1-vinyl-2-pyrrolidone-co-N,N'-methylenebisacrylamide), 391
  potassium chloride, 291, 349
  potassium hydroxide, 308
  potassium triflate, 290
  propylene carbonate, 151, 152, 166, 169, 173, 176, 184, 186, 187, 188, 197, 199, 205, 356, 384, 419
  solid, 96
  stibdic acid polymer, 421T
  sulfuric acid, 82, 86, 149, 178, 259, 349, 409, 420
  tantalum oxide, 150, 420, 421T
  thickeners, 419
   acrylic powder, 419
   poly(ethylene oxide), 419
   poly(vinylbutyral), 419
   silica, 419
  tin phosphate, 154
  titanium dioxide, 421T
  triflic acid, 150, 421
  viscosity, 417
  whiteners, 159, 384, 418, 422, 424
  zinc iodide, 408
  zirconium dioxide, 421T
encapsulation, 424–5, 448
  Surlyn, 425
first patents, 27
flexible, 129, 423
illumination of, 417
large-area, 141, 332, 447
memory effect, 15, 53–4, 152, 153, 403
sealing, 362
self bleaching, 15, 54, 150, 153
  and memory effect, 15, 53–4, 152, 153, 403
  radical annihilation, 386
type-I electrochromes, 77
substrates, 422–4
trichromic, 384
ultra fast, viologens, 363
EDAX, Prussian blue, 288
EDOT, 323, 325, 326
  polymers of, 325
EIC laboratories, 409
Einstein transition probability, 147
electric field, 43, 44, 138
electric paint, ECD application, 364
electroactive material, definition, 1
electroactive polymers, 9
electrochemical cells, 417
electrochemical formation of colour, 52–3
electrochemical impedance spectroscopy (EIS), see *impedance*
electrochemical quartz-crystal microbalance (EQCM), 88, 89, 90, 130, 142, 163, 284, 288, 289, 330, 331
electrochemical titration, 104
electrochemistry, 11
  dynamic, 46–8
  equilibrium, 34–9
  electrochromes
   electrodeposition, bismuth, 304, 305
   hexacyanoferrates, Prussian blue, 285–9
   metal oxides, 138
    cerium oxide, 167
    chromium oxide, 167
    cobalt oxide, 168–9
    copper oxide, 172
    iridium oxide, 157
    iron oxide, 173

Index

manganese oxide, 175–6
molybdenum trioxide, 152–3
nickel oxide, 161–3
niobium pentoxide, 177–8
palladium oxide, 178–9
praseodymium oxide, 178
rhodium oxide, 180
ruthenium oxide, 181
tantalum oxide, 183
titanium dioxide, 184
tungsten trioxide, 142
vanadium pentoxide, 186–8
metal oxyfluorides
   titanium oxyfluoride, 205
   tungsten oxyfluoride, 206
mixtures of metal oxide
   indium–tin oxide, 196–7
phthalocyanines, 262
   lutetium phthalocyanine, 260
polymers, 60
   poly(aniline), 329–30, 331
thermodynamics of, 34–9
viologens, 342, 353, 354–5
electrochromes
  changes in film thickness, 51
  colours of, 2
  decomposition, 49
  laboratory examples of, 3
  memory effect, 15, 53–4, 152, 153, 403
  metal-oxide systems and insertion coefficient, 61
  metal-oxide systems, intervalence of, 61
  photodegradation of, 54
  type, 7–9
electrochromic colours
  counter electrode, see *secondary electrochrome*
  device, see *ECD*
  electrodes, 40–1
  extrinsic intensity of, 52–3
  intensity of, 3
electrochromic hosts
  metal oxides, 190–206
   antimony oxide, 193
   cerium oxide, 193–5
   cobalt oxide, 195–6
   indium oxides, 196–7
   iridium oxide, 198
   iron oxide, 198–9
   molybdenum oxide, 199
   nickel oxide, 200
   niobium pentoxide, 200–1
   titanium dioxide, 201–2
   tungsten trioxide, 191–3, 407
   vanadium pentoxide, 202
   zirconium oxide, 203
  polymers, Nafion as, 405
electrochromic modulation, 3, 53
electrochromic paper, ECD application, 405–6
electrochromic probes, 3
electrochromic–photochromic systems, 438–9
electrochromism
  chemical, 3

complementary, 290
definitions, x, 1, 3
fax transmissions, 26
first use of term, 25
history of, 25–30
ligand-based, 255
near infrared, 165, 183, 253, 254, 265–74
electrode
  as conductor, 37
  ECD, 422–4
  interphase, 43
  kinetics, 46–8
  potential, 35, 39, 48, 75, 91, 93, 104
  reactions, 52
  reactions, under diffusion control, 47
  substrate, see entries listed under *substrate*
electrodeposition of metals, type-II electrochromes, 303, 305
electrodeposition
  forming hexacyanoferrates
   Prussian blue, 283, 284
  forming metals, 303, 305–7
   bismuth, 27, 304, 305–6
   lead, 306–7
   silver, 27, 307
  forming metal oxides, 132–4
   cobalt oxide, 132
   copper oxide, 171
   iron oxide, 173
   manganese oxide, 175
   molybdenum trioxide, 151, 152
   nickel oxide, 132, 160–1
   oxide mixtures, 133
   ruthenium oxide, 181
   tungsten trioxide, 140, 141
   vanadium pentoxide, 186
  forming mixtures of metal oxide
   molybdenum–tungsten oxide, 199
   nickel–titanium oxide, 201
   titanium–tungsten oxide, 202
   tungsten–molybdenum oxide, 199
  forming oxyhydroxides
   cobalt oxyhydroxide, 168
   nickel oxyhydroxide, 161
   nitrate forming metal hydroxide, 132
  forming viologen radicals, 354
  potentiostatic, 133
  precursors, peroxo species, 133
  yields oxyhydroxide, 132
electrokinetic colloids, 5
electroless deposition, of Prussian blue, 283
electroluminescent organic light-emitting diodes, conducting polymers, 312
electrolyte fillers to enhance contrast ratio, 445
electrolyte, ECD
  chemical systems, see *ECD, electrolyte*
  dissolves ITO, 444
  durability, 445
  failure of, 443

electrolyte, ECD (cont.)
  fillers, 445
  organic polymers, 445
  photochemical stability, 445
  semi-solid, 446
electrolytic capacitor, 52
electrolytic side reactions, 43
electrolytic writing paper, 405
electron conduction, through bands, 81
electron donors, photochromism, 438
electron hopping, 81, 99
electron mediation
  bismuth electrodeposition, 305
  mediators
    cerous ion as, 359
    copper as, 305, 306
    ferrocene as, 359
    ferrocyanide as, 342, 350, 358, 359
    ferrous ion as, 359
    hydroquinone as, 359
electron mobility, tungsten–molybdenum oxide, 192
electron transfer
  energy barrier to, 42–3, 47
  fast, 102
  rate of, 33, 34, 42–3, 46, 75
  standard rate constant of, 47
electron-beam evaporation, of chromium oxide, 167
electron-beam sputtering
  forming metal oxides
    manganese oxide, 137, 175
    molybdenum trioxide, 137
    vanadium pentoxide, 138–206
  forming metal oxide mixtures
    indium–tin oxide, 137, 196
electroneutrality, need for, 8
electronic bands, 127
electronic charge, 42
electronic conductivity, 42, 113
  in metal oxides
    nickel oxide, 162
    tungsten trioxide, 99
  in metal oxide mixtures
    indium–tin oxide, 445
  in phthalocyanine complexes, 259
  in polymers
    poly(acetylene), 312
    poly(aniline), 101
  rate of, 42
electronic motion, 81–2
electronic paper, ECD applications, 363
electron–ion pair, see redox pair
electron-transfer rate, viologens, 359
electron-transfer reaction, 75
electrophotography, of tungsten trioxide, 28
electropolychromism, 17–18
  graphite electrochromes, 303, 304
  poly(aniline), 329, 331
  Prussian blue, 287
  seven-colours, 254
  polypyridyl complexes, 255–6

quinones, 384
viologens, 365
electroreduction, of ITO, 444
electroreversibility poor, indium–tin oxide, 197
electrostriction, 51, 129, 445
  definition, 87
  of iridium oxide, 130
  of nickel oxide, 130
  of tungsten trioxide, 87, 129, 445
  of vanadium pentoxide, 87, 129
element, circuit, 50
ellipsometry, 17, 50–1, 81, 109–10
  and film thickness, 50
  and interfaces, 50
  in situ, 51
  of iridium oxide, 157
  of molybdenum trioxide, 109, 153
  of phthalocyanine complexes, 263
  of Prussian blue, 284, 287, 288, 289
  of titanium dioxide, 184
  of tungsten trioxide, 81, 109, 143
  of vanadium pentoxide, 109, 187
emeraldine, 329, 331
emf, 33, 34, 39–40, 94, 95, 104, 143
encapsulation, ECD, 424–5, 448
energetics, 86
  of ion movement through solid oxides, 89–90
energy barrier, 93, 95
  to electron transfer, 47
enhancement factor W, 83, 84
entropy, 88
environmentalism, 398
epoxy resin, 362
equilibrium potential, 35, 41
equivalent circuit, 447
equivalent circuit, impedance, 447
erbium laser, 267
ESCA, 103
ESR
  of methyl viologen, 356
  of molybdenum trioxide, 153
  of tungsten trioxide, 145
  of viologens, 352, 356
ethyl viologen, 344T, 352T, 438
  di-reduced, 358
ethylanthraquinone, 2-, 384
N-ethylcarbazole, carbazoles, 379T, 381T
ethylene glycol, ECD electrolyte, 259
ethylenedioxythiophene, 3,4-, 313, 321–7
evaporated metal-oxide films, water in, 89
evaporation, vacuum
  of bismuth oxide, 166
  of metal-oxide films, 89
  of molybdenum trioxide, 151
  of nickel oxide, 160
  of tantalum oxide, 182
  of tungsten trioxide, 140–1, 147, 150
  of vanadium pentoxide, 185, 186
Eveready, battery charge indicator, 408, 443
Everitt's salt, see Prussian white

EXAFS, of cerium–titanium oxide, 194
exchange current, 47, 95
extinction coefficient, 53, 55, 60, 61, 113, 269, 274, 294, 343, 344T, 349
   and coloration efficiency, 55
extrinsic colour, 52–3
eye, human, see *human eye*
eye wear, see *applications, ECD, eye wear*

Faradaic current, 45, 52, 76
Faraday constant, 34
Faraday's laws, 46, 52
fax transmission, using electrochromism, 26
F-centres, 28
ferric ion, diffusion coefficient, 77T
ferricyanide, 342
   as oxidant, 342
ferrocene
   derivatives, 330, 331
   electron mediator, 359
ferrocene–naphthalimide dyads, 309
ferrocyanide
   charge transfer complexation, 359
   electron mediator, 342, 350, 358, 359
   incorporation into
      nickel oxide, 200
      titanium dioxide, 201
   mediating viologen comproportionation, 358
ferroin, 253
ferrous ion, as electron mediator, 359
fibre-optics, ECD, applications, 265, 404
Fick's laws, 44, 45, 50, 76, 111
   approximation, 45
   first law, 44
   second law, 45, 95
field-effect transistors, conducting polymers, and 312
fillers, ECD electrolyte, 445
film thickness, and ellipsometry, 50
Flabeg Gmbh, ECD, windows, 400
flash evaporation, of vanadium pentoxide, 185
flat-panel screens, TV, 402
flexible ECD, 129, 423
   on indium–tin oxide, 423
fluoreneones, 18, 379, 380, 387
   quasi reversibility of, 380
   2,4,5,7-tetranitro-9-fluorenone, 387
   2,4,7-trinitro-9-fluorenylidene malononitrile, 387
fluorescence, 5
fluorine-doped tin oxide, as substrate, 139, 166, 168, 171, 196, 205, 292, 362, 400, 406, 409, 422
fluoroanil, *p*-, 382
flux, 44, 97
   and colour formation, 75
formation of colour, electrochemical, 52
Fox Talbot, 26
frequency, and impedance, 50
frequency response analysis FRA, see *impedance spectroscopy*
fullerene electrochromes, 303

coloration efficiency, 304, 305–6
degradation of, 303, 305
formation via Langmuir–Blodgett, 304, 305
quasi-reversiblity, 303, 305
near-infrared absorbance, 399
furan, 313
fused bithiophenes, conducting polymers of, 316

gallium hexacyanoferrate, 295
galvanostatic coloration, 96–8, 104, 417
games, ECD applications, 363
gamma rays, 110
gasochromism, 5, 406–7, 407T
   materials
      chromium oxide, 407T
      cobalt oxide, 406
      metalloporphyrin, 407T
      nickel oxide, 407T
      phthalocyanine, 407T
      tungsten trioxide, 407T
   sensors
      for chloride ion, 406
      for nitrate ion, 406
      nitric oxide, 407
      phosphate ion, 406
      toluene, 407
gelled ECD electrolyte, 305, 349, 350, 351, 384
   using agar, 349, 351
   using silica, 348
Gentex Corporation, 376, 385, 396, 398, 417, 425, 447
   aircraft windows, 400
   mirrors (Night-Vision System), ix, 44, 356, 385–7
   cycle life, 356
   memory effect, 387
   radical annihilation, 386
   type-I electrochrome, 396
Gibbs energy, 34–9
   and *emf*, 34
glassy carbon, substrate, 294, 358
gold, 150, 153, 159
   additive
      in cobalt oxide, 204T
      in iridium oxide, 204T
      in molybdenum trioxide, 204T
      in nickel oxide, 200, 204, 204T
      in tungsten trioxide, 204, 204T
      in vanadium pentoxide, 204, 204T
   as substrate, 285
   overlayer of, 446–7
gold nanoparticles, overlayer of, 446
graft copolymer, poly(aniline), 333
grain boundaries, 88, 98, 146
graphite
   electrochromes, 303, 304–5
   electropolychromic, 303, 304
   substrate, 424
Grötthus, conduction in metal oxides, 90
Gyridon 'electrochromic paper', 5

half reaction, 35
Hall effect, 113
hematite, 173
Henderson–Hasselbalch equation, 4
He–Ne laser, 438
heptyl viologen, 8, 9, 11, 14, 28, 190, 344T, 346, 348, 349, 351, 352T, 352–3, 354–5, 356, 357, 359
  anion effects, 353T
  as primary electrochrome, 356, 375, 385
  charge transfer complexation, 359
  critical micelle concentration, 355
  di-reduced, 358
  ECDs of, 360
  incorporated in paper, 365
  morphology of, 355
  power consumption, 14
  radical of, 357, 359
    aging effects, 357
  recrystallisation of, 357
  reduction potentials, 353T
  solubility constant, 351
hexacarbonyl, as CVD precursor, 131, 135–6, 397
hexacyanoferrate(II), see *ferrocyanide*
hexacyanoferrate(III), see *ferricyanide*
hexacyanoferrate of
  copper, 294
  gallium, 295
  indium, 295, 437
  iron, see *Prussian blue*
  miscellaneous, 295–6
  mixed-metal, 296
  nickel, 293–4
  palladium, 294–5
  vanadium, 292–3
hexyl viologen, 352T
history
  of conducting polymers, 312
  of electrochromism, 25–30
  of Prussian blue, 282
history effect, 131
hopping
  electron, 81, 99, 127
  polarons, 143
hue, and colour analysis, 56T, 62, 64, 70
human eye, spectral response, 62
hydride, electrochromic, 307–8
  Anderson transition in, 307
  cycle life, 308
  durability, 307, 308
  ECD, 308
  electrochromic alloys, 308
    lanthanum–magnesium, 308
    samarium–magnesium, 308
  electrochromic metal
    lanthanum, 307
    yttrium, 307
  mirrors, 307
  palladium overlayer on, 307
  response time, 307
  switchable mirrors, 307

hydrogen
  electrode, 36, 37, 40
  evolution at molybdenum oxide, 199
  evolution at tungsten oxide, 89, 102, 104, 445
  uranyl phosphate, ECD electrolyte, 421T
hydrogen peroxide, 135, 308
hydroquinone, electron mediator, 359
hygroscopicity, of metal oxides, 89
hysteresis, 104, 157

IBM Laboratories, 28, 30, 355, 360, 403, 405
ICI Plc, 28, 341, 349–51, 352
illumination
  back-wall, 433, 435
  front-wall, 433
  light sources, 64
  of ECDs, 417
imaginary, impedance 50
immitance, 50
immobilised viologens, see *derivatised electrodes*
impedance spectroscopy, 50, 83, 85, 333
  and frequency, 50
  equivalent circuit, 447
  imaginary, 50
  real, 50
incident light, 50
Inconel-600, oxide on, 203
Indigo Blue, coloration efficiency, 57T
Indigo Carmine, within poly(pyrrole), 333
indium hexacyanoferrate, 295, 437
indium nitride, 309
indium oxide, as electrochromic host, 196–7
indium–tin oxide
  chemical stability, 423
    cycle life, 197
    degradation of, 444–5
  composition of, 196
    containing silver, 204T
  ECDs of, 197
  electrochemistry of, 196–7
    electroreduction of, 444
    electro-reversibility poor, 197
  formation via
    CVD, 132
    dc magnetron sputtering, 136
    electron-beam deposition, 137, 196
    laser ablation, 196
    rf sputtering, 196
    sol–gel, 196
    spin coating, 135, 196
  kinetics of, chemical diffusion coefficient, 197
  optical properties
    as secondary electrochrome, 197
    coloration efficiency, 197, 199, 199T
    contrast ratio, 197
    optical properties, 197
    optically passive, 17, 197, 199
  flexible ECDs, 423
    on Mylar, 423
    on PET, 129, 423
    on polyester, 423

mechanical stability, 129
resistance, effect of, 349
   electronic conductivity, 422, 445
   substrate, 17, 70T, 86, 96, 128, 129, 135, 138, 139, 141, 150, 151, 152, 156, 158, 159, 164, 166, 167, 181, 182, 191, 257, 284, 293, 294, 305, 306, 326, 330, 331, 333, 349, 375, 382, 385, 404, 417, 422–3, 444–5, 447
water sensitivity, 444
XPS of, 197, 445
indole, 313
inert electrode, 38
infra red spectroscopy, 103, 358
inorganic–organic, dual ECD, 333–4
insertion coefficient, 9, 41, 53, 61, 81, 83, 87, 90–1, 92, 95, 96, 101, 104, 108, 113–14, 143, 146, 186, 188, 191, 192, 193, 303, 305
   effect on diffusion coefficient, 90–1
   effect on electroreversibility, 82
   effect on wavelength maximum, 53
   high at grain boundaries, 104
   metal-oxide systems, 61
intensity
   and colour analysis, 63
   of electrochromic colours, 3
interactions, counter ion with water, 89
interfaces, between films, 50
international meetings on electrochromism (IME), x
interphase, electrode, 43
intervalence charge transfer, 61, 102, 125, 127, 145, 153, 188, 192, 253, 267, 284
   heteronuclear, 127
   homonuclear, 127
intrinsic coloration efficiency, 54–60
iodine, 27, 437
iodine laser, 266
ion-conductive electrolyte, tantalum oxide, 181, 183
ion–electron pair, see *redox pairs*
ionic interactions, 36
ionic mobility, 99, 303, 305–7
ionisation, of water, 89
   during coloration, 81
iPod screen, ECD application, 363, 402
IR drop, 153, 405, 422–3
iridium oxide, 10, 125, 155–9, 198
   annealing of, 158
   as secondary electrochrome, 16, 149, 444T
      coloration efficiency, 56T, 70, 158
      reflectance of, 400
   coloration mechanism, 157
   containing
      aramid resin, 198
      gold, 204T
      water, 156
   electrostriction of, 130
   ECDs, 159
   electrochemistry, 157
   electrochromic host, 198
   ellipsometry, 157
   formation via

      anodically grown on Ir, 155–6
      dip coating, 135
      $IrCl_3$, 156
      iridium–carbon composite, 156, 158
      peroxo species, 156
      sol–gel, 156
      spray deposition, 158
      sputtering, 155, 156
   hysteresis of, 157
   mechanical stability, 129
   optical properties, 158
   phase changes, 157
   response time, 156, 159
   specular reflectance, 407T
   water content, 156
   write–erase efficiency, 156
   XPS of, 157
iridium trichloride, 156
iridium trihydroxide, 157
iridium–carbon composite, 156, 158
iridium–cobalt–nickel oxide, 203
iridium–magnesium oxide, 198
iridium–ruthenium oxide, 203
iridium–silicon oxide, 198
   formation via sol–gel, 198
iridium–tantalum oxide, chemical diffusion coefficient, 198
iridium–titanium oxide, 198
   formation via sol–gel, 198
iron acetylacetonate, 174
iron hexacyanoferrate, see *Prussian blue*
iron oxide, 172–5, 201
   annealing, 173, 175
   electrochemistry, 173
   formation via
      CVD, 174, 175T
      dip coating, 135
      electrodeposition, 173, 175T
      oxidised film on Fe metal, 172
      sol–gel, 173, 174, 175T
      spin coating, 135, 174
   mixtures, coloration efficiency, 198
      as electrochromic host, 198–9
   optical properties, 175
      coloration efficiency, 56T, 70, 175, 175T, 201T
      secondary electrochrome, 174
iron oxyhydroxide, 173
iron perchlorate, 173
iron phthalocyanine, 261
iron polypyridyl complexes, 256
iron vanadate, 202
iron–molybdenum oxide, 199
iron–nickel–chromium oxide, 203
iron–niobium oxide, 200, 201
   coloration efficiency, 201T
   formation via sol–gel, 200
iron–titanium oxide
   charge transfer of, 202
   formation via dip coating, 202
irreversibility, when oxidising Li_xWO_3, 82
iso-pentyl viologen, 352T

IUPAC, 55, 147, 161
IVCT, see *intervalence charge transfer*

J-aggregates, 265
junction potential, 39

K-glass, substrate, 422
kinetics
  bleaching
    models
      Faughnan and Crandall, 105–8
      Green, 108, 109
    of metal oxides
      nickel oxide, 164
      vanadium pentoxide, 188
    potentiostatic, 105–8
    type-II, 79–115
    type-III, 79–115
  coloration, 75–115
    of amorphous oxides, 88
    electron as rate limiting, 99
    faster in damp films, 90
    of lutetium phthalocyanine, 260
    type-I, 75–9
    type-II, 75–9
    type-III, 91–115
  effect of counter-ion size on, 87–8
  effect of morphology on, 88
  effect of high resistance of polymers, 11
  effect of water on, 89
  electrochrome transport, 75
  electron transfer, 33, 34
  rate-limiting process, 33, 83, 87, 92
  write–erase efficiency and, 11
Kosower, solvent Z-scale, 343

$L^*a^*b^*$ colour space, 64–71
  data for Prussian blue → Prussian white, 70T
$L^*u^*v^*$ colour space, 64–71
laboratory examples, of electrochromes, 3
Langmuir–Blodgett deposition, 138–206
  forming fullerene electrochromes, 304, 305
  forming phthalocyanine, 262–3, 407
lanthanide hydride, see *hydride, lanthanide*
lanthanum–nickel oxide, 200
large-area ECDs, 141, 332, 447
laser ablation, forming
  indium–tin oxide, 196
  tantalum oxide, 183
  titanium dioxide, 184
  vanadium pentoxide, 185
laser, 404
  Q-switching, 267
  types
    erbium, 267
    He–Ne, 438
    iodine, 266
    YAG, 266, 267
laser-beam deflection, 87, 130, 157
lattice constants, 129
lattice energy, 93

Prussian blue, 289
lattice defects, nickel oxide, 163
lattice stabilisation, 112
layer-by-layer deposition
  of PEDOT:PSS, 329, 331
  of poly(aniline), 329–30, 331
  of poly(viologen), 328–9, 331
LCD, see *liquid crystal display*
lead, electrodeposition of, 8, 306–7
lead fluoride, as ECD electrolyte, 159
lead tetrafluorostannate, as ECD electrolyte, 159
leakage current, 52
leucoemeraldine, 329, 331
LFER, see *linear free-energy relationships*
ligand based, electrochromism, 255
ligand-to-metal charge transfer, 269
light modulation, ECD application, 404–5
light-emitting diodes, 363, 402
lightness, and colour analysis, 63, 64, 66
limiting current, 76
linear diffusion, 76
linear free-energy relationships, 343
liquid electrolytes, transport through, 75
liquid-crystal display, ix, 53, 351, 360, 363, 402, 403, 404, 406, 408, 425
  power consumption of, 15
lithiation, dry, 418
lithium chromate, 167
lithium deficient, cobalt oxide, 167
lithium niobate, ECD electrolyte, 421T
lithium pentafluoroantimonate, ECD electrolyte, 150
lithium perchlorate, ECD electrolyte, 82, 106, 150, 151, 152, 163, 166, 167, 169, 173, 176, 184, 186, 188, 197, 199, 205, 362, 408, 421
lithium phosphorous oxynitride, ECD electrolyte, 363
  overlayer of, 446
lithium pnictide, specular reflectance of, 407T
lithium tetrafluoroaluminate, electrolyte ECD, 150, 152, 421T, 436
  overlayer of, 446
lithium tin oxide, 197
lithium tungsten bronze, 191
  electro-irreversibility of, 82
lithium vanadate, 190
  vanadate, thermochromic, 190
Lucent, 5
luminance, and colour analysis, 56T, 63, 64, 66, 70
lutetium phthalocyanine, 259–60, 261
  cation-free not electrochromic, 260
  chemical diffusion coefficient, 85T
  coloration kinetics, 260
  degradation, 260
  ECDs, 259, 260
  electrochemistry, 260
  protonated, 259
  response times, 260, 261
  formation via sublimation, 259
  write–erase cycles, 259

Madelung constant, 112
maghemite, 173
magnesium fluoride, terminal effect suppressor, 423
magnesium OEP, 265
magnesium phthalocyanine, 260
magnesium–iridium oxide, 198
magnesium–nickel, 200
magnesium–nickel–vanadium oxide, 203
magnetic susceptibility, 113, 345
magnetite, 168, 173
manganese oxide, 175–6, 446
   as secondary electrochrome, 165, 176
   ECDs, 176
   electrochemistry, 175–6
   optical properties, 176
      coloration efficiency, 176
   rechargeable batteries, 176
   formation via
      anodising Mn metal, 175
      electrodeposition, 175
      electron-beam sputtering, 137, 175
      rf sputtering, 175
      sol–gel, 175, 176
   XPS, 176
manganese phthalocyanine, 261
mass balance, 81
   nickel oxide, 162
mass transport, 33, 43–5, 75
mechanical stability, metal oxides, 129–30
medicine, ECD, applications, 265
melamine, plus vanadium pentoxide, 190, 202
membrane potentials, 4
   biological, 3
memory, 15, 53–4, 152, 153, 403
   and Gentex ECD, 387
   and molybdenum trioxide ECD, 152, 153
   and tungsten trioxide ECD, 149, 150
   and viologens ECD, 348, 362
   ECD self-erasure, 54, 387
metal hexacyanomellates, 282–96
metal oxidation to form oxide
   cobalt, 168, 169
   iron, 172
   manganese, 175
   niobium, 177
   rhodium, 179
   ruthenium, 181
   tantalum 182
   titanium, 184
   tungsten, 81, 150
   vanadium, 185, 186, 187
metal oxide, 125–206
   amorphous, 132, 139
   bronzes of, 61, 81, 82, 103
   coloration efficiency, 56T
      insertion coefficient and, 61
   doped, 266
   effect of moisture on, 128–9
   electrochemistry of, 138
   intervalence of, 61
metal oxide

optical properties
   as primary electrochromes, 139–65
   optical passivity, 125
   neutral colours, mixtures, 399
photochemical stability, 125
preparation, 130–8
   oxide formed by chemical vapour deposition, 131–2
   oxide formed by dip coating, 135
   oxide formed by electrodeposition, 132–4
   oxide formed by evaporation, 89
   oxide formed by Langmuir–Blodgett deposition, 138–206
   oxide formed by oxidising alloy, Inconel-600, 203
   oxide formed by oxidising metal
      cobalt, 168, 169
      iron, 172
      manganese, 175
      niobium, 177
      rhodium, 179
      ruthenium, 181
      tantalum, 182
      titanium, 184
      tungsten, 81, 150
      vanadium, 185, 186, 187
   oxide formed by sol–gel deposition, 134–6
   oxide formed by spin coating, 131, 135–6
   oxide formed by spray pyrolysis, 135
stability, 128–30
   mechanical, 129–30
   photochemical, 128, 129
metal-oxide mixtures, 190–206
   colour manipulation, 190
   containing precious metal, 204
   formation via
      dip coating, 135
      rf sputtering, 204
      sol–gel, 204
   neutral colour, 190
   site-saturation model, 190, 192
metal oxyfluorides, 203
metal–insulator transition, see *Anderson transition*
metallic substrates, 423–4
metalloporphyrin, gasochromic, 407T
metals, electrodeposition, 303, 305–7
metal-to-ligand charge transfer, 262, 293
   coordination complexes, 253
methanol, electro-oxidation, 409
methoxybiphenyls, 30, 379–80
   electrode potentials, 379T, 381T, 381T
   optical properties, 379T, 379T, 381T, 381T
   steric effects, 380
methoxyfluorene, 8
methyl viologen, 7, 11, 17, 341, 344T, 346, 352T, 353, 436
   charge-transfer complexation, 359
   coloration efficiency, 57T
   diffusion coefficient, 77T
   di-reduced, 358
   electropolychromic, 17

methyl viologen (cont.)
    ESR, 356
    in paper, 365
        follows Langmuir adsorption isotherm, 365
        mixed-valence salt, 356
methyl–benzyl viologen, in paper, 365
Methylene Blue, 8, 437, 438
    coloration efficiency, 57T
    immobilised, 391, 391T
        in Nafion, 405
methylthiophene, 3-, 318
    oligomers, of 320
micellar, viologens, 355–6
microbalance, see *electrochemical quartz crystal microbalance*
migration, 43, 44, 75, 96–7
    diffusion concurrent, 83
    temperature dependence of, 83
mirror, ECD, see *applications, ECD, mirrors*
mixed valency, methyl viologen, 356
    dinuclear ruthenium complexes, 268T
    Robin–Day classification, 142, 283
    tungsten trioxide, 142
    viologens, 356
mixtures, of metal oxide, see *metal-oxide mixtures*
MLCT, see *metal-to-ligand charge transfer*
mobility
    ionic, 99, 104, 115, 303, 305–7
    proton, 106, 108
modulation, electrochromic, 3, 53
molar absorptivity, 53
mole fraction x, 37
molybdenum ethoxide, forming molybdenum trioxide, 152
molybdenum sulfide, 152
molybdenum trioxide, 11, 27, 28, 103, 109, 125, 130, 151–5, 187
    annealing of, 152, 154
    bronze, 103, 151
    chemical diffusion coefficient, 153
    coloration *in vacuo*, 89
        requires water, 89
    containing
        gold, 204T
        platinum, 204T
    crystal phases
        α phase, 152
        monoclinic, 152
        orthorhombic, 152
    ECD, 154–5, 397
    effect of water on, 89
    electrochromic host, 199
    ellipsometry of, 153
    ESR of, 153
    formation via
        alkoxides, 152
        CVD, 131, 151, 397
        dc magnetron sputtering, 136, 151
        electrodeposition, 132, 151, 152
        electron-beam sputtering, 137
        evaporation, 137–8, 151, 155T
        $Mo(CO)_6$, 397
        molybdenum ethoxide, 152
        molybdenum sulphide, 152, 155T
        organometallic precursors, 152
        oxidation of Mo metal, 151
        peroxo species, 151
        rf sputtering, 151
        sol–gel, 135, 152
        spin coating, 135
        spray pyrolysis, 152
    hydrogen evolution at, 199
    in paper, 27, 405
    memory effect, 152, 153
    optical properties, 153–4
        coloration efficiency, 56T, 154, 155T, 199, 199T
    oxygen deficient, 103, 151, 153
    response time, 154
    self bleaching of, 153
    stability, mechanical, 129
    UV irradiation of, 28
    XPS, 152, 153
    XRD, 153
molybdenum–chromium oxide, 199
molybdenum–iron oxide, 199
molybdenum–niobium oxide, 200
    formation by sol–gel, 200
molybdenum–tin oxide, 199
    coloration efficiency, 199, 199T–201T
molybdenum–titanium oxide, 199
    coloration efficiency, 199
molybdenum–tungsten oxide, see *tungsten–molybdenum oxide*
molybdenum–vanadium oxide, 199
    formation via peroxo species, 199
Moonwatch, ECD display, 402, 403
Mössbauer, 197
    Prussian blue, 282, 283
    tin oxide, 184
motorcycle helmet, ECD application, 398
Mylar, indium–tin oxide substrate, 423

Nafion, 159, 290, 366, 385
    as electrochromic host, 405
    ECD electrolyte, 421T
    incorporating
        Methylene Blue, 405
        phenolsafranine dye, 405
        viologen, 405
    overlayer of, 150, 446
Nanochromic (NTera) displays, 347, 362, 363, 402, 406
naphthalimide–ferrocene dyads, 309
naphthalocyanine complexes, 263–4
1, 4-naphthaquinone, 384
    cyclic voltammetry, 384
N-carbazylcarbazole, carbazoles, 379T
NCD, see *Nanochromic display*
near infrared, electrochromism, 165, 253, 254, 265–74, 303–4, 317, 319, 327, 377T–378T, 399
    of aromatic amines, 376

of diamond, 399
of dinuclear ruthenium complexes, 268T
of fullerene, 399
neodymium–vanadium pentoxide, 202
Nernst equation, 36, 38, 40, 75, 90
Nernst–Planck equation, 43
Nerstian systems, 77
neutral colour, 399
   metal-oxide mixtures, 190, 399
   tungsten–vanadium oxide, 399
neutron diffraction, 144
nickel, underlayer of, 86, 164
nickel dithiolene, 266
nickel hexacyanoferrate, 293–4
nickel hydroxide, 129, 161
   formation via sonication, 133, 134
nickel oxide, 9, 125, 130, 159–65, 167, 200
   activation energy, 111T
   annealing of, 161
   as primary electrochrome, 16, 149, 165, 176
   as secondary electrochromes, 444T, 446, 447
   bleaching of, 164
   chemical diffusion coefficient, 85T
   degradation of, 163
   ECDs, 164–5, 397
   electrochemical quartz microbalance, 163
   electrochemistry of, 161–3
   electrochromic host, 200
      containing
         cobalt metal, 164
         ferrocyanide, 200
         gold, 200, 204, 204T
         lanthanum, 164
         organometallics, 200
   electronic conductivity, 162
   electrostriction of, 130
   formation via
      CVD, 36, 161
      dc magnetron sputtering, 136, 165T
      dip coating, 135, 161, 165T
      electrodeposition, 132, 160–1, 165T
      evaporation, 160, 165T
      plasma oxidation of Ni–C, 161
      rf sputtering, 160, 162, 163, 164
      sol–gel, 135, 161–3, 165T
      sonication, 133, 134, 165T
      spray pyrolysis, 135, 160, 161, 165T
   gasochromic, 407T
   ionic movement rate, 162
   mass balance, 162
      defects lattice, 163
      oxygen deficiency, 16, 159–60
   mechanical stability, 129
   optical properties, 163–4
      coloration efficiency, 56T, 70, 165T
   phases, 162
      crystallites in amorphous NiO, 88
   response times, 12, 164
   thermal instability, 160
   water occluded, 163
   write–erase cycles, 164

nickel oxyhydroxide, 129
   as secondary electrochrome, 400
   formed via electrodeposition, 161
nickel tungstate, 200
nickel–aluminium alloy, 200
nickel–aluminium oxide, 200
nickel–cerium oxide, 200
nickel–chromium oxide, 200
nickel–chromium–iron oxide, 203
nickel-doped tin oxide, 196
nickel–iridium–cobalt oxide, 203
nickel–lanthanum oxide, 200
nickel–magnesium, 200
nickel–titanium oxide, 201
   coloration efficiency, 202
   formation via electrodeposition, 201
nickel–tungsten oxide, 200, 436
   coloration efficiency, 200
   formed via sol–gel, 200
nickel–vanadium pentoxide, 202
nickel–vanadium–magnesium oxide, 203
nickel–yttrium oxide, 200
Nikon, ECD sunglasses, 401
Nile Blue, coloration efficiency, 57T
niobium ethoxide, sol–gel precursor, 134
niobium pentoxide, 17, 125, 176, 201
   annealing, 177
   as secondary electrochrome, 149, 178
   chemical diffusion coefficient, 85T
   cycle life, 178
   cyclic voltammetry, 178
   ECDs, 178
   electrochemistry, 177–8
   electrochromic host, 200–1
      coloration efficiency of mixtures, 201
   optical properties, 178
      coloration efficiency, 56T, 178, 181T, 199T–201T, 200
      optically passive, 17, 178
   redox pairs, 102
   formation via
      dc magnetron sputtering, 136, 177
      dip coating, 135
      oxidising Nb metal, 177
      rf sputtering, 181T
      sol–gel, 134, 176, 178, 181T, 200
      spin coating, 135, 177
      spray pyrolysis, 181T
niobium–iron oxide, 200, 201
   coloration efficiency, 201T
   formation via sol–gel, 200
niobium–molybdenum oxide, 200
   via sol–gel, 200
niobium–silicone oxide, 200
niobium–titanium oxide, 200
niobium–tungsten oxide, 201
   coloration efficiency, 201
Nippon Mitsubishi Oil Corporation, 446–7
NIR, see *near infrared*
nitrate ion, gasochromic, sensor for, 406
nitric oxide, gasochromic, sensor for, 407

nitroaminostilbene, 4
nitrogen-15, see *nuclear reaction analysis*
nitrosylmolybdenum complexes, 270
N-methyl PProDOP, 328
N-methylpyrrolidone, 330, 331
NMP, see *N-methylpyrrolidone*
non-faradaic current, 45, 52, 76
non-linear optical effects, 4
non-redox electrochromism, 3
non-volatile memory, see *memory effect*
N-phenylcarbazole, carbazoles, 379T, 381T
N-PrS PProDOP, 328
NREL Laboratories, 398, 433, 436, 447
NTera, ECD, 6, 10, 347, 348, 362, 363, 402, 404, 405, 406; see also *NanoChromic (NTera) displays*
   phosphonated viologen, 10, 362, 406
     as primary electrochrome, 363, 365
     coloration efficiency, 362, 363
     cycle life, 362
     ECD, response time, 363, 364
nuclear reaction analysis, 16, 110, 111, 162
nucleation, of hydrogen gas, 100
nucleation, of viologen reduction, 354
NVS, see *Gentex Corporation*

occlusion, of water during deposition, 89
octyl viologen, 344T, 352T
Ohm's law, 44
Ohmic migration, 386
oligomers
   of 3-methylthiophene, 320
   of thiophene, 320, 321
   of viologens, 364
opaque substrates, 423–4
optical absorbance, 52
optical analyses, water effect of, 89
optical attenuator, ECD, application, 270
optical charge transfer, see *charge transfer*
Optical Coating Laboratory, Santa Rosa, 448
optical data storage, ECD applications, 265
optical path length, 55
optical properties
   metal oxides
     cerium oxide, 166
     cobalt oxide, 168, 169–70
     iridium oxide, 158
     iron oxide, 175
     manganese oxide, 176
     molybdenum trioxide, 153–4
     nickel oxide, 163–4
     niobium pentoxide, 178
     tantalum oxide, 183
     tin oxide, 184
     titanium dioxide, 184
     tungsten trioxide, 144–9
     vanadium pentoxide, 188–9
   mixtures of metal oxide, indium–tin oxide, 197
   methoxybiphenyls, 379T, 381T
   oligothiophenes, 321T
   pyrazolines, 388T
   quinones, 383T
   tetracyanoquinonedimethanide, 389T
   tetrathiafulvalene, 390T
   viologens, 344T
optical response, deconvolution of, 17
optically passive, 16, 125
   metal oxides, 125
     cerium oxide, 166
     nickel–vanadium oxide, 202
     niobium pentoxide, 178
     titanium dioxide, 184
   mixtures of metal oxide
     antimony–tin oxide, 193
     cerium–titanium oxide, 194
     indium–tin oxide, 197, 199
     titanium–vanadium oxide, 202
     vanadium–nickel oxide, 202
     vanadium–titanium oxide, 202
optically transparent electrode OTE, 62, 129–30, 141, 156, 417, 444, see also *substrates*
optically transparent thin-layer electrode OTTLE, 255
orbitals, and charge transfer, 61
Orgacon EL-350, 332
organic, ECD electrolytes, 420–2
organic electrochromes, coloration efficiency, 57T
organic–inorganic, dual ECD, 333–4
organic, polymers, ECD electrolyte, 445
organometallic
   precursors, of molybdenum trioxide, 152
   in nickel oxide, 200
Orgatron, 323
oscillator strength, 147, 191, 206
osmium dithiolene complexes, 270–4
OTTLE, see *optically transparent thin-layer electrode*
overlayers
   gold, 446–7
     gold nanoparticles, 446
   lithium phosphorus oxynitride, 446
   lithium tetrafluoroaluminate, 446
   Nafion, 446
   palladium, 307
   poly(*o*-phenylenediamine), 446
   tantalum oxide, 150
   tungsten oxyfluoride, 446
   tungsten trioxide, 446
overpotential, 36, 42, 46, 76, 93, 96, 199
oxidation, chemical
   ferricyanide, 342
   oxygen gas, 342
   periodate, 342
oxidation number, 35
oxidation potential, 318, 320
oxidative polymerisation, conducting polymers, 313–14
   of pyrrole, 313
oxide ions, charge transfer with, 85
oxide mixtures, coloration rate enhancement, 200
oxidising metal, to form metal oxide film
   cobalt, 168, 169

iron, 172
manganese, 175
niobium, 177
rhodium, 179
ruthenium, 181
tantalum, 182
titanium, 184
tungsten, 81, 150
vanadium, 185, 186, 187
oxyfluoride, metal, see *metal oxyfluoride*
oxygen
  as oxidant, 342
  molecular, 59
oxygen backfilling, 141
oxygen bridges, in solid metal oxides, 85
oxygen deficiency
  in molybdenum trioxide, 103, 151, 153
  in nickel oxide, 159–60
  in praseodymium oxide, 179
  in tungsten trioxide, 102, 103, 140, 147
oxyhydroxide, via electrodeposition, 132, 133

PAH, see *poly(allylamine hydrochloride)*
paints and pigments of, Prussian blue, 282
palladium, overlayer of, 307
palladium dithiolene, 266
palladium hexacyanoferrate, 294–5
palladium oxide, 150, 178
  electrochemistry, 178–9
paper
  containing hexacyanoferrates, 405
  containing metal oxides
    molybdenum trioxide, 27, 405
    tungsten trioxide, 27, 405
  containing viologens, 365, 366, 405
    heptyl viologen 365
    methyl viologen, 365
    methyl–benzyl viologen, 365
paraquat, 341
parasitic currents, 52
passive, optical, see *optically passive*
patents, 395
PBEDOT-Pyr, 326
PBEDOT-PyrPyr(Ph)$_2$, 326
PBuDOP, 328
PEDOP, 328
PEDOT, 10, 60, 319, 332, 437
  as photoconductor, 436
  as primary electrochrome, 190, 291, 334
  as secondary electrochromes, 149
  band gap of, 322
  coloration efficiency, 437
  colour analysis of, 70, 71
  ECDs of, 409
  specular reflectance, 407T
PEDOT:PSS, 323, 329, 330, 331
  formed via layer-by-layer deposition, 329, 331
PEDOT-S, 330, 331
  formed via spin coating, 330, 331
  self-doped polymers, 330, 331
pentyl viologen, 351, 352T

perchloric acid, as ECD electrolyte, 150, 157
percolation threshold, 99, 100, 101, 113, 114, 115
periodate, as oxidant, 342
permanganate, 60
permittivity, 106, 108, 112
Pernigraniline, 329, 331–2
Perovskite, tungsten trioxide, 127, 140
peroxo species, 10, 133, 135, 141, 151, 156, 168, 184, 186
  electrodeposition with, 133
  forming
    cobalt oxide, 168
    iridium oxide, 156
    molybdenum oxide, 151, 269
    molybdenum–tungsten oxide, 199
    molybdenum–vanadium pentoxide, 199
    titanium dioxide, 184
    tungsten–molybdenum oxide, 199
    tungsten trioxide, 10, 133, 135, 141
    vanadium–molybdenum–oxide, 199
    vanadium pentoxide, 186
Perspex, plus tungsten oxide, 193
PET, indium–tin oxide substrate, 423
phenanthrenes, 379
phenanthroline, 3,8-, pseudo viologen, 360
Phenolsafranine dye, in Nafion, 405
phenothiazines, as secondary electrochromes, 362, 363
phenylenediamine, 386
Philips, 5, 27, 349, 354
phosphate ion, gasochromic sensor for, 406
phosphomolybdic acid, 152
phosphonated viologen, see *NTera*
phosphoric acid, ECD electrolyte, 167, 421, 438
phosphotungstic acid, 150, 192, 309
  in titanium dioxide, formed via sol–gel, 201
photo-activated ECD cells, 433
photo-activity, 129
  titanium dioxide, 445
photocells, photoelectrochromism, 433, 434
photochemistry, metal-oxide stability, 125, 128, 129
photochromic–electrochromic systems, 438–9
photochromism, 28, 404
  electron donors, 438
  of MoO_3, 28
  of $SrTiO_3$, 28
  of WO_3, 103
photoconductors, 433, 434–7
  amorphous silicon, 436
  PEDOT, 436
  poly(3-methylthiophene), 436
  poly(aniline), 436, 439
  poly(*o*-methoxyaniline), 436
  poly(pyrrole), 436
  silicon carbide, 436
  titanium dioxide, 437, 438
photodegradation, 54
photo-driven ECD cells, photoelectrochromism, 433
photoelectrochemistry, 361, 362
  viologens, 362

photoelectrochromism, 129, 421T, 423, 433–9
  beam direction, 433
    back-wall illumination, 433, 435
    front-wall illumination, 433
  of Prussian blue, 267, 437
  photo-activated ECD cells, 433
  photocells, 433, 434
  photoconductors, 433, 434–7
    amorphous silicon, 436
      PEDOT, 436
      poly(aniline), 436, 439
      poly(o-methoxyaniline), 436
      poly(3-methylthiophene), 436
      poly(pyrrole), 436
      silicon carbide, 436
      titanium dioxide, 437, 438
  photo-driven ECD cells, 433
  poised cells, 434
  response time, 436
photogalvanic, 438
  vanadium pentoxide, 438
photography, and Prussian blue, 25
photosensitising, ruthenium tris(2,2′-bipyridyl), 436, 437
photovoltaic, 437–8
  cadmium sulfide, 437
  strontium titanate, 437
  titanium dioxide, 437
phthalocyanine complexes, 9, 258
  conductivity electronic, 263
  ECDs of, 263
  electrochemistry, 262
  electro quasi-reversibility, 261
  electronic conductivity, 259
  ellipsometry, 263
  formation, via electrochemistry, 261–2
  Langmuir–Blodgett, 262–3, 407
  gasochromic, 407T
  including aniline moieties, 261
  mixed cation, 261
  requires central cation, 260
  response times, 261
  tetrasulfonated, 261
physical vapour deposition, of cerium oxide 166
pigments, industrial, 259
Pilkington Glass, 141, 400
pixels, 360, 385, 402, 403
plasma oxidation, of Ni–C, forming nickel oxide, 161
plasma screens, television, 402
platinum
  as substrate, 153, 284, 326, 409, 422, 423
  black, 133
  incorporated into
    molybdenum trioxide, 204T
    ruthenium dioxide, 204T
    tantalum pentoxide, 204T
    tungsten trioxide, 204T
platinum dithiolene, 266
poised cells, photoelectrochromism, 434
polarisation
  of electrode, 76
  of light, 50, 51
polaron, 145, 183, 316
  hopping, 127, 143
  in tungsten trioxide, 147
  polaron–polaron interactions, in WO_3, 88
polished metal, substrates, 418
poly(acetylene), 312, 315
  air sensitive, 312
  electronic conductivity, 312
poly(acrylate), composite
  formed via spin-coating, 333
  with poly(aniline), 333; composite with silica and poly(aniline), via sol–gel, 333
poly(acrylic acid), as ECD electrolyte, 150, 421T
poly(alkenedioxypyrrole)s, 327
poly(allylamine hydrochloride), 330, 331
poly(AMPS), as ECD electrolyte, 12, 150, 157, 260, 330, 331, 348, 366, 391, 391T, 395–410, 420
  immobilising electrochromes, 391
poly(aniline), 9, 11, 30, 101, 313, 329–30, 331, 333, 384, 438, 439
  as photoconductor, 436, 439
  as secondary electrochrome, 149, 290–1, 333, 334, 444T
  castable films, 332–3
  composites
    with cellulose acetate, 333
    with poly(acrylate), 333
    with poly(styrene sulfonic acid), 333
  containing vanadium pentoxide, 190, 202
  cyclic voltammetry, 333
  electrochemistry of, 329–30, 331
  electropolychromic, 329, 331
  encapsulating dyes, 333
  formation via electropolymerisation, 329–30, 331; layer-by-layer deposition, 329–30, 331
  graft copolymer of, 333
  immobilising electrochromes, 391
  poly(acrylate)–silica composite, formed via sol–gel, 333
  redox states, 329, 331
poly(aniline)s, 328–30
  ECDs, 330, 331
  protonation reactions, 328–9, 331
  response times, 330, 331
  spectroelectrochemistry of, 333
poly[3,4-(butylenedioxy)pyrrole], 328
poly(carbazole), chemical diffusion coefficient, 85T
poly(CNFBS), 316
polycrystalline, metal oxides, made by sputtering, 81
poly(DDTP), 326
poly(diphenylamine), specular reflectance, 407T
polyelectrochromism, see *electropolychromism*
polyelectrolytes, ECD electrolyte, 420–2
polyester, indium–tin oxide substrate, 423
poly(ethylene imine), 329, 331
poly(ethylene oxide)
  as ECD electrolyte, 150, 290, 408, 421, 438, 444

as thickener in ECD electrolyte, 419
poly(3,4-ethylenedioxy
   thiophenedidodecyloxybenzene),
   coloration efficiency, 57T
poly(ethylene terephthalate), 332
   ITO on, 27
poly(*iso*-thianaphthene), chemical diffusion
   coefficient, 85T
polymer electrolytes, ECD electrolytes, 421–2
   conducting, 9, 11
   electrolyte, 44
   of EDOT, 325
      polypyridyl complex, via spin-coating, 254–6
      TTF species, ion movement rate limiting, 390
      viologens, 347
   photostability,151
poly(*m*-toluidine), 330, 331–2
poly(methyl methacrylate) blend, as ECD
   electrolyte, 291, 334
poly(3-methylthiophene) 320, 322T
   as photoconductor, 436
   as primary electrochrome, 197
poly(*o*-methoxyaniline), 332
   as photoconductor, 436
poly(*o*-phenylenediamine), overlayer of, 446
poly(*o*-toluidine), 159, 330, 331
poly(oligothiophene)s, 321T, 323T
poly(1,3,5-phenylene), 327
poly(*p*-phenylene terephthalate), 198
   as secondary electrochrome, 159
poly[3,4-(propylenedioxy)pyrrole], 328
   coloration efficiency, 57T
poly(3,4-propylenedioxythiophene), coloration
   efficiency, 57T
poly(propylene glycol), ECD electrolyte, 421
poly(pyrrole), 101, 313, 314, 315, 316, 317, 327
   as photoconductor, 436
   as primary electrochrome, 165
   as secondary electrochrome, 149
   containing dodecylsulfonate, 333
   containing Indigo Carmine, 333
   electro-synthesis of, 30
   specular reflectance, 407T
   viologens of, 346
poly(pyrrole)s, 327–8
   ECDs of, 328
   N-Gly PProDOP, 328; PBuDOP, 328
   PEDOP, 328
   PProDOP, 328
poly(siloxane), immobilising electrochromes, 391
poly(styrene sulfonic acid), 333, 347
   composite with poly(aniline), 333
poly(thiophene), 9, 11, 313, 315, 321
   as primary electrochrome, 165
   star polymers, 327
   viologens of, 347
   PBEDOT-Pyr, 326
   PBEDOT-PyrPyr(Ph)$_2$, 326
   PEDOT, 10, 60, 319, 332, 437
      as photoconductor, 436
      as primary electrochrome, 190, 291, 334
      as secondary electrochrome, 149
      bandgap of, 322
      coloration efficiency, 437
      colour analysis of, 70, 71
      ECDs of, 409
      specular reflectance, 407T
   PEDOT:PSS, 323, 329, 330, 331
      formed via layer-by-layer deposition, 329, 331
   PEDOT-S, 330, 331
      formed via spin coating, 330, 331
      self-doped polymers, 330, 331
poly(thiophene)s, 318–27
   band structure, 152
   Baytron M, 323
   Baytron P, 323
   BEDOT, 326
   BEDOT-*N*-MeCz, 326
   bipolarons in, 320
   DDTP, 326
      dihedral angle, 323T
      formation via spin coating, 327
   PBEDOT
      PBEDOT-B(OC$_{12}$)$_2$, 332
      PBEDOT-*N*-MeCz, 331
      PBEDOT-Pyr, 326
      PBEDOT-PyrPyr(Ph)$_2$, 326
   PProDOT-Me$_2$, 331, 332
   response time, 325
   substituted, 320–1
poly(toluidine)s, 330, 331
poly(triphenylamine), 327
poly(vinyl alcohol), ECD electrolyte, 421
poly(vinyl butyral), ECD electrolyte thickener, 419
poly(1-vinyl-2-pyrrolidone-co-
   N,N'-methylenebisacrylamide), 391, 391T,
   395–410
   ECD electrolyte, 391
poly(viologen), 328–9, 331
   formation via layer-by-layer deposition,
   328–9, 331
Polyvision, 306
porphyrin complexes, 258, 264–5
potassium chloride, ECD electrolyte, 291, 349
potassium triflate, ECD electrolyte, 290
potential, equilibrium, 41
potential, sweep, 48
potential step
   and cycle life, 12
   and coloration, 354–5
potentiostat, 48, 62
potentiostatic
   coloration, 99, 417
   electrodeposition, 133
   interrupted coloration, see *pulsed potentials*
   three-electrode, 443
powder abrasion, of Prussian blue, 283
power consumption, 13–15
   different types of display, 15
poly(methylthiophene), 165
PPG Aerospace, ECD, windows, 400
PPG Industries, 425

praseodymium oxide, 178–9
  cycle life, 179
  electrochemistry of, 178
  oxygen deficiency, 179
  containing cerium oxide, 179
  as secondary electrochrome, 179
  formation via,
    CVD, 178, 179
    dc magnetron sputtering, 136, 178
    XRD, 179
praseodymium phthalocyanine, 263
precious metal, in metal oxide, 204
  formation via
    rf-sputtering, 204
    sol–gel, 204
preparation, of metal oxides, chemical vapour deposition, 131–2
primary and secondary electrochromism, 16–17; see also *complementary electrochromism*
primary electrochromism, 45, 165, 418, 421, 445
  hexacyanoferrates as, Prussian blue, 333
  metal oxides as, 139–65
    nickel oxide, 165, 176
    tungsten trioxide, 149, 154, 165, 170, 178, 179, 184, 190, 197, 290, 291, 333, 334, 400, 421, 436, 438, 444T, 446, 447
  polymers as
    PEDOT, 190, 291, 334
    poly(3-methylthiophene), 165, 197
    poly(pyrrole), 165
    poly(thiophene), 165
  viologens as
    heptyl viologen, 356, 375, 385
    NTera viologen, 363, 365
primary reference electrode, see *standard hydrogen electrode*
probe molecules, 5
propyl viologen, 352T
propylene carbonate, as ECD electrolyte, 100, 106, 151, 152, 166, 169, 173, 176, 184, 186, 187, 188, 197, 199, 205, 356, 384, 419
proton, conductivity in metal oxides, 89
  in tantalum oxide, 183
  mobility, 4, 86, 106, 108
proton transfer, across solution–oxide interface, 86
protonation reactions, poly(aniline)s, 328–9, 331
Prussian blue, 9, 25–6, 41, 57, 61, 446
  and blueprints, 26, 405
  and cyanotype photography, 26
  and drawing, 26
  and photography, 25
  as secondary electrochromes, 149, 290, 333, 334, 363, 365, 444T
  bulk properties, 282–3
  chronoamperometry, 284
  colour analysis of, 62, 70
  cyclic voltammetry, 286, 287
  ECD, 289–91
    comprising single film of, 290
  EDAX of, 288
  electrochemistry of, 58–60, 285–9

electropolychromism of, 287
ellipsometry of, 284, 287, 288, 289
formation via, 283–5
  catalytic silver paint, 283
  directed assembly, 285
  electrodeposition, 283, 284
  electroless deposition, 283
  photolysis, 26
  powder abrasion, 283
  redox cycling, 283
  sacrificial anode methods, 283
history, 282
in paper, 405
'insoluble', 282
lattice energy, 289
Mössbauer of, 282, 283
paints and pigments of, 282
pH effect of, 289
photochargeable battery of, 437
photoelectrochromism of, 267, 437
preparation
'soluble', 283
write–erase efficiency, 286
XPS, 288
XRD, 283
Prussian brown, 26, 285
Prussian green, 285, 334
Prussian white, 57, 286
pseudo viologen, see *viologen, pseudo*
pulsed potential, 104–5, 303, 305
  coloration, 87
  viologens, 365
  enhanced ECD durability, 104
  response time acceleration, 11
purity, and colour analysis, 63, 65, 66
purple line, and colour analysis, 64
PVPD, immobilising electrochromes, 391, 391T, 395–410
PVPD, see *poly(1-vinyl-2-pyrrolidone-co-N,N-methylenebisacrylamide)*
PXDOP, 328, 356
PXDOT, 328
pyrazolines, 387–8
  optical properties, 388T
  response times, 388T
pyridinoporphyrazine complexes, 264
pyrrole, 313; for polymers of pyrrole, see *poly(pyrrole)*
  oxidative polymerisation of, 313

Q-switching, of lasers, 267
quantum-mechanical effects, tunnelling, 81
quartz-crystal microbalance, see *electrochemical quartz-crystal microbalance*
quasi-electrochromism, 406–7
quasi-reference electrodes, 40
quasi-reversibility
  fullerene electrochromes, 303, 305
  phthalocyanine electrochromes, 261
  viologen electrochromes, 358
quaternary oxides, 203

quinhydrone, 384
quinones, 4, 30, 256, 381–5
   amino-4-bromoanthaquinone-2-sulfonate, 384
   aminonaphthaquinone, 384
   benzoquinones
      o-, 381; *p*-, 381, 382
   bis(dimethylamino)diphenylamine, 4,4′-, 384
   bromoanil, *o*-, 382, 383
      solubility product, 383
   catechole, 270–4
   chloranil
      o-, 382, 383
      p-, 382
      type-II electrochrome, 382
   contrast ratio, 348, 384, 385
   ECDs of, 384
   electrode potentials, 383T
   electropolymerisation, 384
   2-ethylanthraquinone, 384
   fluoroanil, *p*-, 382
   naphthaquinone, 1,4-, 384
      cyclic voltammetry, 384
      type-I electrochromes, 384
   optical properties, 383T
   quinhydrone, 384

radical annihilation, and ECD self-erasure, 386
   Gentex mirror, 386
radical, viologen, see *viologen, radical*
radii, ionic, 112
Raman spectroscopy, 86, 88, 103, 130, 357
Randles–Sevčik equation, 49, 83
rate
   of cell operation, 41–6
   of coloration, 33, 139
   of charge transfer, 95
   of electron transfer, 42–3, 75
   of electronic conduction, 42
   of mass transport, 42
rate constant, electron transfer, 34, 46, 102
rate limiting kinetics, 83
   crystal structure changes, 87
   electronic motion, 99, 101, 115, 143
   ionic motion, 92, 163, 188, 390
      diffusion, 97
RBS, see *Rutherford backscattering*
RCA Laboratories, 29
real, impedance, 50
rear-view mirrors, see *applications, ECD, mirrors*
rechargeable batteries, manganese oxide, 176
redox couple, 35, 37
redox cycling, of Prussian blue, 283
redox electrode, 50
redox indicators, 374
redox pairs, 101, 112–13, 115
   niobium pentoxide, 102
redox potential, see *electrode potential*
redox reaction, 35, 54
redox states, of poly(aniline), 329, 331
reference electrode 40, 48, 58, 70T, 149, 155, 157
   primary standard, 40

quasi, 40
saturated calomel electrode, 40, 48, 149, 155, 157, 169, 199, 262, 382, 406, 410
secondary, 40
silver–silver chloride, 58, 70T, 349, 350
silver–silver oxide, 40
reflective, 81, 146, 148, 149T, 303, 305, 399, 407T
   metal oxides
      copper oxide, 407T
      iridium oxide, 400, 407T
      rhodium oxide, 188
      tungsten trioxide, 148–9, 149T, 400, 407T
   miscellaneous
      lithium pnictide, 407T
      tungsten oxyfluoride, 407T
   polymers
      PEDOT, 407T
      poly(diphenylamine), 407T
      poly(pyrrole), 407T
Resazurin, coloration efficiency, 57T
Research Frontiers, 398
resistance, 50
   of electrode substrate, 11
      ITO, effect of, 349
   to charge transfer, 86, 105
Resorufin, coloration efficiency, 57T
response time, 10–11, 86, 98, 139, 141, 150, 268, 274
   metal oxides
      bismuth oxide, 166
      iridium oxide, 156, 159
      molybdenum trioxide, 154
      nickel oxide, 164
      tungsten trioxide, 149, 150
   mixtures of metal oxide, tungsten–cerium oxide, 193
   organic monomers
      aromatic amines, 375
      pyrazolines, 387, 388T
   photoelectrochromism, 436
   phthalocyanine complexes, 261
      lutetium phthalocyanine, 260, 261
   polymers
      poly(aniline)s, 330, 331
      poly(thiophene)s, 325
   pulsed potentials acceleration, 11
   tetrathiafulvalenes, 390T
   viologens, 346, 349, 351, 361, 363
      NTera viologen, 363, 364
reversibility, 39
rf sputtering, 137
   metal oxides
      bismuth oxide, 166
      chromium oxide, 167
      cobalt oxide, 167
      manganese oxide, 175
      molybdenum trioxide, 151
      nickel oxide, 160, 162, 163, 164
      tantalum oxide, 182, 183–4
      tin oxide, 183
      titanium dioxide, 184
      tungsten trioxide, 140, 141, 148

rf sputtering (cont.)
    vanadium pentoxide, 185, 187, 188
    mixtures of metal oxide
        indium–tin oxide, 196
        molybdenum–tungsten oxide, 199
        titanium dioxide mixtures, 201
        tungsten–molybdenum oxide, 199
    precious metal incorporation, 204
rhodium oxide, 125, 179–81
    annealing of, 181
    coloration efficiency, 56T, 181
    cyclic voltammetry of, 181
    electrochemistry of, 180
    formation via
        anodising Rh metal, 179
        sol–gel, 180, 181
    hydrated, 180
    reflective, 188
Robin–Day classification, 142, 283
rocking-chair mechanism, 16
rotated ring-disc electrode, 358
ruthenium complexes, 309
    dinuclear, 267–8
    trinuclear, 268
ruthenium dioxide, 181
    electrochemistry, 181
    formation via electrodeposition, 181; oxidising Ru metal, 181
    incorporating platinum, 204T
ruthenium dithiolene complexes, 270–4, 309
ruthenium hexacyanoferrate, see ruthenium purple
ruthenium polypyridyl complexes, 256
ruthenium purple, 285, 292
    XRD, 292
ruthenium tris(2,2′-bipyridyl), 265, 266
    photosensitiser, 436, 437
    effects of ligands, 255T
ruthenium–iridium oxide, 203
Rutherford backscattering, 103, 160, 205
rutiles, 127

sacrificial anode methods, of Prussian blue, 283
Safranin O, coloration efficiency, 57T
SAGE Incorporated, 398, 447
salt bridge, 39
salvation stabilisation, 89
samarium–vanadium oxide, 202
    coloration efficiency, 202
sapphire, 202
saturated calomel electrode, 40, 48, 149, 155, 157, 169, 199, 262, 382, 406, 410
saturation, and colour analysis, 56T, 63, 65, 66, 70
scan rate, 48
scanning tunnelling microscope, 263, 284
SCE, see saturated calomel electrode
Schott Glass, ECD, 400
SchottDonnelly mirror, 397
screen printing
    carbon black, 424
    carbon ink, 303, 305

tungsten trioxide, 140
sealing, ECD, see encapsulation, ECD
secondary battery, ECD like, 54
secondary electrochromism, 16–17, 165–90, 418, 421
    hexacyanoferrates, Prussian blue, 149, 290, 334, 363, 365, 444T
    metals, bismuth, 444T
    metal oxides
        cobalt oxide, 170
        copper oxide, 165
        iridium oxide, 149, 444T
        iron oxide, 174
        manganese oxide, 165, 176
        nickel oxide, 149, 165, 444T, 446, 447
        niobium pentoxide, 149, 178
        praseodymium oxide, 179
        tin oxide, 165
        titanium dioxide, 184
        tungsten trioxide, 421
        vanadium pentoxide, 149, 190, 438, 444T
    mixtures of metal oxide
        cerium–titanium oxide, 444T
        indium–tin oxide, 197
        titanium–cerium oxide, 444T
    organic monomers
        phenothiazines, 362, 363
        tetramethyl phenylenediamine, as, 356, 375, 385
    oxyhydroxides, nickel, 400
    polymers
        PEDOT, 149
        poly(aniline), 149, 290–1, 333, 334, 444T
        poly(p-phenylene terephthalate), 159
secondary reference electrodes, 40
second-harmonic effects, 4
Seebeck coefficient, 113
self bleaching, of ECD, 15, 54, 150, 153
self-doped polymers, PEDOT-S, 330, 331
self-erasing ECD mirrors, 387
semiconductor theory, 317
semi-solid, ECD electrolyte, 446
sensors, conducting polymers, and, 312
SERS, see surface-enhanced Raman spectroscopy
SHE, see standard hydrogen electrode
shear planes, 103
shutters, ECD applications, 363, 404–5
side reactions, 43, 54, 76, 199
    hydrogen evolution at MoO_3, 199
silica, ECD electrolyte thickener, 348, 419
silicon carbide, as photoconductor, 436
silicon phthalocyanine, 264
silicon–cobalt–aluminium oxide, 204
silicon–iridium oxide, 198
silicone–niobium oxide, 200
silver
    conductive paint, 349
    electrodeposition of, 8, 27, 307

incorporated into
  indium–tin oxide, 204T
  tungsten trioxide, 204T
  vanadium pentoxide, 204T
silver oxide, 40
silver–silver chloride, reference electrode, 58, 70T, 349, 350
silver–silver oxide, reference electrode, 40
SIMS, 110, 111–12
SIROF, 155
site-saturation model, metal-oxide mixtures, 190, 192
ski goggles, ECD application, 398
smart cards, ECD applications, 363
smart glass, 397; see also *ECD, windows*
  non-electrochromic, 5
smart windows, see *ECD applications, windows*
SmartPaper, 5
sodium tungsten bronze, 27
solar energy storage, ECD applications, 265, 266
solar-energy conversion, conducting polymers, and, 312
solar-powered cells, 15
sol–gel
  formation of phosphotungstic acid, in titanium dioxide, 201
  formation of poly(acrylate)–silica composite with poly(aniline), 333
  forming metal oxides, 134–6
    cobalt oxide, 135, 168, 195
    copper oxide, 170
    iridium oxide, 156
    iron oxide, 173, 174
    manganese oxide, 175, 176
    molybdenum trioxide, 135, 152
    nickel oxide, 135, 161–3
    niobium pentoxide, 134, 176, 178, 200
    rhodium oxide, 180, 181
    titanium dioxide, 135, 184
    tungsten trioxide, 135, 141, 149
    vanadium pentoxide, 135, 185
  forming mixtures of metal oxides
    cobalt–aluminium oxide, 195
    indium–tin oxide, 196
    iridium–titanium oxide, 198
    iridium–silicon oxide, 198
    iron–niobium pentoxide, 200
    molybdenum–niobium oxide, 200
    molybdenum–tungsten oxide, 199
    nickel–tungsten oxide, 200
    niobium–iron oxide, 200
    niobium–molybdenum oxide, 200
    titanium dioxide mixtures, 201
    titanium dioxide, plus phosphotungstic acid, 201
    tungsten–molybdenum oxide, 199
  with precious metals, 204
  with titanium butoxide, 135
solid polymer matrix, mirror of, 397
solid solution electrodes, 41
solubility product
  bromoanil, *o*-, 383

viologens, 351, 359
solvatochromism, 3
sonication, 133–4
Sony Corporation, 351
space charges, 105
SPD, see *suspended-particle device*
speciation analyses, 133
spectral locus, 64, 65
spectroelectrochemistry, poly(aniline)s, 333
spectroscopy, impedance, see *impedance*
specular reflectance, see *reflective*
spillover, 407
spin coating
  annealing of product, 135
  formation of metal oxides, 131, 135–6
    cerium oxide, 135
    cobalt oxide, 135
    iron oxide, 135, 174
    molybdenum trioxide, 135
    niobium pentoxide, 135, 177
    tantalum oxide, 135
    titanium dioxide, 135, 184
    tungsten trioxide, 135, 141
    vanadium pentoxide, 135, 185, 186
  formation of mixtures of metal oxide, 136
    indium–tin oxide, 135, 196
  formation of polymers
    PEDOT-S, 330, 331
    poly(acrylate)–poly(aniline) composite, 333
    polymeric polypyridyl complex, 254
    poly(thiophene)s, 327
spirobenzopyran, 438
spiropyrans, 376
SPM, see *solid polymer matrix*
spray pyrolysis
  annealing of product, 135
  formation of metal oxides, 135
    cerium oxide, 135, 166
    cobalt oxide, 135, 168, 169
    iridium oxide, 158
    molybdenum trioxide, 152
    nickel oxide, 135, 160
    tungsten trioxide, 135, 141
sputtering
  product oxide is polycrystalline, 81
sputtering in vacuo, 136–8; see also *dc magnetron sputtering, electron-beam sputtering, evaporation and rf sputtering*
stability
  metal oxide, 128–30
  photochemical, 125, 128, 129
    electrolyte, 445
  tungsten trioxide, 143
Stadsparkasse Bank, ECD, windows, 400
standard electrode potential, 36, 37, 40
  of hydrogen electrode, 40
standard exchange current, 47
standard exchange current density, 47
standard hydrogen electrode (SHE), 40
standard observer, in colour analysis, 63, 64
standard rate constant, of electron transfer, 47

star polymers, of poly(thiophene)s, 327
Stark effects, 4, 25, 61
stibdic acid polymer, ECD electrolyte, 421T
STM, see *scanning tunnelling microscope*
stress, in crystal lattice, 130
strontium titanate, 28
   nickel doped, 309
   photovoltaic, 437
sublimation, of lutetium phthalocyanine, 259
substituted poly(thiophene)s, 320–1
sub-stoichiometry, see *oxygen deficient*
substrates
   antimony–copper alloy, 423
   antimony-doped tin oxide, 362
   carbon, 424
      glassy carbon, 358
      graphite, 424
   durability of, 444–5
   ECD, 422–4
   fluorine-doped tin oxide, 139, 166, 168, 171, 196, 205, 292, 362, 400, 406, 409, 422
   gold, 285
   indium–tin oxide, 17, 70T, 86, 96, 128, 129, 135, 138, 139, 141, 150, 151, 152, 156, 158, 159, 164, 166, 167, 181, 182, 191, 257, 284, 293, 294, 305, 306, 326, 330, 331, 333, 349, 375, 382, 385, 404, 417, 422–3, 444–5, 447
   K-glass, 422
   metallic, 423–4
   opaque, 423–4
   platinum, 153, 284, 312, 326, 409, 418, 422, 423
   resistance of, 11
   tin oxide, 289, 354
   titanium dioxide, 406
   viologens and effect of, 353
sulfuric acid, ECD electrolyte, 82, 86, 149, 178, 259, 349, 409, 420
   degradation by, 420
sunglasses, ECD application, 401
supporting electrolyte, 44
surface enhanced Raman spectroscopy
      viologens, 357
surface potentials, 4
surface states, 86
surfactants, voltammetry and, 356
Surlyn, ECD encapsulation, 425
suspended-particle device, SPD, 5
swamping electrolyte, 75, 76
swapping, of counter ions, 87
sweep rate, see *scan rate*
switchable mirrors, metal hydrides, 307
symmetry factor, 95

Tafel region, 47, 48
Tafel's law, 43, 46, 47
   deviations from, 46
tailoring, of colours, 334
tantalum oxide, 181–3, 198, 446
   as ECD electrolyte, 150, 420, 421T, 447
   electrochemistry, 183
   ion-conductive electrolyte, 181, 183

   mechanical stability, 129
   optical properties, 183
   tantalum oxide, coloration efficiency, 56T, 183
   overlayer of, 150
   plus platinum, 204T
   protonic motion, 183
   formed via
      anodising a metal, 182
      CVD, 132, 182
      dc magnetron sputtering, 136, 182
      dip-coating, 182
      evaporation, 182
      laser ablation, 183
      rf sputtering, 182, 183–4
      spin coating, 135
   water adsorbed on, 183
tantalum–zirconium oxide, 203
   coloration efficiency, 203
TCNQ, see *tetracyanoquinodimethanide*
Teflon, 409
television
   flat-panel screens, 402
   pixels, 402, 403
   plasma screen, 402
temperature management, ECD, applications, 265
terminal effects, 86, 164, 423
   suppressors
      chromium oxide, 423
      magnesium fluoride, 423
tethered electrochromes, see *derivatised electrodes*
tetracyanoquinonedimethanide species, 387, 388–9
   optical properties, 389T
   reversibility, 389
   write–erase efficiency, 388
tetrahydrofuran, 327
tetramethylphenylenediamine, 356
   as secondary electrochrome, 356, 375, 385
tetrathiafulvalene species, 30, 387, 389–90
   ion hopping, 390
   ion tunnelling, 390
   optical properties, 390T
   response times, 390T
Texas Instruments, 28, 352
thermal evaporation, see *evaporation*
thermal instability, nickel oxide, 160
thermoelectrochromism, 408
   lithium vanadate, 190
thermodynamic enhancement, 83–5, 112
   enhancement factor W, 83, 84
thiazines, 385–7
   ECDs, 385
   Methylene Blue, see *Methylene Blue*
thickener, see *electrolyte thickener*
thickness, changes in electrochrome, 51; see also
      electrostriction
thiophene, 313
   oligomers, 321T
thiophene acetic acid, 3-, 320
three-electrode, potentiostatic coloration, 443
tin oxide, 183
   as ionic conductor, 159

as secondary electrochrome, 165
doped
  antimony-doped, see *antimony-doped tin oxide*
  fluorine-doped, see *fluorine-doped tin oxide*
  nickel-doped, 196
  electrochromic host, 201
  formation via rf sputtering, 183
  Mössbauer spectroscopy, 184
  optical properties, 184
    infrared $_{max}$, 183
  substrate, 289, 354
tin oxyfluoride, 205
tin phosphate, as ECD electrolyte, 154
tin–cerium oxide, 201
tin–molybdenum oxide, 199
  coloration efficiency, 199, 199T–201T
titanium alkoxides, 184
titanium butoxide, sol–gel precursor, 135
titanium dioxide, 10, 11, 12, 125, 130, 184, 194
  anatase, 437
  as secondary electrochrome, 184
  coloured with pulsed current, 184
  diffusion coefficient, 184
  ECD electrolyte, 421T, 445
    electrolyte filler, 421
  electrochemistry, 184
  electrochromic host, 201–2
    formed via
      sol–gel, 201
      sputtering, 201
    plus ferrocyanide, 201
    plus phosphotungstic acid, 201
  mechanical stability, 129
  ellipsometry, 184
  formation via
    alkoxides, 184
    dip coating, 135, 184
    evaporation, 8, 185T
    laser ablation, 184
    oxidation of Ti, 184
    peroxo species, 184
    rf sputtering, 184, 185T
    sol–gel, 135, 184, 185T
    spin coating, 135, 184
    thermal evaporation, 184
  nanostructured, 360–4
  optical properties, 184
    coloration efficiency, 56T, 184, 185T
    optically passive, 184
  photo-activity, 445
    photoconductor, 437, 438
    photostability, 1153
    photovoltaic, 437
  substrate, 406
titanium oxyfluoride, 205
  coloration efficiency, 205
  cycle life, 205
  electrochemistry, 205
  formation via dc-sputtering, 205
titanium oxynitride, 184

titanium propoxide, 201
titanium–cerium oxide
  as secondary electrochrome, 444T
  formed via dc magnetron sputtering, 136
titanium–cerium–titanium oxide, 203
titanium–cerium–vanadium oxide, 203
titanium–iridium oxide, 198
titanium–iron oxide
  charge transfer, 202
  formed via dip-coating, 201, 202
titanium–molybdenum oxide, 199
  coloration efficiency, 199
titanium–nickel oxide, 201
  formed via electrodeposition, 201
titanium–niobium oxide, 200
titanium–tungsten oxide, 202
titanium–zirconium–cerium oxide, 203
  formed via electrodeposition, 202
titanium–tungsten–vanadium oxide, 203
titanium–vanadium oxide, 202
  optically passive, 202
titanium–zirconium–cerium oxide, 203
titanium–zirconium–vanadium oxide, 203
titration, electrochemical, 104
tolidine, *o*-, 77, 78
toluene, gasochromic, sensor for, 407
Toluylene Red, coloration efficiency, 57T
tone, and colour analysis, 63
toys, as ECD application, 363
transfer coefficient, 47
transmittivity, 62
transport number, 44, 83
transport, through liquid electrolytes, 75
triflic acid, as ECD electrolyte, 150, 421
trimethoxysilyl viologen, 346
tris(pyrazolyl)borato-molybdenum complexes, 269–70
tris-isocyanate complexes, 268
tristimulus, and colour analysis, 63, 67
TTF, see *tetrathiafulvalene*
tungstate ion, from degradation of WO_3, 89
tungsten hexacarbonyl, 131, 141
  forming tungsten trioxide, 397
tungsten oxyfluoride, 153, 205–6, 446
  coloration efficiency, 205
  electrochemistry, 206
  overlayer of, 150, 446
  specular reflectance, 407T
  formed via dc magnetron sputtering, 205
tungsten trioxide, 9, 10, 11, 25, 27, 28, 35, 40, 79, 81, 103, 109, 110, 111, 115, 125, 130, 139–51, 156, 187, 191, 200, 201, 206, 303, 305, 308, 399, 410, 419, 436, 437, 438, 443, 446
  activation energy, 111T
  amorphous, 81, 88, 113
  Anderson transition, 81, 99, 142, 149
  annealing of, 88, 140, 148
  as primary electrochrome, 16, 149, 154, 159, 165, 170, 178, 179, 184, 190, 197, 290, 291, 333, 334, 400, 421, 436, 438, 444T, 446, 447
  as secondary electrochromes, 421
  bleaching, 106

tungsten trioxide (cont.)
  bronze, 81, 113, 144
  charge transport through, 60, 85
  chemical reduction of, 25, 89, 109
    chemical degradation, 149
    dissolution in acid, 89
  chemical diffusion coefficient, 83, 84T, 85T, 195T
  colour, source of
    F-centres, 145
    intervalence, 145
    oxygen extraction, 145
    polarons, 145
  coloration mechanism
    a two-electron process, 103
    'complicated', 80
    involves W^{IV}, 103
  coloration, without electrolyte, 28
  conductivity, 142
    dielectric properties, 143
    electron localisation, 142
    electronic, 99
    electrons are rate limiting, 143
    insulator at $x =$ 143
    ionic, 83
    low conductivity of, 81
    metallic at high x, 143
    polaron–polaron interactions in, 88
  dry lithiation of, 418
  ECD, first, 27
    in paper, 27, 405
  ECDs of, 28, 29, 61, 104, 139, 149, 397, 399, 402, 408, 409, 410
  ECD applications
    display devices, 149
    mirrors, 149
    sunglasses, 401
    watch displays, 149
    windows, 149, 400
  formation via, 16
    colloidal tungstate, 141
    CVD, 131, 141, 148, 150, 397
    dc magnetron sputtering, 136, 141
    deposition in vacuo, 129
    dip coating, 135, 141, 148T
    electrodeposition, 132, 140, 141, 148T
    evaporation, 81, 99, 140–1, 147, 148T, 150
    organometallic precursors, 141
    oxidising W metal, 81, 150
    peroxo species, 10, 133, 135, 141
    rf sputtering, 140, 141, 146, 147, 148, 148T
    screen printing, 140
    sol–gel, 135, 141, 148T, 149
    spin coating, 135, 141, 148T
    spray pyrolysis, 135, 141
  electrochemistry, 142
    electrophotography, 28
    electrostriction, 87, 129, 130, 445
  ellipsometry of, 81, 143
  ferroelectric properties, 143
  gasochromic, 407T
  mechanical stability, 129
  memory effect, 149, 150
  mixtures of, 88, 191–3, 407
    plus bismuth, 140; gold, 204T; indium, 140; Perspex, 193; platinum, 204T; silver, 140, 204T
  neutron diffraction, 144
  optical effects, 144–9
    colour, by reflection, 149T
    coloration efficiency, 56T, 148T, 191, 193, 201
    spectrum, 144
  overlayer of, 446
  photo-chargeable battery, 437
  photochromism, 103
  proton-free layers while bleaching, 106
  reflective effects, 143, 148–9, 400, 407T
  response time, 149, 150
  structure
    crystal phases, 86
    crystalline, 98, 104
    cubic phase, 89
    morphology, 140
    oxygen deficiency, 102, 103, 140, 147
    perovskite, 140
    polycrystalline, 81
    structural changes, 143–4
  stability, 129, 143
  water and, 89, 145, 150
    hydrated, 115, 150
tungsten–cerium oxide, 193, 195
  formation via dc magnetron sputtering, 136
  response time, 193
tungsten–cobalt oxide, chemical diffusion coefficient, 195T
tungsten–molybdenum oxide, 88, 191, 192, 199
  amorphisation, 193
  coloration efficiency, 56T, 192
  ECD, 397
  electron mobility, 192
  formation via
    CVD, 397
    electrodeposition, 199
    peroxo species, 199
    rf sputtering, 199
    sol–gel, 199
  intervalence, 192
tungsten–nickel oxide, 130, 200, 436
  coloration efficiency, 200
  formation via sol–gel, 200
tungsten–niobium oxide, 201
  coloration efficiency, 201
tungsten–titanium oxide, 202
tungsten–vanadium oxide, 202
  neutral colour, 399
tungsten–vanadium–titanium oxide, 203
tunnelling, 81
Tyndall effect, 134
type-I electrochromes, 33, 43, 45, 46, 54, 328, 346, 354, 356, 359, 396, 403, 417, 419, 425
  aromatic amines, 375
  coloration kinetics, 75–9, 403
  Gentex mirror, 396, 398, 400

naphthaquinones, 384
type-II electrochromes, 33, 45, 46, 78, 79–115, 417, 419, 425
  aromatic amines, 375
  bleaching, 79–115
  carbazoles, 376
  chloranil, 382
  coloration kinetics, 75–9
  electrodeposition of metals, 303, 305
  viologens, 346, 348–9, 351, 354
type-III electrochromes, 45, 46, 54, 79, 397, 403, 407, 417, 445
  bleaching of, 79–115
  coloration, 79–115, 403
  concentration gradients, 303, 305
  diffusion coefficients through, 83
  formation via chemical tethering, 346, 361
  kinetic modelling, 91–115; see also *coloration models*
  viologens, 346, 361
  viscous solvents immobilising, 391
types, of electrochrome, 7–9

$u'v'$ uniform colour space, 67, 70, 71
Ucolite, ECD, 400
underlayers, nickel, 86, 164
uniform colour space, 66, 67, 70, 71
UV electrochromism, 165

vacuum evaporation, product oxide is amorphous, 81
value, and colour analysis, 63
vanadium dioxide, 190
vanadium ethoxide, 131
vanadium hexacyanoferrate, 292–3
  cyclic voltammetry, 292
  XPS, 293
vanadium pentoxide, 16, 56T, 87, 109, 130, 156, 185–90, 399, 446
  annealing, 185
  anodising vanadium metal, 185, 186, 187
  as secondary electrochrome, 16, 149, 190, 438, 444T
  bleaching rate, 188
  chemical diffusion coefficient, 85T
  coloration rate, 188
  cycle life, 188
  cyclic voltammetry, 187
  dissolution in acid, 186
  ECDs of, 189–90
  electrochemistry, 186–8
    quasi-reversible, 188
  electrostriction of, 87, 129
  ellipsometry, 187
  formation via
    cathodic arc deposition, 185
    CVD, 132, 190T
    dc sputtering, 136, 185
    dip coating, 135
    electrodeposition, 186
    electron-beam sputtering, 138–206
    evaporation, 185, 186
    flash evaporation, 185
    laser ablation, 185
    peroxo species, 186
    rf sputtering, 185, 187, 188, 190T
    sol–gel, 135, 185, 190T
    spin coating, 135, 185, 186
    vanadium propoxide, 185
    xerogel, 185
  intervalence effects, 188
  mixtures
    as electrochromic host, 202
    composites
      with gold, 204, 204T
      with melamine, 190, 202
      with poly(aniline), 190, 202
      with silver, 204T
  optical properties, 188–9
    coloration efficiency, 56T, 189, 190T
  structure, 186, 188
    monoclinic, 186
  write–erase efficiency, 188
  xerogel, 202
  XPS, 189
  XRD, 185, 202
vanadium propoxide, forming vanadium pentoxide, 185
vanadium–dysprosium oxide, 202
vanadium–magnesium–nickel oxide, 203
vanadium–molybdenum oxide, 199
vanadium–neodymium oxide, 202
vanadium–nickel oxide, 202
  optically passive, 202
vanadium–samarium oxide, 202
  coloration efficiency, 202
vanadium–titanium oxide, 202
  optically passive, 202
vanadium–titanium–cerium oxide, 203
vanadium–titanium–tungsten oxide, 203
vanadium–titanium–zirconium oxide, 203
vanadium–tungsten oxide, 202
  coloration efficiency, 202
  neutral colour, 399
video display units, 402, 403
violenes, 374
viologens, 12, 17, 341–66, 385
  asymmetric, 355, 360
  bleaching, chemical, 359
  chain length, see *viologens, substituent*
  charge movement through solid layers of, 81
  charge transfer complexation, 342–5, 353, 359
    potentiostatic, 358
    via pulsed potentials, 365
  comproportionation of, 357–8, 365
  contrast ratio, 346, 349, 352
  counter ions, effect of, 352–4
  covalently tethered, 10, 12
  cycle life, 362
  cyclic voltammetry, 352, 355, 356, 357, 359
  degradation of, 357
    oiling, 351

viologens (cont.)
  derivatised electrodes, 348
  di-reduced, 343, 357, 358
  ECDs, 346–8, 349, 352, 357
    five-colour, 385
    memory, 348, 362
    paper quality, 362
    see also *cyanophenyl paraquat, heptyl viologen, Nanochromics* and *NTera*
    ultra fast, 363
  electrochemistry, 342, 353, 354–5
    electrochemistry, quasi-reversibility, 358
    electrodeposition, 354
    electron transfer rate, 359
  electropolychromic, 365
  ESR, 352, 356
  in Nafion, 405
  in paper, 365, 366, 405
  infrared spectroscopy of, 358
  memory effect, 348, 362
  micellar, 355–6
    critical micelle concentration, 355, 356
  mixed valency of, 356
  modified, 360
  optical properties, 344T
    coloration efficiency, 349, 361, 362, 363
    colours of, 343, 351
    extinction coefficient, 343, 344T, 349
  polymers of, 328–9, 331, 347
    poly(pyrrole), 346
    poly(thiophene), 347
    oligomers, 364
  photoelectrochemistry, 362
    photostability, 129
  pseudo
    bipyridine, 2,2′-, 364
    phenanthroline, 3,8-, 360
  radicals of
    aging effects, 355, 357; see also *recrystallisation*
    chemical oxidation of, 352, 359
    dimerisation, 351, 355, 357, 358
    radical, nucleation, 358
    radical, recrystallisation, 357–8, 359
    radical, stability, 352T
  reduction, multi-step, 353, 354–5
    occurs via nucleation, 354
  response time, 346, 349, 351, 361, 363
  solubility product, 351, 359
  substituent, 349, 351–2
    alkyl, 359
    aryl, cyanophenyl, see *cyanophenyl paraquat*
    benzyl, 8, 344T, 346, 352T, 356, 358
    butyl, 352T
    ethyl, 344T, 352T, 438
    heptyl, see *heptyl viologen*
    hexyl, 352T
    methyl, see *methyl viologen*
    pentyl, 351, 352T
    propyl, 352T
    octyl, 344T, 352T
  substrates
    effect of, 353
    on nanostructured titania, 360–4
    tethered, 361
  type
    type-I electrochrome, 328, 346, 354, 356, 359
    type-II electrochrome, 346, 348–9, 351, 354
    type-III electrochrome, 346, 361
  write–erase efficiency, 348, 351, 356–60
viscous solvents
  forming type-III electrochromes, 391
  immobilised electrochromes
    carbazoles in, 391
    diacetylbenzene, *p*-, 390T, 391T
    diethyl terephthalate, 391T
    dimethyl terephthalate, 391T
    Methylene Blue, 391, 391T
  thickeners
    poly(AMPS), 391
    poly(aniline), 391
    poly(siloxane), 391
    PVPD, 391T, 395–410
visors, ECD application, 401
volatile memory, 54
voltammetry, cyclic, see *cyclic voltammetry*
voltmeters, 39

watch face, application, ECD, 443
  of tungsten trioxide, 149
water
  adsorbed, 89, 183
  and molybdenum trioxide, 89
  and tungsten trioxide, 145, 150
  coloration, acceleration, 90
  counter-ion interaction, 89
  degrades metal-oxide films, 89, 128–9
    dissolves ITO, 444
  ionisation of, 89
  occluded, 87, 89, 96–7, 156, 163
  solid oxide films, effect on, 89
wavelength maximum, 53
  change with insertion coefficient for WO_3, 53
Wien effect, 4
white point, and colour analysis, 64
whitener, in ECD electrolyte, 159, 384, 418, 422
windows, ECD, see *applications, ECD windows*
working electrode, 48
write–erase efficiency, 11–12, 129, 144–9, 156, 164, 259, 286, 348, 351, 356–60
  and tethered electrochromes, 12, 346

xerogel, 161, 185
  vanadium pentoxide, 202
Xerox, 5
XPS of
  indium–tin oxide, 197, 445
  iridium oxide, 26
  manganese oxide, 176
  molybdenum trioxide,1529, 153
  Prussian blue, 288
  tungsten trioxide, 103
  vanadium pentoxide, 189

X-ray reflector, ECD application, 397
XRD of
   molybdenum trioxide, 153
   praseodymium oxide, 179
   Prussian blue, 283
   ruthenium purple, 292
   tungsten trioxide, 140, 141
   vanadium pentoxide, 185, 202
XYZ-tristimulus, and colour analysis, 63

YAG laser, 266, 267
ytterbium phthalocyanine, 261, 291
   colour source, 261
   formed via plasma polymerisation, 291
yttrium–nickel oxide, 200

zinc iodide, ECD electrolyte, 408
zinc phthalocyanine, chemical diffusion
     coefficient, 85T
zinc TPP, 264
zirconium dioxide, ECD electrolyte, 421T
   electrochromic host, 203
   electro-inert, 203
zirconium–cerium oxide, 203
zirconium–cerium–titanium
     oxide, 203
zirconium–titanium–vanadium
     oxide, 203
zirconium–tantalum oxide, 203
   coloration efficiency, 203
Z-scale, Kosower, 343